ANALYSIS OF STRUCTURES

ANALYSIS OF STRUCTURES

AN INTRODUCTION INCLUDING NUMERICAL METHODS

Joe G. Eisley
Anthony M. Waas
College of Engineering
University of Michigan, USA

A John Wiley & Sons, Ltd., Publication

This edition first published 2011
© 2011 John Wiley & Sons, Ltd

Registered office
John Wiley & Sons Ltd, The Atrium, Southern Gate, Chichester, West Sussex, PO19 8SQ, United Kingdom

For details of our global editorial offices, for customer services and for information about how to apply for permission to reuse the copyright material in this book please see our website at www.wiley.com.

Library of Congress Cataloging-in-Publication Data

Eisley, Joe G.
  Analysis of structures : an introduction including numerical methods / Joe G. Eisley, Anthony M. Waas.
    p. cm.
  Includes bibliographical references and index.
  ISBN 978-0-470-97762-0 (cloth)
  1. Structural analysis (Engineering)–Mathematics.   2. Numerical analysis.   I. Waas, Anthony M.   II. Title.
  TA646.W33 2011
  624.1′71–dc22

                                                                                    2011009723

A catalogue record for this book is available from the British Library.

Print ISBN: 9780470977620
E-PDF ISBN: 9781119993285
O-book ISBN: 9781119993278
E-Pub ISBN: 9781119993544
Mobi ISBN: 9781119993551

Typeset in 9/11pt Times by Aptara Inc., New Delhi, India
Printed and bound in Singapore by Markono Print Media Pte Ltd

We would like to dedicate this book to our families.

To Marilyn, Paul and Susan
—Joe

To Dayamal, Dayani, Shehara and Michael
—Tony

Contents

About the Authors

Joe G. Eisley received degrees from St. Louis University, BS (1951), and the California Institute of Technology, MS (1952), PhD (1956), all in the field of aeronautical engineering. He served on the faculty of the Department of Aerospace Engineering from 1956 to 1998 and retired as Emeritus Professor of Aerospace Engineering in 1998. His primary field of teaching and research has been in structural analysis with an emphasis on the dynamics of structures. He also taught courses in space systems design and computer aided design. After retirement he has continued some part time work in teaching and consulting.

Anthony M. Waas is the Felix Pawlowski Collegiate Professor of Aerospace Engineering and Professor of Mechanical Engineering, and Director, Composite Structures Laboratory at the University of Michigan. He received his degrees from Imperial College, Univ. of London, U.K., B.Sc. (first class honors, 1982), and the California Institute of Technology, MS (1983), PhD (1988) all in Aeronautics. He joined the University of Michigan in January 1988 as an Assistant Professor, and is currently the Felix Pawlowski Collegiate Professor. His current teaching and research interests are related to lightweight composite aerostructures, with a focus on manufacturability and damage tolerance, ceramic matrix composites for "hot" structures, nano-composites, and multi-material structures. Several of his projects have been funded by numerous US government agencies and industry. In addition, he has been a consultant to several industries in various capacities. At Michigan, he has served as the Aerospace Engineering Department Graduate Program Chair (1998–2002) and the Associate Chairperson of the Department (2003–2005). He is currently a member of the Executive Committee of the College of Engineering. He is author or co-author of more than 175 refereed journal papers, and numerous conference papers and presentations.

Preface

This textbook is intended to be an introductory text on the mechanics of solids. The authors have targeted an audience that usually would go on to obtain undergraduate degrees in aerospace and mechanical engineering. As such, some specialized topics that are of importance to aerospace engineers are given more coverage. The material presented assumes only a background in introductory physics and calculus. The presentation departs from standard practice in a fundamental way. Most introductory texts on this subject take an approach not unlike that adopted by Timoshenko, in his 1930 *Strength of Materials* books, that is, by primarily formulating problems in terms of forces. This places an emphasis on statically determinate solid bodies, that is, those bodies for which the restraint forces and moments, and internal forces and moments, can be determined completely by the equations of static equilibrium. Displacements are then introduced in a specialized way, often only at a point, when necessary to solve the few statically indeterminate problems that are included. Only late in these texts are distributed displacements even mentioned. Here, we introduce and formulate the equations in terms of distributed displacements from the beginning. The question of whether the problems are statically determinate or indeterminate becomes less important. It will appear to some that more time is spent on the slender bar with axial loads than that particular structure deserves. The reason is that classical methods of solving the differential equations and the connection to the rational development of the finite element method can be easily shown with a minimum of explanation using the axially loaded slender bar. Subsequently, the development and solution of the equations for more advanced structures is facilitated in later chapters.

Modern advanced analysis of the integrity of solid bodies under external loads is largely displacement based. Once displacements are known the strains, stresses, strain energies, and restraint reactions are easily found. Modern analysis solutions methods also are largely carried out using a computer. The direction of this presentation is first to provide an understanding of the behavior of solid bodies under load and second to prepare the student for modern advanced courses in which computer based methods are the norm.

Analysis of Structures: An Introduction Including Numerical Methods is accompanied by a website (www.wiley.com/go/waas) housing exercises and examples that use modern software which generates color contour plots of deformation and internal stress. It offers invaluable guidance and understanding to senior level and graduate students studying courses in stress and deformation analysis as part of aerospace, mechanical and civil engineering degrees as well as to practicing engineers who want to re-train or re-engineer their set of analysis tools for contemporary stress and deformation analysis of solids and structures.

We are grateful to Dianyun Zhang, Ph.D candidate in Aerospace Engineering, for her careful reading of the examples presented.

Corrections, comments, and criticisms are welcomed.

Joe G. Eisley
Anthony M. Waas
June 2011
Ann Arbor, Michigan

1

Forces and Moments

1.1 Introduction

Mechanics of solids is concerned with the analysis and design of solid bodies under the action of applied forces in order to ensure "acceptable" behavior. These solid bodies are the components and the assemblies of components that make up the structures of aircraft, automobiles, washing machines, golf clubs, roller blades, buildings, bridges, and so on, that is, of many manufactured and constructed products. If the solid body is suitably restrained to exclude "rigid body" motion it will deform when acted upon by applied forces, or loads, and internal forces will be generated in the body. For "acceptable" behavior:

1. Internal forces must not exceed values that the materials can withstand.
2. Deformations must not exceed certain limits.

In later chapters of this text we shall identify, define, and examine the various quantities, such as internal forces, stresses, deformations, and material stress-strain relations, which determine acceptable behavior. We shall study methods for analyzing solid bodies and structures when loaded and briefly study ways to design solid bodies to achieve a desired behavior.

All solid bodies are three dimensional objects and there is a general theory of mechanics of solids in three dimensions. Because understanding the behavior of three dimensional objects can be difficult and sometimes confusing we shall work primarily with objects that have simplified geometry, simplified applied forces, and simplified restraints. This enables us to concentrate on the process instead of the details. After we have a clear understanding of the process we shall consider ever increasing complexity in geometry, loading, and restraint.

In this introductory chapter we examine three categories of force. First are *applied forces* which act on the surface or the mass of the body. Next are *restraint forces*, that is, forces on the surfaces where displacement is constricted (or restrained). Thirdly, *internal forces* generated by the resistance of the material to deformation as a result of applied and restraint forces.

Forces can generate moments acting about some point. For the most part we carefully distinguish between forces and moments; however, it is common practice to include both forces and moments when referring in general terms to the *forces acting on the body* or the *forces at the restraints*.

1.2 Units

The basic quantities in the study of solid mechanics are length (L), mass (M), force (F), and time (t). To these we must assign appropriate units. Because of their prominent use in every day life in the

Analysis of Structures: An Introduction Including Numerical Methods, First Edition. Joe G. Eisley and Anthony M. Waas.
© 2011 John Wiley & Sons, Ltd. Published 2011 by John Wiley & Sons, Ltd.

United States, the so-called English system of units is still the most familiar to many of us. Some engineering is still done in English units; however, global markets insist upon a world standard and so a version of the International Standard or SI system (from the French Système International d'Unités) prevails. The standard in SI is the meter, m, for length, the Newton, N, for force, the kilogram, kg, for mass, and the second, s, for time. The Newton is defined in terms of mass and acceleration as

$$1N = 1 \ kg \cdot 1\frac{m}{s^2} \tag{1.2.1}$$

For future reference the acceleration due to gravity on the earth's surface, g, in metric units is

$$g = 9.81\frac{m}{s^2} \tag{1.2.2}$$

The standard English units are the foot, ft, for length, the pound, lb, for force, the slug, $slug$, for mass, and the second, s, for time. The pound is defined in terms of mass and acceleration as

$$1 \ lb = 1 \ slug \cdot 1\frac{ft}{s^2} \tag{1.2.3}$$

The acceleration due to gravity on the earth's surface, g, in English units is

$$g = 32.2\frac{ft}{s^2} \tag{1.2.4}$$

We shall use SI units as much as possible.

Most of you are still thinking in English units and so for quick estimates you can note that a meter is approximately 39.37 inches; there are approximately 4.45 Newtons in a pound; and there are approximately 14.59 kilograms in a slug. But since you are not used to thinking in slugs it may help to note that a kilogram of mass weighs about 2.2 pounds on the earth's surface. For those who must convert between units there are precise tables for conversion. In time you will begin to think in SI units.

Often we obtain quantities that are either very large or very small and so units such as millimeter are defined. One millimeter is one thousandth of a meter, or $1 \ mm = 0.001 \ m$, and, of course, one kilogram is one thousand grams, or $1 \ kg = 1000 \ g$. The following table lists the prefixes for different multiples:

Multiple	Prefix	Symbol
10^9	giga	G
10^6	mega	M
10^3	kilo	k
10^{-3}	milli	m
10^{-6}	micro	μ
10^{-9}	nano	n

One modification of SI is that it is common practice in much of engineering to use the millimeter, mm, as the unit of length. Thus force per unit length is often, perhaps usually, given as Newtons per millimeter or N/mm. Force per unit area is given as Newtons per millimeter squared or N/mm^2. One N/m^2 is called a Pascal or Pa, so the unit of $1 \ N/mm^2$ is called 1 mega Pascal or $1 \ MPa$. Mass density has the units of kilograms per cubic millimeter or kg/mm^3. Throughout we shall use millimeter, Newton, and kilogram in all examples, discussions, and problems.

As noted in the above table: Only multiples of powers of three are normally used; thus, we do not use, for example, centimeters, decimeters, or other multiples that are the power of one or two. These are conventions, of course, so in the workplace you will find a variety of practices.

1.3 Forces in Mechanics of Materials

There are several types of forces that act on solid bodies. These consist of forces applied to the mass of the body and to the surface of the body, forces at restraints, and internal forces.

In Figure 1.3.1 we show a general three dimensional body with forces depicted acting on its surface and on its mass.

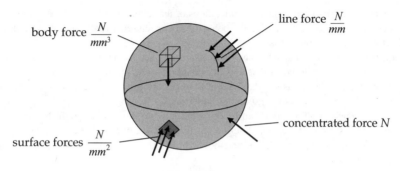

Figure 1.3.1

Forces that are volume or mass related are called *body forces*. In the system of units we are using they have the units of Newtons per cubic millimeter (N/mm^3). Gravity forces are a good example. Inertia forces generated by accelerations are another.

Surface forces can be specified in terms of force per unit area distributed over a surface and have the units of Newtons per square millimeter (N/mm^2). As noted one Newton per square millimeter is also called one mega Pascal (*MPa*).

If a force is distributed along a narrow band it is specified as a *line force*, that is, a force per unit length or Newtons per millimeter (N/mm).

If the force acts at a point it is a *concentrated force* and has the units of Newtons (N). Concentrated forces and line forces are usually idealizations or resultants of distributed surface forces. We can imagine an ice pick pushing on a surface creating a concentrated force. More likely the actual force acts on a small surface area where small means the size of the area is very small compared to other characteristic dimensions of the surface. Likewise a line force may be the resultant of a narrow band of surface forces.

When a concentrated, line, surface, or body force acts on the solid body or is applied to the body by means of an external agent it is called an *applied force*. When the concentrated, line, or surface force is generated at a point or region where an external displacement is imposed it is called a *restraint force*. In addition, for any body that is loaded and restrained, a force per unit area can be found on any internal surface. This particular distributed force is referred to as *internal* or simply as *stress*.

Generally, in the initial formulation of a problem for analysis, the geometry, applied forces, and physical restraints (displacements on specified surfaces) are known while the restraint forces and internal stresses are unknown. When the problem is formulated for design, the acceptable stress limits may be specified in advance and the final geometry, applied forces, and restraints may initially be unknown. For the most part the problems will be formulated for analysis but the subject of design will be introduced from time to time.

The analysis of the interaction of these various forces is a major part of the following chapters. For the most part we shall use rectangular Cartesian coordinates and resolve forces into components with respect to these axes. An exception is made for the study of torsion in Chapter 6. There we use cylindrical coordinates.

In the sign convention adopted here, applied force components and restraint force components are positive if acting in the positive direction of the coordinate axes. Positive stresses and internal forces will be defined in different ways as needed.

We start first with a discussion of concentrated forces.

1.4   Concentrated Forces

As noted, concentrated forces are usually idealizations of distributed forces. Because of the wide utility of this idealization we shall first examine the behavior of concentrated forces. In all examples we shall use the Newton (N) as our unit of force.

Force is a vector quantity, that is, it has both magnitude and direction. There are several ways of representing a concentrated force in text and in equations; however, the pervasive use of the digital computer in solving problems has standardized how forces are usually represented in formulating and solving problems in the behavior of solid bodies under load.

First, we shall consider a force that can be oriented in a two dimensional right handed rectangular Cartesian coordinate system and we shall define positive unit vectors $\mathbf{i}$ and $\mathbf{j}$ in the x, and y directions, respectively, as shown in Figure 1.4.1. Using boldface has been a common practice in representing vectors in publications.

Figure 1.4.1

A force is often shown in diagrams as a line that starts at the point of application and has an arrowhead to show its direction as shown in Figure 1.4.2.

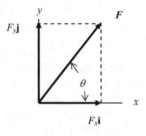

Figure 1.4.2

The concentrated force, $\boldsymbol{F}$, can be represented by its components in the x and y directions.

$$\boldsymbol{F} = F_x\mathbf{i} + F_y\mathbf{j} \tag{1.4.1}$$

In keeping with the notation most commonly used for later computation we represent this force vector by a column matrix $\{F\}$ as shown in Equation 1.4.2.

$$\{F\} = \begin{bmatrix} F_x \\ F_y \end{bmatrix} \tag{1.4.2}$$

In matrix notation the unit vector directions are implied by the component subscripts. From the properties of a right triangle the magnitude of the vector is given by

$$F = \sqrt{F_x^2 + F_y^2} \qquad (1.4.3)$$

The orientation of the force can be represented by the angle between the force and either axis. For example, with respect to the x axis

$$\tan \theta = \frac{F_y}{F_x} \quad \rightarrow \quad \theta = \tan^{-1} \frac{F_y}{F_x} \qquad (1.4.4)$$

Quite often we must sum two or more forces such as those shown in Figure 1.4.3 as solid lines.

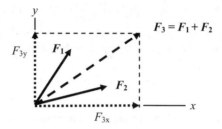

Figure 1.4.3

To add or subtract vectors is simply to add or subtract components. For example,

$$\{F_3\} = \{F_1\} + \{F_2\} = \begin{bmatrix} F_{1x} \\ F_{1y} \end{bmatrix} + \begin{bmatrix} F_{2x} \\ F_{2y} \end{bmatrix} = \begin{bmatrix} F_{1x} + F_{2x} \\ F_{1y} + F_{2y} \end{bmatrix} = \begin{bmatrix} F_{3x} \\ F_{3y} \end{bmatrix} \qquad (1.4.5)$$

The sum is shown by the dashed line and its components in the two coordinate directions by the dotted lines.

###########

Example 1.4.1

Problem: Two forces are acting at a point at the origin of the coordinate system as shown in Figure (a). Sum the two to find the resultant force and its direction.

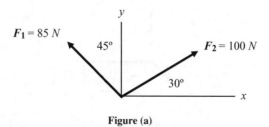

Figure (a)

Solution: Resolve the forces into components and sum. Solve for the resultant force and its orientation. The components are

$$\{F_1\} = \begin{bmatrix} F_{1x} \\ F_{1y} \end{bmatrix} = \begin{bmatrix} -85\sin 45° \\ 85\cos 45° \end{bmatrix} = \begin{bmatrix} -60.1 \\ 60.1 \end{bmatrix} \quad \{F_2\} = \begin{bmatrix} F_{2x} \\ F_{2y} \end{bmatrix} = \begin{bmatrix} 100\cos 30° \\ 100\sin 30° \end{bmatrix} = \begin{bmatrix} 86.6 \\ 50 \end{bmatrix} N \quad (a)$$

The sum is

$$\{F_3\} = \{F_1\} + \{F_2\} = \begin{bmatrix} -60.1 \\ 60.1 \end{bmatrix} + \begin{bmatrix} 86.6 \\ 50 \end{bmatrix} = \begin{bmatrix} 26.5 \\ 110.1 \end{bmatrix} N \qquad (b)$$

The total magnitude of the force is

$$F_3 = \sqrt{F_x^2 + F_y^2} = \sqrt{(26.5)^2 + (110.1)^2} = 113.2\ N \qquad (c)$$

The resultant force vector makes an angle with respect to the *x* axis,

$$\theta = \tan^{-1}\frac{F_y}{F_x} = \tan^{-1}\frac{110.1}{26.5} = \tan^{-1} 4.15 = 76.5° \qquad (d)$$

The resultant force is shown as a dashed line and its components as dotted lines in Figure (b).

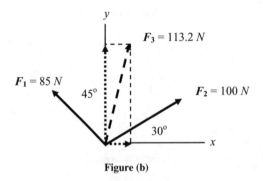

Figure (b)

##########

This can be extended to three dimensions. We shall define positive unit vectors **i, j, k** in the *x, y, z* directions, respectively, as shown in Figure 1.4.4.

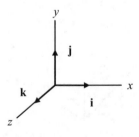

Figure 1.4.4

The concentrated force, F, can be represented by its components in the x, y, and z directions as in Equation 1.4.6.

$$F = F_x\mathbf{i} + F_y\mathbf{j} + F_z\mathbf{k} \qquad (1.4.6)$$

This is shown graphically in Figure 1.4.5.

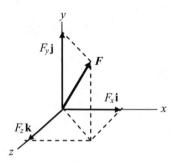

Figure 1.4.5

The components in matrix form are

$$\{F\} = \begin{bmatrix} F_x \\ F_y \\ F_z \end{bmatrix} \qquad (1.4.7)$$

The magnitude of the vector F is given by

$$F = \sqrt{F_x^2 + F_y^2 + F_z^2} \qquad (1.4.8)$$

The angular orientation of the force F with respect to each axis is given by

$$\cos\alpha = \frac{F_x}{F} \qquad \cos\beta = \frac{F_y}{F} \qquad \cos\gamma = \frac{F_z}{F} \qquad (1.4.9)$$

The angle between the force, F, and the x axis is α, between the force F and the y axis is β, and between the force F and the z axis is γ. The quantities in Equation 1.4.9 are called the *direction cosines*.

As noted in the two dimensional case, to add or subtract vectors is simply to add or subtract components. For example, given three forces acting at a point the force representing the sum is

$$\{F_4\} = \{F_1\} + \{F_2\} - \{F_3\} = \begin{bmatrix} F_{1x} \\ F_{1y} \\ F_{1z} \end{bmatrix} + \begin{bmatrix} F_{2x} \\ F_{2y} \\ F_{2z} \end{bmatrix} - \begin{bmatrix} F_{3x} \\ F_{3y} \\ F_{3z} \end{bmatrix} = \begin{bmatrix} F_{1x} + F_{2x} - F_{3x} \\ F_{1y} + F_{2y} - F_{3y} \\ F_{1z} + F_{2z} - F_{3z} \end{bmatrix} = \begin{bmatrix} F_{4x} \\ F_{4y} \\ F_{4z} \end{bmatrix}$$

$$(1.4.10)$$

###########

Example 1.4.2

Problem: Two forces act in perpendicular planes as shown in Figure (a). Sum the two to find the resultant force and its direction.

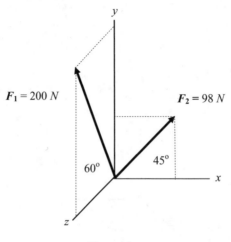

Figure (a)

Solution: Resolve the forces into components and sum. Solve for the value of the resultant force and its orientation.

The components of the forces are

$$\{F_1\} = \begin{bmatrix} F_{1x} \\ F_{1y} \\ F_{1z} \end{bmatrix} = \begin{bmatrix} 0 \\ 200\sin 60° \\ 200\cos 60° \end{bmatrix} = \begin{bmatrix} 0 \\ 173.2 \\ 100 \end{bmatrix} N \qquad \{F_2\} = \begin{bmatrix} F_{2x} \\ F_{2y} \\ F_{2z} \end{bmatrix} = \begin{bmatrix} 98\cos 45° \\ 98\sin 45° \\ 0 \end{bmatrix} = \begin{bmatrix} 69.3 \\ 69.3 \\ 0 \end{bmatrix} N$$

(a)

The sum of the forces is

$$\{F\} = \{F_1\} + \{F_2\} = \begin{bmatrix} 0 \\ 173.2 \\ 100 \end{bmatrix} + \begin{bmatrix} 69.3 \\ 69.3 \\ 0 \end{bmatrix} = \begin{bmatrix} 69.3 \\ 242.5 \\ 100 \end{bmatrix} N \qquad (b)$$

The magnitude of the total force is

$$F = \sqrt{F_x^2 + F_y^2 + F_z^2} = \sqrt{(69.3)^2 + (242.5)^2 + (100)^2} = 271.3 \ N \qquad (c)$$

The direction cosines are

$$\cos \alpha = \frac{F_x}{F} = \frac{69.3}{271.3} = 0.255 \quad \rightarrow \quad \alpha = 75.2°$$

$$\cos \beta = \frac{F_y}{F} = \frac{242.5}{271.3} = 0.662 \quad \rightarrow \quad \beta = 26.6° \qquad (d)$$

$$\cos \gamma = \frac{F_z}{F} = \frac{100}{271.3} = 0.369 \quad \rightarrow \quad \gamma = 68.4°$$

The final result is shown in Figure (b).

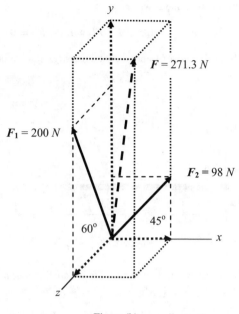

Figure (b)

############

Another property of a matrix that we shall use shortly is multiplication of a matrix by a scalar. It is simply

$$a\{F\} = a \begin{bmatrix} F_x \\ F_y \\ F_z \end{bmatrix} = \begin{bmatrix} aF_x \\ aF_y \\ aF_z \end{bmatrix} \tag{1.4.11}$$

Additional matrix operations will be introduced as needed. They are summarized in Appendix A.

1.5 Moment of a Concentrated Force

A concentrated force can produce a moment about any given axis. In all examples we shall use Newton millimeter $(N \cdot mm)$ as our unit for moments. Consider the force applied to the rigid bar at point B as shown in Figure 1.5.1.

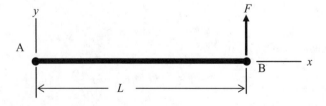

Figure 1.5.1

If we take moments about points A and B we get

$$M_A = FL \qquad M_B = M - FL = FL - FL = 0 \tag{1.5.1}$$

Now consider the force has been moved to point A and a concentrated moment equal to FL is added at point A as shown in Figure 1.5.2.

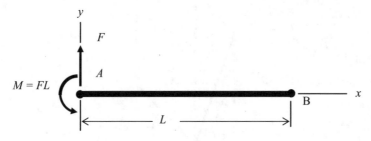

Figure 1.5.2

The moments about points A and B in this new configuration are the same as for the first configuration. Summing moments about each point we get

$$M_A = F \cdot 0 + M = M = FL \qquad M_B = M - FL = FL - FL = 0 \tag{1.5.2}$$

We can, in fact, take moments about any point in the xy plane and get the same result for both configurations. For example, take moments about the point C as shown in Figure 1.5.3 located at

$$x_C = \frac{L}{2} \qquad y_C = \frac{L}{4} \tag{1.5.3}$$

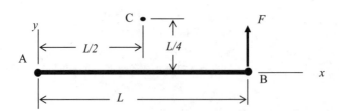

Figure 1.5.3

From the configuration in Figure 1.5.3 we get

$$M_C = F\frac{L}{2} \tag{1.5.4}$$

From the configuration in Figure 1.5.2 we get

$$M_C = M - F\frac{L}{2} = FL - F\frac{L}{2} = F\frac{L}{2} \tag{1.5.5}$$

The problem can be posed in another way: If you move a force, what moment must be added to achieve an equivalent balance of moments?

Let us consider the rigid bar in Figure 1.5.4 with the force initially at point B. We shall call this configuration 1. It is then moved to point C (shown by a dashed line). This we call configuration 2. What moment must be added (and at what location) to provide equivalence?

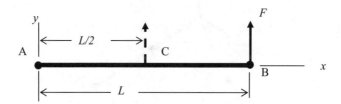

Figure 1.5.4

Let us take moments about point A for configuration 1 and 2.

$$M_{A1} = F \cdot L \qquad M_{A2} = F \cdot \frac{L}{2} \tag{1.5.6}$$

For equivalence we must add a moment that is equal to the difference in the two values or

$$M = M_{A1} - M_{A2} = F \cdot L - F \cdot \frac{L}{2} = F \cdot \frac{L}{2} \tag{1.5.7}$$

Now where should it be added? The answer is anywhere. In this case anywhere along the bar, for example, at point A, or point C, or point B, or any point in between.

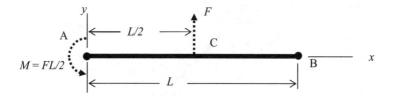

Figure 1.5.5

Just to be sure let us place the new applied moment at point A and sum moments about points A, B, and C in Figure 1.5.5.

$$M_A = \frac{FL}{2} + \frac{FL}{2} = FL \qquad M_B = -\frac{FL}{2} + \frac{FL}{2} = 0 \qquad M_C = \frac{FL}{2} \tag{1.5.8}$$

If you compare this with the original configuration in Figure 1.5.1 you will see that the moments about points A, B, and C agree.

The use of the half circle symbol in Figures 1.5.2 and 1.5.5 is one way of representing a concentrated moment in diagrams. It is used when the moment is about an axis perpendicular to the plane of the page. A common practice is to use a vector with a double arrowhead shown here in an isometric view to represent a moment. The vector is parallel to the axis about which the moment acts. The right hand rule of the thumb pointed in the vector direction and the curve fingers of the right hand showing the direction of the moment is implied here. The moment of Figure 1.5.2 is repeated in Figure 1.5.6 using a double arrowhead notation.

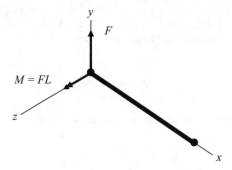

Figure 1.5.6

############

Example 1.5.1

Problem: A force is applied to a rigid body at point A as shown in Figure (a). If the force is moved to point B what moment must be applied at point C (origin of coordinates) to produce the same net moment about all points in space?

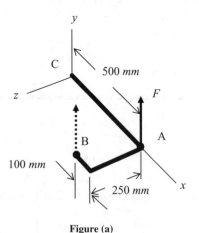

Figure (a)

Solution: Find the moment components at point C due to the force at point B and add the necessary moments so the total is equivalent to the moment components generated by the force at point A.

The force at point A produces the following moment components about the origin (point C)

$$M_{CAx} = 0 \qquad M_{CAy} = 0 \qquad M_{CAz} = 500 \cdot F \qquad \text{(a)}$$

This can be written in matrix form as a column vector.

$$\{M_{CA}\} = F \begin{bmatrix} 0 \\ 0 \\ 500 \end{bmatrix} \qquad \text{(b)}$$

The same force at point B would produce the following moment components at point C.

$$M_{CBx} = -250 \cdot F \qquad M_{CBy} = 0 \qquad M_{CBz} = 400 \cdot F \tag{c}$$

or

$$\{M_{CB}\} = F \begin{bmatrix} -250 \\ 0 \\ 400 \end{bmatrix} \tag{d}$$

The needed moment components would be

$$\{M_C\} = \begin{bmatrix} M_{Cx} \\ M_{Cy} \\ M_{Cz} \end{bmatrix} = \{M_{CA}\} - \{M_{CB}\} = F \begin{bmatrix} 0 \\ 0 \\ 500 \end{bmatrix} - F \begin{bmatrix} -250 \\ 0 \\ 400 \end{bmatrix} = F \begin{bmatrix} 250 \\ 0 \\ 100 \end{bmatrix} \tag{e}$$

This is illustrated in Figure (b).

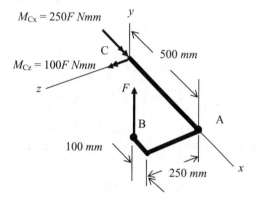

Figure (b)

Let us check our answer by summing moments about the origin. Using Figure (b)

$$\sum M_x = 250F - 250F = 0 \qquad \sum M_y = 0 \qquad \sum M_z = 400F + 100F = 500F \tag{f}$$

In matrix form we get the same answer as Equation (a)

$$\{M\} = F \begin{bmatrix} 0 \\ 0 \\ 500 \end{bmatrix} \tag{g}$$

Summing moments about any point in space will prove that the answer is always the same.

##########

So far we have considered a single force parallel to one of the axes. Consider now a force with components in all three dimensions acting at a point in space. We select a point about which we wish to find the moment components of this force with respect to a set of rectangular Cartesian coordinate axes.

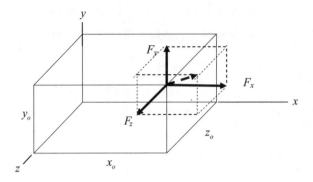

Figure 1.5.7

In Figure 1.5.7 we show the components of a force $\{F\}$ and the axes about which we wish to find the moment. If we take the moment at the origin of the coordinate system about each axis in turn we get

$$\sum M_x = F_z y_o - F_y z_o \qquad \sum M_y = F_x z_o - F_z x_o \qquad \sum M_z = F_y x_o - F_x y_o \qquad (1.5.9)$$

The moment components in matrix form about all three axes are

$$\{M\} = \begin{bmatrix} M_x \\ M_y \\ M_z \end{bmatrix} = \begin{bmatrix} F_z y_o - F_y z_o \\ F_x z_o - F_z x_o \\ F_y x_o - F_x y_o \end{bmatrix} \qquad (1.5.10)$$

This same information is often presented in the form of vector notation. The cross product of two vectors is stated as

$$C = A \times B \qquad (1.5.11)$$

The magnitude of the cross product is

$$C = AB \cos \theta \qquad (1.5.12)$$

where θ is the angle between the two vectors. The vector C has a direction that is perpendicular to the plane containing A and B and its direction is defined by the right hand rule. Since the angle between unit vectors is either $90°$, $-90°$, or $0°$ the cross product of unit vectors is found to be

$$
\begin{array}{lll}
\mathbf{i} \times \mathbf{j} = \mathbf{k} & \mathbf{i} \times \mathbf{k} = -\mathbf{j} & \mathbf{i} \times \mathbf{i} = 0 \\
\mathbf{j} \times \mathbf{k} = \mathbf{i} & \mathbf{j} \times \mathbf{i} = -\mathbf{k} & \mathbf{j} \times \mathbf{j} = 0 \\
\mathbf{k} \times \mathbf{i} = \mathbf{j} & \mathbf{k} \times \mathbf{j} = -\mathbf{i} & \mathbf{k} \times \mathbf{k} = 0
\end{array}
\qquad (1.5.13)
$$

With this definition in mind the moment of the force in Figure 1.5.7 is represented as the cross product of a position vector and the force. Thus

$$M = d \times F \qquad (1.5.14)$$

If the moment is taken with respect to the origin of the coordinates in Figure 1.5.7 the position vector is

$$d = x_o \mathbf{i} + y_o \mathbf{j} + z_o \mathbf{k} \qquad (1.5.15)$$

The moment is then

$$M = d \times F = (x_o i + y_o j + z_o k) \times (F_x i + F_y j + F_z k)$$
$$= x_o F_x (i \times i) + x_o F_y (i \times j) + x_o F_z (i \times k)$$
$$+ y_o F_x (j \times i) + y_o F_y (j \times j) + z_o F_z (j \times k) \qquad (1.5.16)$$
$$+ z_o F_x (k \times i) + z_o F_x (k \times j) + z_o F_x (k \times k)$$

By combining terms this reduces to

$$M = d \times F = (y_o F_z - z_o F_y)i + (x_o F_z - z_o F_x)j + (x_o F_y - y_o F_x)k \qquad (1.5.17)$$

This is often presented in the form of a determinant.

$$M = \begin{vmatrix} i & j & k \\ x_o & y_o & z_o \\ F_x & F_y & F_z \end{vmatrix} \qquad (1.5.18)$$

Equations 1.5.17 and 1.5.18 convey exactly the same information as that contained in Equation 1.5.10.
The matrix formulation of vectors very often has replaced the boldfaced vector representation more common in past treatises. We shall discontinue the use of boldface in representing vectors since we shall be using the matrix representation in all future work. The representation of any vector quantity will be clear from the context of its use.

As noted before we denote a moment using a vector with a double arrowhead as shown in Figure 1.5.8.

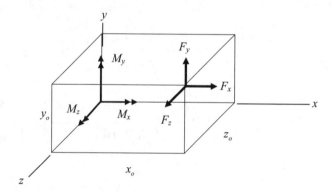

Figure 1.5.8

The vector components can be combined to obtain the total value of the moment.

$$M = \sqrt{M_x^2 + M_y^2 + M_z^2} \qquad (1.5.19)$$

and the orientation can be given by the direction cosines.

$$\cos \alpha = \frac{M_x}{M} \qquad \cos \beta = \frac{M_y}{M} \qquad \cos \gamma = \frac{M_z}{M} \qquad (1.5.20)$$

In all our deliberations applied forces are positive if they are in the positive directions of the axes and applied moments resulting from applied forces are positive by the right hand rule. Right handed rectangular Cartesian coordinate systems are used for the most part. Cylindrical coordinates will be used when we study torsion in Chapter 6.

###########

Example 1.5.2

Problem: Find the moment components of the force shown in Figure (a) about the origin of the coordinate axes and the total value of the moment.

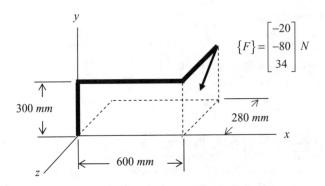

Figure (a)

Solution: Use Equations 1.5.18–20.

The components are

$$\{F\} = \begin{bmatrix} -20 \\ -80 \\ 34 \end{bmatrix} N \qquad \{d\} = \begin{bmatrix} x_o \\ y_o \\ z_o \end{bmatrix} = \begin{bmatrix} 600 \\ 300 \\ -280 \end{bmatrix} mm \tag{a}$$

The moment can be obtained from

$$M = \begin{vmatrix} \mathbf{i} & \mathbf{j} & \mathbf{k} \\ 600 & 300 & -280 \\ -20 & -80 & 34 \end{vmatrix} \quad \rightarrow \quad \{M\} = \begin{bmatrix} M_x \\ M_y \\ M_z \end{bmatrix} = \begin{bmatrix} 34 \cdot 300 - 80 \cdot 280 \\ -34 \cdot 600 + 20 \cdot 280 \\ -80 \cdot 600 + 20 \cdot 300 \end{bmatrix} = \begin{bmatrix} -12200 \\ -14800 \\ -42000 \end{bmatrix} N \cdot mm \tag{b}$$

The total value of the moment is

$$M = \sqrt{M_x^2 + M_y^2 + M_z^2} = \sqrt{(12200)^2 + (14800)^2 + (42000)^2} = 46172.3 \; Nmm \tag{c}$$

The orientation of the resultant moment is given by the direction cosines.

$$\cos\alpha = \frac{M_x}{M} = \frac{-12200}{46172.3} = -0.2642 \quad \rightarrow \quad \alpha = 105.32°$$

$$\cos\beta = \frac{M_y}{M} = \frac{-14800}{46172.3} = -0.3205 \quad \rightarrow \quad \beta = 108.7° \tag{d}$$

$$\cos\gamma = \frac{M_z}{M} = \frac{-42000}{46172.3} = -0.9096 \quad \rightarrow \quad \gamma = 155.46°$$

###########

To find the moment about some point other than the origin of the coordinate system requires only defining a new position vector. For example suppose we wish to find the moment components about point A as shown in Figure 1.5.9.

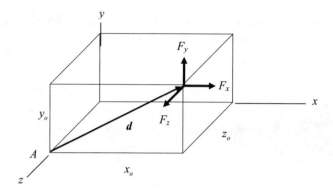

Figure 1.5.9

The components of the new position vector would be

$$\{d\} = \begin{bmatrix} x_o \\ y_o \\ 0 \end{bmatrix}$$

(1.5.21)

A special case of a moment caused by forces occurs when there are two parallel forces of equal and opposite direction separated by a distance a. This might occur, for example, with the loads applied to a member as shown in Figure 1.5.10.

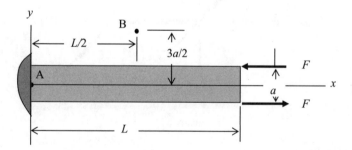

Figure 1.5.10

The resultant force of the two forces is zero. The resultant moment about any point in the plane is

$$M = Fa$$

(1.5.22)

To illustrate that the location of the point in the plane is of no effect take moments about point A which is at the origin of the coordinates and about point B which is at $x = L/2$ and $y = 3d/2$.

$$M_A = F\frac{a}{2} + F\frac{a}{2} = Fa \qquad (1.5.23)$$

$$M_B = -\left(\frac{3a}{2} - \frac{a}{2}\right)F + \left(\frac{3a}{2} + \frac{a}{2}\right)F = Fa$$

Such a force combination as shown in Figure 1.5.10 is called a *couple*. When d is small it can often be represented as a concentrated moment as in Figure 1.5.11.

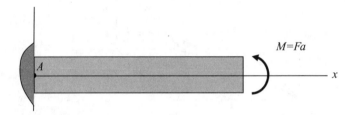

Figure 1.5.11

############

Example 1.5.3

Problem: Find the moment components about the axes of the set of couples shown in Figure (a).

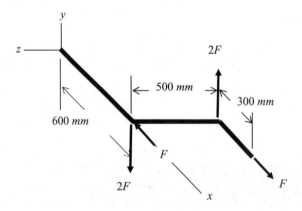

Figure (a)

Solution: Use the definition of a couple.
    The moment components about the axes are simply

$$M_x = 500 \cdot 2F \qquad M_y = -500 \cdot F \qquad M_z = 0 \qquad (a)$$

Example 1.5.4

Problem: A force system consists of a couple and another force as shown in Figure (a). Find the moment about the z axis at point A.

Solution: Sum moments about the z axis.

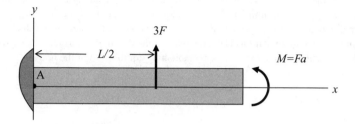

Figure (a)

The sum of moments about point A is

$$M_A = 3F \cdot \frac{L}{2} + M = 3F \cdot \frac{L}{2} + Fa = \left(\frac{3L}{2} + a\right) F \tag{a}$$

###########

1.6 Distributed Forces—Force and Moment Resultants

Forces may be distributed along a line, on a surface, or throughout a volume. It is often necessary to find the total force resultant of the distributed force and also find through what point it is acting. Consider the area shown in Figure 1.6.1 and the force per unit area acting upon it. We have chosen a planar rectangular area and a particular distribution for ease of explanation. Real surfaces with loads will be found in many shapes and sizes and can be external or internal surfaces.

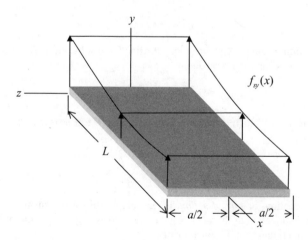

Figure 1.6.1

This particular surface loading acts in the y direction, varies in the x direction, and is uniform in the z direction. We label this force $f_{sy}(x, z)$ where the subscript s denotes a surface force and the y its component direction. We note that its units are force per unit area (N/mm^2) and since it is uniform in the z direction we can multiply the surface force by the width, a, of the planar surface and obtain an equivalent line force with units of force per unit length (N/mm) as shown in Figure 1.6.2. This line force acts in the plane of symmetry, that is, in the xy plane at $z = 0$. Our coordinate system was selected conveniently with this in mind. Bear in mind that this is the resultant of the distributed surface force and not the actual force acting on the surface. It may be used for establishing equilibrium of the body.

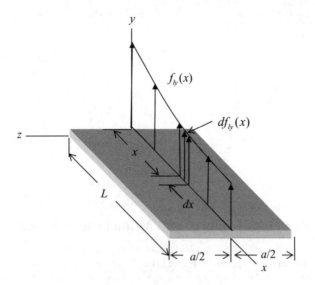

Figure 1.6.2

We label this force $f_{ly}(x)$ where the subscript l denotes a line force and note that

$$f_{ly}(x) = af_{sy}(x, z) \tag{1.6.1}$$

Now consider an infinitesimal length, dx, along the line force at location x, as shown in Figure 1.6.2. On the length, dx, the force in the y direction is

$$dF_y = df_{ly}(x) \tag{1.6.2}$$

If we sum the forces on all such dx lengths, ranging for $x = 0$ to $x = L$, we obtain the total resultant of the distributed force

$$F_y = \int_0^L f_{ly}(x)dx = a \int_0^L f_{sy}(x)dx \tag{1.6.3}$$

We can find the location, or *line of action*, of the force resultant by equating the moments of the distributed force to the moment of the force resultant as follows. Again, the sum of all the moments of all the forces on all the infinitesimal elements dx is

$$\bar{x}F_y = \int_0^L xf_{ly}(x)dx \tag{1.6.4}$$

And the location of the resultant is

$$\bar{x} = \frac{1}{F_y} \int_0^L x f_{ly}(x) dx = \frac{\displaystyle\int_0^L x f_{ly}(x) dx}{\displaystyle\int_0^L f_{ly}(x) dx} \tag{1.6.5}$$

The use of symmetry to locate the line force and the resultant force at $z = 0$ can be confirmed by equating the moments of the distributed force to the moment of the resultant force about the z axis.

$$\bar{z} F_y = \int_0^L \int_{-\frac{a}{2}}^{\frac{a}{2}} z f_{sy}(x) dz dx \quad \rightarrow \quad \bar{z} = \frac{1}{F_y} \int_{-\frac{a}{2}}^{\frac{a}{2}} z dz \int_0^L f_{sy}(x) dx \tag{1.6.6}$$

Clearly the integral

$$\int_{-\frac{a}{2}}^{\frac{a}{2}} z dz = 0 \tag{1.6.7}$$

and therefore $\bar{z} = 0$.

The line force and the resultant force and its location are depicted in Figure 1.6.3.

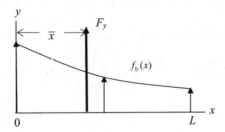

Figure 1.6.3

###########

| **Example 1.6.1** |

Problem: A distributed force per unit area is applied to the surface as shown in Figure (a). The force is uniform in the z direction.

$$f_{sy}(x, z) = f_0 \frac{x}{L} \tag{a}$$

Find the total value and its line of action.

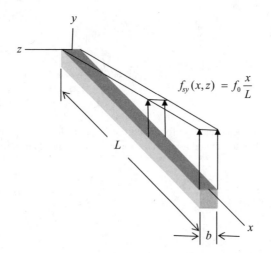

Figure (a)

Solution: Convert the surface force (N/mm^2) to a line force (N/mm) and integrate to find the total resultant force (N). Then equate moments to find its line of action.

The distributed line force and resultant force are

$$f_{ly}(x) = bf_0 \frac{x}{L} \qquad F_y = bf_0 \int_0^L \frac{x}{L} dx = bf_0 \left. \frac{x^2}{2L} \right|_0^L = bf_0 \frac{L}{2} \tag{b}$$

Find the line of action by equating moments.

$$\bar{x} F_y = \int_0^L x bf_0 \frac{x}{L} dx \quad \rightarrow \quad \bar{x} = \frac{1}{F_y} \int_0^L x bf_0 \frac{x}{L} dx = \frac{1}{F_y} \left. \frac{bf_0 x^3}{3L} \right|_0^L = \frac{2}{3} L \tag{c}$$

The line force and the location of the force resultant are shown in Figure (b).

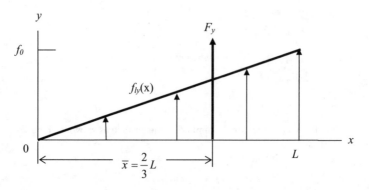

Figure (b)

###########

Symmetry may be used to identify quickly the location of a resultant force. The two line forces shown in Figure 1.6.4 are symmetrical about their midpoints and so the location of the resultant is known instantly as shown.

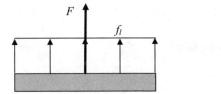
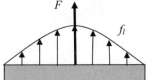

Figure 1.6.4

Distributed surface forces may be functions of two variables; for example, as shown in Figure 1.6.5.

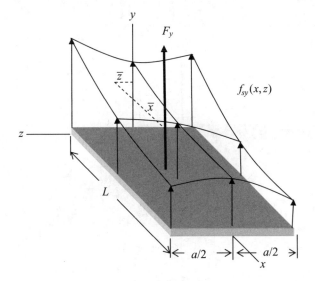

Figure 1.6.5

Then the resultant force for the area shown, using the same procedure as before, is

$$F_y = \int_{-\frac{a}{2}}^{\frac{a}{2}} \int_0^L f_{sy}(x,z) dx dz \qquad (1.6.6)$$

and its line of action is found by

$$\bar{x} F_y = \int_{-\frac{a}{2}}^{\frac{a}{2}} \int_0^L x f_{sy}(x,z) dx dz \rightarrow \bar{x} = \frac{1}{F_y} \int_{-\frac{a}{2}}^{\frac{a}{2}} \int_0^L x f_{sy}(x,z) dx dz = \frac{\int_{-\frac{a}{2}}^{\frac{a}{2}} \int_0^L x f_{sy}(x,z) dx dz}{\int_{-\frac{a}{2}}^{\frac{a}{2}} \int_0^L f_{sy}(x,z) dx dz}$$

$$\bar{z} F_y = \int_{-\frac{a}{2}}^{\frac{a}{2}} \int_0^L z f_{sy}(x,z) dx dz \rightarrow \bar{z} = \frac{1}{F_y} \int_{-\frac{a}{2}}^{\frac{a}{2}} \int_0^L z f_{sy}(x,z) dx dz = \frac{\int_{-\frac{a}{2}}^{\frac{a}{2}} \int_0^L z f_{sy}(x,z) dx dz}{\int_{-\frac{a}{2}}^{\frac{a}{2}} \int_0^L f_{sy}(x,z) dx dz}$$

$$(1.6.7)$$

The location of the resultant force is also depicted in Figure 1.6.5.

Of course, surfaces are not always planar and rectangular and surface forces may have components in all coordinate directions. We shall be satisfied with simplified geometry and forces until and unless the need arises for more complicated cases.

##########

Example 1.6.2

Problem: A distributed force per unit area is applied to the surface shown in Figure (a). The surface force is represented by the function in Equation (a). Find the resultant force and its line of action.

$$f_{sy}(x, z) = f_0 \left(1 - \frac{x^2}{a^2}\right) \left(\frac{1}{2} + \frac{z}{2b}\right) \tag{a}$$

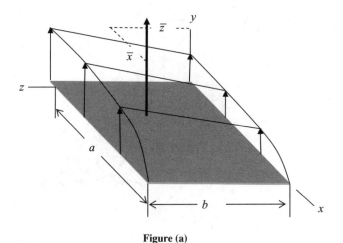

Figure (a)

Solution: Integrate to find the resultant force. Equate moments to find its line of action.

The resultant force is given by Equation (b)

$$F_y = \int_0^a \int_0^b f_{sy}(x, z) dz dx = f_0 \int_0^a \int_0^b \left(1 - \frac{x^2}{a^2}\right) \left(\frac{1}{2} + \frac{z}{2b}\right) dz dx$$

$$= f_0 \left(x - \frac{x^3}{3a^2}\right) \Big|_0^a \left(\frac{z}{2} + \frac{z^2}{4b}\right) \Big|_0^b = f_0 \frac{ab}{2} \tag{b}$$

Its location is given by Equations (c) and (d).

$$\bar{x} F_y = \int_0^a \int_0^b x f_{sy}(x, z) dx dz \quad \rightarrow \quad \bar{x} = \frac{1}{F_y} f_0 \int_0^a \int_0^b x \left(1 - \frac{x^2}{a^2}\right) \left(\frac{1}{2} + \frac{z}{2b}\right) dz dx$$

$$= \frac{2}{ab} \left(\frac{x^2}{2} - \frac{x^4}{4a^2}\right) \Big|_0^a \left(\frac{z}{2} + \frac{z^2}{4b}\right) \Big|_0^b = \frac{3a}{8} \tag{c}$$

$$\bar{z}F_y = \int_0^a \int_0^b z f_{sy}(x, z)dxdz \quad \rightarrow \quad \bar{z} = \frac{1}{F_y}f_0 \int_0^a \int_0^b z\left(1 - \frac{x^2}{a^2}\right)\left(\frac{1}{2} + \frac{z}{2b}\right)dzdx$$

$$= \frac{2}{ab}\left(x - \frac{x^3}{3a^2}\right)\Big|_0^a \left(\frac{z^2}{4} + \frac{z^3}{6b}\right)\Big|_0^b = \frac{5b}{9} \tag{d}$$

############

We may also need to find the resultants of body forces and their location. Body forces have the units of force per unit volume (N/mm^3) and may be distributed over the volume. Consider the general solid shown in Figure 1.6.6.

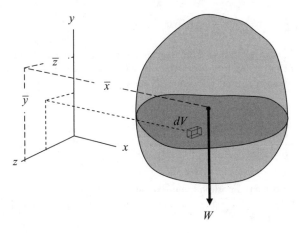

Figure 1.6.6

The most common body force is that due to gravity, that is, weight. The weight (dW) of a infinitesimal volume dV is given by

$$dW = \rho(x, y, z)g dV \tag{1.6.8}$$

where $\rho(x, y, z)$ is the mass density (kg/mm^3) and g is the acceleration due to gravity (mm/s^2). (See Section 1.2, Equation 1.2.2.)

If the y axis is oriented parallel to the gravity vector then the total weight is given by

$$W = g \int_V \rho(x, y, z)dV \tag{1.6.9}$$

and its line of action, or, in this case, the point through which it acts is given by

$$\bar{x} = \frac{\displaystyle\int_V x\rho(x, y, z)dV}{\displaystyle\int_V \rho(x, y, z)dV} \qquad \bar{y} = \frac{\displaystyle\int_V y\rho(x, y, z)dV}{\displaystyle\int_V \rho(x, y, z)dV} \qquad \bar{z} = \frac{\displaystyle\int_V z\rho(x, y, z)dV}{\displaystyle\int_V \rho(x, y, z)dV} \tag{1.6.10}$$

The location of this resultant force is called the *center of gravity* and is commonly abbreviated as C.G. Notice that the acceleration due to gravity is a constant that cancels out of the integrals for finding the C.G. and the resulting equations depended only upon the mass density and the volume. This is also called the *center of mass*.

In many situations the solid body is homogenous, that is, the mass density is a constant. In such cases the mass density terms also cancel. Equations 1.6.10 then become

$$\bar{x} = \frac{\displaystyle\int_V x\,dV}{\displaystyle\int_V dV} \qquad \bar{y} = \frac{\displaystyle\int_V y\,dV}{\displaystyle\int_V dV} \qquad \bar{z} = \frac{\displaystyle\int_V z\,dV}{\displaystyle\int_V dV} \tag{1.6.11}$$

This locates the *centroid* of the volume. In such cases the centroid, the center of mass, and the center of gravity are the same point.

When a homogenous body can be divided into sub volumes with simple geometry so that the centroids of the sub volumes are known we can find the centroid of the total using the following formulas.

$$\bar{x} = \frac{\sum x_s V_s}{\sum V_s} \qquad \bar{y} = \frac{\sum y_s V_s}{\sum V_s} \qquad \bar{z} = \frac{\sum z_s V_s}{\sum V_s} \tag{1.6.12}$$

The quantities x_s, y_s, and z_s represent the distances from the base axes to the centroids of the sub volumes V_s. For a body with uniform mass density you can replace the volume in Equation 1.6.12 with the mass or the weight to find the center of mass or the center of gravity. All are at the same location.

###########

Example 1.6.3

Problem: A cylindrical bar has a portion hollowed out as shown in Figure (a). It is made of aluminum which has a mass density of 2.72 *E-06 kg/mm³*. Find its total weight and center of gravity. The y axis is aligned with the gravity vector.

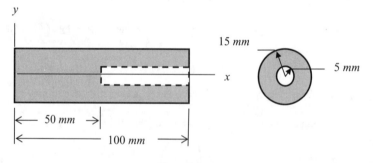

Figure (a)

Solution: The total weight is the volume times the mass density times the acceleration due to gravity. From axial symmetry we know the center of gravity will lie on the centerline of the cylinder. We find the x location by summing moments about the z axis.

To find the total weight we find the weight of the outer cylinder and subtract the weight of the inner cylinder.

$$W_{total} = W_{outer} - W_{inner} = \rho g\left(\pi r_{outer}^2 L - \frac{\pi r_{inner}^2 L}{2}\right)$$

$$= 2.72\cdot 10^{-6}\cdot 9.81\left(\pi\,(15)^2\,100 - \pi\,(5)^2\,50\right) = (1.886 - 0.105)\,N = 1.781N \tag{a}$$

Using symmetry the center of gravity location is located on the centerline of the cylinder and at an x position given by

$$\bar{x}W_{total} = 50W_{outer} - 75W_{inner} = 50 \cdot 1.886 - 75 \cdot 0.105 = 86.425$$

$$\rightarrow \bar{x} = \frac{86.425}{1.781} = 48.526 \; mm \qquad \text{(b)}$$

$$\bar{y} = \bar{z} = 0$$

###########

1.7   Internal Forces and Stresses—Stress Resultants

As we have said, when forces are applied to a solid body that is suitably restrained to eliminate rigid body motions it will deform and internal forces will be generated. It has been found convenient on any internal surface to define *stress* as the force per unit area (N/mm^2). The stress on any internal surface is usually divided into components normal to that surface and tangent to it. The stress and stress components are depicted in Figure 1.7.1.

Consider a force, Δp, acting on an element of area, ΔA, in the interior of the solid. If we resolve the force Δp into components normal and tangential to the surface, Δp_n and Δp_t, respectively, we can define a *normal* component of stress, σ, and a *tangential* or *shearing* (or *shear*) stress component, τ, as shown in Figure 1.7.1.

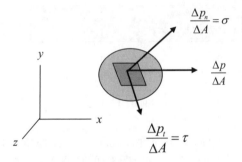

Figure 1.7.1

If we allow ΔA to shrink to an infinitesimal size, then,

$$\lim_{\Delta A \to 0} \frac{\Delta p_n}{\Delta A} = \sigma \qquad \lim_{\Delta A \to 0} \frac{\Delta p_t}{\Delta A} = \tau \qquad (1.7.1)$$

To illustrate internal stresses let us first consider a uniform slender bar in equilibrium with equal but opposite distributed loads on each end as shown in Figure 1.7.2. These distributed forces have units of force per unit area and act on the end surfaces. There are no force components in the y and z directions. The applied force resultants in the x direction are

$$F_0 = \int_A f_x(0, y, z)dA = -F \qquad F_L = \int_A f_x(L, y, z)dA = F \qquad (1.7.2)$$

Figure 1.7.2

Let us choose distributed forces for which the applied moment resultants are zero, that is

$$M_{y0} = \int_A f_x(0, yz)z\, dA = 0 \qquad M_{yL} = \int_A f_x(L, yz)z\, dA = 0$$

$$M_{z0} = \int_A f_x(0, yz)y\, dA = 0 \qquad M_{zL} = \int_A f_x(L, yz)y\, dA = 0$$

(1.7.3)

This is illustrated in Figure 1.7.3.

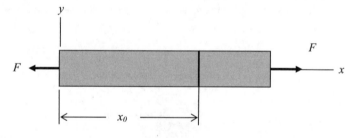

Figure 1.7.3

Now let us ask what the distributed internal force or stress is on an interior surface. Let us choose a flat surface normal to the x axis at, say, $x = x_0$ and examine the stress on that surface by considering the equilibrium of the segment of the bar between $x = x_0$ and $x = L$, as shown in Figure 1.7.4. As noted above, in the usual notation for stress, we designate the normal stress component with the symbol σ and the tangential or shearing stress component with the symbol τ. To further specify the specific components of stress we use a double subscript notation. The first subscript refers to the direction of the normal to the surface and the second to the direction of the stress.

Thus we show the two possible stress components on this particular surface and label them σ_{xx} and τ_{xy}. Since normal stress components always have the same subscript repeated we usually use only one and so we use σ_x for σ_{xx} in subsequent applications.

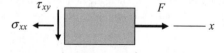

Figure 1.7.4

The sign convention for stresses is different from that for external and restraint forces; the normal and shear stress components are positive as shown in Figure 1.7.4. Sign conventions for stress will be discussed in greater detail shortly.

At this time we do not know the distribution of these stresses on that surface; however, we can define stress resultants on the surface as

$$P = \int_A \sigma_x dA \qquad V = \int_A \tau_{xy} dA \qquad (1.7.4)$$

This is shown in Figure 1.7.5.

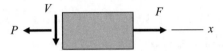

Figure 1.7.5

If we sum the forces acting on the segment in the x and y directions we get

$$\sum F_x = F - P = 0 \quad \rightarrow \quad P = F \qquad \sum F_y = V = 0 \quad \rightarrow \quad V = 0 \qquad (1.7.5)$$

While we do not yet know the actual distribution of stress on this surface we can define an average stress as

$$(\sigma_x)_{ave} = \frac{F}{A} \qquad (\tau_{xy})_{ave} = \frac{V}{A} = 0 \qquad (1.7.6)$$

Now consider that the chosen internal surface is not normal to the axis but is defined by a local coordinate system, tn, at an angle α to the xy coordinate system as shown in Figure 1.7.6.

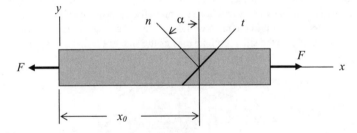

Figure 1.7.6

The stresses on this surface, that we have labeled A_α, are shown in Figure 1.7.7.

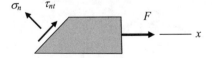

Figure 1.7.7

Once again we can define stress resultants on this surface as

$$P_n = \int_{A_\alpha} \sigma_n dA \qquad V_t = \int_{A_\alpha} \tau_{nt} dA \qquad (1.7.7)$$

This is shown in Figure 1.7.8.

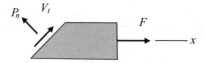

Figure 1.7.8

When we sum the forces acting on the segment in the x and y directions we get

$$\sum F_x = F + V_t \cos \alpha - P_n \sin \alpha = 0 \quad \rightarrow \quad V_t \cos \alpha - P_n \sin \alpha = -F$$
$$\sum F_y = V_t \sin \alpha + P_n \cos \alpha = 0 \quad \rightarrow \quad V_t \sin \alpha + P_n \cos \alpha = 0$$

(1.7.8)

Solving we get

$$P_n = F \sin \alpha \quad \rightarrow \quad V_t = -F \cos \alpha$$

(1.7.9)

And the average stresses are

$$(\sigma_n)_{ave} = \frac{P_n}{A_\alpha} = \frac{F}{A_\alpha} \sin \alpha \qquad (\tau_{nt})_{ave} = \frac{V_t}{A_\alpha} = -\frac{F}{A_\alpha} \cos \alpha$$

(1.7.10)

Stresses are always defined by their magnitude, direction, and the surface upon which they act.

The actual distribution of the stresses on the interior surfaces in this example is discussed in detail in Chapter 4.

To continue our discussion of stress we consider a thin flat plate acted upon by forces in the plane of the plate. We orient this plate in the xy plane of the coordinate system. With forces only in the plane of the plate on the thin plate edge surfaces and no forces on the plate surfaces in the z direction we can assume that the only stress components are those acting in the plane of the plate.

To consider the stress at a point in the plate we take a small rectangular element, $dxdy$, of the plate as shown in Figure 1.7.9 and note the stress components as the size of the element approaches zero.

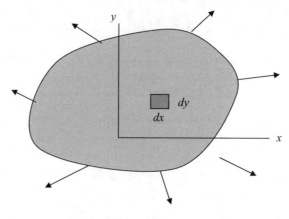

Figure 1.7.9

The stresses on this rectangular element are shown in Figure 1.7.10. This is called a two dimensional state of stress. Remember this is a uniform stress through the thickness at a point where the edges dx and dy approach zero.

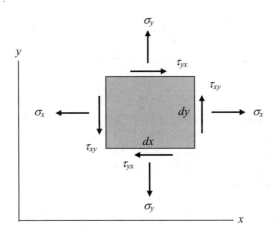

Figure 1.7.10

Stresses are positive as shown. If the normal to the surface upon which the stress acts points in the positive x direction the normal stress is positive in the positive x direction. If the normal points in the negative x direction the normal stress is positive in the negative x direction. If the normal points in the positive x direction the shear stress is positive in the positive y direction. If the normal points in the negative x direction the shear stress is positive in the negative y direction. We see that the forces called stress resultants have a different sign convention than the applied forces.

At this time one would conclude that there are four components of stress in a two dimensional solid - $\sigma_x, \sigma_y, \tau_{xy}, \tau_{yx}$. We note in Figure 1.7.10 that the τ_{xy} and τ_{yx} components each form a couple and as the sides of the rectangle shrink to a point by applying moment equilibrium let us conclude that

$$\tau_{xy} = \tau_{yx} \qquad (1.7.11)$$

This 2D state of stress is often depicted in matrix form as

$$\{\sigma\} = \begin{bmatrix} \sigma_x \\ \sigma_y \\ \tau_{xy} \end{bmatrix} \qquad (1.7.12)$$

To depict a complete state of stress at a point in a three dimensional solid with respect to rectangular Cartesian coordinates consider a rectangular element $dxdydz$ oriented as shown in Figure 1.7.11. Stress components are shown on the three faces with positive normal directions and on the one face with a negative normal (x) direction. Similar components are on the negative y and z directions but are not shown to avoid confusion.

On a face whose normal is in the positive x direction we have three stress components as follows:

$$\sigma_x = \lim_{\Delta A \to 0} \frac{\Delta P_x}{\Delta A} \qquad \tau_{xy} = \lim_{\Delta A \to 0} \frac{\Delta P_y}{\Delta A} \qquad \tau_{xz} = \lim_{\Delta A \to 0} \frac{\Delta P_z}{\Delta A} \qquad (1.7.13)$$

On a face whose normal is in the negative direction of the axis the positive directions of both normal and shearing stresses are in the negative direction of their respective axes.

At this time one would conclude that there are nine components of stress in a three dimensional solid - $\sigma_x, \sigma_y, \sigma_z, \tau_{xy}, \tau_{yx}, \tau_{xz}, \tau_{zx}, \tau_{yz}, \tau_{zy}$. By the same argument used in the two dimensional case each of the three pairs of shear stresses form a couple and as the element shrinks to a point, we can conclude from

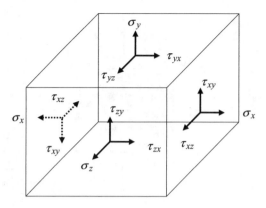

Figure 1.7.11

moment equilibrium that

$$\tau_{xy} = \tau_{yx} \qquad \tau_{yz} = \tau_{zy} \qquad \tau_{zx} = \tau_{xz} \tag{1.7.14}$$

Thus, there are six independent components of stress at a material point within a solid. These are variously arranged in matrix form as a column or square matrix according to their use in equations.

$$\{\sigma\} = \begin{bmatrix} \sigma_x \\ \sigma_y \\ \sigma_z \\ \tau_{xy} \\ \tau_{yz} \\ \tau_{zx} \end{bmatrix} \quad \text{or} \quad [\sigma] = \begin{bmatrix} \sigma_x & \tau_{xy} & \tau_{zx} \\ \tau_{xy} & \sigma_y & \tau_{yz} \\ \tau_{zx} & \tau_{yz} & \sigma_z \end{bmatrix} \tag{1.7.15}$$

At the beginning of our deliberation the stresses, generally, are unknown. Finding stresses in particular situations is a major part of the task before us. The state of stress on surfaces at other orientations than normal to the xyz axis is discussed in great detail in Chapter 9.

Acceptable magnitudes of internal forces are limited by the properties of the material of the solid body. We all know that if they are too great the body will fail either from undesirable permanent deformations or from fracture. Material properties are introduced in Chapter 3 and discussed in greater detail in Chapter 9.

1.8   Restraint Forces and Restraint Force Resultants

Solid bodies can be restrained at a point, they can be restrained along a line, or over a surface. When the body is subjected to applied forces there must be forces at the restraints. In a common statement of the problem these forces are unknown. Part of the purpose of our study in the following chapters is to find the value of these forces. In some circumstances it is possible to predict the value of the resultants of these forces without finding the actual distribution of the forces. This, in fact, is an important lesson of Chapter 2. Just how the restraint forces are distributed will be considered in Chapters 4 and beyond.

1.9 Summary and Conclusions

We have introduced the definition and notation for forces and moments of forces. An important part of the study of the mechanics of material is the interaction among applied forces, restraint forces, and internal forces and the moments generated by them. These forces come in various forms such as concentrated forces with units of Newtons (N), forces per unit length or line forces with units of Newtons per millimeter (N/mm), surface forces with units of Newtons per millimeter squared (N/mm^2), and body forces with units of Newtons per millimeter cubed. (N/mm^3).

We have shown how force resultants of distributed forces are found and how their location is determined. We have defined internal forces and stresses, and restraint forces.

Now we shall put all this to good use in Chapter 2 in establishing static equilibrium, that is, in satisfying Newton's laws as the various forces interact.

2

Static Equilibrium

2.1   Introduction

While all solid bodies deform to some extent under loads these deformations are often so small that the bodies may be considered to be rigid for certain purposes. A rigid body is an idealization that considers a body to have no deformation when subject to loads. When all forces, both known applied forces and unknown restraint forces, are identified and we attempt to sum the forces and moments we can have two possible situations:

1. The number of unknown forces is less than the number of independent equations of motion. Rigid body motion may result. This is the subject of *rigid body dynamics*. We shall not study this case here.
2. The number of unknown forces is equal to or greater than the number of independent equations of motion. There is no rigid body motion and we have *static equilibrium*. There are two subcases:
   a. The body is *statically determinate* when the number of unknown forces is equal to the number of independent equations of static equilibrium. All the unknown forces can be determined by the summation of forces and moments without regard to what the body is made of or to its deformation.
   b. The body is *statically indeterminate* when the number of unknown forces is greater than the number of independent equations of static equilibrium. The summation of forces and moments is not sufficient to find the unknown forces. We need to consider the deformation of the body and the physical properties that contribute to resisting deformation.

In the case of a statically indeterminate body we must introduce additional equations to obtain a solution. In this chapter we shall be interested primarily in statically determinate bodies. The subsequent chapters will contend with both statically determinate and statically indeterminate problems.

2.2   Free Body Diagrams

We shall illustrate each subcase with a simple example in two dimensions using concentrated forces only. In these examples it is assumed that the applied forces initially are known and the restraint forces are unknown.

Consider the *rigid body* shown in Figure 2.2.1. We introduce some common symbols to illustrate forces and restraints. For now we consider only concentrated loads and, in this case, assume they act entirely in the *xy* plane. We add concentrated restraints. The restraint on the lower left prevents movement in both the *x* and *y* directions but does allow rotation about the *z* axis. The restraint on the lower right

Analysis of Structures: An Introduction Including Numerical Methods, First Edition. Joe G. Eisley and Anthony M. Waas.
© 2011 John Wiley & Sons, Ltd. Published 2011 by John Wiley & Sons, Ltd.

prevents movement only in the y direction, since it is assumed to roll freely (without friction) in the x direction. Since it is assumed that there are no forces in the z direction there is no need to consider restraints in the z direction.

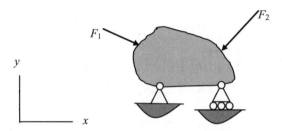

Figure 2.2.1

By replacing the restraint symbols in Figure 2.2.1 with visual representations of the characteristics to which we have assigned them, we create what is called a *free body diagram*. We draw the free body in Figure 2.2.2 showing all the applied forces and restraint forces. Unless otherwise stated we shall use the symbol F for concentrated applied forces and R for concentrated restraint forces. Appropriate subscripts will be added to identify individual forces and components. It is common to represent both the applied forces and the restraint forces in terms of their components in the appropriate coordinate directions. For the applied forces these components are

$$F_{1x} = F_1 \cos \alpha \qquad F_{1y} = F_1 \sin \alpha \qquad F_{2x} = F_2 \cos \beta \qquad F_{2y} = F_2 \sin \beta \qquad (2.2.1)$$

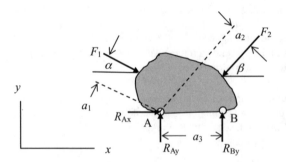

Figure 2.2.2

According to standard sign convention, restraint force components generally are drawn directed along the positive x and y directions, as defined by the coordinate system. If a force component acts in the direction it is drawn, it has a positive value. If it acts in the opposite direction, it has a negative value.

We have static equilibrium between the known applied forces and the unknown restraint forces. There are four known applied force components (F_{1x}, F_{1y}, F_{2x}, and F_{2y}), three unknown restraint force components (R_{Ax}, R_{Ay}, and R_{By}) and three equations of static equilibrium (summation of forces in the x and y directions and summation of moments about the z axis). The summation of moments about the z axis may be taken about any point in the xy plane. In the following equations the summation of moments is taken about point A.

$$\sum F_x = R_{Ax} + F_{1x} - F_{2x} = 0 \quad \rightarrow \quad R_{Ax} = F_{2x} - F_{1x} = F_2 \cos \beta - F_1 \cos \alpha$$

$$\sum M_z = R_{By}d_3 - F_1 a_1 - F_2 a_2 = 0 \quad \rightarrow \quad R_{By} = \frac{F_1 a_1 + F_2 a_2}{a_3}$$

$$\sum F_y = R_{Ay} + R_{By} - F_{1y} - F_{2y} = 0 \quad \rightarrow \quad R_{Ay} = F_{1y} + F_{2y} - \frac{F_1 a_1 + F_2 a_2}{a_3} \qquad (2.2.2)$$

$$= F_1 \sin \alpha + F_2 \sin \alpha - \frac{F_1 a_1 + F_2 a_2}{a_3}$$

In this case, the unknown restraint forces can be found by solving the equations of *static equilibrium*. Bodies that satisfy this condition are called *statically determinate*. These bodies will play a prominent role in subsequent chapters.

The statically determinate case is considered in some detail in this text. Many structures are naturally, or are purposely made to be, statically determinate; however, that will not be the case for many of our problems in later chapters. Indeterminate, also called *redundant*, structures have important positive characteristics and are widely used.

Consider the body in Figure 2.2.3. Suppose we remove the rollers from the bottom right restraint and add a third restraint at the top of the body as shown.

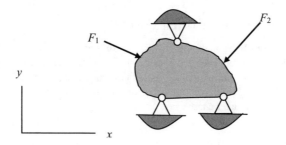

Figure 2.2.3

The free body diagram is shown in Figure 2.2.4.

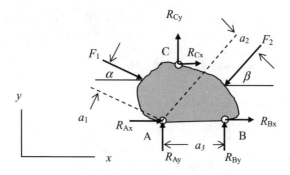

Figure 2.2.4

Just by removing the rollers on the bottom right support we add a fourth restraint force component R_{Ax}. This alone would make the number of unknowns greater than the number of equations of static equilibrium. Add the third support and we then have six unknown restraint force components. We can write down the three equations of static equilibrium and solve them for three of the unknowns in terms of the remaining three unknowns, but, we shall need additional information in order to find all the restraint forces. You shall soon see that, for a deformable body, equations based on distributed displacements and material properties will supply the additional necessary information. Distributed displacements of the deformed body will play a prominent role in subsequent chapters.

The focus of most of this text will be the subject of *statics*. The bodies that we study will be in static equilibrium. Accordingly, all forces and all moments acting on those bodies will always sum to zero. In *dynamics*, on the other hand, bodies often experience rigid body motion and/or nonzero net forces. Some dynamic considerations are introduced in Chapter 15.

So far our examples have been two dimensional. In three dimensions, when presented in rectangular Cartesian coordinates, the equations of statics are

$$\sum F_x = 0 \qquad \sum F_y = 0 \qquad \sum F_z = 0$$
$$\sum M_x = 0 \qquad \sum M_y = 0 \qquad \sum M_z = 0 \tag{2.2.3}$$

where the moment summations are taken about each axis at any point in space. For a statically determinate system, these equations will be sufficient to find the restraint forces and internal forces. For a statically indeterminate system, additional equations based on deformations will be needed.

2.3   Equilibrium—Concentrated Forces

Concentrated forces occur in the form of point loads, restraints at a point, resultants of distributed forces, and stress resultants. It is often very useful to consider a deformable, restrained body under the influence of applied loading to be rigid and to replace its actual applied loads and restraints with equivalent concentrated applied and restraint force components. Several such cases will be introduced here and will have significant applications in later chapters.

2.3.1   Two Force Members and Pin Jointed Trusses

If a structural member has only equal and opposite point forces applied at its ends, such as those examples shown in Figure 2.3.1, it is called a *two force member*. The forces, which could be applied via pins inserted through holes at each end of the member, would, in reality, be distributed, that is, where the pins bear against the inside surfaces of the holes. For simplicity of analysis, however, the actual distributed forces will be represented by their concentrated force resultants acting through the centers of the holes. Exactly how the forces are applied is not now of concern as long as no moments are generated. Since this is the study of *statics*, the member must be in equilibrium, and thus the two forces must be equal and opposite and must act along the direction defined by the dotted lines below.

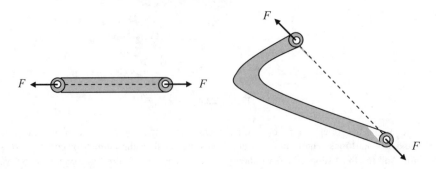

Figure 2.3.1

Structures are often constructed by joining the ends of several two force members with pins. This is called a *pin jointed truss*. A 2D example is shown in Figure 2.3.2a. Both 2D and 3D truss structures are used as load carrying components.

Several conventions are used in illustrating pin jointed trusses. Each of the two force members is shown as a solid line. The actual geometry of the cross section of the member is not of concern at this time. Only the cross sectional area is relevant. The pinned joints are sometimes shown as small circles (Fig. 2.3.2a). Other times the pinned joints are implied and no circles are shown (Fig. 2.3.2b).

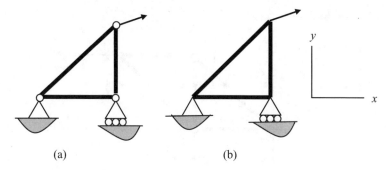

(a)   (b)

Figure 2.3.2

Pinned joints are free to translate in directions in which they are not restrained. In Figure 2.3.2 the lower left joint is restrained in both the x and y directions, and thus is fixed. These two restraints prevent rigid body motion of the truss as a whole. The lower right joint is restrained in the y direction but is free to move in the x direction. The upper right joint is free to move in both the x and y directions. Because two of the joints are able to translate, the individual members are able to deform and rotate.

If the assumption of two force members is to remain valid, forces can be applied only at the joints of the truss. Accordingly, member weights, the resultants of which act through the members' respective centers of gravity, must be able to be neglected. Since the applied forces are frequently much larger than the member weights, this is often the case. In some cases the member weights are divided up as equivalent forces at the joints. Exceptions will be identified explicitly.

############

Example 2.3.1

Problem: Find the restraint forces and the forces acting on each member of the pin jointed truss shown in Figure (a). Neglect the weight of the members.

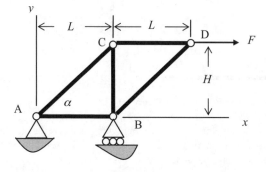

Figure (a)

Solution: We shall first draw a "global" free body diagram for the entire truss to identify the restraint forces generated at its supports. Those restraint forces will then be found using the equations of static equilibrium. Once the restraint forces are known, free body diagrams will be drawn for each member and each joint so that the force on each member can be determined.

Step 1: Draw the global free body diagram (Figure (b). According to standard sign convention, the restraint force components are drawn to have positive values when acting in the positive directions of the coordinate axes.

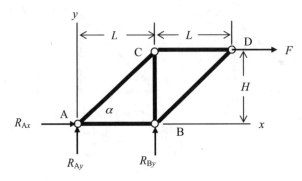

Figure (b)

To find the restraint forces sum all forces in the x and y directions and sum moments about the z axis. Specifically, we shall take moments about point A.

$$\sum F_x = F + R_{Ax} = 0 \quad \rightarrow \quad R_{Ax} = -F$$
$$\sum M_z = R_{By}L - FH = 0 \quad \rightarrow \quad R_{By} = \frac{FH}{L} \tag{a}$$
$$\sum F_y = R_{Ay} + R_{By} = 0 \quad \rightarrow \quad R_{Ay} = -R_{By} \rightarrow R_{Ay} = -\frac{FH}{L}$$

Step 2: We can now draw free body diagrams for each member and each pinned joint (Figures (c) and (d)).

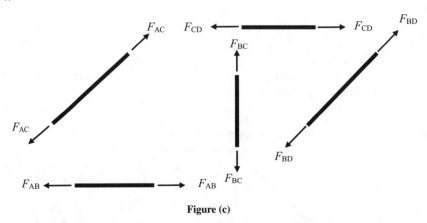

Figure (c)

Free body diagrams for each member are shown in Figure (c). According to standard sign convention, forces acting on each member are drawn to have positive values when putting the member in tension. Thus, compressive forces will have negative values.

Free body diagrams for each pinned joint are shown in Figure (d). According to standard sign convention, joint forces are drawn to have positive values when directed away from the joint. This is consistent with the sign convention used for member forces in that forces directed away from the joint are tensile forces.

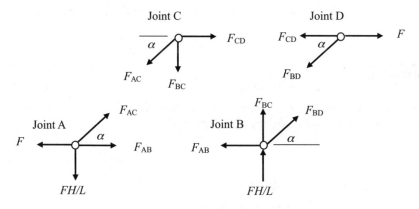

Figure (d)

Because the entire truss is in equilibrium, each member and each joint is also in equilibrium. Accordingly, equations of static equilibrium must be satisfied. Since we are dealing with two force members, the moment equilibrium equation is identically satisfied for each member and each joint. In summing forces at each joint we assume positive forces are in the direction of the positive coordinate axes.

At Joint D:

$$\sum F_y = -F_{BD} \sin \alpha = 0 \quad \rightarrow \quad F_{BD} = 0$$
$$\sum F_x = F - F_{CD} - F_{BD} = 0 \quad \rightarrow \quad F_{CD} = F$$
(b)

At Joint B:

$$\sum F_y = F_{BD} \sin \alpha + F_{BC} + \frac{FH}{L} = 0 \quad \rightarrow \quad F_{BC} = -\frac{FH}{L}$$
$$\sum F_x = F_{BD} \cos \alpha - F_{AB} = 0 \quad \rightarrow \quad F_{AB} = 0$$
(c)

At Joint A:

$$\sum F_y = F_{AC} \sin \alpha - \frac{FH}{L} = 0 \quad \rightarrow \quad F_{AC} = \frac{FH}{\sin \alpha L}$$
$$\sum F_x = F_{AC} \cos \alpha + F_{AB} - F = 0 \quad \rightarrow \quad F_{AC} = \frac{F}{\cos \alpha}$$
(d)

It would appear that at Joint A we have found two different values for F_{AC}; however, if we note that $H/L = \tan \alpha$ then both values are the same.

We have found the values for all the forces on the members and so it is not necessary to sum forces at Joint C; however, we can use Joint C to confirm that the above answers are correct. It is always desirable to have an independent check of any analysis.

At Joint C:

$$\sum F_x = F_{CD} - F_{AC} \cos\alpha = 0 \quad \rightarrow \quad F - \frac{F}{\cos\alpha} \cos\alpha = 0$$

$$\sum F_y = -F_{AC} \sin\alpha - F_{BC} = 0 \quad \rightarrow \quad \frac{F}{\cos\alpha} \sin\alpha - F \tan\alpha = 0$$

(e)

Example 2.3.2

Problem: Find the support reactions (restraint forces) and the forces acting on each member for the pin jointed truss in Figure (a). Neglect the weight of the members.

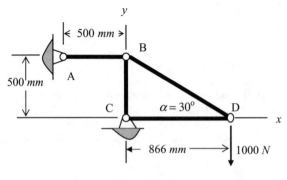

Figure (a)

Solution: This time we shall first set up a free body diagram of one joint, solve for the internal forces in the members connected by that joint, and then proceed from joint to joint without first finding the support reactions.

Step 1: The free body diagram for Joint D is shown in Figure (b).

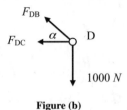

Figure (b)

From summation of forces we get

$$\sum F_y = F_{DB} \sin\alpha - 1000 = 0 \quad \rightarrow \quad F_{DB} = \frac{1000}{\sin\alpha} = \frac{1000}{\sin 30°} = 2000N$$

$$\sum F_x = -F_{DC} - F_{DB} \cos\alpha = 0 \quad \rightarrow \quad F_{DC} = -2000 \cos\alpha = -2000 \cos 30° = -1732N$$

(a)

Step 2: The free body diagram for Joint B is shown in Figure (c).

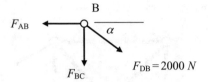

Figure (c)

From summation of forces we get

$$\sum F_x = 2000\cos\alpha - F_{AB} = 0 \quad\rightarrow\quad F_{AB} = 2000\cos\alpha = 2000\cos 30° = 1732N$$
$$\sum F_y = -F_{BC} - F_{DB}\sin\alpha = 0 \quad\rightarrow\quad F_{BC} = -F_{DB}\sin\alpha = -2000\sin 30° = -1000N$$

(b)

As defined previously, a positive value indicates a tensile force and a negative value indicates a compressive force.

Step 3: The free body diagram for Joint A is shown in Figure (d)

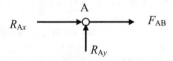

Figure (d)

From summation of forces we get

$$\sum F_x = R_{Ax} + F_{AB} = 0 \quad\rightarrow\quad R_{Ax} = -F_{AB} = -1732N$$
$$\sum F_y = R_{Ay} = 0 \quad\rightarrow\quad R_{Ay} = 0$$

(c)

The free body diagram for Joint C is shown in Figure (e).

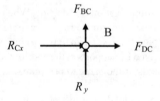

Figure (e)

From summation of forces we get

$$\sum F_x = R_{Cx} + F_{DC} = 0 \quad\rightarrow\quad R_{Cx} = -F_{DC} = 1732N$$
$$\sum F_y = R_{Cy} + F_{BC} = 0 \quad\rightarrow\quad R_{Cy} = -F_{BC} = 1000N$$

(d)

The support reactions are

$$R_{Ax} = -1732N \qquad R_{Ay} = 0 \qquad R_{Cx} = 1732N \qquad R_{Cy} = 1000N$$

(e)

##########

Not all pin jointed trusses are statically determinate. The truss shown in Figure 2.3.3 has more unknown forces than available equations of static equilibrium and therefore is statically indeterminate. Note that the two interior members are not joined where they cross.

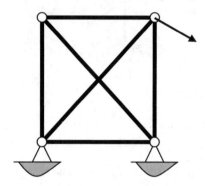

Figure 2.3.3

The method used to find the additional equations required to solve a statically indeterminate truss problem is considered in Chapters 4 and 5.

As stated earlier trusses, both determinate and indeterminate, can also be three dimensional. In such cases, we will have one more relevant force equation and two more relevant moment equations, raising our number of relevant static equilibrium equations from three to six. We will deal with these cases in later chapters.

2.3.2   Slender Rigid Bars

Slender bars that can carry lateral and rotational loads as well as axial loads are one structural form that has received a lot of specialized attention. A slender bar is defined as a solid body having one dimension that is much larger than the other two. Two force members used in trusses are a special case of slender bars where the bar deformation is restricted to be along the axis, that is, a one dimensional axially deformable member.

Using slender bars allows certain simplifications to be made. When forces and moments are distributed over sufficiently small areas, they can be approximated as concentrated forces and moments. For a slender bar, the surface areas of the ends of the bar are very small when compared to the surface area of the entire bar. Accordingly, restraint forces created by supports at the ends of the bar, which are actually distributed over the small surface area of the ends of the bar, can be approximated as concentrated restraint force and moment resultants.

All bodies deform when subjected to loads and restraints; however, in many instances the amount of deformation is very small. In those cases the loaded bodies very nearly maintain their unloaded sizes and shapes; therefore, their dimensions and location of loads and restraints in the unloaded state can be used when establishing, for example, equilibrium in the loaded state. This is an important assumption made in the analysis of structures that undergo small deformations. We did this with two force members in trusses. Thus, we extend this assumption when we examine the equilibrium of deformable bars in this section.

Initially we restrict discussion to two dimensional cases, in which all forces lie in the xy plane and only moments about the z axis can be non zero. This keeps the problems simple and allows us to solve them without becoming overwhelmed by complications in geometry. There are several conventions for showing restraints and depicting restraint forces in drawings and diagrams. The special case of using concentrated fixed restraint force resultants for a slender, rigid bar is shown in Figure 2.3.4 for three different types of supports. Once again we shall use R to designate concentrated restraint forces and use Q to designate concentrated restraint moments, with subscripts used to show the coordinate direction.

While the forces are, in fact, distributed over the surface area of the end of the bar, they are represented here as concentrated resultant forces. In later chapters we shall consider distributed force restraints as well.

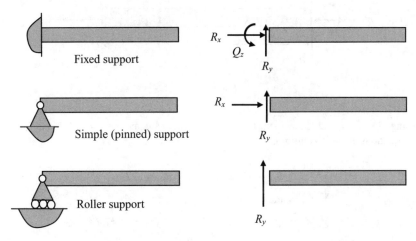

Figure 2.3.4

The forces and moments are shown in what is the convention used in this text for their positive directions. At the beginning of a problem the values of these forces and moments will not be known. To find out more about them is part of what we are about to study.

############

Example 2.3.3

Problem: Consider the rigid bar fixed at the left end and free at the right end as shown in Figure (a). This is known as a *cantilever* support. It has two concentrated applied forces as shown. Find the restraint forces and moments. Neglect the weight of the bar.

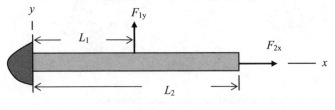

Figure (a)

Solution: Draw a free body diagram and sum forces and moments.

The free body diagram is shown in Figure (b).

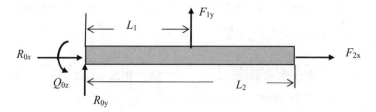

Figure (b)

Assuming F_{1y} and F_{2x} are known, the three restraint forces and moments can be found from the three available equations of static equilibrium.

$$\sum F_x = R_{0x} + F_{2x} = 0 \quad \rightarrow \quad R_{0x} = -F_{2x}$$
$$\sum F_y = R_{0y} + F_{1y} = 0 \quad \rightarrow \quad R_{0y} = -F_{1y} \tag{a}$$
$$\sum M_z = Q_{0z} + F_{1y}L_1 = 0 \quad \rightarrow \quad Q_{0z} = -F_{1y}L_1$$

Example 2.3.4

Problem: Consider the rigid bar with pinned supports at both ends as shown in Figure (a). This is known as a *simply supported* beam. It has two concentrated applied forces as shown. Rollers are added at the right end to allow the force F_{2x} to be transmitted to the bar. Find the restraint forces. Neglect the weight of the bar.

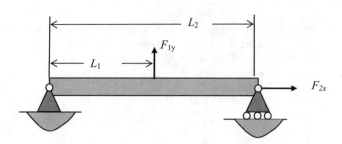

Figure (a)

Solution: Draw a free body diagram and sum forces and moments.

The free body diagram is shown in Figure (b).

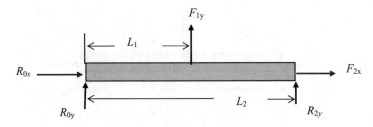

Figure (b)

Assuming F_{1y} and F_{2x} are known, the three restraint forces and moments can be found from the three available equations of static equilibrium.

If we sum moments about the left end and then sum forces we have

$$\sum M_z = F_{1y}L_1 + R_{2y}L_2 = 0 \quad \rightarrow \quad R_{2y} = -\frac{F_{1y}L_1}{L_2}$$

$$\sum F_x = R_{0x} + F_{2x} = 0 \quad \rightarrow \quad R_{0x} = -F_{2x} \tag{a}$$

$$\sum F_y = R_{0y} + F_{1y} + R_{2y} = 0 \quad \rightarrow \quad R_{0y} = -F_{1y} - R_{2y} = -F_{1y} + \frac{F_{1y}L_1}{L_2} = F_{1y}\left(\frac{L_1}{L_2} - 1\right)$$

Example 2.3.5

Problem: Consider the rigid bar fixed at one end and pinned at the other as shown in Figure (a). It has two concentrated applied forces as shown. Find the restraint forces. Neglect the weight of the bar.

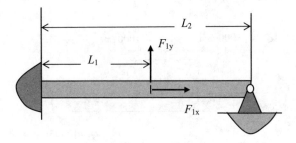

Figure (a)

Solution: Draw a free body diagram and sum forces and moments.

The free body diagram is shown in Figure (b).

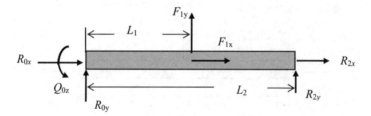

Figure (b)

Assuming F_{1y} and F_{2x} are known, there are now five unknown restraint forces but just three available equations of static equilibrium.

$$\sum F_x = R_{0x} + F_{1x} + R_{2x} = 0$$
$$\sum F_y = R_{0y} + F_{1y} + R_{2y} = 0 \qquad\qquad (a)$$
$$\sum M_z = Q_{0z} + F_{1y}L_1 + R_{2y}L_2 = 0$$

We can reduce the number of unknowns to two by solving the equations but need additional equations to obtain a full solution. This is a statically indeterminate problem. Where and how to find the additional equations is the subject of later chapters.

Example 2.3.6

Problem: Find the support reactions for the built up bar structure shown in Figure (a).
 Neglect the weight of the structure.

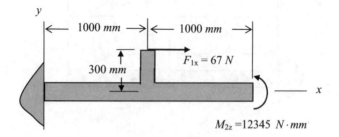

Figure (a)

Solution: Set up the free body diagram and sum forces and moments. Note that we have a concentrated applied moment at the right end.

The free body diagram is shown in Figure (b).

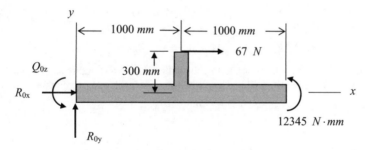

Figure (b)

And the solution is

$$\sum F_x = R_{0x} + 67 = 0 \quad \rightarrow \quad R_{0x} = -67N$$

$$\sum F_y = R_{0y} = 0 \qquad\qquad\qquad\qquad\qquad\qquad \text{(a)}$$

$$\sum M_z = Q_{0z} + 12345 - 67 \cdot 300 = 0 \quad \rightarrow \quad Q_{0z} = 7755N \cdot mm$$

###########

2.3.3 Pulleys and Cables

Flexible cables combined with pulleys enable convenient changes in the direction of tensile forces. Consider the case shown in Figure 2.3.5. Neglect the weight of the cable, pulley, and pin jointed truss support.

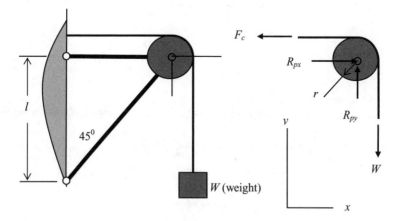

Figure 2.3.5

Using the pulley free body diagram in Figure 2.3.5 we can obtain the force in the cable, F_c, by summing moments about the center of the pulley and the reactions on the pulley hub by summing forces.

Note that the radius of the pulley is r.

$$\sum M_z = F_c r - Wr = 0 \quad \rightarrow \quad F_c = W$$
$$\sum F_x = R_{px} - F_c = 0 \quad \rightarrow \quad R_{px} = F_c = W \qquad\qquad (2.3.1)$$
$$\sum F_y = R_{py} - W = 0 \quad \rightarrow \quad R_{py} = W$$

The free body diagram of the body supporting the pulley is shown is Figure 2.3.6. Note that the pulley and the truss supporting it apply equal and opposite forces on each other through the pin that connects them.

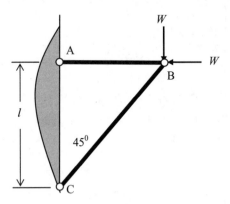

Figure 2.3.6

A free body diagram of the pinned joint supporting the pulley is shown in Figure 2.3.7

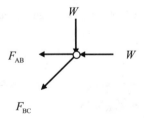

Figure 2.3.7

The two equations of static equilibrium give us the forces in the truss members.

$$\sum F_y = -F_{BC} \cos 45^o - W = 0 \quad \rightarrow \quad F_{BC} = -\frac{W}{\cos 45^o} = -\frac{2W}{\sqrt{2}} \qquad (2.3.2)$$
$$\sum F_x = -F_{AB} - F_{BC} \sin 45^o - W = 0 \quad \rightarrow \quad F_{AB} = -F_{BC} \sin 45^o - W = W - W = 0$$

###########

Example 2.3.7

Problem: Find the applied forces and the support reactions for the structure shown in Figure (a). The radius of the pulley is 100 *mm*. Neglect the weights of the structure and the pulley.

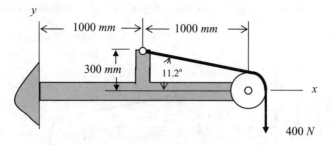

Figure (a)

Solution: To find the forces transmitted to the bar by the pulley set up the free body diagram of the pulley and sum forces and moments. To find the support reactions set up a free body diagram of the bar.

The free body diagram of the pulley is shown in Figure (b).

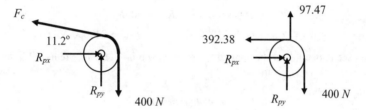

Figure (b)

From moments about the hub of the pulley we have

$$\sum M_z = F_c \cdot 100 - 400 \cdot 100 = 0 \quad \rightarrow \quad F_c = 400N \tag{a}$$

From summation of forces on the pulley we have

$$\sum F_x = R_{px} - 400\cos 11.2° = R_{px} - 392.38 = 0 \quad \rightarrow \quad R_{px} = 392.38N$$
$$\sum F_y = R_{py} + 400\sin 11.2° - 400 = R_{py} - 322.31 = 0 \quad \rightarrow \quad R_{py} = 322.31N \tag{b}$$

Then set up the free body diagram of the structure as shown in Figure (c) and solve.

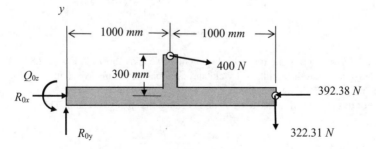

Figure (c)

Actually, it is simpler to use a global free body diagram to find the support reactions. Here it is in Figure (d).

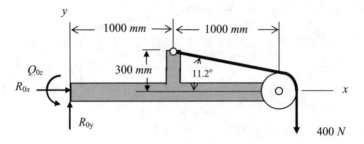

Figure (d)

From Figure (d) we can quickly determine that

$$\sum F_x = R_{0x} = 0$$
$$\sum F_y = R_{0y} - 400 = 0 \quad \rightarrow \quad R_{0y} = 400N$$
$$\sum M_z = Q_{0z} - 400 \cdot 2100 = 0 \quad \rightarrow \quad Q_{0z} = 840000 \ N \cdot mm \tag{c}$$

As we shall see in detail later the applied forces as shown in Figure (c) are useful in finding internal forces and stresses for the structure.

$$\#\#\#\#\#\#\#\#\#\#$$

We can, of course, have components of applied forces and restraints in the z direction and applied moments and restraints about the y axis as well. We shall consider such cases in Section 2.5. Moments about the x axis will be considered separately.

2.3.4 Springs

Springs are often used to support solid bodies. Their behavior is expressed in terms of overall displacement, d, and force, F. Consider a spring acted upon by equal and opposite forces (it is a two force member actually) or restrained at one end and loaded by a force at the other as shown in Figure 2.3.8.

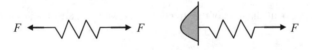

Figure 2.3.8

For a linearly elastic spring the force displacement curve is as shown in Figure 2.3.9. At some value of force and displacement the spring will yield and eventually fracture. We shall limit discussion to the linear elastic range.

The slope of this curve is designated as, k, and is called the *spring constant*. Thus

$$\frac{F}{d} = k \quad \rightarrow \quad F = kd \tag{2.3.3}$$

The spring constant k has units of Newtons per millimeter (N/mm).

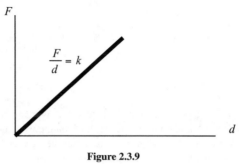

Figure 2.3.9

##########

Example 2.3.8

Problem: A rigid bar of length, L, is supported at the ends by two springs, both having spring constant, k, as shown in Figure (a). Find the forces in and elongations of the springs due to the weight of the bar. Neglect the weight of the springs.

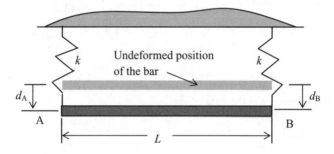

Figure (a)

Solution: Draw a free body diagram. Then solve for the forces in the springs. Finally, use Equation 2.3.3 to find the spring elongations (d_A and d_B). The free body diagram is shown in Figure (b).

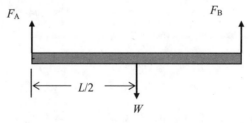

Figure (b)

By summing forces and taking the moment about the left end we get

$$\sum F_y = F_A + F_B - W = 0 \quad \rightarrow \quad F_A + F_B = W$$

$$\sum M_z = F_B \cdot L - W\frac{L}{2} = 0 \quad \rightarrow \quad F_B \cdot L = W\frac{L}{2}$$

(a)

and the solution is

$$F_A = F_B = \frac{W}{2} \quad \rightarrow \quad d_A = d_B = \frac{W}{2k} \qquad \text{(b)}$$

Example 2.3.9

Problem: A weight equal to the weight of the bar is added to the right side of the bar, one fourth of the length of the bar from its right end as shown in Figure (a). Given the value of the spring constant at the left end (k_A) what should be the value of the spring constant at the right end (k_B) for the bar to remain level? Include the weight of the bar but neglect the weight of the springs.

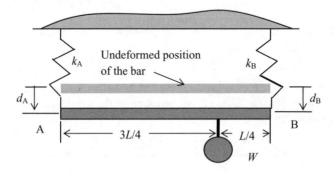

Figure (a)

Solution: Draw a free body diagram. Then solve for the spring forces (F_A and F_B). Finally, use Equation 2.3.3 to find the spring constant of the right spring (k_B).

The free body diagram is shown in Figure (b).

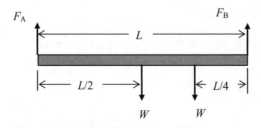

Figure (b)

First find the moment about the left end.

$$\sum M_z = F_B \cdot L - W\frac{L}{2} - W\frac{3L}{4} = 0 \quad \rightarrow \quad F_B = \frac{5W}{4} \qquad \text{(a)}$$

Next sum forces in the y direction.

$$\sum F_y = F_A + F_B - 2W = 0 \quad \rightarrow \quad F_A = 2W - F_B = \frac{3W}{4} \qquad \text{(b)}$$

Let $d_B = d_A$, where d_A and d_B represent the elongations of the respective springs.

Then

$$d_A = \frac{3W}{4k_A} \quad \rightarrow \quad k_B = \frac{F_B}{d_A} = \frac{5W}{4} \cdot \frac{4k_A}{3W} = \frac{5}{3}k_A \tag{c}$$

###########

2.4 Equilibrium—Distributed Forces

In Section 1.6 we noted that distributed forces can be resolved into force resultants which act at a point with a particular line of action. It is often the case that the forces per unit area applied to the surface of a bar, such as those we considered in Section 1.6, are constant across the width of the bar (see Example 1.6.1). These can be resolved into forces per unit length along the bar. Consider a cantilever bar with a distributed load $f_y(x)$ as shown in Figure 2.4.1. Neglect the weight of the bar.

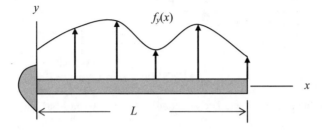

Figure 2.4.1

The distributed force has units of force per unit length (*N/mm*). The force resultant can be found.

$$F_R = \int_0^L f_y(x)dx \tag{2.4.1}$$

The centroid, or location of the line of action, of the distributed force is

$$\bar{x} = \frac{\int_0^L x f_y(x)dx}{\int_0^L f_y(x)dx} \tag{2.4.2}$$

A free body diagram of the beam is shown in Figure 2.4.2 with the force resultant applied in place of the actual distributed force.

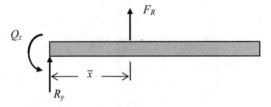

Figure 2.4.2

From summation of forces and moments

$$\sum F_y = R_y + F_R = 0 \quad \rightarrow \quad R_y = -F_R$$
$$\sum M_z = Q_z + F_R \bar{x} = 0 \quad \rightarrow \quad Q_z = -F_R \bar{x}$$

(2.4.3)

############

Example 2.4.1

Problem: The bar in Figure 2.4.1 is given a uniform distributed load (*N/mm*) as shown in Figure (a). Find the support reactions. Neglect the weight of the bar.

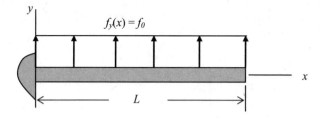

Figure (a)

Solution: Find the force resultant and line of action of the distributed load. Then set up a free body diagram and solve.

The force resultant can be found to be

$$F_R = \int_0^L f_y(x)dx = f_0 \int_0^L dx = f_0 \, x|_0^L = f_0 L$$

(a)

By inspection, the location of the centroid of the distributed load, and thus the location of the line of action of the distributed load, is at $L/2$. Or, if you wish to find it formally by integration.

$$\bar{x} = \frac{f_0 \int_0^L x dx}{f_0 \int_0^L dx} = \frac{f_0 \frac{x^2}{2}\Big|_0^L}{f_0 L} = \frac{L^2}{2L} = \frac{L}{2}$$

(b)

The free body diagram is shown in Figure (b)

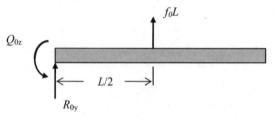

Figure (b)

From summation of forces and moments

$$\sum F_y = R_{0y} + f_0 L = 0 \quad \rightarrow \quad R_{0y} = -f_0 L$$

$$\sum M_z = Q_{0z} + f_0 L \cdot \frac{L}{2} = 0 \quad \rightarrow \quad Q_{0z} = -\frac{f_0 L^2}{2}$$ (c)

Example 2.4.2

Problem: A cantilever bar is loaded with a triangularly distributed line load (*N/mm*) as shown in Figure (a). Find the support reactions. Neglect the weight of the bar.

Solution: Find the force resultant and line of action of the distributed load. Then set up a free body diagram and sum forces and moments.

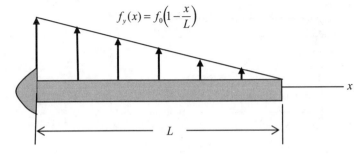

$$f_y(x) = f_0\left(1 - \frac{x}{L}\right)$$

Figure (a)

The resultant of the distributed load is

$$F_y = \int_0^L f_y(x)dx = \int_0^L f_0\left(1 - \frac{x}{L}\right)dx = \frac{f_0 L}{2}$$ (a)

and its line of action passes through the centroid of the load distribution.

$$\bar{x} = \frac{\int_0^L x f_y(x)dx}{\int_0^L f_y(x)dx} = \frac{f_0 \int_0^L x\left(1 - \frac{x}{L}\right)dx}{f_0 \int_0^L \left(1 - \frac{x}{L}\right)dx} = \frac{L}{3}$$ (b)

The free body diagram is shown in Figure (b).

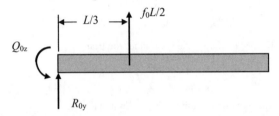

Figure (b)

The support reactions are

$$\sum F_y = R_{0y} + \frac{f_0 L}{2} = 0 \quad \rightarrow \quad R_{0y} = -\frac{f_0 L}{2}$$

$$\sum M_z = Q_{0z} + \frac{f_0 L}{2} \cdot \frac{L}{3} = 0 \quad \rightarrow \quad Q_{0z} = -\frac{f_0 L^2}{6}$$

(c)

Example 2.4.3

Problem: A rigid bar one meter long is loaded as shown in Figure (a). What is the force in the spring and what is the total displacement at the right end of the bar in terms of f_0? Neglect the weights of the bar and the spring.

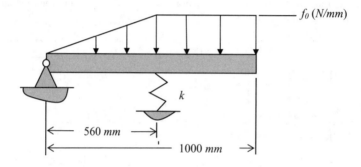

Figure (a)

Solution: First, replace the load distribution with equivalent force resultants. Then draw a free body diagram and sum forces and moments. Finally, calculate the spring deflection (d_s) and use that to find the displacement of the right end (d_L).

It is convenient to split the load distribution into a triangularly distributed load and a uniformly distributed load. The functions for the triangularly distributed load $f_1(x)$ and the uniformly distributed load $f_2(x)$ are

$$f_1(x) = f_0 \frac{x}{560} \quad 0 \le x \le 560$$

$$f_2(x) = f_0 \quad 560 \le x \le 1000$$

(a)

The resultant force, F_1, of the triangularly distributed region is found by integration.

$$F_1 = \int_0^{560} f_0 \frac{x}{560} dx = f_0 \frac{x^2}{1120} \Big|_0^{560} = 280 f_0$$

(b)

You could also find F_1 by calculating the area under the triangular region

$$F_1 = \frac{560}{2} f_0 = 280 f_0$$

(c)

From the area under the rectangular uniformly distributed region

$$F_2 = 440 f_0$$

(d)

The line of action of F_1 is

$$\bar{x}_1 = \frac{1}{F_1} \int_0^{560} x f_0 \frac{x}{560} dx = \frac{1}{F_1} f_0 \frac{x^3}{3 \cdot 560}\Big|_0^{560} = \frac{1120}{3} mm \qquad (e)$$

or, from the equation for the centroid of the triangle, two thirds the distance from the left end.
The line of action of F_2 is at the midpoint of the region of uniformly distributed load.

$$\bar{x}_2 = 560 + 220 = 780 \, mm \qquad (f)$$

The free body diagram is shown in Figure (b).

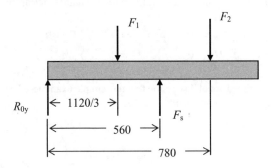

Figure (b)

By summing moments about the left end we find the force in the spring.

$$\sum M_z = 560 F_s - F_1 \frac{1120}{3} - F_2 780 = 0 \quad \rightarrow \quad F_s = \frac{1}{560}\left(280 f_0 \frac{1120}{3} + 440 f_0 780\right) = 799.5 f_0 \qquad (g)$$

The deflection of the spring is

$$d_s = \frac{F_s}{k} = 799.5 \frac{f_0}{k} \qquad (h)$$

The displacement at the right end (d_L) is found by the simple ratio

$$\frac{d_L}{1000} = \frac{d_s}{560} \quad \rightarrow \quad d_L = \frac{1000}{560} d_s = \frac{1000}{560} 799.5 \frac{f_0}{k} = 1428 \frac{f_0}{k} \qquad (i)$$

The restraint force at the left end is found by summing forces in the y direction.

$$\sum F_y = R_{0y} + F_s - F_1 - F_2 = 0$$
$$\rightarrow R_{0y} + 799.5 f_0 - 280 f_0 - 440 f_0 = 0 \qquad (j)$$
$$\rightarrow R_{0y} = -79.5 f_0$$

############

2.5   Equilibrium in Three Dimensions

All the explanations and examples up to now have been limited to two dimensions, that is, to static equilibrium in a plane. We have arbitrarily chosen that plane to be the xy plane and have presented all cases in terms of rectangular Cartesian coordinates. In such cases there are three equations of static equilibrium.

$$\sum F_x = 0 \qquad \sum F_y = 0 \qquad \sum M_z = 0 \qquad\qquad (2.5.1)$$

We are fortunate that a number of real problems can be formulated and solved in this way. Many other real problems must be formulated and solved in terms of three dimensions.

For these cases the equations of static equilibrium in rectangular Cartesian coordinates are

$$\begin{aligned} \sum F_x = 0 \qquad \sum F_y = 0 \qquad \sum F_z = 0 \\ \sum M_x = 0 \qquad \sum M_y = 0 \qquad \sum M_z = 0 \end{aligned} \qquad (2.5.2)$$

We shall limit our three dimensional discussion to a few simple examples. The principles are the same but the application can get quite complicated.

##########

Example 2.5.1

Problem: Find the internal forces in the members of the three dimensional truss shown in Figure (a) in terms of the applied weight, W. Neglect the weight of the truss members.

Solution: Set up free body diagrams and sum forces.

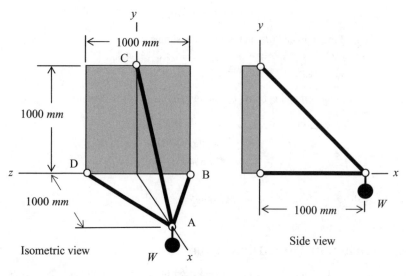

Figure (a)

The free body diagram at Joint A is shown in Figure (b).

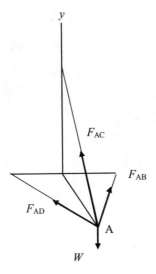

Figure (b)

We sum forces in all three coordinate directions.

$$\sum F_y = F_{AC} \sin 45° - W = 0 \quad \rightarrow \quad F_{AC} = \sqrt{2}W$$

$$\sum F_z = F_{AD} \sin 26.565° - F_{AB} \sin 26.565° = 0 \quad \rightarrow \quad F_{AD} = F_{AB}$$

$$\sum F_x = F_{AD} \cos 26.565° + F_{AB} \cos 26.565° + F_{AC} \cos 45° = 0$$

(a)

$$\rightarrow \quad F_{AD} + F_{AB} = -\frac{W}{\cos 26.565°}$$

$$\rightarrow \quad F_{AD} = F_{AB} = -\frac{W}{2\cos 26.565°} = -0.559W$$

The internal forces are the same as the forces at the joints.

$$P_{AB} = F_{AB} = -0.559W \qquad P_{AC} = F_{AC} = \sqrt{2}W \qquad P_{AD} = F_{AD} = -0.559W \qquad \text{(b)}$$

Example 2.5.2

Problem: Find the support reactions for the rigid bar shown in Figure (a). A free body diagram is shown in Figure (b). Restraint moments are indicated by double arrowheads. Neglect the weights of the bar and the flanges.

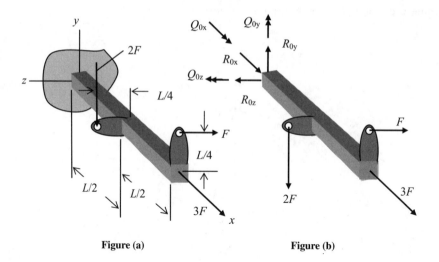

Figure (a)   **Figure (b)**

Solution: Sum forces and moments in all six degrees of freedom.

The sum of the forces and moments and the solutions are

$$\sum F_x = R_{0x} + 3F = 0 \quad \rightarrow \quad R_{0x} = -3F$$

$$\sum F_y = R_{0y} - 2F = 0 \quad \rightarrow \quad R_{0y} = 2F$$

$$\sum F_z = R_{0z} - F = 0 \quad \rightarrow \quad R_{0z} = F$$

$$\sum M_x = Q_{0x} + 2F \cdot \frac{L}{4} - F \cdot \frac{L}{4} = 0 \quad \rightarrow \quad Q_{0x} = -\frac{FL}{4} \tag{a}$$

$$\sum M_y = Q_{0y} + F \cdot L = 0 \quad \rightarrow \quad Q_{0y} = -FL$$

$$\sum M_z = Q_{0z} - 2F \cdot \frac{L}{2} = 0 \quad \rightarrow \quad Q_{0z} = FL$$

##########

2.6   Equilibrium—Internal Forces and Stresses

Up to now we have been concerned with the equilibrium of the entire body or the separate members under the actions of applied loads and restraints. Now let us consider what is happening in the interior of the body. Any portion of the interior must also be in equilibrium. A free body diagram can be drawn for a portion of the solid body so that the summation of forces may consist of both external (applied) forces and internal forces (stress resultants). In Section 2.3 we found the forces acting on a member of a determinate pin jointed truss. A simple free body diagram of a portion of that member shows that the internal force is the same as the applied force (Fig. 2.6.1).

We label the internal force, P, and note that it must be equal to F.

$$\sum F_x = F - P = 0 \quad \rightarrow \quad P = F \tag{2.6.1}$$

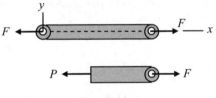

Figure 2.6.1

P is the force resultant of the stress distribution over the cross section upon which it acts. For the given coordinate system, the normal stress on that particular surface would be labeled σ_x. Thus

$$P = \int_A \sigma_x dA \qquad (2.6.2)$$

We know that the value of P is equal to the integral of the stress, σ_x, acting on the cross section surface but we do not yet know how that stress is distributed. The actual distribution of that stress is considered in detail in Chapter 4.

Similarly we can make a free body diagram of the portion of the bar in Figure 2.6.2 that is to the right of a coordinate x_0 on the bar.

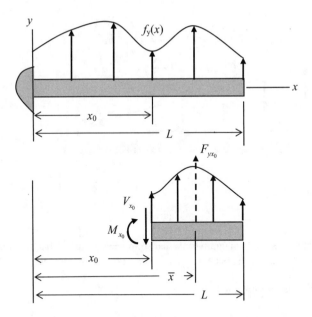

Figure 2.6.2

We show an internal transverse force, V_{x_0}, and an internal moment, M_{x_0}, on the bar at $x = x_0$. We choose the notation V and M to distinguish the internal quantities from the restraint quantities which are generally designated by R and Q, for quantities of force and moment, respectively. We also choose a sign convention that will be explained in detail in subsequent chapters. Shown are the directions of positive values for internal force resultants in a slender bar according to said sign convention.

To find the internal force and moment we first find the resultant of the applied distributed load and its location for the portion of the bar to the right of x_0.

$$F_{yx_0} = \int_{x_0}^{L} f_y(x)\, dx \quad \rightarrow \quad \bar{x} = \frac{\int_{x_0}^{L} x f_y(x)}{F_{yx_0}} \qquad (2.6.3)$$

Then we can calculate V and M at x_0 by summation of forces and moments.

$$\sum F_y = F_{yx_0} - V_{x_0} = 0 \quad \rightarrow \quad V_{x_0} = F_{yx_0}$$
$$\sum M_z = (\bar{x} - x_0) F_{yx_0} - M_{x_0} = 0 \quad \rightarrow \quad M_{x_0} = (\bar{x} - x_0) F_{yx_0} \qquad (2.6.4)$$

Once again we can find the internal stress resultants but we do not yet know how the stress is distributed. There would be both a shearing stress, τ_{xy}, and a normal stress, σ_x, acting on a surface normal to the x axis. Thus the stress resultants are

$$V_{x_0} = \int_A \tau_{xy}\, dA \qquad M_{x_0} = \int_A y\sigma_x\, dA \qquad (2.6.5)$$

The actual distribution of the stresses is considered in detail in Chapter 7.

##########

Example 2.6.1

Problem: For the bar in Figure (a) find the internal forces at $L/2$.

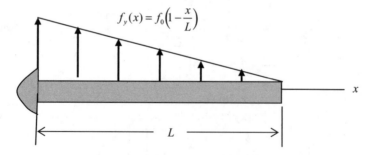

$$f_y(x) = f_0\left(1 - \frac{x}{L}\right)$$

Figure (a)

Solution: First set up a free body diagram of the bar from $x = L/2$ to L. Then find the force resultant of the applied distributed load and its location. Finally, sum forces and moments.

The free body diagram is shown in Figure (b).

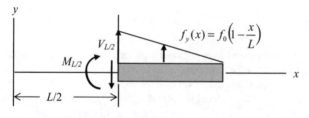

$$f_y(x) = f_0\left(1 - \frac{x}{L}\right)$$

Figure (b)

The applied force resultant for the portion of the beam from $x = L/2$ to L is found by taking the area of the triangle.

$$F_R = \frac{1}{2} \cdot \frac{f_0}{2} \cdot \frac{L}{2} = \frac{f_0 L}{8} \tag{a}$$

The location of the resultant is at the centroid of the triangle which is one third the distance from its left side. As measured from the origin the line of action is

$$\bar{x} = \frac{L}{2} + \frac{1}{3}\frac{L}{2} = \frac{2L}{3} \tag{b}$$

The free body diagram with force resultants for the portion of the beam outboard of $x = L/2$ is shown in Figure (c).

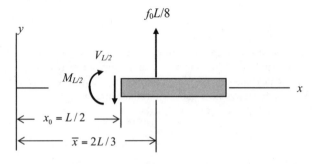

Figure (c)

The internal force and moment are

$$\sum F_y = \frac{f_0 L}{8} - V_{L/2} = 0 \quad \rightarrow \quad V_{L/2} = \frac{f_0 L}{8}$$

$$\sum M_z = \frac{f_0 L}{8} \cdot \frac{L}{6} - M_{L/2} = 0 \quad \rightarrow \quad M_{L/2} = \frac{f_0 L^2}{48} \tag{c}$$

###########

Now let us go completely into the interior of the body and look at equilibrium there. This time we deal directly with the stresses rather than with the stress resultants.

2.6.1 Equilibrium of Internal Forces in Three Dimensions

Consider a small rectangular element of infinitesimal dimensions dx by dy by dz somewhere inside the boundaries of a three dimensional solid body as shown in Figure 2.6.3. The body is restrained and has applied loads.

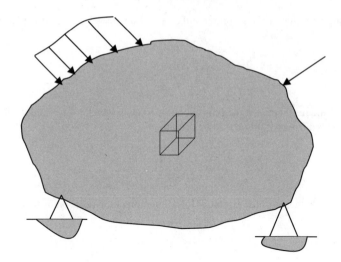

Figure 2.6.3

We wish to place this internal element in equilibrium. Using a Taylor series expansion we can write the change in stress from one face of the element to another face dx away as shown in Figure 2.6.4. We show the stress vectors as solid lines on faces with normals in the positive coordinate directions and as dotted lines on faces with normals in the negative coordinate directions.

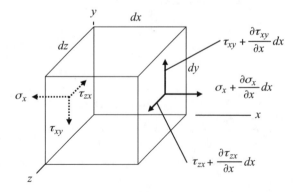

Figure 2.6.4

For example, on the left face of the cube we have $\sigma_x(0, y, z)$ and on the right face we have $\sigma_x(0 + dx, y, z)$. We do a Taylor's series expansion and neglect higher order terms.

$$\sigma_x(0 + dx, y, z) = \sigma_x(0, y, z) + \frac{\partial \sigma_x}{\partial x} dx + \text{higher order terms} \qquad (2.6.6)$$

Also on those same two faces we have

$$\tau_{zx}(0 + dx, y, z) = \tau_{zx}(0, y, z) + \frac{\partial \tau_{zx}}{\partial x} dx \quad \tau_{xy}(0 + dx, y, z) = \tau_{xy}(0, y, z) + \frac{\partial \tau_{xy}}{\partial x} dx \qquad (2.6.7)$$

A body force, that is, a force per unit volume, may also be acting on the body. It is designated by $f_b(x, y, z)$. An example of a body force is the force due to gravity or an inertial force caused by acceleration. In many, but not all, of our examples this body force is small enough compared to the other forces present that it may be ignored.

Similar expansions can be written for the faces of constant y and z as well. In Figure 2.6.5 we show all of the components in the x direction.

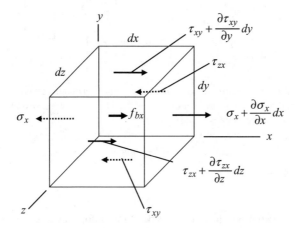

Figure 2.6.5

These include the expansions

$$\tau_{xy}(x, 0 + dy, z) = \tau_{xy}(x, 0, z) + \frac{\partial \tau_{xy}}{\partial y} dy \qquad (2.6.8)$$

and

$$\tau_{zx}(x, y, 0 + dz) = \tau_{zx}(x, y, 0) + \frac{\partial \tau_{zx}}{\partial z} dz \qquad (2.6.9)$$

For equilibrium, summation of forces in the x direction results in

$$\sum F_x = \left(\sigma_x + \frac{\partial \sigma_x}{\partial x} dx\right) dydz - \sigma_x dydz + \left(\tau_{xy} + \frac{\partial \tau_{xy}}{\partial y} dy\right) dxdz - \tau_{xy} dxdz$$
$$+ \left(\tau_{zx} + \frac{\partial \tau_{zx}}{\partial z} dz\right) dxdy - \tau_{zx} dxdy + f_{bx} dxdydz = 0 \qquad (2.6.10)$$

This simplifies to

$$\frac{\partial \sigma_x}{\partial x} + \frac{\partial \tau_{xy}}{\partial y} + \frac{\partial \tau_{zx}}{\partial z} + f_{bx} = 0 \qquad (2.6.11)$$

Similar expansions and summations in the y and z directions result in

$$\frac{\partial \tau_{xy}}{\partial x} + \frac{\partial \sigma_y}{\partial y} + \frac{\partial \tau_{yz}}{\partial z} + f_{by} = 0$$
$$\frac{\partial \tau_{zx}}{\partial x} + \frac{\partial \tau_{yz}}{\partial y} + \frac{\partial \sigma_z}{\partial z} + f_{bz} = 0 \qquad (2.6.12)$$

This may be put in matrix form as shown in Equation 2.6.13.

$$[E]\{\sigma\} = \begin{bmatrix} \dfrac{\partial}{\partial x} & 0 & 0 & \dfrac{\partial}{\partial y} & 0 & \dfrac{\partial}{\partial z} \\[2mm] 0 & \dfrac{\partial}{\partial y} & 0 & \dfrac{\partial}{\partial x} & \dfrac{\partial}{\partial z} & 0 \\[2mm] 0 & 0 & \dfrac{\partial}{\partial z} & 0 & \dfrac{\partial}{\partial y} & \dfrac{\partial}{\partial x} \end{bmatrix} \begin{bmatrix} \sigma_x \\ \sigma_y \\ \sigma_z \\ \tau_{xy} \\ \tau_{yz} \\ \tau_{zx} \end{bmatrix} = - \begin{bmatrix} f_{bx} \\ f_{by} \\ f_{bz} \end{bmatrix} = -\{f_b\} \qquad (2.6.13)$$

These are the three dimensional *differential equations of equilibrium* in terms of the stress components in the interior of the body. These equations must be satisfied to obtain what we call an exact solution for the stress distribution in a three dimensional solid body.

############

Example 2.6.2

Problem: A cube of material is subject on all sides to a uniform hydrostatic pressure as shown in Figure (a). The same pressures are on all six faces. What are the stresses in the cube?

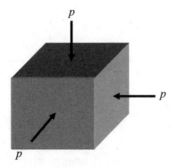

Figure (a)

Solution: A possible solution would be uniform normal stress in all directions equal to the applied pressure. Give it a try.

Substitute this assumed solution in the three dimensional equations of equilibrium to see if it satisfies equilibrium. Let

$$\sigma_x = \sigma_y = \sigma_z = -p \qquad \tau_{xy} = \tau_{yz} = \tau_{zx} = 0 \qquad\qquad (a)$$

Substitute the values in Equation (a) into Equation 2.6.13.

$$\begin{bmatrix} \dfrac{\partial}{\partial x} & 0 & 0 & \dfrac{\partial}{\partial y} & 0 & \dfrac{\partial}{\partial z} \\[2mm] 0 & \dfrac{\partial}{\partial y} & 0 & \dfrac{\partial}{\partial x} & \dfrac{\partial}{\partial z} & 0 \\[2mm] 0 & 0 & \dfrac{\partial}{\partial z} & 0 & \dfrac{\partial}{\partial y} & \dfrac{\partial}{\partial x} \end{bmatrix} \begin{bmatrix} -p \\ -p \\ -p \\ 0 \\ 0 \\ 0 \end{bmatrix} = \begin{bmatrix} 0 \\ 0 \\ 0 \end{bmatrix} = - \begin{bmatrix} f_{bx} \\ f_{by} \\ f_{bz} \end{bmatrix} \qquad\qquad (b)$$

Since the derivative of a constant is zero it is seen to be a solution if no body force is applied. And the stress state does match the applied loads at the boundary faces. Both equilibrium and boundary conditions are satisfied.

###########

2.6.2 Equilibrium in Two Dimensions—Plane Stress

In many cases the applied forces and restraint forces all lie within the same plane. This often happens when a thin sheet is loaded by in-plane edge loads. This state of stress is called *plane stress*. The plane stress equations of equilibrium are obtained from Figure 2.6.2, in which a small element of dimensions dx by dy and thickness h is shown under a state of plane stress.

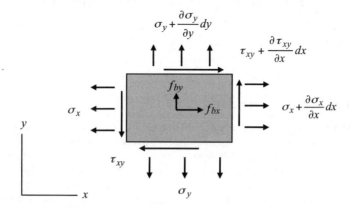

Figure 2.6.2

Equilibrium, from summation of forces in the x and y directions, results in

$$\sum F_x = \left(\sigma_x + \frac{\partial \sigma_x}{\partial x}dx\right)hdy - \sigma_x hdy + \left(\tau_{xy} + \frac{\partial \tau_{xy}}{\partial y}dy\right)hdx - \tau_{xy}hdx + f_{bx}hdxdy = 0$$

$$\sum F_y = \left(\sigma_y + \frac{\partial \sigma_y}{\partial y}dy\right)hdx - \sigma_y hdx + \left(\tau_{xy} + \frac{\partial \tau_{xy}}{\partial x}dx\right)hdy - \tau_{xy}hdy + f_{by}hdxdy = 0$$

(2.6.14)

This reduces to

$$\frac{\partial \sigma_x}{\partial x} + \frac{\partial \tau_{xy}}{\partial y} + f_{bx} = 0$$

$$\frac{\partial \tau_{xy}}{\partial x} + \frac{\partial \sigma_y}{\partial y} + f_{by} = 0$$

(2.6.15)

The plane stress equilibrium equations can be written in matrix form as

$$[E]\{\sigma\} = \begin{bmatrix} \dfrac{\partial}{\partial x} & 0 & \dfrac{\partial}{\partial y} \\ 0 & \dfrac{\partial}{\partial y} & \dfrac{\partial}{\partial x} \end{bmatrix} \begin{bmatrix} \sigma_x \\ \sigma_y \\ \tau_{xy} \end{bmatrix} = -\begin{bmatrix} f_{bx} \\ f_{by} \end{bmatrix} = -\{f_b\}$$

(2.6.16)

One may conclude that a uniform stress on the boundary would produce a uniform stress in the interior. We shall examine this in more detail in Chapter 4.

2.6.3   Equilibrium in One Dimension—Uniaxial Stress

In an even more simple case, all of the applied forces and restraint forces lie along a single line and there is stress in only one direction. This stress state is called *uniaxial stress* and is shown in Figure 2.6.3.

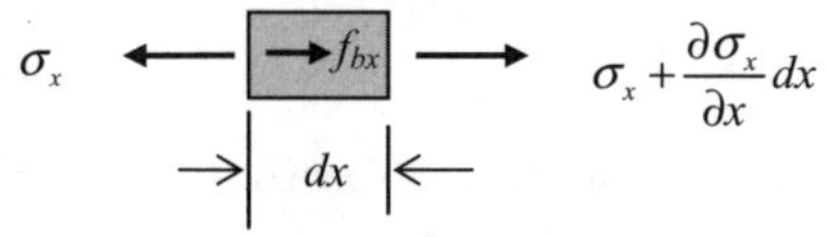

Figure 2.6.3

The one dimensional equation of equilibrium is found by summing forces.

$$\sum F_x = \left(\sigma_x + \frac{\partial \sigma_x}{\partial x} dx \right) h dy - \sigma_x h dy + f_{bx} h dx dy = 0 \qquad (2.6.17)$$

This reduces to

$$\frac{\partial \sigma_x}{\partial x} = -f_{bx} \qquad (2.6.18)$$

2.7   Summary and Conclusions

We have applied the equations of static equilibrium to a number of cases. We have limited the geometry to simplified or idealized bodies. Geometry simplification has come in the form of two forces members, cables, springs, and slender, rigid bars. These topics will be expanded in subsequent chapters. Finally, we have examined the equilibrium of internal forces in terms of the six components of stress.

The task before us is to find a connection between the external forces and restraints acting on the body and the stress state within the body. Trying to satisfy the equations of equilibrium for the stresses within the body for different combinations of external forces and restraints can be quite difficult. We understand the relationship between external forces and force resultants, but we don't yet understand the exact stress distribution itself.

Fortunately, by simplifying geometry, loading, and restraints it is possible to work with one and two dimensional equations in many practical situations. Examples of this simplified approach make up most of the rest of this text; however, we shall touch base with some three dimensional examples to justify the simplifying assumptions. Before trying to solve any problems of deformable bodies we need some further definitions and equations. That will be the subject of the next chapter.

3

Displacement, Strain, and Material Properties

3.1 Introduction

We now provide a brief look at the representation of displacements and deformation of the solid body. We know from experience that solid bodies are not truly rigid. The deformation may be small and not visually perceptible, such as when we walk across a floor, or quite perceptible, such as when we jump from a diving board. Our use of rigid body equilibrium equations assumes that the displacements are so small they do not affect the equations, that is, we can calculate the effect of the loads using the undeformed geometry. This is a central assumption in problems of linear kinematics, which is relaxed when we consider buckling problems in later chapters. We shall soon learn that knowing information about displacements is essential to solving indeterminate structures. The remaining chapters treat this in great detail.

In the process of finding displacements it is desirable to define a quantity called *strain*. Strain is essentially a measure of the rate of change of displacement with respect to position. At first you may not see why it is necessary to define strain but it soon becomes obvious when material properties are discussed.

For a given geometry we are well aware that the size of the displacement is related to the size of the load. We are also aware that the size of the displacement for a given load and geometry will depend upon the material. Under a given tensile load a rubber band will stretch much more than a steel wire of the same cross section area. How we represent material properties is introduced later in this chapter.

3.2 Displacement and Strain

When a restrained body is acted upon by external forces it deforms and internal forces and stresses are generated. When a body deforms the distance between any two particles within the body changes, that is, the positions of particles within the body, in general, will change by different amounts. If all points in the body change their location by the same amount the body displaces without deforming. This is referred to as *rigid body* displacement.

To solve certain problems it is necessary to find relations between the displacements and the internal stresses. To do this directly is not easy. An analysis of the deformation of the body through the use of strain facilitates the connection between stress and displacement.

Analysis of Structures: An Introduction Including Numerical Methods, First Edition. Joe G. Eisley and Anthony M. Waas.
© 2011 John Wiley & Sons, Ltd. Published 2011 by John Wiley & Sons, Ltd.

3.2.1   Displacement

First consider a three dimensional solid body at rest and unloaded but suitably restrained from having any rigid body motion. Then, as it is loaded, it will deform. Consider a point within the body with coordinates (x, y, z) in the undeformed state. After deformation this point will have a new position coordinate (x^*, y^*, z^*). If the displacement components $u(x,y,z)$, $v(x,y,z)$, and $w(x,y,z)$, are introduced in the x, y, and z directions then

$$x^* = x + u \qquad y^* = y + v \qquad z^* = z + w \tag{3.2.1}$$

Throughout this text we shall assume that the displacements are so small that they will have negligible effect on the location of loads and restraints, that is, changes in dimensions due to displacements caused by external loads are ignored unless otherwise stated. This is typical of most, but not all, situations that we shall consider.

At the beginning of our deliberations the displacements, generally, are unknown. Finding them in particular situations is a major part of the task before us.

3.2.2   Strain

The amount of deformation for any given applied load and set of restraints depends upon the properties of the material of the solid body. Some materials resist deformation more than others, for example, steel compared to soft rubber. To discuss this resistance to deformation it is convenient, indeed necessary, to define a deformation related quantity called *strain*. So far the quantities we have discussed – force, force per unit area (stress or at least pressure), and displacement are familiar from studies in physics. Unless you have had previous study in the behavior of solid bodies you may find strain less intuitive. Its value will soon become clear.

For an introductory and elementary look at strain we find it convenient to consider a geometrically simple solid body, namely, a slender bar with a uniform cross section shown unloaded and undeformed by the shaded image in Figure 3.2.1. Away for the ends of the bar we scribe two lines a distance L apart. Now we load it on the ends by equal and opposite forces and note that it elongates by the amount shown by the unshaded extensions at each end. Next we measure the new distance between the scribed lines and note that the lines are now an additional distance apart, d.

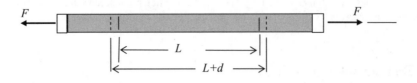

Figure 3.2.1

We call the ratio of this change in length to the original length as the *strain*, ε, that is,

$$\varepsilon = \frac{d}{L} \tag{3.2.2}$$

In a more general case of loading we assume that the distributed displacement in the x direction can be represented by a single independent variable, the displacement $u(x)$. Given a quantity $u(x)$, we can define a quantity called the *normal strain*, ε_x, at any point along the bar. Consider the bar in Figure 3.2.2 and the displacement at x and at a point $x + \Delta x$ as shown in the figure.

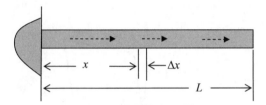

Figure 3.2.2

At any point along the bar the local value of strain can be defined as

$$\varepsilon_x = \lim_{\Delta x \to 0} \frac{u(x + \Delta x) - u(x)}{\Delta x} = \lim_{\Delta x \to 0} \frac{\Delta u}{\Delta x} = \frac{du}{dx} \tag{3.2.3}$$

If we take a more general example, namely, a three dimensional body under the action of normal stress in all three coordinate directions by a similar argument we obtain

$$\varepsilon_x = \lim_{\Delta x \to 0} \frac{\Delta u}{\Delta x} = \frac{\partial u}{\partial x} \qquad \varepsilon_y = \lim_{\Delta y \to 0} \frac{\Delta v}{\Delta y} = \frac{\partial v}{\partial y} \qquad \varepsilon_z = \lim_{\Delta z \to 0} \frac{\Delta w}{\Delta z} = \frac{\partial w}{\partial z} \tag{3.2.4}$$

It is observed that when a body is subjected to shearing stresses it will distort in another way. Consider an infinitesimal cubical element imbedded in a solid body and subjected to shearing stress τ_{xy}. Once again we start with a geometrically simple body. The imposed shear stresses are shown in Figure 3.2.3 only in the xy directions, that is, there are no z direction components of shear stress in this example.

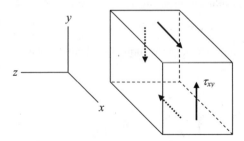

Figure 3.2.3

Such an element will distort under small displacement. Looking at it in the xy plane as shown in Figure 3.2.4 we see the body before and after strain.

The *shear strain* or *shearing strain* is defined as the change in angular orientation of points on the body. In Figure 3.2.4 after deformation the line elements AC and AD have rotated through the angles γ_1 and γ_2 to the positions A'C' and A'D'. Since the derivatives of the displacements are small, these angles are approximately equal to the sine of the angle. Thus

$$\gamma_1 = \frac{\partial u}{\partial y} \qquad \gamma_2 = \frac{\partial v}{\partial x} \tag{3.2.5}$$

It may be seen that the decrease in the right angle formed by AC and AD when the body deforms to the position A'C' and A'D' is equal to the shearing strain.

$$\gamma_{xy} = \gamma_1 + \gamma_2 = \frac{\partial u}{\partial y} + \frac{\partial v}{\partial x} \tag{3.2.6}$$

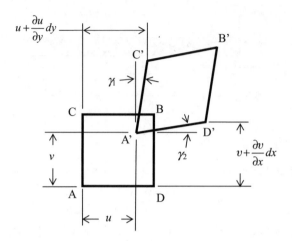

Figure 3.2.4

In a similar way the shearing strains in the yz and zx planes may be found to be

$$\gamma_{yz} = \frac{\partial w}{\partial y} + \frac{\partial v}{\partial z} \qquad \gamma_{zx} = \frac{\partial u}{\partial z} + \frac{\partial w}{\partial x} \qquad (3.2.7)$$

Equations 3.2.4, 3.2.6, and 3.2.7 collectively are called the *strain displacement relations*. There are six independent components of strain since

$$\gamma_{xy} = \gamma_{yx} \qquad \gamma_{yz} = \gamma_{zy} \qquad \gamma_{zx} = \gamma_{xz} \qquad (3.2.8)$$

The strains are variously arranged in matrix form as a column or square matrix according to their use in equations.

$$\{\varepsilon\} = \begin{bmatrix} \varepsilon_x \\ \varepsilon_y \\ \varepsilon_z \\ \gamma_{xy} \\ \gamma_{yz} \\ \gamma_{zx} \end{bmatrix} \quad \text{or} \quad [\varepsilon] = \begin{bmatrix} \varepsilon_x & \gamma_{xy} & \gamma_{zx} \\ \gamma_{xy} & \varepsilon_y & \gamma_{yz} \\ \gamma_{zx} & \gamma_{yz} & \varepsilon_z \end{bmatrix} \qquad (3.2.9)$$

The matrix form of the strain displacement relations for a three dimensional solid is

$$\{\varepsilon\} = \begin{bmatrix} \varepsilon_x \\ \varepsilon_y \\ \varepsilon_z \\ \gamma_{xy} \\ \gamma_{yz} \\ \gamma_{zx} \end{bmatrix} = \begin{bmatrix} \dfrac{\partial}{\partial x} & 0 & 0 \\ 0 & \dfrac{\partial}{\partial y} & 0 \\ 0 & 0 & \dfrac{\partial}{\partial z} \\ \dfrac{\partial}{\partial y} & \dfrac{\partial}{\partial x} & 0 \\ 0 & \dfrac{\partial}{\partial z} & \dfrac{\partial}{\partial y} \\ \dfrac{\partial}{\partial z} & 0 & \dfrac{\partial}{\partial x} \end{bmatrix} \begin{bmatrix} u \\ v \\ w \end{bmatrix} = [D]\{u\} \qquad (3.2.10)$$

When a thin sheet is acted upon only by in plane loads there are no stresses in, say, the z direction. As noted in Section 2.6 this is called plane stress. The corresponding strain displacement equations are

$$\{\varepsilon\} = \begin{bmatrix} \varepsilon_x \\ \varepsilon_y \\ \gamma_{xy} \end{bmatrix} = \begin{bmatrix} \dfrac{\partial}{\partial x} & 0 \\ 0 & \dfrac{\partial}{\partial y} \\ \dfrac{\partial}{\partial y} & \dfrac{\partial}{\partial x} \end{bmatrix} \begin{bmatrix} u \\ v \end{bmatrix} = [D]\{u\} \qquad (3.2.11)$$

For a slender bar loaded in one dimension the equations become

$$\varepsilon_x = \frac{du}{dx} \qquad (3.2.12)$$

##########

<h2>Example 3.2.1</h2>

Problem: A 100 *mm* square thin steel plate is deformed from the shape ABCD to A'B'C'D' indicated by the dotted lines as shown in Figure (a). Determine ε_x, ε_y, γ_{xy}.

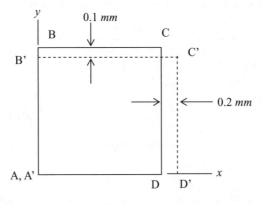

Figure (a)

Solution: The normal strain is the change in length divided by the length. The shearing strain is the change in angle between two lines at right angles.

Comparing the geometry of the deformed part to the original we obtain

$$\varepsilon_x = \frac{A'D' - AD}{AD} = \frac{0.2}{100} = 0.002$$

$$\varepsilon_y = \frac{A'B' - AB}{AB} = -\frac{0.1}{100} = -0.001 \qquad (a)$$

$$\gamma_{xy} = \text{angle } BAB' + \text{angle } DAD' = 0$$

Example 3.2.2

Problem: A 100 *mm* square thin steel plate is deformed from the shape ABCD to A'B'C'D' indicated by the dotted lines as shown in Figure (a). Determine ε_x, ε_y, γ_{xy}.

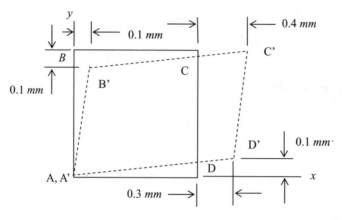

Figure (a)

Solution: The normal strain is the change in length divided by the length. The shearing strain is the change in angle between two lines at right angles.

Comparing the geometry of the deformed part to the original we obtain

$$\varepsilon_x = \frac{A'D' - AD}{AD} = \frac{\sqrt{(100.3)^2 + (0.1)^2} - 100}{100} = 0.003$$

$$\varepsilon_y = \frac{A'B' - AB}{AB} = \frac{\sqrt{(99.9)^2 + (0.1)^2} - 100}{100} = -0.001 \tag{a}$$

$$\gamma_{xy} = \text{angle BAB'} + \text{angle DAD'} = \tan^{-1}\left(\frac{0.1}{99.9}\right) + \tan^{-1}\left(\frac{0.1}{100.3}\right) = 0.002 \text{ rad}$$

###########

3.3 Compatibility

There are six components of strain defined in term of three components of displacement. From that you may conclude that the strain components are not independent of each other.

Consider the two dimensional case expressed in Equation 3.2.12. Expanded we have

$$\varepsilon_x = \frac{\partial u}{\partial x} \qquad \varepsilon_y = \frac{\partial v}{\partial y} \qquad \gamma_{xy} = \frac{\partial u}{\partial y} + \frac{\partial v}{\partial x} \tag{3.3.1}$$

Take the second derivatives of each strain component in the following way.

$$\frac{\partial^2 \varepsilon_x}{\partial y^2} = \frac{\partial^3 u}{\partial x \partial y^2} \qquad \frac{\partial^2 \varepsilon_y}{\partial x^2} = \frac{\partial^3 v}{\partial x^2 \partial y} \qquad \frac{\partial^2 \gamma_{xy}}{\partial x \partial y} = \frac{\partial^3 u}{\partial x \partial y^2} + \frac{\partial^3 v}{\partial x^2 \partial y} \tag{3.3.2}$$

By summing the first two equations in Equation 3.3.2 we note from the third equation that the following relation exists among the strain components.

$$\frac{\partial^2 \varepsilon_x}{\partial y^2} + \frac{\partial^2 \varepsilon_y}{\partial x^2} = \frac{\partial^2 \gamma_{xy}}{\partial x \partial y} \tag{3.3.3}$$

This is called the *compatibility* equation in two dimensions. When applied to the three dimensional case additional relationships are found. We shall not go through the steps but only record that in three dimensions there are five more relations that must be satisfied.

$$\frac{\partial^2 \varepsilon_y}{\partial z^2} + \frac{\partial^2 \varepsilon_z}{\partial y^2} = \frac{\partial^2 \gamma_{yz}}{\partial y \partial z}$$

$$\frac{\partial^2 \varepsilon_z}{\partial x^2} + \frac{\partial^2 \varepsilon_x}{\partial z^2} = \frac{\partial^2 \gamma_{xz}}{\partial z \partial x}$$

$$2\frac{\partial^2 \varepsilon_x}{\partial y \partial z} = \frac{\partial}{\partial x}\left(-\frac{\partial \gamma_{yz}}{\partial x} + \frac{\partial \gamma_{xz}}{\partial y} + \frac{\partial \gamma_{xy}}{\partial z}\right) \tag{3.3.4}$$

$$2\frac{\partial^2 \varepsilon_y}{\partial z \partial x} = \frac{\partial}{\partial y}\left(\frac{\partial \gamma_{yz}}{\partial x} - \frac{\partial \gamma_{xz}}{\partial y} + \frac{\partial \gamma_{xy}}{\partial z}\right)$$

$$2\frac{\partial^2 \varepsilon_x}{\partial x \partial y} = \frac{\partial}{\partial z}\left(\frac{\partial \gamma_{yz}}{\partial x} + \frac{\partial \gamma_{xz}}{\partial y} - \frac{\partial \gamma_{xy}}{\partial z}\right)$$

We note that in both two and three dimensions the compatibility equations consist of relations among the second derivatives of the strain components with respect to the coordinates. It follows that any state of strain involving constant values of strain components satisfies compatibility and so do cases of linearly varying strain components.

Whatever the state of strain in a body, to be a valid solution of a given problem, compatibility must be satisfied. We shall use this fact to advantage in several instances in later chapters.

3.4 Linear Material Properties

There are two physical laws that govern the behavior of solid bodies under load. One of these is static equilibrium, discussed in detail in Chapter 2. The other is based on the properties of materials. When a material is *homogeneous* it has properties that do not change from point to point. When it is *isotropic* it has the same properties in all possible coordinate directions and it often has the simple properties that are about to be defined. These properties apply to many, but not all, materials that solid bodies used in structures are made of. The three most commonly used aerospace metals, aluminum, steel, and titanium, are both homogenous and isotropic; however, modern materials like carbon reinforced composites have more involved properties and will be dealt with later in this chapter.

3.4.1 Hooke's Law in One Dimension—Tension

Material properties are determined initially by a simple experiment. Consider that same slender bar in Figure 3.2.1 used to define strain. The force is applied in a manner yet to be determined but assume, for the moment, that it acts through the centroid of the cross section. If we examine a cross section normal to the axis at, say, position $0 < x < L$, there must be an internal force P acting on that cross section that satisfies Newton's law of equilibrium, namely, that the summation of forces is zero.

We illustrate this with a *free body diagram*, that is, we consider a portion of the bar in Figure 3.2.1 as shown in Figure 3.4.1.

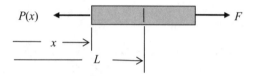

Figure 3.4.1

As noted before, it is common to define applied force components and restraint force components as positive if acting in the positive directions of the axes. So F is shown as a positive force. Two different conventions are used for internal forces. The most common has been to define normal force components acting away from the surface as positive, as P is shown in Figure 3.4.1. By this convention positive stresses produce positive internal force resultants. The occasion will arise when we shall use another convention but for now we shall use the one just stated.

Another convention relates to the direction of the y axis in the figures. Civil engineers, in particular, are inclined to show the axis pointing down. Since applied loads in civil engineering tend to be gravity loads they can then be positive pointing down in the same direction of the positive y axis. We have chosen to have it pointing up, perhaps because of our aerospace background. Airplanes are particularly pleased when some of the forces acting upon them are acting up.

From Newton's laws the summation of forces in the x-direction must be zero.

$$\sum F_x = F - P(x) = 0 \quad \rightarrow \quad P = F \tag{3.4.1}$$

In this case $P(x)$ is a constant for all values of x. The force P, however, must be distributed over the cross sectional area. Let us assume, for now, that it is uniformly distributed over that area. This is justified by experimental evidence. We can denote the force per unit area or stress by σ_x. The stress is

$$\sigma_x = \frac{P}{A} \quad \rightarrow \quad P = \sigma_x A \tag{3.4.2}$$

Note that it makes no difference what the material of the bar is. For any applied force F there will be an internal force P and for any area A the stress will be given by Equation 3.4.2.

Note again that this applies only if the force is acting through the centroid of the cross sectional area.

We would expect, and we can verify by experiment, that the bar will deform. In this case the ends would displace some distance. Just what distance would depend upon the material. You can readily agree that if the bar was made of soft rubber it would displace much more than if made of steel for a given value of F and A. To find the displacement then we must find some way to describe the properties of the material being used. We might, for example, actually construct a bar of a material of interest and measure the displacement of two scribe marks just as we did in defining strain. For many materials, such as those commonly used in structures of many kinds (for example, steel, aluminum, or titanium) we would obtain a plot similar to that shown in Figure 3.4.2.

There would be a straight portion, that is, a linear relation between force and displacement, followed by a region in which the displacement increases more rapidly with respect to an increase in force. For many materials used in real structures the deflection returns to zero if the load is reduced to zero provided the maximum load does not exceed a certain value. Quite often this value is very near where the curve deviates from a straight line. Materials that behave this way are called *elastic,* or more precisely *linearly elastic*, up to the point where the curve is no longer straight.

The straight portion of this plot up to a value of the force F_0 has the slope k where

$$k = \frac{F}{d} \tag{3.4.3}$$

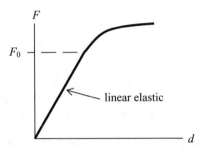

Figure 3.4.2

The quantity k is called the *stiffness* of the bar. Given that $F \le F_0$ and a value for k we can say

$$d = \frac{F}{k} \tag{3.4.4}$$

The greater the stiffness the less is the displacement for a given force. The same concept is used to define the stiffness of springs where k is called the *spring constant*.

For the moment we shall restrict our interest to loads that do not exceed F_0. In a later chapter we shall discuss what happens when larger loads are applied.

This defines the principal material property over the linearly elastic range for this particular structure; however, if we change the cross sectional area A or the length L of the bar between scribe marks the test results would provide a different value of k. So we need a more general way to define the properties of the material that is independent of the geometry of the solid bar. Fortunately, a way has been found. We define the quantity

$$\varepsilon_x = \frac{du}{dx} = \frac{d}{L} \tag{3.4.5}$$

This is our old friend the *normal strain*. If the stress is uniform across the cross section we can define the normal stress as

$$\sigma_x = \frac{P}{A} = \frac{F}{A} \tag{3.4.6}$$

We will show this to be true in Chapter 4.

Now the same data used to plot the curve in Figure 3.4.2 can be represented as shown in Figure 3.4.3.

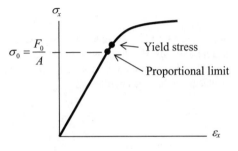

Figure 3.4.3

The slope of the straight portion is designated by E. It follows that

$$E = \frac{\sigma_x}{\varepsilon_x} \quad \text{or} \quad \sigma_x = E\varepsilon_x \quad \text{or} \quad \varepsilon_x = \frac{\sigma_x}{E} \tag{3.4.7}$$

In this case the plot is independent of A and L and is representative of the material and independent of the particular geometry of the slender bar. Thus it has general validity for that material for any value of the length between scribe marks, L, or of the cross section area, A. The straight portion is an example of *Hooke's law* for elastic materials. The quantity E is called *Young's modulus*. We will study this in more detail later.

The quantity σ_0 is called the *proportional limit*. Beyond this point the material may or may not be elastic. For many metals a short distance beyond the proportional limit the material starts to have permanent deformation. That point it called the *yield point* or *yield stress*.

Often the two points are so close together that the material is assumed to be elastic up to the yield stress and it is taken as the value at which our assumption of linear elasticity no longer applies.

If the material we pulled on was amorphous glass or carbon fiber the stress-strain curve would be a straight line up to the point where it fractures with no nonlinear portion as shown in Figure 3.4.4. This is characteristic of *brittle* materials while Figure 3.4.3 is characteristic of *ductile* materials.

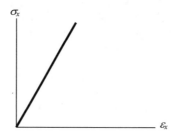

Figure 3.4.4

When the stress is positive it is called a *tensile* stress, when negative a *compressive* stress. For many materials of interest the slope of the curve, E, is the same in the negative stress and strain region as in the positive region.

For future reference we note that the total displacement in our test specimen between scribe marks is

$$d = \varepsilon_x L = \frac{\sigma_x L}{E} = \frac{PL}{AE} = \frac{FL}{AE} \tag{3.4.8}$$

############

Example 3.4.1

Problem: Consider that the bar in Figure 3.2.1 is made of aluminum. For the aluminum alloy 6010-T6 the proportional limit is approximately 250 N/mm^2 *(MPa)* and $E = 68950\ N/mm^2$. The dimension between scribe marks is $L = 200\ mm$ and the cross section area is $A = 4\ mm^2$. The applied load is $F = 500\ N$. Find the stress and the total displacement between marks.

Solution: Use Equations 3.4.6 – 3.4.8.

The stress is

$$\sigma_x = \frac{P}{A} = \frac{500}{4} = 125\ N/mm^2 \tag{a}$$

Since this value is below the proportional limit the displacement would be

$$d = \frac{FL}{AE} = \frac{500 \cdot 200}{4 \cdot 68950} = 0.3626 \; mm \tag{b}$$

The fact that the displacement is a small quantity when compared to the length of the bar is important in certain assumptions to be made in the development of the equations of mechanics of materials. In fact, this simple example points out what we are usually looking for in an analysis – the size of the stress and of the displacement. Most structures are said to fail when either one or both of these exceed acceptable limits.

###########

3.4.2 Poisson's Ratio

A careful observer of the experimental determination of Young's modulus will note that the slender bar's cross section narrows as the bar elongates under load. This is shown in side view in Figure 3.4.5(a) and in end view in Figure 3.4.5b with the change in dimensions greatly exaggerated. The shaded bar shows the original dimensions. The bar has a width b and a height h in the unloaded state.

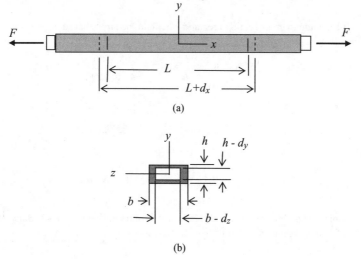

Figure 3.4.5

So far we have neglected this effect. We designate the amount of elongation as d_x and of lateral displacement d_y in the y direction and d_z in the z direction. The corresponding strains are

$$\varepsilon_x = \frac{d_x}{L} \qquad \varepsilon_y = -\frac{d_y}{h} \qquad \varepsilon_z = -\frac{d_z}{b} \tag{3.4.9}$$

It may be noted that for a linear elastic, homogenous, and isotropic material experimental evidence will confirm that

$$-\frac{\varepsilon_y}{\varepsilon_x} = -\frac{\varepsilon_z}{\varepsilon_x} = v \tag{3.4.10}$$

where v is a constant called *Poisson's ratio*, named after the person who defined it. The negative sign accounts for lateral contractions in the y and z directions for a positive elongation in the x direction when a positive tensile force is applied.

From this we may conclude that if

$$\varepsilon_x = \frac{\sigma_x}{E} \tag{3.4.11}$$

then from Equation 3.4.10

$$\varepsilon_y = -v\varepsilon_x = -\frac{v\sigma_x}{E} \qquad \varepsilon_z = -v\varepsilon_x = -\frac{v\sigma_x}{E} \tag{3.4.12}$$

If there is a contraction in y and z directions the area of the cross section must be smaller than the nominal value that we have been using to determine the stress. We turn to an example to explore this effect.

############

Example 3.4.2

Problem: Find the Poisson's ratio effect on the magnitude of the stress for the bar in Example 3.4.1. In a typical material, such as aluminum, the value of Poisson's ratio is approximately $v = 0.3$. Width and height of the cross section is $b = h = 2\ mm$.

Solution: Find the change in cross section area and calculate the stress based on the new area.

The new cross section dimensions are

$$b_{new} = b + \varepsilon_z b = b - \frac{v\sigma_x}{E}b = b - \frac{vP}{AE}b = 2 - \frac{0.3 \cdot 500}{4 \cdot 68950} \cdot 2 = 1.9989 = h_{new} \tag{a}$$

Thus the new area is

$$A_{new} = b_{new} \cdot h_{new} = (1.9989)^2 = 3.9956\ mm^2 \tag{b}$$

and the stress based on this area is

$$\sigma_{xnew} = \frac{P}{A_{new}} = \frac{500}{3.9956} = 125.138\ N/mm^2 \tag{c}$$

This may be compared to the stress value of $125\ N/mm^2$ found in Example 3.4.1. Such a small change justifies the use of nominal dimensions in nearly all analysis using most common structural materials.

############

3.4.3   Hooke's Law in One Dimension—Shear in Isotropic Materials

When an experiment is performed that produces a pure shear stress it produces a curve similar to the one in Figure 3.4.2. When the shearing stress, τ, is plotted versus the shearing strain, γ, a curve very similar to Figure 3.4.3 is produced. The slope of the stress strain curve is designated, G, and is called the *shear modulus*. The corresponding Hooke's law for shear is then

$$\tau = G\gamma \tag{3.4.13}$$

We shall show in Chapter 9 that G is not independent of E and ν, in fact,

$$G = \frac{E}{2(1+\nu)} \tag{3.4.14}$$

For now we shall accept the experimental evidence confirms both Equations 3.4.13 and 3.4.14.

3.4.4 Hooke's Law in Two Dimensions for Isotropic Materials

Some structural components consist of thin flat plates with loads and restraints applied in the plane of the plate. This is called a state of *plane stress*. The stress components at a point in the interior of the plate are as shown in Figure 3.4.6.

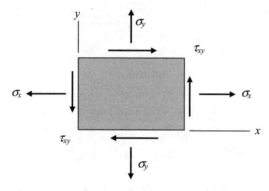

Figure 3.4.6

In such cases all z component of stresses are zero.

$$\sigma_z = \tau_{yz} = \tau_{zx} = 0 \tag{3.4.15}$$

The σ_x normal stress produces the same strains as given in Equations 3.4.11 and 3.4.12, namely,

$$\varepsilon_x = \frac{\sigma_x}{E} \qquad \varepsilon_y = -\frac{\nu\sigma_x}{E} \qquad \varepsilon_z = -\frac{\nu\sigma_x}{E} \tag{3.4.16}$$

Similarly the σ_y normal stress produces

$$\varepsilon_y = \frac{\sigma_y}{E} \qquad \varepsilon_x = -\frac{\nu\sigma_y}{E} \qquad \varepsilon_z = -\frac{\nu\sigma_y}{E} \tag{3.4.17}$$

These two can be combined to provide

$$\varepsilon_x = \frac{1}{E}\left[\sigma_x - \nu\sigma_y\right]$$

$$\varepsilon_y = \frac{1}{E}\left[\sigma_y - \nu\sigma_x\right] \tag{3.4.18}$$

$$\varepsilon_z = \frac{-\nu}{E}\left[\sigma_y + \sigma_x\right]$$

and to this we add the shearing strains.

$$\gamma_{xy} = \frac{1}{G}\tau_{xy} \tag{3.4.19}$$

The first two of Equations 3.4.18 and Equation 3.4.19 can be put in the matrix form

$$
\begin{bmatrix} \varepsilon_x \\ \varepsilon_y \\ \gamma_{xy} \end{bmatrix} = \begin{bmatrix} \dfrac{1}{E} & \dfrac{-\nu}{E} & 0 \\ \dfrac{-\nu}{E} & \dfrac{1}{E} & 0 \\ 0 & 0 & \dfrac{1}{G} \end{bmatrix} \begin{bmatrix} \sigma_x \\ \sigma_y \\ \tau_{xy} \end{bmatrix} = [\Gamma]\{\sigma\}
\tag{3.4.20}
$$

The matrix $[\Gamma]$ can be inverted to put these equations in a more useful form.

$$
[\Gamma]^{-1} = \begin{bmatrix} \dfrac{1}{E} & \dfrac{-\nu}{E} & 0 \\ \dfrac{-\nu}{E} & \dfrac{1}{E} & 0 \\ 0 & 0 & \dfrac{1}{G} \end{bmatrix}^{-1} = \begin{bmatrix} \dfrac{E}{1-\nu^2} & \dfrac{\nu E}{1-\nu^2} & 0 \\ \dfrac{\nu E}{1-\nu^2} & \dfrac{E}{1-\nu^2} & 0 \\ 0 & 0 & G \end{bmatrix} = [G]
\tag{3.4.21}
$$

Hooke's law for plane stress can now be written as

$$
\{\sigma\} = \begin{bmatrix} \sigma_x \\ \sigma_y \\ \tau_{xy} \end{bmatrix} = \begin{bmatrix} \dfrac{E}{1-\nu^2} & \dfrac{\nu E}{1-\nu^2} & 0 \\ \dfrac{\nu E}{1-\nu^2} & \dfrac{E}{1-\nu^2} & 0 \\ 0 & 0 & G \end{bmatrix} \begin{bmatrix} \varepsilon_x \\ \varepsilon_y \\ \gamma_{xy} \end{bmatrix} = [G]\{\varepsilon\}
\tag{3.4.22}
$$

The notation here can be confusing. The shear modulus is labeled G and is a scalar quantity. The matrix $[G]$ relates all components of stress to strain.

We note that ε_z is not zero; therefore the thickness of the plate at any point is dependent upon the state of stress at that point.

$$
\varepsilon_z = -\frac{\nu}{E}\left(\sigma_x + \sigma_y\right)
\tag{3.4.23}
$$

For most structural materials the effect is small so that the nominal thickness of the plate is used in all calculations with minimal effect, that is, the z component of strain is neglected.

3.4.5 Generalized Hooke's Law for Isotropic Materials

If a material is subjected at a point to triaxial stress components we add the z components of strain to the components listed in Equations 3.4.16-17.

$$
\varepsilon_z = \frac{\sigma_z}{E} \qquad \varepsilon_y = -\frac{\nu\sigma_z}{E} \qquad \varepsilon_x = -\frac{\nu\sigma_z}{E}
\tag{3.4.24}
$$

When all three stress components are acting we can superimpose the above to obtain

$$
\varepsilon_x = \frac{1}{E}\left[\sigma_x - \nu\left(\sigma_y + \sigma_z\right)\right]
$$

$$
\varepsilon_y = \frac{1}{E}\left[\sigma_y - \nu\left(\sigma_x + \sigma_z\right)\right]
\tag{3.4.25}
$$

$$
\varepsilon_z = \frac{1}{E}\left[\sigma_z - \nu\left(\sigma_y + \sigma_x\right)\right]
$$

When a homogenous, isotropic, and linear elastic material is subjected at a point to all three shearing stress components we have

$$\gamma_{xy} = \frac{1}{G}\tau_{xy} \qquad \gamma_{yz} = \frac{1}{G}\tau_{yz} \qquad \gamma_{zx} = \frac{1}{G}\tau_{zx} \qquad (3.4.26)$$

This can be put in matrix form

$$\{\varepsilon\} = \begin{bmatrix} \varepsilon_x \\ \varepsilon_y \\ \varepsilon_z \\ \gamma_{xy} \\ \gamma_{yz} \\ \gamma_{zx} \end{bmatrix} = \begin{bmatrix} \dfrac{1}{E} & \dfrac{-\nu}{E} & \dfrac{-\nu}{E} & 0 & 0 & 0 \\ \dfrac{-\nu}{E} & \dfrac{1}{E} & \dfrac{-\nu}{E} & 0 & 0 & 0 \\ \dfrac{-\nu}{E} & \dfrac{-\nu}{E} & \dfrac{1}{E} & 0 & 0 & 0 \\ 0 & 0 & 0 & \dfrac{1}{G} & 0 & 0 \\ 0 & 0 & 0 & 0 & \dfrac{1}{G} & 0 \\ 0 & 0 & 0 & 0 & 0 & \dfrac{1}{G} \end{bmatrix} \begin{bmatrix} \sigma_x \\ \sigma_y \\ \sigma_z \\ \tau_{xy} \\ \tau_{yz} \\ \tau_{zx} \end{bmatrix} = [\Gamma]\{\sigma\} \qquad (3.4.27)$$

where

$$G = \frac{E}{2(1+\nu)} \qquad (3.4.28)$$

The matrix $[\Gamma]$ can be inverted to put these equations in a more useful form.

$$[G] = [\Gamma]^{-1} \qquad (3.4.29)$$

Hooke's law for a 3D solid can now be written as

$$\{\sigma\} = \begin{bmatrix} \sigma_x \\ \sigma_y \\ \sigma_z \\ \tau_{xy} \\ \tau_{yz} \\ \tau_{zx} \end{bmatrix} = \begin{bmatrix} \lambda+2G & \lambda & \lambda & 0 & 0 & 0 \\ \lambda & \lambda+2G & \lambda & 0 & 0 & 0 \\ \lambda & \lambda & \lambda+2G & 0 & 0 & 0 \\ 0 & 0 & 0 & G & 0 & 0 \\ 0 & 0 & 0 & 0 & G & 0 \\ 0 & 0 & 0 & 0 & 0 & G \end{bmatrix} \begin{bmatrix} \varepsilon_x \\ \varepsilon_y \\ \varepsilon_z \\ \gamma_{xy} \\ \gamma_{yz} \\ \gamma_{zx} \end{bmatrix} = [G]\{\varepsilon\} \qquad (3.4.30)$$

where

$$\lambda = \frac{\nu E}{(1+\nu)(1-2\nu)} \qquad (3.4.31)$$

3.5 Some Simple Solutions for Stress, Strain, and Displacement

To find a valid solution for the stress strain and displacement in a solid body in the linear elastic range we must satisfy

1. The equations of equilibrium.
2. The strain displacement equations consistent with compatibility.
3. Hooke's law for material properties.
4. Boundary conditions on loading and restraint.

We shall examine some simple examples here.

###########

| Example 3.5.1 |

Problem: A thin plate has two sets of uniform edge loads as shown in Figures (a) and (b). The coordinate system is attached to the center of the plate. Find the stresses and displacements.

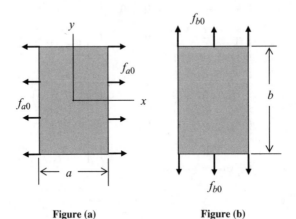

Figure (a)   Figure (b)

The loading in Figure (a) is

$$f_{ax}\left(\pm\frac{a}{2}, y\right) = \pm f_{a0} \ N/mm^2 \tag{a}$$

The loading in Figure (b) is

$$f_{by}\left(x, \pm\frac{b}{2}\right) = \pm f_{b0} \ N/mm^2 \tag{b}$$

Solution: Assume the stresses are the same constant values as the edge loads and check to see if the stress components satisfy the equilibrium equations, match the loads at the boundaries, satisfy Hooke's law, satisfy the compatibility equations, and then integrate the strains to find the displacements.

Since this is a thin plate and all forces lie in the xy plane this may be treated as a plane stress problem. It is suggested that the stress components in the interior of the plate are the same as on the edges where the load is applied in each case.

For the loading in Figure (a) try this for the stress.

$$\sigma_x(x, y) = f_{a0} \qquad \sigma_y(x, y) = \tau_{xy}(x, y) = 0 \tag{c}$$

Substitute these values in the equilibrium equations.

$$\begin{bmatrix} \dfrac{\partial}{\partial x} & 0 & \dfrac{\partial}{\partial y} \\[2mm] 0 & \dfrac{\partial}{\partial y} & \dfrac{\partial}{\partial x} \end{bmatrix} \begin{bmatrix} f_{a0} \\ 0 \\ 0 \end{bmatrix} = \begin{bmatrix} 0 \\ 0 \end{bmatrix} \tag{d}$$

In the absence of body forces they are satisfied. The stresses at the boundaries match the applied load. From Hooke's law the strains are

$$\varepsilon_x = \frac{\sigma_x}{E} = \frac{f_{a0}}{E} \qquad \varepsilon_y = -v\frac{\sigma_x}{E} = -v\frac{f_{a0}}{E} \qquad \gamma_{xy} = 0 \qquad\qquad (e)$$

The compatibility equation is satisfied as shown in Equation (f).

$$\frac{\partial^2 \varepsilon_x}{\partial y^2} + \frac{\partial^2 \varepsilon_y}{\partial x^2} = \frac{\partial^2 \gamma_{xy}}{\partial x \partial y}$$

$$\rightarrow \quad \frac{\partial^2}{\partial y^2}\left(\frac{f_{a0}}{E}\right) + \frac{\partial^2}{\partial x^2}\left(-v\frac{f_{a0}}{E}\right) = \frac{\partial^2}{\partial x \partial y}(0) \quad \rightarrow \quad 0 + 0 = 0 \qquad (f)$$

To find the displacements integrate the strains.

$$\varepsilon_x = \frac{du}{dx} = \frac{f_{a0}}{E} \quad \rightarrow \quad u(x, y) = \frac{f_{a0}}{E}\int dx = \frac{f_{a0}}{E}x + a$$

$$\rightarrow \quad u(0, 0) = \frac{f_{a0}}{E}0 + a = 0 \quad \rightarrow \quad a = 0$$

$$\rightarrow \quad u(x, y) = \frac{f_{a0}}{E}x$$

$$\varepsilon_y = \frac{dv}{dy} = -v\frac{f_{a0}}{E} \quad \rightarrow \quad v(x, y) = -v\frac{f_{a0}}{E}\int dy = -v\frac{f_{a0}}{E}y + b \qquad (g)$$

$$\rightarrow \quad v(0, 0) = -v\frac{f_{a0}}{E}0 + b = 0 \quad \rightarrow \quad b = 0$$

$$\rightarrow \quad v(x, y) = -v\frac{f_{a0}}{E}y$$

For the loading in Figure (b), following the same steps, we find that

$$\sigma_y(x, y) = f_{b0} \qquad \sigma_x(x, y) = \tau_{xy}(x, y) = 0 \qquad\qquad (h)$$

satisfies equilibrium and the stresses match the applied loads at the edges.
   The strains are

$$\varepsilon_y = \frac{\sigma_y}{E} = \frac{f_{b0}}{E} \qquad \varepsilon_x = -v\frac{\sigma_y}{E} = -v\frac{f_{b0}}{E} \qquad \gamma_{xy} = 0 \qquad\qquad (i)$$

and they satisfy compatibility. The displacements are found to be

$$v(x, y) = \frac{f_{b0}}{E}y \qquad u(x, y) = -v\frac{f_{b0}}{E}x \qquad\qquad (j)$$

If both loads are applied at the same time the stresses and displacements are simply the sum of those found, that is, the principle of superposition applies.

Example 3.5.2

Problem: The same thin plate has varying edge loads as shown in Figure (a).

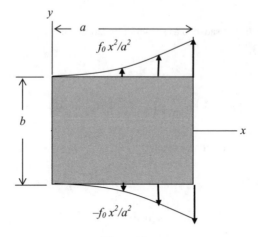

Figure (a)

These loads are

$$f_y\left(x, \pm\frac{b}{2}\right) = \pm f_0\frac{x^2}{a^2} \ N/mm^2 \tag{a}$$

Once again it is suggested that the stresses in the interior be the same as those on the boundaries, that is,

$$\sigma_y(x, y) = f_0\frac{x^2}{a^2} \qquad \sigma_x = \tau_{xy} = 0 \tag{b}$$

Are these correct values for the stress components?

Solution: As noted, to be valid the stress components must satisfy the equilibrium equations, match the loads at the boundaries, satisfy Hooke's law, and the compatibility equations.

Insert the suggested stress components into the equilibrium equations and evaluate.

$$\begin{bmatrix} \dfrac{\partial}{\partial x} & 0 & \dfrac{\partial}{\partial y} \\[2mm] 0 & \dfrac{\partial}{\partial y} & \dfrac{\partial}{\partial x} \end{bmatrix} \begin{bmatrix} 0 \\[2mm] f_0\dfrac{x^2}{b^2} \\[2mm] 0 \end{bmatrix} = \begin{bmatrix} 0 \\ 0 \end{bmatrix} \tag{c}$$

The left hand side is identically zero in the absence of body forces. The equilibrium equations are satisfied. Great! And the stresses at the boundaries do match the applied loads.

Now find the strain components.

$$\varepsilon_y = \frac{\sigma_y}{E} = \frac{f_0}{E}\frac{x^2}{a^2} \qquad \varepsilon_x = -v\frac{\sigma_y}{E} = -v\frac{f_0}{E}\frac{x^2}{a^2} \qquad \gamma_{xy} = 0 \tag{d}$$

Insert these in the compatibility equation and evaluate.

$$\frac{\partial^2}{\partial y^2}\left(\frac{f_0}{E}\frac{x^2}{a^2}\right) + \frac{\partial^2}{\partial x^2}\left(-\nu\frac{f_0}{E}\frac{x^2}{a^2}\right) = \frac{\partial^2}{\partial x \partial y}(0) \quad \rightarrow \quad 0 - \frac{2\nu}{Ea^2} \neq 0 \tag{e}$$

Compatibility is not satisfied. This cannot be the stress in the plate. The actual solution is quite difficult to obtain and is beyond our scope at this time, but we shall find a way to solve this problem later on.

##########

3.6   Thermal Strain

A change in temperature from an initial state can cause a material to expand or contract. An increase in temperature will cause an expansion and for many materials this change is linear with respect to temperature. For such materials we denote a *coefficient of thermal expansion* with the symbol α. It has units of strain per degree of temperature.

In one dimension for a change in temperature ΔT the thermal strain is

$$\varepsilon_{xT} = \alpha \Delta T \tag{3.6.1}$$

When there is a temperature change in addition to a strain caused by a mechanical load Equation 3.2.12 for the one dimensional case can be written as the superposition of the mechanical and thermal strains.

$$\varepsilon_x = \varepsilon_x^{mech} + \alpha \Delta T = \frac{du}{dx} \tag{3.6.2}$$

Since the mechanical strain in one dimension is given by Equation 3.4.7, that is,

$$\varepsilon_x^{mech} = \frac{\sigma_x}{E} \tag{3.6.3}$$

we have

$$\frac{du}{dx} = \frac{\sigma_x}{E} + \alpha \Delta T \tag{3.6.4}$$

In two dimensions the mechanical strain in terms of stress is given by Equation 3.4.20. The effect of adding thermal strain results in

$$[D]\{u\} = \begin{bmatrix} \dfrac{\partial}{\partial x} & 0 \\[2mm] 0 & \dfrac{\partial}{\partial y} \\[2mm] \dfrac{\partial}{\partial y} & \dfrac{\partial}{\partial x} \end{bmatrix} \begin{bmatrix} u \\ v \end{bmatrix} = \begin{bmatrix} \dfrac{1}{E} & \dfrac{-\nu}{E} & 0 \\[2mm] \dfrac{-\nu}{E} & \dfrac{1}{E} & 0 \\[2mm] 0 & 0 & \dfrac{1}{G} \end{bmatrix} \begin{bmatrix} \sigma_x \\ \sigma_y \\ \tau_{xy} \end{bmatrix} + \alpha \Delta T \begin{bmatrix} 1 \\ 1 \\ 0 \end{bmatrix} \tag{3.6.5}$$

In three dimensions the mechanical strain in terms of stress is given by Equation 3.4.27. The effect of adding thermal strain results in

$$[D]\{u\} = \begin{bmatrix} \dfrac{\partial}{\partial x} & 0 & 0 \\[6pt] 0 & \dfrac{\partial}{\partial y} & 0 \\[6pt] 0 & 0 & \dfrac{\partial}{\partial z} \\[6pt] \dfrac{\partial}{\partial y} & \dfrac{\partial}{\partial x} & 0 \\[6pt] 0 & \dfrac{\partial}{\partial z} & \dfrac{\partial}{\partial y} \\[6pt] \dfrac{\partial}{\partial z} & 0 & \dfrac{\partial}{\partial x} \end{bmatrix} \begin{bmatrix} u \\ v \\ w \end{bmatrix} = \begin{bmatrix} \dfrac{1}{E} & \dfrac{-\nu}{E} & \dfrac{-\nu}{E} & 0 & 0 & 0 \\[6pt] \dfrac{-\nu}{E} & \dfrac{1}{E} & \dfrac{-\nu}{E} & 0 & 0 & 0 \\[6pt] \dfrac{-\nu}{E} & \dfrac{-\nu}{E} & \dfrac{1}{E} & 0 & 0 & 0 \\[6pt] 0 & 0 & 0 & \dfrac{1}{G} & 0 & 0 \\[6pt] 0 & 0 & 0 & 0 & \dfrac{1}{G} & 0 \\[6pt] 0 & 0 & 0 & 0 & 0 & \dfrac{1}{G} \end{bmatrix} \begin{bmatrix} \sigma_x \\ \sigma_y \\ \sigma_z \\ \tau_{xy} \\ \tau_{yz} \\ \tau_{zx} \end{bmatrix} + \alpha \Delta T \begin{bmatrix} 1 \\ 1 \\ 1 \\ 0 \\ 0 \\ 0 \end{bmatrix}$$

$$(3.6.6)$$

Some examples of the use of these equations that have been extended to include changes in temperature will be given in later chapters.

3.7   Engineering Materials

Steel and aluminum are metallic materials often used in structures. Titanium is used in high performance structures where high temperatures are important. The properties depend upon the particular alloy selected. For exact values for these and other materials we must refer to handbooks and manufacturer's data that list properties for particular alloys. For use in example problems in the following chapters we list properties in Table 3.7.1 for a generic steel, aluminum, and titanium that are representative of these three metals.

Table 3.7.1

Material	Density kg/mm^3	Young's Modulus MPa	Shear Modulus MPa	Yield Stress MPa	Poisson's Ratio	Thermal Expansion $(10^{-6})/°F$
Steel	0.0078	206800	79500	240	0.3	16
Aluminum	0.0028	68950	26500	360	0.3	24
Titanium	0.0044	120000	46000	900	0.3	9

In particular non metallic and composite materials are coming into increasing use in many types of structures. A brief introduction to composite material properties is given in the next section.

3.8   Fiber Reinforced Composite Laminates

Modern aerospace structures use a significant amount of fiber reinforced polymer (FRP) matrix composite materials. There are many constructions of FRP laminates that are made by stacking together several lamina and then curing them in an autoclave under specified pressure and temperature histories. Each lamina of a FRP may be woven (like a textile), braided (a more complex weave pattern), or simply consist of a collection straight fibers held within a matrix, such as shown in Figure 3.8.1. We designate the direction of the fiber as 1 and the two perpendicular directions as 2 and 3.

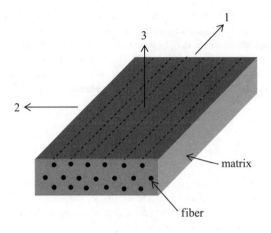

Figure 3.8.1

Using straight fiber stacked lamina to make a laminate is popular in modern aerospace construction. In particular the Airbus 380, Boeing 777 and 787, and most military aircraft use a substantial amount of laminate materials. Typically, the lamina shown in Figure 3.8.1 is about 0.13 *mm* thick. Carbon fibers, which are most popular in the aerospace industry, are about 5-7 μm in diameter, and within each lamina fibers occupy about 50 percent of the volume.

We will focus here on understanding the stress-strain behavior of a single lamina. Later, when we analyze plate problems we shall derive the equations to characterize the stress-strain behavior of a collection of stacked laminae, or a laminate.

3.8.1   Hooke's Law in Two Dimensions for a FRP Lamina

Consider a FRP lamina with loads and restraints applied in the plane of the lamina as shown in Figure 3.8.2.

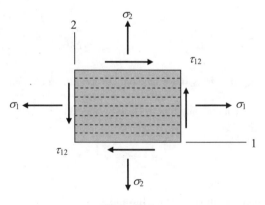

Figure 3.8.2

This is a state of plane stress as we noted before. We shall treat the lamina as a homogeneous but orthotropic layer, with "1" denoting the fiber direction and "2" the transverse direction. The assumption about homogeneity implies that we are treating the lamina as a single material with regard to the discrete fibers and matrix. The reason is that the dimension of the lamina we are interested in is several times greater than the fiber diameter or the spacing between fibers. The fiber diameter is usually about 5-7 μm for carbon fibers and about 20 μm for glass fibers. The spacing between fibers is usually about 5-7μm for the carbon fibers and about 15-20 μm for the glass fibers. If one has equal amounts of fiber and matrix by volume in a lamina we say the lamina has a 50 percent volume fraction (V_f) of fiber.

For plane stress

$$\sigma_3 = \tau_{23} = \tau_{31} = 0 \qquad (3.8.1)$$

Consider what happens when we have σ_1 acting alone. Then, unlike isotropic materials, the amounts by which the lamina will strain in the 2 and 3 directions, that is, the Poison's ratio effect, will be different from the 1 direction. This is because the properties of the lamina in the fiber direction will be different than those perpendicular to the fiber direction.

So we can write

$$\varepsilon_1 = \frac{\sigma_1}{E_1} \qquad \varepsilon_2 = -\nu_{12}\frac{\sigma_1}{E_1} \qquad \varepsilon_3 = -\nu_{13}\frac{\sigma_1}{E_1} \qquad (3.8.2)$$

However, for a unidirectional lamina the 2 and 3 directions are identical in constitution and so $\nu_{12} = \nu_{13}$. Equation 3.8.2 becomes

$$\varepsilon_1 = \frac{\sigma_1}{E_1} \qquad \varepsilon_2 = -\nu_{12}\frac{\sigma_1}{E_1} \qquad \varepsilon_3 = -\nu_{12}\frac{\sigma_1}{E_1} \qquad (3.8.3)$$

The notation for ν_{12} signifies that the loading is in the "1" direction and the contraction occurs in the "2" direction. Young's modulus E_1 is in the fiber or 1 direction.

We can generalize Equation 3.8.3 to include other stress components, like we did before for isotropic materials, to obtain

$$\varepsilon_1 = \frac{\sigma_1}{E_1} - \nu_{21}\frac{\sigma_2}{E_2} \qquad \varepsilon_2 = -\nu_{12}\frac{\sigma_1}{E_1} + \frac{\sigma_2}{E_2} \qquad \gamma_{12} = \frac{\tau_{12}}{G_{12}} \qquad (3.8.4)$$

Equations 3.8.4 are the plane stress-strain relations for an orthotropic lamina. Notice that we have used 5 material constants, $E_1, E_2, G_{12}, \nu_{12}$ and ν_{21} to describe the plane stress –strain relation for an orthotropic material. It turns out that only four of these are independent, because,

$$\frac{E_1}{\nu_{12}} = \frac{E_2}{\nu_{21}} \qquad (3.8.5)$$

which is called the reciprocity relation. In general, with i, j ranging from 1 to 3

$$\frac{E_i}{\nu_{ij}} = \frac{E_j}{\nu_{ji}} \qquad (3.8.6)$$

When we generalize Equation 3.8.4 to 3D without the plane stress assumption but still with fibers in only the 1 direction we get

$$
\begin{bmatrix} \varepsilon_1 \\ \varepsilon_2 \\ \varepsilon_3 \\ \gamma_{23} \\ \gamma_{31} \\ \gamma_{12} \end{bmatrix} =
\begin{bmatrix}
\dfrac{1}{E_1} & -\dfrac{\nu_{21}}{E_2} & -\dfrac{\nu_{21}}{E_2} & 0 & 0 & 0 \\[2mm]
-\dfrac{\nu_{12}}{E_1} & \dfrac{1}{E_2} & -\dfrac{\nu_{32}(=\nu_{23})}{E_2} & 0 & 0 & 0 \\[2mm]
-\dfrac{\nu_{12}}{E_1} & -\dfrac{\nu_{23}}{E_2} & \dfrac{1}{E_2} & 0 & 0 & 0 \\[2mm]
0 & 0 & 0 & \dfrac{2(1+\nu_{23})}{E_2} & 0 & 0 \\[2mm]
0 & 0 & 0 & 0 & \dfrac{1}{G_{12}} & 0 \\[2mm]
0 & 0 & 0 & 0 & 0 & \dfrac{1}{G_{12}}
\end{bmatrix}
\begin{bmatrix} \sigma_1 \\ \sigma_2 \\ \sigma_3 \\ \tau_{23} \\ \tau_{31} \\ \tau_{12} \end{bmatrix} \qquad (3.8.7)
$$

There are five independent constants. These are E_1, E_2, ν_{12}, G_{12}, and ν_{23}. This is referred to as a transversely isotopic material.

###########

Example 3.8.1

Problem: A thin unidirectional lamina has uniform edge loads (along the fiber direction) as shown in Figure (a).

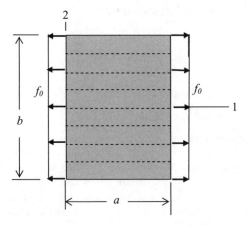

Figure (a)

These loads are

$$ f_1 = f_0 \ N/mm^2 \qquad (a) $$

Since this is a thin sheet and all forces lie in the 1,2 plane this may be treated as a plane stress problem. It is suggested that the stress components in the interior of the plate are

$$ \sigma_1(1,2) = f_0 \qquad \sigma_2 = \tau_{12} = 0 \qquad (b) $$

Are these correct values for the stress components?

Solution: Check to see if the stress components satisfy the equilibrium equations, match the loads at the boundaries, satisfy Hooke's law, and satisfy the compatibility equations. Notice that we are using 1 and 2 as coordinate directions instead of x and y.

The equilibrium equations in two dimensions are given in Section 2.6, Equation 2.6.16. Insert the suggested stress components and evaluate.

$$
\begin{bmatrix}
\dfrac{\partial}{\partial x_1} & 0 & \dfrac{\partial}{\partial x_2} \\[2ex]
0 & \dfrac{\partial}{\partial x_2} & \dfrac{\partial}{\partial x_1}
\end{bmatrix}
\begin{bmatrix} 0 \\ f_0 \\ 0 \end{bmatrix}
= -\begin{bmatrix} f_{b1} \\ f_{b2} \end{bmatrix}
\tag{c}
$$

The left hand side is identically zero. In the absence of body forces the equilibrium equations are satisfied.

The stresses at the boundaries do match the applied loads.

From Hooke's law the strain components are

$$
\varepsilon_1 = \frac{f_0}{E_1} \qquad \varepsilon_2 = -v_{12}\frac{f_0}{E_1} \qquad \gamma_{12} = 0
\tag{d}
$$

Insert these in the compatibility equation and evaluate.

$$
\frac{\partial^2 \varepsilon_1}{\partial x_2^2} + \frac{\partial^2 \varepsilon_2}{\partial x_1^2} = \frac{\partial^2 \gamma_{12}}{\partial x_1 \partial x_2} \quad \rightarrow \quad \frac{\partial^2}{\partial x_2^2}\left(\frac{f_0}{E_1}\right) + \frac{\partial^2}{\partial x_1^2}\left(-v_{12}\frac{f_0}{E_1}\right) = \frac{\partial^2}{\partial x_1 \partial x_2}(0) \quad \rightarrow \quad 0+0=0
$$
$$\tag{e}$$

All necessary equations are satisfied. These stress values are correct.

###########

3.8.2 Properties of Unidirectional Lamina

The most popular unidirectional material in the aerospace industry is graphite/epoxy. There are several suppliers of this material, which consists of graphite filaments (fibers) that are held in a pre-impregnated (pre-preg) polymer matrix. The lamina are supplied as rolls. The manufacturer purchases these lamina and lays them up (stacks them one on top of another) to form a laminate by applying pressure and temperature to the stack in a specified manner. After curing, a laminate is made. AS4 fiber (a commercial grade of graphite fiber) in a 3501-6 thermoset polymer is widely used. The designation is AS4/3501-6. Another popular graphite/epoxy is T300/5208. The auto industry and some in the civil engineering sector use glass fiber/epoxy. A popular glass fiber pre-preg is S2 glass/epoxy. The lamina elastic properties for these three materials are listed below (from Herakovich, 1998)[1] in Table 3.8.1

In a later chapter we will examine these and other materials in much greater detail including additional properties.

---

[1] Herakovich, CT. *Mechanics of Fibrous Composites*, John Wiley & Sons, Ltd, 1998.

Table 3.8.1

Material	AS4/3501	T300/5208	S2 Glass/Epoxy
Density, g/cm^2	1.52	1.54	2.0
Axial Modulus, E_{11}, GPa	148	132	43.5
Transverse Modulus, E_{22}, GPa	10.50	10.8	11.5
Poissons Ratio, ν_{12}	0.30	0.24	0.27
Poissons Ratio, ν_{23}	0.59	0.59	0.40
Shear Modulus, G_{12}	5.61	5.65	3.45
Axial Coeff. Of thermal expansion, α_1, $(m/^\circ C)$	-0.8	-0.77	6.84
Transverse Coeff. of Thermal Expansion, α_2, $(m/^\circ C)$	29	25	29
Fiber Volume Fraction, V_f	0.62	0.62	0.60
Lamina Thickness, mm	0.127	0.127	varies

###########

Example 3.8.2

Problem: Problem: For the plate in Example 3.8.1 assign these initial dimensions before the load is applied.

$$a = 300 \; mm \qquad b = 200 \; mm \qquad h = 2 \; mm \qquad \text{(a)}$$

where h is the thickness. It is made of graphite/epoxy (AS4/3501-6) with $E_{11} = 148 \times 10^3 \; MPa$, $E_{22} = 10.5 \times 10^3 \; MPa$, $G_{12} = 5.61 \times 10^3 \; MPa$ and $\nu_{12} = 0.3$. If $f_0 = 200 \; N/mm^2$. What are its dimensions after it is loaded?

Solution: Find the strains and multiply by the original lengths to find the change in dimensions.

From, 3.8.4, we have

$$\varepsilon_1 = \frac{\sigma_1}{E_1} - \nu_{21}\frac{\sigma_2}{E_2}$$
$$\varepsilon_2 = -\nu_{12}\frac{\sigma_1}{E_1} + \frac{\sigma_2}{E_2} \qquad \text{(b)}$$
$$\gamma_{12} = \frac{\tau_{12}}{G_{12}}$$

Thus

$$\varepsilon_1 = \frac{\sigma_1}{E_1} = \frac{200}{148000} = 0.0014, \quad \varepsilon_2 = -\nu_{12}\frac{\sigma_1}{E_1} = -0.3\frac{200}{148000} = -0.00041, \quad \gamma_{12} = 0 \quad \text{(c)}$$

To these, we must add the three components of the original dimensions. From Equation 3.8.3, $\varepsilon_3 = \varepsilon_2$. All shear strains are zero.

Since the strains are constant the new dimensions can be found as follows:

$$a_{new} = 300\,(1 + \varepsilon_1) = 300\,(1 + 0.0014) = 300.42 \; mm$$
$$b_{new} = 200\,(1 + \varepsilon_2) = 200\,(1 - 0.00042) = 199.92 \; mm \qquad \text{(d)}$$
$$h_{new} = 2\,(1 + \varepsilon_3) = 2\,(1 - 0.00042) = 1.9992 \; mm$$

###########

3.9   Plan for the Following Chapters

Our goal in the following chapters is to use the physical laws of equilibrium (Equations 2.6.13, 2.6.16, 2.6.18) and material properties (Equations 3.4.7, 3.4.22, 3.4.30) and the strain displacement relations (Equations 3.2.10, 3.2.11, 3.2.12) to derive equations connecting the known quantities such as geometry, applied loads, restraints, and material properties with the unknown quantities of displacement, strain, stress, internal forces, and restraint forces.

Symbolically, for homogenous and isotropic materials

$$\{F\} \quad \rightarrow \quad \{f_b\} \quad \{\sigma\} \quad \{\varepsilon\} \quad \{u\} \quad \leftarrow \quad \{\rho\}$$
$$\{\varepsilon\}\,[D]\,\{u\}$$
$$\{\sigma\} = [G]\,\{\varepsilon\}$$
$$\{f_b\} = [E]\,\{\sigma\}$$

$$(3.9.1)$$

The quantities in the first row of Equation 3.9.1 are the matrix components of the unknown quantities for which we shall seek answers

$\{\sigma\}$ stresses

$\{\varepsilon\}$ strains

$\{u\}$ displacements

and the known quantities that are given in the statement of the problem

$\{F\}$ applied concentrated and surface forces ($\{F\} = \{F_c\} + \{f_s\}$)

$\{f_b\}$ body forces

$\{\rho\}$ geometric restraints

These quantities are connected by the following equations where the matrices in the second, third, and fourth rows are the differential, integral, and algebraic operators which define the following equations.

$\{\varepsilon\} = [D]\,\{u\}$ Strain displacement

$\{\sigma\} = [G]\,\{\varepsilon\}$ Hooke's law

$\{f_b\} = [E]\,\{\sigma\}$ Equilibrium (E is the equilibrium operator matrix)

The arrows indicate that $\{F\}$ and $\{\rho\}$ appear in the boundary conditions.

The particular formulation of these equations depends on the geometry of the solid body. There are three classes of bodies that are usually considered:

- One dimensional (function of 1 independent variable):
    Slender bars – axial, torsional, bending
- Two dimensional (function of 2 independent variables):
    Plane stress, plates, and shells
- Three dimensional (function of 3 independent variables):
    General solid bodies

These are considered in order in the following chapters. In the one dimensional case the equations of equilibrium (Equation 2.6.18), material properties (Equation 3.4.7), and strain displacement

Equation (3.2.12) in order are summarized here as Equation 3.9.2.

$$\frac{\partial \sigma_x}{\partial x} = -f_{bx} \qquad \sigma_x = E\varepsilon_x \qquad \varepsilon_x = \frac{du}{dx} \tag{3.9.2}$$

In two dimensions we have for equilibrium (Equation 2.6.16)

$$[E]\{\sigma\} = \begin{bmatrix} \dfrac{\partial}{\partial x} & 0 & \dfrac{\partial}{\partial y} \\[2mm] 0 & \dfrac{\partial}{\partial y} & \dfrac{\partial}{\partial x} \end{bmatrix} \begin{bmatrix} \sigma_x \\ \sigma_y \\ \tau_{xy} \end{bmatrix} = -\begin{bmatrix} f_{bx} \\ f_{by} \end{bmatrix} = -\{f_b\} \tag{3.9.3}$$

for material properties (Equation 3.4.22)

$$\{\sigma\} = \begin{bmatrix} \sigma_x \\ \sigma_y \\ \tau_{xy} \end{bmatrix} = \begin{bmatrix} \dfrac{E}{1-v^2} & \dfrac{vE}{1-v^2} & 0 \\[2mm] \dfrac{vE}{1-v^2} & \dfrac{E}{1-v^2} & 0 \\[2mm] 0 & 0 & G \end{bmatrix} \begin{bmatrix} \varepsilon_x \\ \varepsilon_y \\ \gamma_{xy} \end{bmatrix} = [G]\{\varepsilon\} \tag{3.9.4}$$

and for strain displacement (Equation 3.2.12)

$$\{\varepsilon\} = \begin{bmatrix} \varepsilon_x \\ \varepsilon_y \\ \gamma_{xy} \end{bmatrix} = \begin{bmatrix} \dfrac{\partial}{\partial x} & 0 \\[2mm] 0 & \dfrac{\partial}{\partial y} \\[2mm] \dfrac{\partial}{\partial y} & \dfrac{\partial}{\partial x} \end{bmatrix} \begin{bmatrix} u \\ v \end{bmatrix} = [D]\{u\} \tag{3.9.5}$$

In three dimensions for equilibrium (Equation 2.6.13)

$$[E]\{\sigma\} = \begin{bmatrix} \dfrac{\partial}{\partial x} & 0 & 0 & \dfrac{\partial}{\partial y} & 0 & \dfrac{\partial}{\partial z} \\[2mm] 0 & \dfrac{\partial}{\partial y} & 0 & \dfrac{\partial}{\partial x} & \dfrac{\partial}{\partial z} & 0 \\[2mm] 0 & 0 & \dfrac{\partial}{\partial z} & 0 & \dfrac{\partial}{\partial y} & \dfrac{\partial}{\partial x} \end{bmatrix} \begin{bmatrix} \sigma_x \\ \sigma_y \\ \sigma_z \\ \tau_{xy} \\ \tau_{yz} \\ \tau_{zx} \end{bmatrix} = -\begin{bmatrix} f_{bx} \\ f_{by} \\ f_{bz} \end{bmatrix} = -\{f_b\} \tag{3.9.6}$$

For material properties (Equation 3.4.30)

$$\{\sigma\} = \begin{bmatrix} \sigma_x \\ \sigma_y \\ \sigma_z \\ \tau_{xy} \\ \tau_{yz} \\ \tau_{zx} \end{bmatrix} = \begin{bmatrix} \lambda+2G & \lambda & \lambda & 0 & 0 & 0 \\ \lambda & \lambda+2G & \lambda & 0 & 0 & 0 \\ \lambda & \lambda & \lambda+2G & 0 & 0 & 0 \\ 0 & 0 & 0 & G & 0 & 0 \\ 0 & 0 & 0 & 0 & G & 0 \\ 0 & 0 & 0 & 0 & 0 & G \end{bmatrix} \begin{bmatrix} \varepsilon_x \\ \varepsilon_y \\ \varepsilon_z \\ \gamma_{xy} \\ \gamma_{yz} \\ \gamma_{zx} \end{bmatrix} = [G]\{\varepsilon\} \tag{3.9.7}$$

and for strain displacement (Equation 3.2.10)

$$\{\varepsilon\} = \begin{Bmatrix} \varepsilon_x \\ \varepsilon_y \\ \varepsilon_z \\ \gamma_{xy} \\ \gamma_{yz} \\ \gamma_{zx} \end{Bmatrix} = \begin{bmatrix} \dfrac{\partial}{\partial x} & 0 & 0 \\ 0 & \dfrac{\partial}{\partial y} & 0 \\ 0 & 0 & \dfrac{\partial}{\partial z} \\ \dfrac{\partial}{\partial y} & \dfrac{\partial}{\partial x} & 0 \\ 0 & \dfrac{\partial}{\partial z} & \dfrac{\partial}{\partial y} \\ \dfrac{\partial}{\partial z} & 0 & \dfrac{\partial}{\partial x} \end{bmatrix} \begin{Bmatrix} u \\ v \\ w \end{Bmatrix} = [D]\{u\} \tag{3.9.8}$$

The material properties for laminate materials will also be introduced at an appropriate point.

The study of the three dimensional Equations 3.9.6, 3.9.7, and 3.9.8 and their two and one dimensional cousins is called the *theory of elasticity*. Satisfying these equations consistent with their boundary conditions of applied load and restraints for a given geometry has been proved to be a very accurate description of the behavior of many load carrying solid bodies. Finding exact analytical solutions, however, has proved quite difficult except in certain cases of simplified geometry, loading, and restraint. Fortunately, simplifying assumptions can provide equations that can be solved for a wide variety of cases with great practical value.

The rest of this text deals with the equations and their solutions made possible by these simplifying assumptions. On occasion we even find exact solutions to the complete equations of elasticity.

3.10 Summary and Conclusions

The various quantities that make up the study of mechanics of solids are identified and defined. In the remaining chapters we apply external forces on solid bodies which are restrained and find the internal forces or stresses and the displacements. To derive suitable equations for analysis we invoke two physical laws that are defined here – equilibrium and materials properties. To properly define material properties it is necessary to introduce the quantity called strain and connect it to the stresses via Hooke's law. Strain is defined in terms of the displacements.

In the past many solution methods were based on finding a state of stress that first satisfied equilibrium. Then the strains were found using Hooke's law. Finally the displacements were found from the strain displacement equations. Because the six components of strain are defined in terms of three components of displacement the strain components are not independent of each other. Thus the compatibility equations also must be satisfied for the displacements to be valid.

In more recent times equilibrium equations usually are formulated in terms of displacements. From the displacements we determine the strains, then the stresses, and compatibility is satisfied implicitly.

This introduction is deliberately brief. It is believed that the things studied so far will become familiar and more meaningful in the process of solving a variety of problems that are introduced in the subsequent chapters.

4

Classical Analysis of the Axially Loaded Slender Bar

4.1 Introduction

In Chapters 1–3 we learned that in a typical problem in the analysis of solid bodies the known quantities are the geometry, applied forces or loads, restraints, and material properties. We note that the unknown quantities usually are restraint forces, internal forces and stresses, displacements, and strains. To have a successful structure the stresses and displacements must fall within acceptable limits. To define these quantities we have presented the two physical laws, equilibrium and material properties, and one geometric relationship, strain-displacement, that must be satisfied. Now we must put them together in a form that obtains solutions for the displacements, strains, stresses, and internal forces. To illustrate this in the clearest way we shall use some examples that have very simple geometry, loading, and restraint. In that way we can concentrate on the process and not be diverted from understanding by the details of more complicated cases.

In this chapter we consider the axially loaded slender bar. We shall assume it is made of a homogenous and isotropic material and that its strain and displacement follow Hooke's law within the load limits imposed.

4.2 Solutions from the Theory of Elasticity

To set the stage for developing the equations for an axially loaded slender bar consider a rectangular solid body that is loaded by a uniform force per unit area, f_0, in the x direction on each end as shown in Figure 4.2.1. There are no loads on any other surface, no body forces, and no restraints. The coordinate axes originate at the centroid of the body and are aligned as principal axes.

The body is in equilibrium as a rigid body. We wish to find solutions for the stress and deformation that satisfy equilibrium in the interior, match the loads on the boundaries, and satisfy the compatibility equations.

It is intuitive that the normal stress, σ_x, shall be constant throughout the body and all other stress components be zero, that is,

$$\sigma_x(x, y, z) = f_0 \qquad \sigma_y = \sigma_z = \tau_{xy} = \tau_{yz} = \tau_{zx} = 0 \qquad (4.2.1)$$

In fact we showed this to be true for the plane stress case in Example 3.5.1.

Analysis of Structures: An Introduction Including Numerical Methods, First Edition. Joe G. Eisley and Anthony M. Waas.
© 2011 John Wiley & Sons, Ltd. Published 2011 by John Wiley & Sons, Ltd.

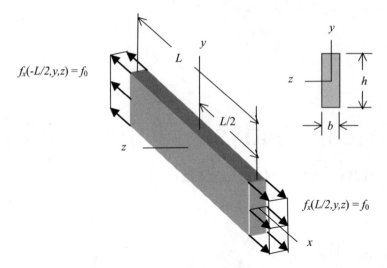

Figure 4.2.1

Let us see if these values do satisfy the three dimensional equations of equilibrium in Section 2.6, Equations 2.6.13. In fact, upon substitution all equations are satisfied exactly when there are no body forces.

$$
\begin{bmatrix}
\dfrac{\partial}{\partial x} & 0 & 0 & \dfrac{\partial}{\partial y} & 0 & \dfrac{\partial}{\partial z} \\[2mm]
0 & \dfrac{\partial}{\partial y} & 0 & \dfrac{\partial}{\partial x} & \dfrac{\partial}{\partial z} & 0 \\[2mm]
0 & 0 & \dfrac{\partial}{\partial z} & 0 & \dfrac{\partial}{\partial y} & \dfrac{\partial}{\partial x}
\end{bmatrix}
\begin{bmatrix}
f_0 \\ 0 \\ 0 \\ 0 \\ 0 \\ 0
\end{bmatrix}
=
\begin{bmatrix}
0 \\ 0 \\ 0
\end{bmatrix}
\tag{4.2.2}
$$

The stresses match the applied loads at all boundaries.

$$
\sigma_x\left(\pm\frac{L}{2}, y, z\right) = f_0 \qquad \sigma_y\left(x, \pm\frac{h}{2}, z\right) = 0 \qquad \sigma_z\left(x, y, \pm\frac{b}{2}\right) = 0
\tag{4.2.3}
$$

and all shear stresses on the boundaries are zero.

The state of strain represented by this state of stress is obtained from Hooke's law.

$$
\varepsilon_x = \frac{\sigma_x}{E} = \frac{f_0}{E} \qquad \varepsilon_y = -\frac{\nu\sigma_x}{E} = -\frac{\nu f_0}{E} \qquad \varepsilon_z = -\frac{\nu\sigma_x}{E} = -\frac{\nu f_0}{E} \qquad \gamma_{xy} = \gamma_{yz} = \gamma_{zx} = 0
\tag{4.2.4}
$$

It is noted in Section 3.3 that the strain components are not independent of each other. The conditions that must be met are called conditions of compatibility. It is noted there that compatibility is satisfied if the strain components are constant. Let us check Equation 3.3.3 which is repeated here as Equation 4.2.5.

$$\frac{\partial^2 \varepsilon_x}{\partial y^2} + \frac{\partial^2 \varepsilon_y}{\partial x^2} = \frac{\partial^2 \gamma_{xy}}{\partial x \partial y} \tag{4.2.5}$$

Inserting the strain components we see that compatibility is satisfied.

$$\frac{\partial^2}{\partial y^2}\left(\frac{f_0}{E}\right) + \frac{\partial^2}{\partial x^2}\left(-\frac{\nu f_0}{E}\right) = \frac{\partial^2}{\partial x \partial y}(0) \quad \rightarrow \quad 0 + 0 = 0 \tag{4.2.6}$$

The displacements are found by integrating the strain components.

$$\varepsilon_x = \frac{f_0}{E} \quad \rightarrow \quad \frac{\partial u}{\partial x} = \frac{f_0}{E} \quad \rightarrow \quad u(x, y, z) = \frac{f_0}{E}x + a(y, z)$$

$$\varepsilon_y = -\frac{\nu f_0}{E} \quad \rightarrow \quad \frac{\partial v}{\partial y} = -\frac{\nu f_0}{E} \quad \rightarrow \quad v(x, y, z) = -\frac{\nu f_0}{E}y + b(x, z) \tag{4.2.7}$$

$$\varepsilon_z = -\frac{\nu f_0}{E} \quad \rightarrow \quad \partial\frac{\partial w}{dz} = -\frac{\nu f_0}{E} \quad \rightarrow \quad w(x, y, z) = -\frac{\nu f_0}{E}z + c(x, y)$$

If we assign zero displacement at the origin of the axes we have the following boundary conditions

$$u(0, y, z) = \frac{f_0}{E}\cdot 0 + a(y, z) = 0 \quad \rightarrow \quad a(y, z) = 0$$

$$v(x, 0, z) = -\frac{\nu f_0}{E}\cdot 0 + b(x, z) = 0 \quad \rightarrow \quad b(x, z) = 0 \tag{4.2.8}$$

$$w(x, y, 0) = -\frac{\nu f_0}{E}\cdot 0 + c(x, y) = 0 \quad \rightarrow \quad c(x, y) = 0$$

The displacements are

$$u(x, y, z) = \frac{f_0}{E}x \quad v(x, y, z) = -\frac{\nu f_0}{E}y \quad w(x, y, z) = -\frac{\nu f_0}{E}z \tag{4.2.9}$$

Note that the displacement in the x direction is independent of y and z and linear in x. All points lying on any yz plane move an equal amount in the x direction. This is referred to as *plane sections remain plane*.

There is also a displacement in the y and z directions even though there are no forces in the y and z directions. This was noted in Section 3.4 as the Poisson's ratio effect. Since for many materials the value of Poisson's ratio is approximately $\nu = 0.3$ we can see that the displacements in the y and z directions can be quite sizeable compared to the displacement in the x direction for a three dimensional body. If the body is much longer in the x direction compared to its height and thickness the displacements in the y and z directions are relatively small in comparison. We noted in Section 3.4 that for a slender bar these displacements can often be ignored. Example 3.4.2 also confirms that for a typical structural material, in that case aluminum, all displacements will be quite small.

The solution for this combination of geometry, loading, and restraint is one of the few exact solutions available in the theory of elasticity. If we modify the end loading to be non uniform an exact solution can become quite difficult to obtain. Refer again to Chapter 3, Section 3.5. To illustrate consider a parabolic end load such as that shown in Figure 4.2.2. Let the loading at each end be

$$f_x(\pm L/2, y, z) = \pm f_0\left(1 - \frac{4y^2}{h^2}\right)$$ (4.2.10)

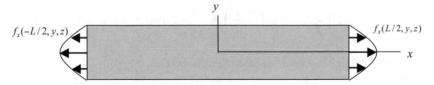

Figure 4.2.2

Let us assume that the stress in the interior has the same form as the loads on the boundaries, that is,

$$\sigma_x(x, y, z) = f_0\left(1 - \frac{4y^2}{h^2}\right) \qquad \sigma_y = \sigma_z = \tau_{xy} = \tau_{yz} = \tau_{zx} = 0$$ (4.2.11)

Let us check to see if Equation 4.2.11 is a possible solution. First let us check to see if this state of stress satisfies equilibrium.

$$\begin{bmatrix} \frac{\partial}{\partial x} & 0 & 0 & \frac{\partial}{\partial y} & 0 & \frac{\partial}{\partial z} \\ 0 & \frac{\partial}{\partial y} & 0 & \frac{\partial}{\partial x} & \frac{\partial}{\partial z} & 0 \\ 0 & 0 & \frac{\partial}{\partial z} & 0 & \frac{\partial}{\partial y} & \frac{\partial}{\partial x} \end{bmatrix} \begin{bmatrix} f_0\left(1 - \frac{4y^2}{h^2}\right) \\ 0 \\ 0 \\ 0 \\ 0 \\ 0 \end{bmatrix} = \begin{bmatrix} 0 \\ 0 \\ 0 \end{bmatrix}$$ (4.2.12)

Sure enough it does in the absence of body forces. And by inspection we see that it satisfies the boundary conditions on applied loads at the boundaries. On the ends

$$\sigma_x\left(\pm\frac{L}{2}, y, z\right) = f_0\left(1 - \frac{4y^2}{h^2}\right)$$ (4.2.13)

and the stresses are zero on all other faces.

So far so good, but now we must check compatibility. The components of strain are

$$\varepsilon_x = \frac{\sigma_x}{E} = \frac{f_0}{E}\left(1 - \frac{4y^2}{h^2}\right) \qquad \varepsilon_y = -\frac{v\sigma_x}{E} = -\frac{vf_0}{E}\left(1 - \frac{4y^2}{h^2}\right)$$

(4.2.14)

$$\varepsilon_z = -\frac{v\sigma_x}{E} = -\frac{vf_0}{E}\left(1 - \frac{4y^2}{h^2}\right) \qquad \gamma_{xy} = \gamma_{yz} = \gamma_{zx} = 0$$

Insert the components of strain into Equation 4.2.5.

$$\frac{\partial^2 \varepsilon_x}{\partial y^2} + \frac{\partial^2 \varepsilon_y}{\partial x^2} = \frac{\partial^2 \gamma_{xy}}{\partial x \partial y}$$

$$\rightarrow \quad \frac{\partial^2}{\partial y^2}\left(\frac{f_0}{E}\left(1 - \frac{4y^2}{h^2}\right)\right) + \frac{\partial^2}{\partial x^2}\left(-\frac{v f_0}{E}\left(1 - \frac{4y^2}{h^2}\right)\right) = 0 \qquad (4.2.15)$$

$$\rightarrow \quad -\frac{8 f_0}{E h^2} \neq 0$$

However, this equation is not satisfied. The second derivative of the normal strain in the x direction is a non zero constant.

An exact analytical solution has not been found; however, approximate numerical, but highly accurate, solutions for this and other boundary conditions have been found. The validity of these solutions is supported by experimental evidence. We shall now present one of these numerical solutions without proof; however, as this text proceeds we shall develop the methods used here. In particular a numerical solution has been found for a non uniform end force per unit area as shown in Figure 4.2.2. The Unigraphics I-DEAS software was used in the following example.

##########

Example 4.2.1

Problem: A rectangular bar is loaded with a parabolic end load as shown in Figure (a). It has dimensions of $L = 250\ mm$, $h = 50\ mm$, $b = 10\ mm$. It is made of aluminum with $E = 68950\ N/mm^2$. The distributed end loading is

$$f_x\left(\pm\frac{L}{2}, y, z\right) = \pm f_0\left(1 - \frac{4y^2}{h^2}\right) = \pm\frac{3}{2}\left(1 - \frac{4y^2}{50^2}\right)\frac{N}{mm^2} \qquad (a)$$

Find the stresses and displacements.

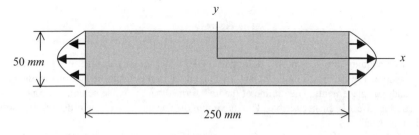

Figure (a)

Solution: Use the I-DEAS software to find a numerical solution based on 3D elasticity equations.

Note that we can determine from symmetry that the resultant forces at the two ends act through the centroids of the end faces.

The value of the resultant force at each end is

$$F = \int_{-\frac{h}{2}}^{\frac{h}{2}} f_0 \left(1 - \frac{4y^2}{h^2}\right) b \, dy = \int_{-25}^{25} \frac{3}{2} \left(1 - \frac{4y^2}{50^2}\right) 10 \, dy = \frac{3}{2} \cdot 10 \left(y - \frac{4y^3}{3 \cdot 50^2}\right)\Bigg|_{-25}^{25} = 500 \, N \quad \text{(b)}$$

A contour plot of the σ_x component of stress from the numerical analysis is shown in Figure (b).

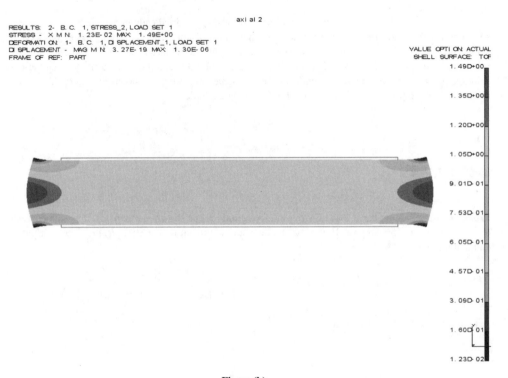

Figure (b)

The stress contours are shown on a deformed model which greatly exaggerates the deformation so that it is visible. The deformed model shows the magnitude of the displacement, that is, the vector sum of the x and y displacements. The outline of the undeformed bar is visible behind the contour plot. The Poisson's ratio effect is clearly visible. The maximum displacement is of the order of 10^{-6} so you can see it is quite small.

The lines between colors are lines of constant stress. (For those looking at a black and white print the lines between shades of gray are lines of constant stress.) Over most of the bar the stress is a uniform $1 \, N/mm^2$. This is the same value you would get if a uniform distributed load was placed on the ends as explained in Equations 4.2.1-3 with a value of $1 \, N/mm^2$ (500 N total force). The surprising fact is that over most of the length of the bar the stress is uniformly distributed no matter how the load is distributed over the ends. Only the value of the resultant applied force and the cross section area are important.

Near the ends the stress distribution is affected. There is a transition from the uniform value to the value of the distributed applied load. We can see better what is happening near the ends if we zoom in as shown in Figure (c).

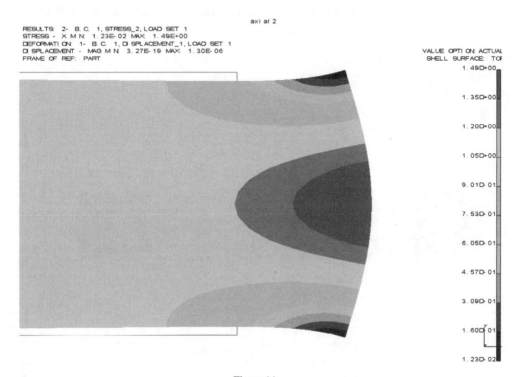

RESULTS: 2- B. C. 1, STRESS_2, LOAD SET 1
STRESS - X M N: 1.23E-02 MAX 1.49E+00
DEFORMATION: 1- B. C. 1, DISPLACEMENT_1, LOAD SET 1
DISPLACEMENT - MAG M N: 3.27E-19 MAX 1.30E-06
FRAME OF REF: PART

axial 2

VALUE OPTION: ACTUAL
SHELL SURFACE: TOP

1.49D+00
1.35D+00
1.20D+00
1.05D+00
9.01D-01
7.53D-01
6.05D-01
4.57D-01
3.09D-01
1.60D-01
1.23D-02

Figure (c)

The black and white printed version, which shows the colors in gray scale, is not so clear but in any case you can see that a portion of the bar is affected significantly at each end. The stress transitions from the value of the applied surface load on the end faces to a uniform value over a distance about equal to the height of the bar. Over the middle 60 percent the stress is very close to uniform as expressed by a single color. The maximum stress at the ends is approximately 50 percent greater than the average uniform stress over the middle portion.

A contour plot of the x displacement is shown in Figure (d). As noted the displacement displayed is greatly exaggerated in order to make it visible. As we noted the actual displacement is quite small. Straight vertical edges of color zones are evidence of constant values of displacement in the bar.

Of particular importance is that the displacement away from the ends is uniform for all values of y and z. The material particles that lie in any yz plane before the load is applied continue to lie on an yz plane after the load is applied except near the ends. This is referred to as *plane sections remain plane*.

   This phenomenon of the localized effect of loading, that is, over most of the length of a slender bar the internal stresses and displacements depend only on the value of the total force and not on its particular distribution, was first discovered by St. Venant who enunciated his now famous *St. Venant's principle*. According to this principle the strains that are produced in any body by an application, to a small part of its surface, of a system of forces statically equivalent to zero force and zero moment, are of negligible magnitude at distances which are large compared to the linear dimensions of the body.

   This principle will aid in developing a simplified theory of analysis for axially loaded slender bars.

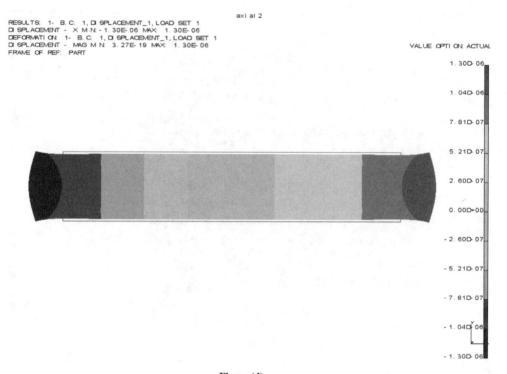

Figure (d)

   Now let us repeat the solution using a more slender bar. This has a length to height ratio of 10 to 1 as shown in Figure (e). The cross section and the loading are the same.

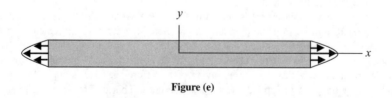

Figure (e)

In Figure (f) we have the contour plot of the stress.

axi al 2

RESULTS: 2- B. C. 1, STRESS_2, LOAD SET 1
STRESS - X M N: 1. 23E- 02 MAX: 1. 49E+00
DEFORMATI ON: 1- B. C. 1, DI SPLACEMENT_1, LOAD SET 1
DI SPLACEMENT - MAG M N: 1. 05E- 17 MAX: 2. 51E- 06
FRAME OF REF: PART

VALUE OPTI ON: ACTUAL
SHELL SURFACE: TOP

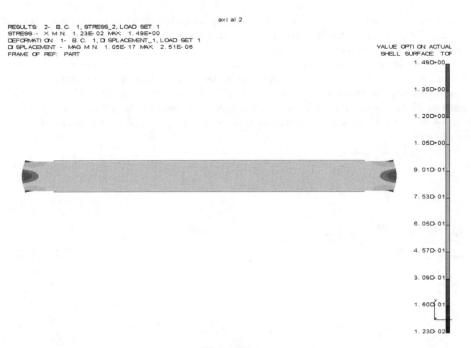

Figure (f)

In Figure (g) we have the contour plot of the x displacement.

axi al 2

RESULTS: 1- B. C. 1, DI SPLACEMENT_1, LOAD SET 1
DI SPLACEMENT - X M N: - 2. 51E- 06 MAX: 2. 51E- 06
DEFORMATI ON: 1- B. C. 1, DI SPLACEMENT_1, LOAD SET 1
DI SPLACEMENT - MAG M N: 1. 05E- 17 MAX: 2. 51E- 06
FRAME OF REF: PART

VALUE OPTI ON: ACTUAL

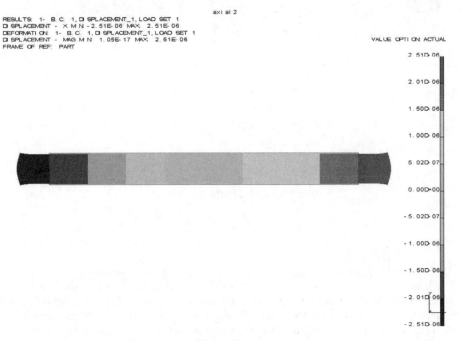

Figure (g)

In this case the same effect is noted; however, only about 10 percent of the region at each end is affected, that is, a length roughly equal to the height. Thus as the bar becomes more slender the end effect is a less significant part of the total.

Before continuing with the development of this particular theory of structures we note that constraints can also affect the stress and displacement distribution. Consider a slender bar built into a wall, or has what is called a fixed end. The fixed end boundary condition consists of a total restraint in all degrees of freedom. In particular we solve the same bar as in Figure (a) with the same load on the left end but with the right end fixed. In Figure (h) we show the results of the stress contour plot.

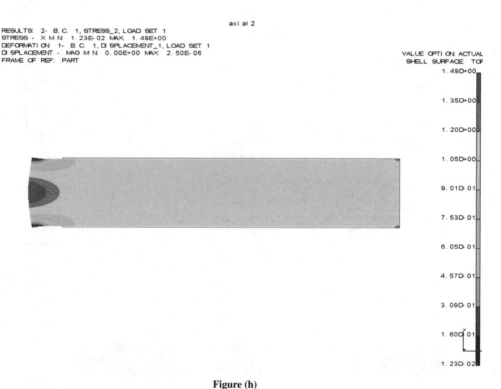

Figure (h)

This is a bit hard to see so we enlarge the image of the right end in Figure (i).

The high stress levels at the corners are due to restraint against the Poisson's ratio effect. There is a decrease in the height of the bar over most of the length due to this effect but at the right end it is restrained by the fixed condition. Once again the effect is localized in the vicinity of the restraint and does not affect the rest of the bar. The actual displacements in the y and z directions are quite small for the slender member.

axi al 2

RESULTS: 2- B. C. 1, STRESS_2, LOAD SET 1
STRESS - X M N: 1.23E- 02 MAX: 1.49E+00
DEFORMATI ON: 1- B. C. 1, DI SPLACEMENT_1, LOAD SET 1
DI SPLACEMENT - MAG M N: 0.00E+00 MAX: 2.50E- 06
FRAME OF REF: PART

VALUE OPTI ON: ACTUAL
SHELL SURFACE: TOP

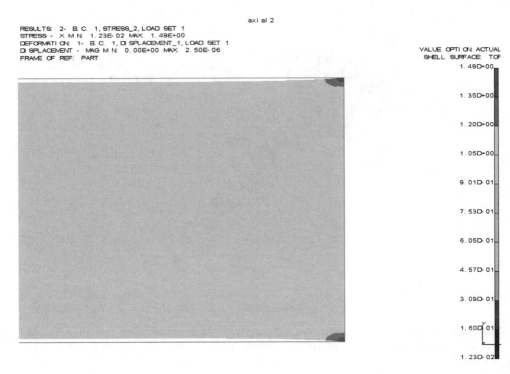

1.49D+00
1.35D+00
1.20D+00
1.05D+00
9.01D- 01
7.53D- 01
6.05D- 01
4.57D- 01
3.09D- 01
1.60D- 01
1.23D- 02

Figure (i)

############

In the meantime we shall use the assumption that plane sections remain plane to learn what we can about the behavior of slender bars by neglecting the end effects.

Many of the great achievements in structural design and analysis have been made possible by the validity of this assumption in appropriate circumstances.

4.3   Derivation and Solution of the Governing Equations

In the examples in Section 4.2 the loads and restraints are applied only at the ends of the body. We wish now to allow loads to be applied along the bar in the x direction as well as at the ends. Consider a slender bar of arbitrary cross section as shown in Figure 4.3.1. It may be restrained at one or both ends and it has some combination of distributed loads, $f_x(x)$, acting along the bar and concentrated loads, F_c, at points along the bar and at its ends.

We pause for a moment to ask how the applied forces are, in fact, applied. A load distributed over the end faces can be resolved into a concentrated resultant force. It must satisfy the requirement of a line of action through the centroid. How distributed and concentrated forces are applied between the ends requires some interpretation. Let us for now assume that some device or substance is used to attach the forces to the cross section or to the surfaces, but always with the condition that the resultant acts through the centroid. Gravity and inertia loads are one way to apply distributed axial loads. Point loads can be applied by drilling a hole and applying loads to a pin or a lug or collar can be fastened to the bar

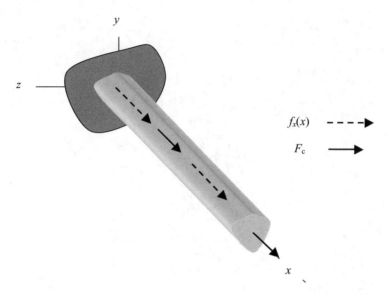

$$f_x(x) \quad \text{- - - -} \blacktriangleright$$
$$F_c \quad \longrightarrow$$

Figure 4.3.1

for purposes of applying loads as illustrated in Figure 4.3.2. We would, of course, rely on St. Venant's principle to be able to ignore local effects.

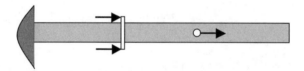

Figure 4.3.2

In whatever manner the forces are applied we assume that their effect on modifying the uniform stress and the plane section displacements is ignored. If the bar is slender enough this localized effect can be ignored by the grace and goodness of St. Venant's principle. We shall assume that plane sections remain plane over the entire length of the bar.

It is assumed that the x axis lies on the loci of centroids of the cross sections in the yz plane. All resultants of distributed forces and all concentrated forces act through the centroids of the cross sections. Finding centroids of areas is discussed in detail in Appendix B. In most of our examples the cross sections will be simple shapes with axes of symmetry; such as, rectangular or circular. The centroids are where the axes of symmetry cross. For more complicated cross sectional shapes refer to Appendix B.

From experimental evidence and our discussion in Section 4.2 we assume that plane sections remain plane and that the stress, σ_x, on a plane cross section surface normal to the x axis is constant. No shearing stresses occur on that surface. Furthermore the displacement is defined by a single component, u, which is constant across the plane and at most is only a function of x, that is, $u(x)$. It follows that there is one component of strain, $\varepsilon_x(x)$. While it is true that the axial stress will generate lateral strain from the Poisson's ratio effect, as we have noted in Section 3.4, this effect is small and will be neglected in the slender bar.

Let us consider that the slender bar in Figure 4.3.1 has constant cross section, A, is fastened at one end, and is loaded with a distributed force, $f_x(x)$, with units of force per unit length (N/mm), and one or more concentrated loads all acting in the direction of and coincident with the x axis. If a body force is present it must be restricted to the same conditions and is resolved into a force per unit length and combined with any distributed surface forces.

We shall denote the distributed displacement as $u(x)$. From our discussion of stress and strain in Chapter 3 the strain displacement equation is

$$\varepsilon_x(x) = \frac{du}{dx} \tag{4.3.1}$$

From Hooke's law we have

$$\sigma_x(x) = E\varepsilon_x = E\frac{du}{dx} \tag{4.3.2}$$

Since the x axis is located at the centroid of the cross section of the bar we can assume that only an axial internal force is generated, that is, there are no internal moments about any axis and no internal lateral forces. We denote the internal stress resultant force with the letter P. The internal force in terms of the stress, strain, and displacement is

$$P = \sigma_x A = EA\varepsilon_x = EA\frac{du}{dx} \tag{4.3.3}$$

Now let us seek the equilibrium of a slice of the bar at x of length dx as shown in Figure 4.3.3.

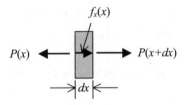

Figure 4.3.3

We are faced with defining a convention for the positive direction of a force. It is common to define applied force components and restraint force components as positive if acting in the positive directions of the axes. So F and $f_x(x)$ in Figure 4.3.3 are shown as positive forces. Two different conventions are used for internal forces. The most common has been to define internal force components acting away from the surface as positive, as P is shown in Figure 4.3.3. Internal force resultants have the same sign conventions as stresses, thus a positive stress component will generate a positive stress resultant.

We shall use this convention until further notice. In some cases later on we shall use the same convention for internal and external forces, that is, positive if acting in the positive direction of the axes, for reasons that will be made clear when we encounter the need.

If the internal force at x is $P(x)$ then at $x + dx$ it is $P(x + dx)$. Using a Taylor's series expansion and neglecting higher order terms, we have

$$P(x + dx) = P(x) + \frac{dP}{dx}dx + \text{higher order terms} \tag{4.3.4}$$

From Newton's laws the summation of forces in the x-direction must be zero.

$$\sum F_x = P(x) + \frac{dP}{dx}dx - P(x) + f_x(x)dx = 0 \tag{4.3.5}$$

which reduces to

$$\frac{dP}{dx} = -f_x(x) \tag{4.3.6}$$

Combining Equations 4.3.3 and 4.3.6 we obtain

$$\frac{dP}{dx} = \frac{d}{dx}EA\frac{du}{dx} = -f_x(x) \tag{4.3.7}$$

The quantities and equations governing the behavior of an axially loaded bar are summarized in Equations 4.3.8. As a result of the assumption of plane sections remaining plane throughout the bar the first row differs from Equation 3.9.1 in that internal force resultants are added explicitly, distributed applied forces are explicit in the equations and any body forces are absorbed into the applied forces. All concentrated applied forces and restraints are introduced in the boundary conditions. Since each quantity is the function of a single variable, the matrices reduce to single components.

In summary we have

$$F_c \quad \rightarrow \quad f_x(x) \qquad P(x) \qquad \sigma_x(x) \qquad \varepsilon_x(x) \qquad u(x) \quad \leftarrow \quad \rho$$

$$\frac{dP}{dx} = -f_x(x) \qquad\qquad\qquad \varepsilon_x(x) = \frac{du}{dx}$$

$$\sigma_x(x) = E\varepsilon_x$$

$$P(x) = \sigma_x A = EA\varepsilon_x = EA\frac{du}{dx}$$

$$\frac{dP}{dx} = \frac{d}{dx}EA\frac{du}{dx} = -f_x(x)$$

(4.3.8)

In the first row of Equation 4.3.8 we have the typical internal unknowns in the formulation of a problem.

$u(x)$ - distributed displacement
$\varepsilon_x(x)$ - normal strain
$\sigma_x(x)$ - normal stress
$P(x)$ - internal force resultant

and the known external quantities

$f_x(x)$ - distributed applied force
F_c - concentrated applied forces
ρ - displacement restraints

The arrows are there to indicate that both F_c and ρ are external *boundary conditions* to be satisfied in the process of solving the equations.

The equations connecting these quantities are in the second and third rows.

$\dfrac{dP}{dx} = -f_x(x)$ - static equilibrium

$\varepsilon_x = \dfrac{du}{dx}$ - strain displacement

$\sigma_x = E\varepsilon_x$ - Hooke's law for material properties

In the fourth row we have combined the three expressions for the internal quantities to get the internal force resultant in terms of the displacement (Equation 4.3.3).

$$P = \sigma_x A = EA\varepsilon_x = EA\frac{du}{dx} \tag{4.3.9}$$

and in the fifth row equilibrium is expressed in terms of the single unknown displacement (Equation 4.3.7).

$$\frac{dP}{dx} = \frac{d}{dx}EA\frac{du}{dx} = -f_x(x) \tag{4.3.10}$$

In the special case of a bar with a constant cross sectional area and constant Young's modulus this equation becomes

$$EA\frac{d^2u}{dx^2} = -f_x(x) \tag{4.3.11}$$

The solution of Equations 4.3.10 and 4.3.11 involve straightforward direct integration. From Equation 4.3.11 we have

$$\frac{du}{dx} = -\frac{1}{EA}\int f_x(x)dx + a \quad \rightarrow \quad u(x) = -\frac{1}{EA}\int\int f_x(x)dx + ax + b \tag{4.3.12}$$

The constants of integration, a and b, are found by assigning boundary conditions. The boundary conditions shall consist of one displacement constraint and one applied force or two displacement constraints. We must have at least one displacement restraint to obtain static equilibrium. This solution works for both statically determinate and indeterminate cases. Having found the displacement we find the stress by differentiating the displacement.

$$\sigma_x = E\varepsilon_x = E\frac{du}{dx} \tag{4.3.13}$$

In the determinate case (one displacement restraint) the internal force, P, can be determined directly from the equations of static equilibrium. Once we know the internal force we know immediately that the stress is

$$P = \sigma_x A \quad \rightarrow \quad \sigma_x = \frac{P}{A} \tag{4.3.14}$$

If we need the displacement in this case we get it from the first order equation, Equation 4.3.9.

$$P = EA\frac{du}{dx} \quad \rightarrow \quad \frac{du}{dx} = \frac{P}{EA} \quad \rightarrow \quad u(x) = \frac{1}{EA}\int P(x)dx + c \tag{4.3.15}$$

where c is the constant of integration to be determined from a displacement boundary condition. We shall illustrate both approaches to a solution in the next simple example. Note that since u is a function of x only compatibility is always satisfied

############

Example 4.3.1

Problem: Consider the simplest problem of them all – a uniform bar with equal and opposite forces at the ends as shown in Figure (a). Find the stresses and displacements.

This is the same problem examined in Section 4.2 but now with the assumption of plane sections remaining plane regardless of how the forces are applied as long as their resultant acts through the centroid of the cross sections.

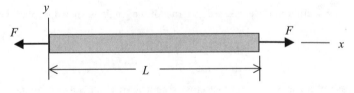

Figure (a)

Solution: First, integrate the second order displacement equation and use two boundary conditions to find the constants of integration. From the displacement find the strain, stress, and internal force. Second, find the internal force and stress using the equations of static equilibrium and then solve the first order displacement equation.

First method:
Since there is no distributed applied load the second order equation and its solution are

$$EA\frac{d^2u}{dx^2} = 0 \quad \rightarrow \quad \frac{d^2u}{dx^2} = 0 \quad \rightarrow \quad \frac{du}{dx} = a \quad \rightarrow \quad u(x) = ax + b \tag{a}$$

The known boundary condition at the right end of the bar yields

$$P(L) = EA\frac{du(L)}{dx} = EAa = F \quad \rightarrow \quad a = \frac{F}{EA} \tag{b}$$

For our other boundary condition we attach the origin of the coordinate system to the left end of the bar. The displacement there will then be zero.

$$u(0) = a \cdot 0 + b = 0 \quad \rightarrow \quad b = 0 \tag{c}$$

The solution is

$$u(x) = \frac{F}{EA}x \tag{d}$$

To find the internal force, P, we use Equation 4.3.9.

$$P = EA\frac{du}{dx} = EA\left(\frac{F}{EA}\right) = F \tag{e}$$

Note well that we find the displacement by directly integrating the second order displacement equation and **then** find the internal force by differentiating the displacement equation. The stress throughout the bar is found to be

$$\sigma_x = \frac{P}{A} = \frac{F}{A} \tag{f}$$

Plots of the displacement and stress are shown in Figure (b).

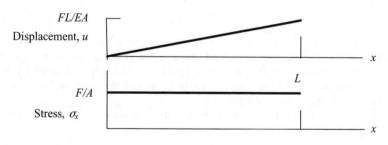

Figure (b)

Second method:

For an alternative approach, since this is a statically determinate case, we can first draw a free body diagram of a portion of the bar from some point x to the right end as shown in Figure (c).

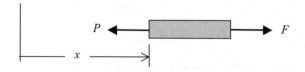

Figure (c)

The internal force may be found immediately from static equilibrium

$$\sum F_x = F - P = 0 \quad \rightarrow \quad P = F \tag{g}$$

The stress is found directly to be

$$\sigma_x = \frac{P}{A} = \frac{F}{A} \tag{h}$$

The governing equation for the displacement and its solution is from Equation 4.3.9

$$P = EA\frac{du}{dx} \quad \rightarrow \quad \frac{du}{dx} = \frac{F}{EA} \quad \rightarrow \quad u(x) = \frac{F}{EA}x + c \tag{i}$$

where c is a constant of integration. To find the value of the constant of integration we invoke the condition of zero displacement at $x = 0$ or $u(0) = 0$.

$$u(0) = \frac{F}{EA}0 + c = 0 \quad \text{thus} \quad c = 0 \tag{j}$$

The displacement $u(x)$ is

$$u(x) = \frac{F}{EA}x \tag{k}$$

Note well that in this second case the internal force is found first using the equations of static equilibrium and from it the stress is found and **then** the displacement is found by integrating the first order equation.

We note that the displacement at $x = L$ is

$$u(L) = \frac{FL}{EA} \qquad (1)$$

This alternative approach is possible only if the bar is statically determinate. The first method is good in all cases, both determinate and indeterminate. Clearly the first approach using the second order differential equation is more general since it applies to all cases but in many statically determinate cases it is faster and easier first to find the internal force and then to solve the first order equation. When the stress is all you are looking for, of course, this is the way to go.

We must note that this is exactly the same problem with exactly the same solutions as that for a bar with the left end fixed as shown in Figure (d).

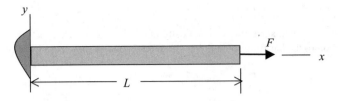

Figure (d)

A free body diagram of the entire bar as shown in Figure (e) quickly confirms that it is the same problem.

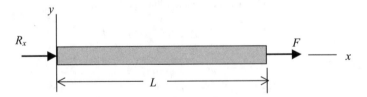

Figure (e)

From static equilibrium

$$\sum F_x = R_x + F = 0 \quad \rightarrow \quad R_x = -F \qquad (m)$$

and the boundary conditions are

$$u(0) = 0 \quad P(L) = EA\frac{du(L)}{dx} = F \qquad (n)$$

###########

4.4   The Statically Determinate Case

We shall provide several more examples of the statically determinate case. Simply put, the number of restraints may not exceed the number of applicable equations of static equilibrium. For an axially loaded slender bar that number is one. Since we have only one constant of integration the displacement restraint is necessarily the boundary condition of record.

Let us summarize the various determinate cases that arise. With $f_x(x)$ and EA as continuous functions of x, with concentrated loads at one or both ends, or with a restraint at one end and a concentrated load at the other we have three possible configurations.

One possibility is for no specified displacement restraint but for all axial forces to be in equilibrium. This is illustrated in Figure 4.4.1. Example 4.4.1 is a special case of this.

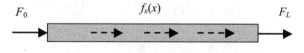

Figure 4.4.1

To be in equilibrium the summation of all concentrated forces and the resultants of all distributed forces must be zero.

$$\sum F_x = F_0 + F_L + \int_0^L f_x(x)dx = 0 \tag{4.4.1}$$

In such cases we assign the origin of the coordinate system to be the point from which displacements are measured. In Section 4.2 the origin was placed at the centroid of the bar and the axes were oriented as principal axes. The origin may be placed at any convenient point along the centroidal axis of the bar including the end points. In Example 4.3.1 the origin was placed at the left end. This is in effect a displacement restraint.

With a restraint at one end, $f_x(x)$ a continuous function of x, and a concentrated load at the other end, the possible configurations are shown in Figure 4.4.2.

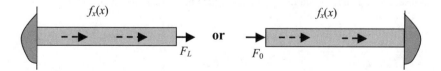

Figure 4.4.2

In all three cases the internal forces $P(x)$ can be found by using static equilibrium. To find the stress

$$\sigma_x = \frac{P(x)}{A} \tag{4.4.2}$$

To find displacements we integrate the first order equation.

$$EA\frac{du}{dx} = P(x) \quad \rightarrow \quad u(x) = \int \frac{P(x)}{EA}dx + c \tag{4.4.3}$$

The constant of integration is found by imposing a displacement restraint, either

$$u(0) = 0 \quad \text{or} \quad u(L) = 0 \tag{4.4.4}$$

###########

Example 4.4.1

Problem: The bar with the left end restrained is given a distributed load $f_x(x) = f_0$, a constant value, as shown in Figure (a). Find the displacement and stress.

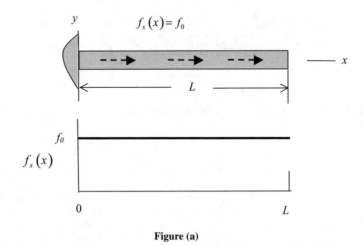

Figure (a)

Solution: Find the internal force and stress using the equations of static equilibrium and then the displacement by integrating the first order equation.

You can set up a free body diagram of a portion of the bar to help find the internal force as we did in Example 4.3.1.

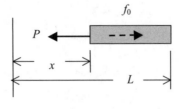

Figure (b)

From static equilibrium the axial internal force is

$$\sum F_x = f_0(L - x) - P(x) = 0 \quad \rightarrow \quad P(x) = f_0(L - x) \tag{a}$$

and we can immediately say that

$$\sigma_x(x) = \frac{P(x)}{A} = \frac{f_0}{A}(L - x) \tag{b}$$

To find the deflection we use

$$P = EA\frac{du}{dx} \tag{c}$$

to get

$$\frac{du}{dx} = \frac{f_0}{EA}(L - x) \quad \rightarrow \quad u(x) = \frac{f_0}{EA}\left(Lx - \frac{x^2}{2}\right) + c \tag{d}$$

The displacement boundary condition determines that

$$u(0) = c = 0 \tag{e}$$

The final answer is

$$u(x) = \frac{f_0}{EA}\left(Lx - \frac{x^2}{2}\right) \qquad u(L) = \frac{f_0 L^2}{2EA} \tag{f}$$

Plots of the displacement and stress are given in Figure (c).

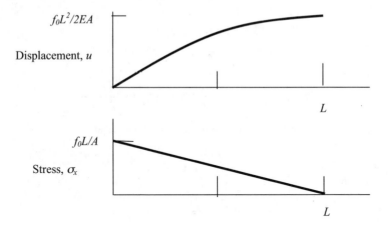

Figure (c)

Example 4.4.2

Problem: Our next example is a bar with a triangular loading and a displacement restraint on the right end as shown in Figure (a). Find the displacement and stress.

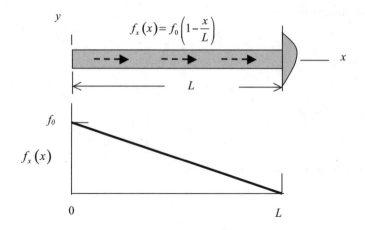

Figure (a)

Solution: As the functional equation for the loading gets more complicated it is convenient to use the equilibrium equation, Equation 4.3.6, to find the internal axial force in place of a free body diagram. Then solve the first order equation for the displacement.

For this case

$$\frac{dP}{dx} = -f_x(x) = -f_0\left(1 - \frac{x}{L}\right) \tag{a}$$

Integrate once.

$$P(x) = -f_0 \int \left(1 - \frac{x}{L}\right)dx = -f_0\left(x - \frac{x^2}{2L}\right) + a \tag{b}$$

The boundary condition at $x = 0$ is that there is no concentrated load, so

$$P(0) = a = 0 \tag{c}$$

We have for the stress

$$\sigma_x(x) = \frac{P(x)}{A} = -\frac{f_0}{A}\left(x - \frac{x^2}{2L}\right) \tag{d}$$

The differential equation for the displacement is

$$EA\frac{du}{dx} = P(x) = -f_0\left(x - \frac{x^2}{2L}\right) \quad \rightarrow \quad u(x) = -\frac{f_0}{EA}\int \left(x - \frac{x^2}{2L}\right)dx$$

$$\rightarrow \quad u(x) = -\frac{f_0}{EA}\left(\frac{x^2}{2} - \frac{x^3}{6L}\right) + c \tag{e}$$

The boundary condition is the restraint at $x = L$, that is, $u(L) = 0$.

$$u(L) = -\frac{f_0}{EA}\left(\frac{L^2}{2} - \frac{L^3}{6L}\right) + c = -\frac{f_0L^2}{3EA} + c = 0 \quad \rightarrow \quad c = \frac{f_0L^2}{3EA} \tag{f}$$

And so the displacement is

$$u(x) = \frac{f_0 L^2}{3EA} - \frac{f_0}{EA}\left(\frac{x^2}{2} - \frac{x^3}{6L}\right) = \frac{f_0}{EA}\left(\frac{L^2}{3} - \frac{x^2}{2} + \frac{x^3}{6L}\right) \tag{g}$$

Plots of the displacement and stress are given in Figure (b).

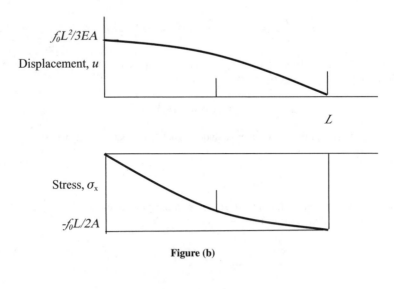

Figure (b)

##########

When distributed loading is discontinuous, and/or concentrated loads are at midpoints, or *EA* is discontinuous, the equations must be written for each segment. Examples of possible configurations are shown in Figure 4.4.3. There are, of course, an infinite variety of loads and *EA* values, but just the three possible end loads or restraints.

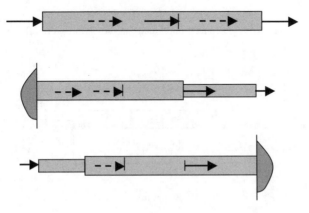

Figure 4.4.3

Let two adjacent segments be identified by the subscripts i and $i+1$ and the coordinate where they meet by x_j. All loads including concentrated loads at a boundary are satisfied by using static equilibrium to find $P(x)$ in each region. To find the stresses

$$\sigma_{xi} = \frac{P_i(x)}{A_i} \tag{4.4.5}$$

To find displacements:

$$E_i A_i \frac{du_i}{dx} = P_i(x) \quad \rightarrow \quad u_i(x) = \int \frac{P_i(x)}{E_i A_i} dx + c_i \tag{4.4.6}$$

For boundary conditions apply whatever restraints and concentrated applied forces occur at the ends of the bar. Then apply continuity of displacement where segments meet. Then

$$u_i(x_j) = u_{i+1}(x_j) \tag{4.4.7}$$

Where necessary apply equilibrium of internal and external forces between segments.

$$- E_i A_i \frac{du_i(x_j)}{dx} + F_j + E_{i+1} A_{i+1} \frac{du_{i+1}(x_j)}{dx} = 0 \tag{4.4.8}$$

For the top example in Figure 4.4.3 to be in equilibrium the summation of all concentrated forces and the resultants of all distributed forces must be zero.

$$\sum F_x = \sum_{i=1}^{n} F_i + \sum_{j=1}^{m} \int f_j(x) dx = 0 \tag{4.4.9}$$

The bar is attached to the origin of the coordinate system. This serves as the restraint boundary condition to prevent rigid body motion.

##########

Example 4.4.3

Problem: Consider the bar with two concentrated loads as shown in Figure (a). Find the displacements and stresses.

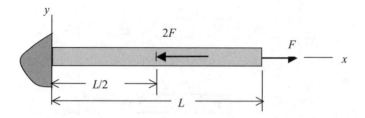

Figure (a)

Solution: Find the internal forces and stresses using the equations of static equilibrium and then find the displacements by integrating the first order equation. Satisfy boundary restraints and forces at the ends and continuity at the mid point.

Region 1 ....

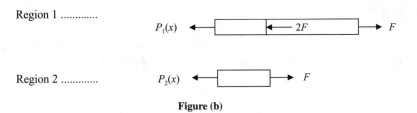

Region 2 ....

Figure (b)

From static equilibrium

Region1: $\sum F_x = F - 2F - P_1(x) = 0 \quad \rightarrow \quad P_1(x) = -F \quad 0 \le x \le \dfrac{L}{2}$

Region2: $\sum F_x = F - P_2(x) = 0 \quad \rightarrow \quad P_2(x) = F \quad \dfrac{L}{2} < x \le L$ (a)

The stresses are

$$\sigma_{x1} = -\frac{F}{A} \quad 0 \le x \le \frac{L}{2} \qquad \sigma_{x2} = \frac{F}{A} \quad \frac{L}{2} < x \le L \tag{b}$$

The deflections are found by integrating and applying boundary conditions.

$$\frac{du_1}{dx} = -\frac{F}{EA} \quad \rightarrow \quad u_1(x) = -\frac{F}{EA}x + c_1 \qquad 0 \le x \le \frac{L}{2}$$

$$\frac{du_2}{dx} = \frac{F}{EA} \quad \rightarrow \quad u_2(x) = \frac{F}{EA}x + c_2 \qquad \frac{L}{2} \le x \le L \tag{c}$$

The boundary condition at the left end is

$$u_1(0) = c_1 = 0 \quad \rightarrow \quad u_1(x) = -\frac{F}{EA}x \quad \rightarrow \quad u_1\left(\frac{L}{2}\right) = -\frac{FL}{2EA} \tag{d}$$

Now we use $u_2(L/2) = u_1(L/2)$ to get

$$u_2\left(\frac{L}{2}\right) = \frac{FL}{2EA} + c_2 = u_1\left(\frac{L}{2}\right) = -\frac{FL}{2EA} \quad \rightarrow \quad u_2(x) = \frac{F}{EA}(x - L) \quad \rightarrow \quad u_2(L) = 0 \quad \text{(e)}$$

Plots of the displacement and stress are shown in Figure (c).

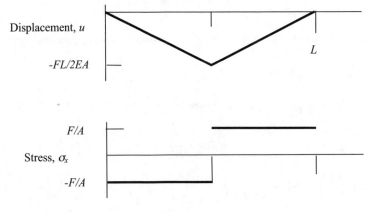

Figure (c)

Example 4.4.4

Problem: Suppose the same bar has both a distributed load over its whole length as given in Example 4.4.1 and two concentrated forces as given in Example 4.4.3 as shown in Figure (a). Find the displacements and stresses.

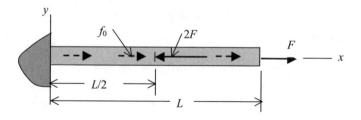

Figure (a)

Solution: We have two choices here. First we could proceed exactly as in the two previous examples and find the axial internal force. From it we can find the stress and solve the differential equation for the displacement. Second we could use superposition of known solutions.

First Method:
Region 1:

$$\sum F_x = f_0(L - x) + F - 2F - P_1(x) = 0$$

$$\rightarrow \quad P_1(x) = f_0(L - x) - F \qquad 0 \le x \le \frac{L}{2} \tag{a}$$

Region 2:

$$\sum F_x = f_0(L - x) + F - P_2(x) = 0$$

$$\rightarrow \quad P_2(x) = f_0(L - x) + F \qquad \frac{L}{2} < x \le L \tag{b}$$

The stress is

$$\sigma_{x1} = \frac{P_1}{A} = \frac{f_0}{A}(L - x) - \frac{F}{A} \qquad 0 \le x \le \frac{L}{2}$$

$$\sigma_{x2} = \frac{P_2}{A} = \frac{f_0}{A}(L - x) + \frac{F}{A} \qquad \frac{L}{2} < x \le L \tag{c}$$

The deflection is found by integrating the following two equations and applying the boundary restraint.

$$\frac{du_1}{dx} = \frac{P_1}{EA} = \frac{f_0}{EA}(L - x) - \frac{F}{EA}$$

$$\rightarrow \quad u_1(x) = \frac{f_0}{EA}\left(Lx - \frac{x^2}{2}\right) - \frac{F}{EA}x + c_1 \quad \rightarrow \quad u_1(0) = c_1 = 0 \quad \cdot \quad 0 \le x \le \frac{L}{2} \tag{d}$$

$$\frac{du_2}{dx} = \frac{P_2}{EA} = \frac{f_0}{EA}(L - x) + \frac{F}{EA} \quad \rightarrow \quad u_2(x) = \frac{f_0}{EA}\left(Lx - \frac{x^2}{2}\right) + \frac{F}{EA}x + c_2$$

$$\rightarrow \quad u_2\left(\frac{L}{2}\right) = u_2\left(\frac{L}{2}\right) = \frac{f_0}{EA}\left(\frac{L^2}{2} - \frac{L^2}{8}\right) + \frac{F}{EA}\frac{L}{2} + c_2 = \frac{f_0}{EA}\left(\frac{L^2}{2} - \frac{L^2}{8}\right) - \frac{F}{EA}\frac{L}{2}$$

$$\rightarrow \quad c_2 = -\frac{FL}{EA} \qquad\qquad \frac{L}{2} < x \le L$$

The resulting displacements are

$$u_1(x) = \frac{f_0}{EA}\left(Lx - \frac{x^2}{2}\right) - \frac{F}{EA}x \qquad\qquad 0 \le x \le \frac{L}{2}$$

$$u_2(x) = \frac{f_0}{EA}\left(Lx - \frac{x^2}{2}\right) + \frac{F}{EA}(x - L) \qquad \frac{L}{2} < x \le L$$

(e)

Second Method:
Actually, there is an easier way. From the principle of superposition we simply add the displacement from previous examples. If we label Example 4.4.1 as case a, Example 4.4.3 as case b, and this example as case c, then

$$EA\frac{du_a}{dx} = P_a \quad \text{and} \quad EA\frac{du_b}{dx} = P_b \tag{f}$$

and since the boundary conditions are the same for both, then for

$$EA\frac{du_c}{dx} = P_a + P_b \quad \rightarrow \quad u_c = u_a + u_b \tag{g}$$

Thus

$$u_{c1}(x) = u_{a1} + u_{b1} = \frac{f_0}{EA}\left(Lx - \frac{x^2}{2}\right) - \frac{F}{EA}x \qquad\qquad 0 \le x \le \frac{L}{2}$$

$$u_{c2}(x) = u_{a2} + u_{b2} = \frac{f_0}{EA}\left(Lx - \frac{x^2}{2}\right) + \frac{F}{EA}(x - L) \qquad \frac{L}{2} \le x \le L$$

(h)

You will note that the stresses can also be found by adding the two solutions.

Example 4.4.5

Problem: The slender bar is given three concentrated loads at equal intervals along the length as shown in Figure a. Let $l = L/3$. Find the displacements and stresses.

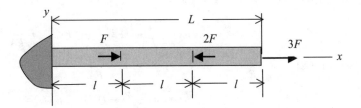

Figure (a)

Solution: Find the internal force and stress using the equations of static equilibrium and then the displacements by integrating the first order equations.

The bar is divided into three regions with separate equations covering each region. Since this is statically determinate we can find the constant axial internal force in each region to be

Region1: $\quad \sum F_x = 3F - 2F + F - P_1 = 0 \quad \rightarrow \quad P_1 = 3F - 2F + F = 2F \qquad 0 \leq x \leq l$

Region2: $\quad \sum F_x = 3F - 2F - P_2 = 0 \quad \rightarrow \quad P_2 = 3F - 2F = F \qquad l \leq x \leq 2l \qquad$ (a)

Region3: $\quad \sum F_x = 3F - P_3 = 0 \quad \rightarrow \quad P_3 = 3F \qquad 2l \leq x \leq L$

The stresses follow immediately.

$$\sigma_{x1} = \frac{2F}{A} \qquad \sigma_{x2} = \frac{F}{A} \qquad \sigma_{x3} = \frac{3F}{A} \qquad \text{(b)}$$

The equations governing the displacement in each region are

$$EA\frac{du_1}{dx} = P_1 = 2F$$

$$EA\frac{du_2}{dx} = P_2 = F \qquad \text{(c)}$$

$$EA\frac{du_3}{dx} = P_3 = 3F$$

The solutions are

$$u_1 = \frac{2F}{EA}x + c_1$$

$$u_2 = \frac{F}{EA}x + c_2 \qquad \text{(d)}$$

$$u_3 = \frac{3F}{EA}x + c_3$$

The boundary conditions are

$$u_1(0) = 0$$

$$u_2(l) = u_1(l) \qquad \text{(e)}$$

$$u_3(2l) = u_2(2l)$$

Applying these boundary conditions:

$$u_1(0) = c_1 = 0 \quad \rightarrow \quad u_1(x) = \frac{2F}{EA}x \quad \rightarrow \quad u_1(l) = \frac{2Fl}{EA}$$

$$u_2(l) = u_1(l) \quad \rightarrow \quad \frac{Fl}{EA} + c_2 = \frac{2Fl}{EA} \quad \rightarrow \quad c_2 = \frac{Fl}{EA}$$

$$\rightarrow \quad u_2(x) = \frac{F}{EA}(x+l) \quad \rightarrow \quad u_2(2l) = \frac{3Fl}{EA} \qquad \text{(f)}$$

$$u_3(2l) = u_2(2l) \quad \rightarrow \quad \frac{6Fl}{EA} + c_3 = \frac{3Fl}{EA} \quad \rightarrow \quad c_3 = -\frac{3Fl}{EA}$$

$$\rightarrow \quad u_3(x) = \frac{3F}{EA}(x-l) \quad \rightarrow \quad u_3(3l) = \frac{6Fl}{EA}$$

We can now say that

$$u(0) = u_0 = 0 \qquad u(l) = u_l = 2\frac{Fl}{EA} \qquad u(2l) = u_{2l} = 3\frac{Fl}{EA} \qquad u(3l) = u_{3l} = 6\frac{Fl}{EA} \qquad \text{(g)}$$

We plot the displacements and internal forces in Figure (b) noting that the displacements are linear between points of load application and the internal forces are constant between points of load application.

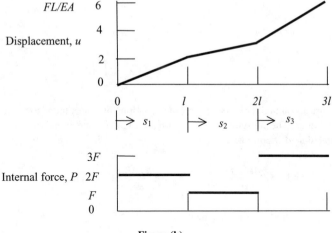

Figure (b)

We have labeled the displacements at the discrete points at which loads are applied as u_l, u_{2l}, and u_{3l}. It has been found useful at times to represent the distributed displacement $u(x)$ in terms of these discrete displacements and the local coordinates s_1, s_2, and s_3, as shown in Figure (b). For example, consider region 2.

$$u(s_2) = u_l + (u_{2l} - u_l)\frac{s_2}{l} = \left(1 - \frac{s_2}{l}\right)u_l + \frac{s_2}{l}u_{2l} \tag{h}$$

This can be put in a standard matrix form.

$$u(s_2) = \left[1 - \frac{s_2}{l} \quad \frac{s_2}{l}\right]\left[\begin{array}{c} u_l \\ u_{2l} \end{array}\right] \tag{i}$$

As a check let us evaluate $u(s_2)$ at $s_2 = 0$ and $s_2 = l$.

$$u(s_2) = \left[1 - \frac{s_2}{l} \quad \frac{s_2}{l}\right]\left[\begin{array}{c} u_l \\ u_{2l} \end{array}\right] = \left[1 - \frac{s_2}{l} \quad \frac{s_2}{l}\right]\left[\begin{array}{c} \dfrac{2Fl}{EA} \\ \dfrac{3Fl}{EA} \end{array}\right] \tag{j}$$

$$\rightarrow \quad u(0) = \frac{2Fl}{EA} \quad \rightarrow \quad u(l) = \frac{3Fl}{EA}$$

In fact this applies to any and all regions, say, region m, when the appropriate local coordinate, s_m, and the discrete displacements at each end of the region are used.

$$u(s_m) = \left[1 - \frac{s_m}{l_m} \quad \frac{s_m}{l_m}\right]\left[\begin{array}{c} u_n \\ u_{n+1} \end{array}\right] \tag{k}$$

where u_n is the displacement at $s_m = 0$ and u_{n+1} is the displacement at $s_m = l_m$. Try it on regions 1 and 3 to satisfy yourself that it works. We shall have an important use for this way of representing the displacements later in the next chapter.

###########

As we have noted determinate problems also can be solved using the second order differential equation. With $f_x(x)$ a continuous function of x and a concentrated load at one end only and a restraint at the other end as shown in Figure 4.4.1:

$$EA\frac{d^2u}{dx^2} = -f_x(x) \quad \rightarrow \quad \frac{du}{dx} = -\frac{1}{EA}\int f_x(x)dx + a$$

$$\rightarrow \quad u(x) = -\frac{1}{EA}\int\int f(x)dxdx + ax + b \tag{4.4.10}$$

Distributed loads are accounted for in the differential equation but concentrated loads must enter through the boundary conditions. The constants of integration are found by one boundary displacement restraint and one concentrated load condition.

$$u(0) = 0 \quad \text{and} \quad EA\frac{du(L)}{dx} = F_L \quad \text{or} \quad EA\frac{du(0)}{dx} = F_0 \quad \text{and} \quad u(L) = 0 \tag{4.4.11}$$

To find the stress differentiate the displacement.

$$EA\frac{du}{dx} = P \quad \rightarrow \quad \sigma_x = \frac{P}{A} \tag{4.4.12}$$

When distributed loading is discontinuous, and/or concentrated loads are at midpoints, or EA is discontinuous the equations must be written for each segment.

$$E_1A_1\frac{d^2u_1}{dx^2} = -f_1(x) \quad \rightarrow \quad \frac{du_1}{dx} = -\frac{1}{E_1A_1}\int f_1(x)dx + a_1$$

$$\rightarrow \quad u_1(x) = -\frac{1}{E_1A_1}\int\int f_1(x)dxdx + a_1x + b_1$$

$$E_2A_2\frac{d^2u_2}{dx^2} = -f_2(x) \quad \rightarrow \quad \frac{du_2}{dx} = -\frac{1}{E_2A_2}\int f_2(x)dx + a_2$$

$$\rightarrow \quad u_2(x) = -\frac{1}{E_2A_2}\int\int f_2(x)dxdx + a_2x + b_2 \tag{4.4.13}$$

$$\cdots$$

$$E_nA_n\frac{d^2u_1}{dx^2} = -f_n(x) \quad \rightarrow \quad \frac{du_n}{dx} = -\frac{1}{E_nA_n}\int f_n(x)dx + a_n$$

$$\rightarrow \quad u_n(x) = -\frac{1}{E_nA_n}\int\int f_n(x)dxdx + a_nx + b_n$$

The constants of integration are found by imposing a displacement restraint, either

$$u_1(0) = 0 \quad \text{or} \quad u_n(L) = 0 \tag{4.4.14}$$

and continuity of displacement between regions of integration.

$$u_1(x_1) = u_2(x_1) \qquad u_2(x_2) = u_3(x_2) \quad \cdots \quad u_{n-1}(x_{n-1}) = u_n(x_{n-1}) \tag{4.4.15}$$

and equilibrium of internal and external forces between regions

$$E_1 A_1 \frac{du_1(x_1)}{dx} - E_2 A_2 \frac{du_2(x_1)}{dx} - F_{x1} = 0$$

$$E_2 A_2 \frac{du_2(x_2)}{dx} - E_3 A_3 \frac{du_3(x_2)}{dx} - F_{x2} = 0$$

$$\cdots$$

$$E_{n-1} A_{n-1} \frac{du_{n-1}(x_{n-1})}{dx} - E_n A_n \frac{du_n(x_{n-1})}{dx} - F_{xn} = 0$$

(4.4.16)

To find stresses differentiate the displacements.

$$E_1 A_1 \frac{du_1}{dx} = P_1(x) \quad \rightarrow \quad \sigma_{x1} = \frac{P_1(x)}{A_1}$$

$$E_2 A_2 \frac{du_2}{dx} = P_2(x) \quad \rightarrow \quad \sigma_{x2} = \frac{P_2(x)}{A_2}$$

$$\cdots$$

(4.4.17)

$$E_n A_n \frac{du_n}{dx} = P_n(x) \quad \rightarrow \quad \sigma_{xn} = \frac{P_n(x)}{A_n}$$

When the second order equation is used the displacements are found first by integration. The constants of integration are found from a combination of displacement boundary conditions and force boundary conditions for statically determinate cases.

The stresses are then found by differentiating the displacements to find the internal axial forces and dividing by the areas.

If the problem is indeterminate you must use the second order equations. If the problem is determinate you may use either the first order or the second order equations.

4.5 The Statically Indeterminate Case

We shall provide several examples of the statically indeterminate case. Simply put, the number of restraints exceeds the number of applicable equations of static equilibrium, that is, there are two displacement restraints and only one equilibrium equation.

Let us summarize the various indeterminate cases that arise. First consider that E, A, and $f_x(x)$ are continuous functions of x and there are displacement restraints at both ends as shown in Figure 4.5.1.

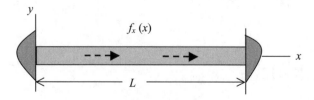

Figure 4.5.1

A free body diagram of the whole bar gives us Figure 4.5.2.

Figure 4.5.2

From the summation of forces

$$\sum F_x = \int_0^L f_x dx + R_{x1} + R_{x2} = 0 \tag{4.5.1}$$

Of course, this is our only equilibrium equation and it has two unknown restraint forces in it. This is a statically *indeterminate* problem. In such circumstances we cannot find the internal force P by means of equilibrium alone.

In this case we must use the second order equation, which combined the equation of equilibrium with the bar strain displacement and material properties relations.

$$\frac{dP}{dx} = \frac{d}{dx} EA \frac{du}{dx} = -f_x(x) \tag{4.5.2}$$

or when *EA* is constant

$$EA \frac{d^2u}{dx^2} = -f_x(x) \tag{4.5.3}$$

###########

<hr>

Example 4.5.1

Problem: Consider the special case represented by Figure 4.5.1 when $f_x(x) = f_0$ is a constant and *EA* is a constant. Find the displacement and stress.

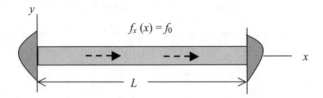

Figure (a)

Solution: Solve the second order equation for the displacement and then differentiate to find the stress. The second order equation and its solution are

$$EA \frac{d^2u}{dx^2} = -f_0 \quad \rightarrow \quad \frac{du}{dx} = -\frac{f_0}{EA}x + a \quad \rightarrow \quad u(x) = -\frac{f_0}{2EA}x^2 + ax + b \tag{a}$$

where a and b are constants of integration. In this case we use two displacement boundary conditions to find the two constants of integration. The bar is restrained on both ends so

$$u(0) = b = 0 \qquad u(L) = -\frac{f_0 L^2}{2EA} + aL + b = 0 \tag{b}$$

From these we obtain

$$b = 0 \qquad a = \frac{f_0 L}{2EA} \tag{c}$$

and so

$$u(x) = \frac{f_0 x}{2EA}(L - x) \tag{d}$$

Now, the equations for strain displacement and Hooke's law are the same as we used before, and so is the stress resultant P in terms of the stress, thus

$$\varepsilon_x = \frac{du}{dx} = \frac{f_0}{EA}\left(\frac{L}{2} - x\right) \qquad \sigma_x = E\varepsilon_x = \frac{f_0}{A}\left(\frac{L}{2} - x\right) \qquad P = \sigma_x A = f_0\left(\frac{L}{2} - x\right) \tag{e}$$

Plots of the displacement and stress are shown in Figure (b).

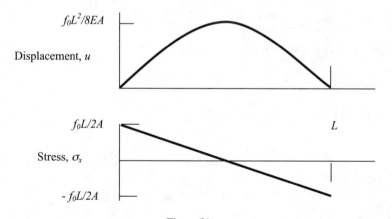

Figure (b)

The support restraint forces are

$$R_{x1} = -P(0) = -f_0\frac{L}{2} \qquad R_{x2} = P(L) = -f_0\frac{L}{2} \tag{f}$$

Note how the sign conventions for restraint forces and internal forces are accommodated in Equation (f).

############

When distributed loading is discontinuous, and/or concentrated loads are at midpoints, or EA is discontinuous, the equations must be written for each segment. An example of a possible configuration is shown in Figure 4.5.3.

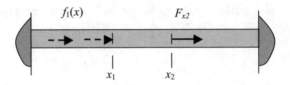

Figure 4.5.3

In this case we must use the second order equations. The constants of integration are found by imposing displacement restraints at both ends.

$$u_1(0) = 0 \quad \text{and} \quad u_n(L) = 0 \tag{4.5.4}$$

and continuity of displacement between regions of integration.

$$u_1(x_1) = u_2(x_1) \qquad u_2(x_2) = u_3(x_2) \quad \dots \quad u_{n-1}(x_{n-1}) = u_n(x_{n-1}) \tag{4.5.5}$$

and equilibrium of internal and external forces between regions

$$E_1 A_1 \frac{du_1(x_1)}{dx} - E_2 A_2 \frac{du_2(x_1)}{dx} = F_{x1}$$

$$E_2 A_2 \frac{du_2(x_2)}{dx} - E_3 A_3 \frac{du_3(x_2)}{dx} = F_{x2} \tag{4.5.6}$$

$$\dots$$

$$E_{n-1} A_{n-1} \frac{du_{n-1}(x_{n-1})}{dx} - E_n A_n \frac{du_n(x_{n-1})}{dx} = F_{xn}$$

To find the stresses differentiate the displacements.

$$E_1 A_1 \frac{du_1}{dx} = P_1(x) \quad \rightarrow \quad \sigma_{x1} = \frac{P_1(x)}{A_1}$$

$$E_2 A_2 \frac{du_2}{dx} = P_2(x) \quad \rightarrow \quad \sigma_{x2} = \frac{P_2(x)}{A_2} \tag{4.5.7}$$

$$\dots$$

$$E_n A_n \frac{du_n}{dx} = P_n(x) \quad \rightarrow \quad \sigma_{xn} = \frac{P_n(x)}{A_n}$$

###########

Example 4.5.2

Problem: A uniform bar has fixed ends and a concentrated load F at $x = x_0$ as shown in Figure (a). Find the displacement and stress.

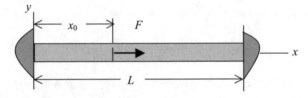

Figure (a)

Solution: Solve the second order equation for the displacement and then find the stress.

We will distinguish the two regions with subscripts where $u_1(x)$ is from $0 \le x \le x_0$ and $u_2(x)$ is from $x_0 \le x \le L$. The solution to the first equation is

$$EA\frac{d^2u_1}{dx^2} = 0 \quad \rightarrow \quad u_1(x) = a_1x + b_1 \qquad 0 \le x \le x_0 \tag{a}$$

The solution to the second equation is

$$EA\frac{d^2u_2}{dx^2} = 0 \quad \rightarrow \quad u_2(x) = a_2x + b_2 \qquad x_0 \le x \le L \tag{b}$$

We must find four boundary conditions to solve for the constants of integration. We have a displacement boundary condition at each end or

$$u_1(0) = a_1 0 + b_1 = 0 \qquad u_2(L) = a_2 L + b_2 = 0 \tag{c}$$

Now it follows that we must have continuity at x_0. First, apply continuity of displacements.

$$u_1(x_0) = u_2(x_0) \quad \rightarrow \quad a_1x_0 + b_1 = a_2x_0 + b_2 \tag{d}$$

Second, apply equilibrium at $x = x_0$. See the free body diagram in Figure (b).

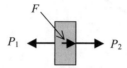

Figure (b)

From summation of forces

$$\sum F_x = P_2 + F - P_1 = 0 \quad \rightarrow \quad P_2 - P_1 = EA\frac{du_2}{dx} - EA\frac{du_1}{dx} = -F \tag{e}$$

or

$$\frac{du_2}{dx} - \frac{du_1}{dx} = a_2 - a_1 = -\frac{F}{EA} \tag{f}$$

We have then four equations to solve for the four constants of integration. As the number of unknowns increase to four and more it becomes more tedious to solve by the usual substitution methods that work so well for two or three unknowns.

We can cast the boundary condition equations in the following format

$$a_1 0 + b_1 = 0 \qquad a_1 x_0 + b_1 - a_2 x_0 - b_2 = 0 \qquad a_1 - a_2 = \frac{F}{EA} \qquad a_2 L + b_2 = 0 \tag{g}$$

and form matrix equations

$$\begin{bmatrix} 0 & 1 & 0 & 0 \\ x_0 & 1 & -x_0 & -1 \\ 1 & 0 & -1 & 0 \\ 0 & 0 & L & 1 \end{bmatrix} \begin{bmatrix} a_1 \\ b_1 \\ a_2 \\ b_2 \end{bmatrix} = \begin{bmatrix} 0 \\ 0 \\ \dfrac{F}{EA} \\ 0 \end{bmatrix} \tag{h}$$

This can be solved with one of the symbolic equation solving software packages, such as, Mathematica, Maple, and others. Appendix C contains some Mathematica instructions for solving sets of equations symbolically and numerically.

By any method the solution is

$$a_1 = \frac{F}{EA}\left(1 - \frac{x_0}{L}\right) \qquad b_1 = 0 \qquad a_2 = -\frac{F}{EA}\frac{x_0}{L} \qquad b_2 = \frac{F}{EA}x_0 \tag{i}$$

The displacement is

$$u_1(x) = \frac{F}{EA}\left(1 - \frac{x_0}{L}\right)x \qquad u_2(x) = \frac{F}{EA}\left(1 - \frac{x}{L}\right)x_0 \tag{j}$$

Also

$$\varepsilon_{1x} = \frac{du_1}{dx} = \frac{F}{EA}\left(1 - \frac{x_0}{L}\right) \qquad \sigma_{1x} = E\varepsilon_{1x} = \frac{F}{A}\left(1 - \frac{x_0}{L}\right) \qquad P_1 = \sigma_{1x}A = F\left(1 - \frac{x_0}{L}\right) \tag{k}$$

$$\varepsilon_{2x} = \frac{du_2}{dx} = -\frac{F}{EA}\frac{x_0}{L} \qquad \sigma_{2x} = E\varepsilon_x = -\frac{F}{A}\frac{x_0}{L} \qquad P_2 = \sigma_{2x}A = -F\frac{x_0}{L}$$

The plots of the displacement and stress are shown in Figure (c)

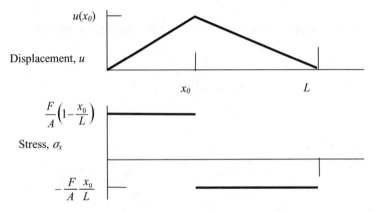

Figure (c)

Note that if we label the nodal displacements

$$u_1(0) = u_0 = 0 \qquad u_1(x_0) = u_2(x_0) = u_{x_0} \qquad u_2(L) = u_L \tag{l}$$

the distributed displacements can be written in terms of local coordinates and the nodal displacements.

$$u_1(s_1) = \begin{bmatrix} 1 - \dfrac{s_1}{x_0} & \dfrac{s_1}{x_0} \end{bmatrix}\begin{bmatrix} 0 \\ u_{x_0} \end{bmatrix}$$

$$u_2(s_2) = \begin{bmatrix} 1 - \dfrac{s_2}{L - x_0} & \dfrac{s_2}{L - x_0} \end{bmatrix}\begin{bmatrix} u_{x_0} \\ u_L \end{bmatrix} \tag{m}$$

This form for the displacements applies to all axial cases with concentrated loads.

###########

This solution can be used to solve for multiple concentrated loads for a given set of displacement boundary conditions. From the principle of superposition, if multiple concentrated loads are applied at various values of x_0, you can substitute the appropriate force and location values and simply add them up to obtain the displacements, strains, stresses, and axial stress resultants.

In all cases so far the displacement constraints have all been zero. It is possible to induce stress and displacement by imposing a non zero displacement constraint. The next example demonstrates this.

##########

Example 4.5.3

Problem: The bar in Example 4.5.1 is modified by imposing a non zero restraint at the right end. This might be accomplished with a screw jack. In this case we create a negative displacement δ as shown in Figure (a).

$$u(L) = -\delta \tag{a}$$

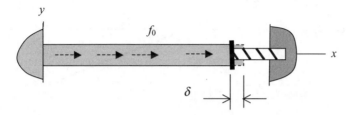

Figure (a)

Solution: This is an indeterminate problem (two displacement restraints) so solve the second order equation.

From Example 4.5.1 we have for the displacement

$$u(x) = -\frac{f_0}{2EA}x^2 + ax + b \tag{b}$$

The boundary conditions are now

$$u(0) = b = 0 \qquad u(L) = -\frac{f_0 L^2}{2EA} + aL + b = -\delta \tag{c}$$

From these we obtain

$$b = 0 \qquad a = -\frac{\delta}{L} + \frac{f_0 L}{2EA} \tag{d}$$

and so

$$u(x) = \frac{f_0 x}{2EA}(L - x) - \frac{\delta}{L}x \tag{e}$$

Now, the equations for strain displacement and Hooke's law are the same as we used before, and so is the stress resultant P in terms of the stress, thus

$$P(x) = EA\frac{du}{dx} = f_0\left(\frac{L}{2} - x\right) - \frac{EA\delta}{L} \qquad \sigma_x = \frac{P}{A} = \frac{f_0}{A}\left(\frac{L}{2} - x\right) - \frac{E\delta}{L} \tag{f}$$

The support restraint forces are

$$R_{x1} = -P(0) = -f_0\frac{L}{2} + \frac{EA\delta}{L} \qquad R_{x2} = P(L) = -f_0\frac{L}{2} - \frac{EA\delta}{L} \qquad (g)$$

Plots of the displacement and stress are shown in Figure (b).

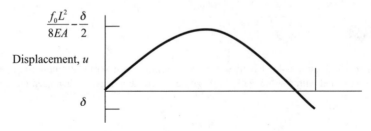

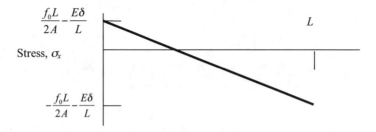

Figure (b)

############

4.6 Variable Cross Sections

Consider an axially loaded slender bar with a cross section that varies with x, that is, $A(x)$. Once again we shall assume that the stress is uniform across the section if the axial load acts through the centroid of the section.

The same rules apply for statically determinate and indeterminate cases. Given the internal force we can find the stresses.

$$\sigma_x = \frac{P(x)}{A(x)} \qquad (4.6.1)$$

Finding the deflection requires integrating the first order equation for the determinate case.

$$EA(x)\frac{du}{dx} = P(x) \quad \rightarrow \quad u = \int \frac{P(x)}{EA(x)}dx + c \qquad (4.6.2)$$

or the second order equation for the indeterminate case

$$\frac{d}{dx}EA(x)\frac{du}{dx} = -f_x(x) \quad \rightarrow \quad u(x) = -\int \frac{1}{EA(x)}\left(\int f_x(x)dx + a\right)dx + b \qquad (4.6.3)$$

An example will help.

############

Example 4.6.1

Problem: An axially loaded slender bar with linear taper, restrained as shown in Figure (a), has a uniform distributed load. Find the stress and displacement. The cross section area is

$$A(x) = A_0\left(1 - \frac{x}{2L}\right) \tag{a}$$

where A_0 is the area at $x = 0$.

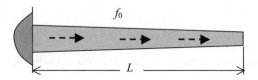

f_0

L

Figure (a)

Solution: Find the internal force and stress and then solve the first order equation for the displacement. The internal force is

$$\sum F_x = f_0(L - x) - P(x) = 0 \quad \rightarrow \quad P(x) = f_0(L - x) \tag{b}$$

The stress is

$$\sigma_x = \frac{P(x)}{A(x)} = \frac{f_0(L - x)}{A_0\left(1 - \frac{x}{2L}\right)} \tag{c}$$

The deflection is

$$u(x) = \frac{1}{E}\int \frac{P(x)}{A(x)}dx = \frac{1}{E}\int \frac{f_0(L - x)}{A_0\left(1 - \frac{x}{2L}\right)}dx = \frac{f_0 L^2}{EA_0}\int \frac{1}{\left(1 - \frac{x}{2L}\right)}d\frac{x}{L} - \int \frac{\frac{x}{L}}{\left(1 - \frac{x}{2L}\right)}d\frac{x}{L} \tag{d}$$

It is convenient to introduce

$$\xi = \frac{x}{L} \tag{e}$$

then the integral may be written and looked up in the integral tables.

$$u(\xi L) = \frac{f_0 L^2}{EA_0}\left\{\int \frac{1}{\left(1 - \frac{\xi}{2}\right)}d\xi - \int \frac{\xi}{\left(1 - \frac{\xi}{2}\right)}d\xi\right\} = \frac{f_0 L^2}{EA_0}\left\{2\ln\left(1 - \frac{\xi}{2}\right) + 2\xi + c\right\} \tag{f}$$

Applying the boundary condition $u(0) = 0$ we get $c = 0$. Thus

$$u(x) = \frac{f_0 L^2}{EA_0}\left\{2\ln\left(1 - \frac{x}{2L}\right) + 2\frac{x}{L}\right\} \tag{g}$$

It is always good practice to find a way to check your answer. Let us compare the displacement at $x = L$ with the uniform bar with the same loading in Example 4.4.1.

If the bar has a constant cross section equal to A_0, the value at $x = 0$, the deflection would be

$$u(L) = 0.5 \frac{f_0 L^2}{EA_0}$$

(h)

On the other hand if the cross section was constant and equal to that at the end of the tapered bar $(A(L) = A_0/2)$, the end displacement would be

$$u(L) = \frac{f_0 L^2}{EA_0}$$

(i)

For this tapered case we have

$$u(L) = 0.6137 \frac{f_0 L^2}{EA_0}$$

(j)

It is logical that the displacement would fall between the two.

Plots of the displacement, internal force, and stress are given in Figure (b).

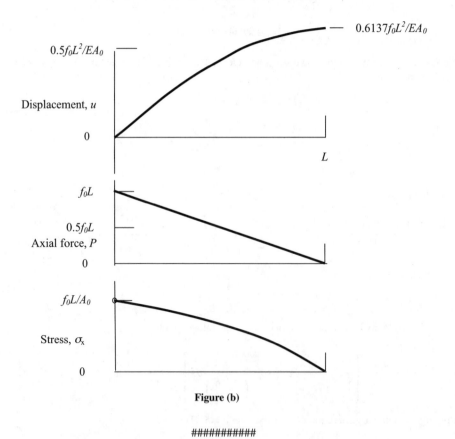

Figure (b)

############

If the restraints are such that the bar is statically indeterminate the second order equation must be used.

$$\frac{d}{dx} EA \frac{du}{dx} = -f_x(x)$$

(4.6.4)

Once we find the displacement $u(x)$ by integrating twice and applying constraints to find the constants of integration we can find the axial internal force

$$P(x) = EA(x)\frac{du}{dx} \tag{4.6.5}$$

and the stress

$$\sigma_x(x) = \frac{P(x)}{A(x)} \tag{4.6.6}$$

###########

Example 4.6.2

Problem: Consider the same slender bar and loading as in Example 4.6.1 but with both ends fixed as shown in Figure (a). Find the displacement and stress.

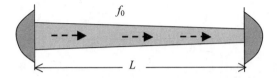

Figure (a)

Solution: Solve the second order equation for the displacement and then find the stress.
Integrate Equation 4.6.4 once

$$\frac{d}{dx}EA\frac{du}{dx} = -f_0 \quad \rightarrow \quad EA\frac{du}{dx} = -f_0 x + a$$

$$\rightarrow \quad \frac{du}{dx} = \frac{1}{EA_0}\left\{ -\frac{f_0 x}{\left(1-\dfrac{x}{2L}\right)} + \frac{a}{\left(1-\dfrac{x}{2L}\right)} \right\} \tag{a}$$

where a is the constant of integration.
The second integration provides

$$u(x) = \frac{1}{EA_0}\left\{ -f_0\int \frac{x}{\left(1-\dfrac{x}{2L}\right)}dx + a\int \frac{1}{\left(1-\dfrac{x}{2L}\right)}dx + b \right\} \tag{b}$$

where b is the second constant of integration.
For convenience let $\xi = \dfrac{x}{L}$, then

$$u(\xi L) = \frac{1}{EA_0}\left\{ -f_0 L^2\int \frac{\xi}{\left(1-\dfrac{\xi}{2}\right)}d\xi + aL\int \frac{1}{\left(1-\dfrac{\xi}{2}\right)}d\xi + b \right\} \tag{c}$$

and from a table of integrals

$$u(\xi L) = \frac{1}{EA_0}\left\{-f_0L^2\left[-2\xi - 4\ln\left(1 - \frac{\xi}{2}\right)\right] - 2aL\ln\left(1 - \frac{\xi}{2}\right) + b\right\}$$ (d)

Now apply the boundary restraints.

$$u(0) = \frac{1}{EA_0}\left\{-f_0L^2\left[0 - 4\ln(1)\right] - 2aL\ln(1) + b\right\} = 0 \rightarrow b = 0$$

$$u(L) = \frac{1}{EA_0}\left\{-f_0L^2\left[-2 - 4\ln(0.5)\right] - 2aL\ln(0.5) + 0\right\} = 0 \rightarrow a = 0.5573f_0L$$ (e)

Finally, the displacement is

$$u(x) = \frac{f_0L}{EA_0}\left\{2.8854L\ln\left(1 - \frac{x}{2L}\right) + 2x\right\}$$ (f)

The internal force in the bar is

$$P(x) = EA(x)\frac{du}{dx} = -f_0x + a = -f_0x + 0.5573f_0L$$ (g)

Let us check to see if the two internal forces at the restraints are in equilibrium with the applied load.

$$P(0) = 0.5573f_0L \quad \rightarrow \quad P(L) = f_0L(-1 + 0.5573) = -0.4427f_0L$$ (h)

Recognizing the definition of positive P we have the values shown in Figure (b).

$P(0) = 0.5573f_0L$ f_0 $P(L) = -0.4427f_0L$

Figure (b)

Summing all forces we get

$$f_0L + P(L) - P(0) = f_0L - 0.4427f_0L - 0.5573f_0L = 0$$ (i)

We see that the bar is in equilibrium. This is a good check on the answer.

The stress is

$$\sigma_x(x) = \frac{P(x)}{A(x)} = E\frac{du}{dx} = E(-f_0x + a) = Ef_0(0.5573L - x) \tag{j}$$

As a check we note the values of the stress at $x = 0$ and $x = L$.

$$\sigma(0) = 0.5573\frac{f_0L}{A_0} \qquad \sigma(L) = -0.8854\frac{f_0L}{A_0} \tag{k}$$

Plots of the displacements, internal forces, and stresses are given in Figure (b).

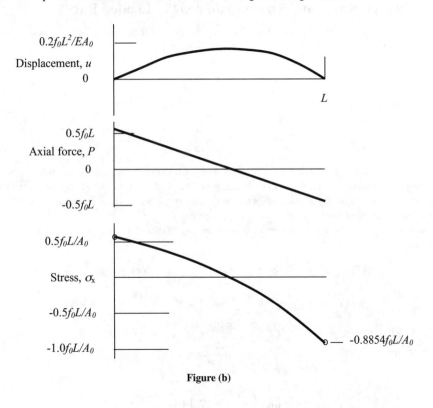

Figure (b)

##########

In real life we may not find that the cross sectional area is so easily represented by a simple functional relationship; therefore, in today's world it is more common to adopt approximate methods. For example, the tapered bar can be divided into a series of uniform segments as shown in Figure 4.6.1. We use an average value of the area for each segment.

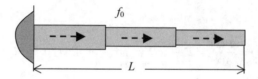

Figure 4.6.1

We shall demonstrate solutions in the next chapter in Section 5.5 and verify the validity of this type of approximation.

4.7   Thermal Stress and Strain in an Axially Loaded Bar

Thermal strain occurs when a bar is heated. Defining the change in temperature from some base value as ΔT we have for the thermal strain (See Chapter 3, Section 3.6)

$$\varepsilon_{xT} = \alpha \Delta T \tag{4.7.1}$$

The strain from applied loads we designate as ε_x^{mech} then the total strain is

$$\varepsilon_x^{Total} = \varepsilon_x^{mech} + \varepsilon_{xT} = \frac{\sigma_x}{E} + \alpha \Delta T = \frac{du}{dx} \tag{4.7.2}$$

If a statically determinate bar is heated a uniform amount it expands thus increasing the displacement. No stresses are induced. When the bar is restrained on both ends, that is, is indeterminate, stresses are induced in the absence of any applied loads but no displacement takes place. These effects can be added to the effects of applied loads by the principle of superposition.

In the presence of a change in temperature the summary of equations in Equation 4.3.8 are modified as follows.

$$F_c \quad \rightarrow \quad f_x(x) \qquad P(x) \qquad \sigma_x(x) \qquad \varepsilon_x^{mech}(x) \qquad \Delta T(x) \qquad u(x) \quad \leftarrow \quad \rho$$

$$\frac{dP}{dx} = -f_x(x) \qquad\qquad\qquad \varepsilon_x^{Total} = \varepsilon_x^{mech} + \varepsilon_{xT} = \frac{\sigma_x}{E} + \alpha \Delta T = \frac{du}{dx}$$

$$\sigma_x(x) = E\frac{du}{dx} - E\alpha \Delta T \tag{4.7.3}$$

$$P(x) = \sigma_x A = EA\left(\frac{du}{dx} - \alpha \Delta T\right)$$

$$\frac{dP}{dx} = \frac{d}{dx}\left(EA\left(\frac{du}{dx} - \alpha \Delta T\right)\right) = -f_x(x)$$

##########

Example 4.7.1

Problem: The two bars shown in Figure (a) are initially at room temperature. Their temperature is then increased uniformly an amount ΔT. What is the displacement and stress in each case?

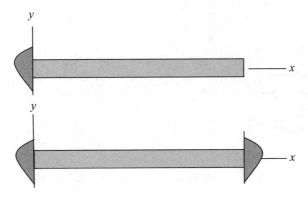

Figure (a)

Solution: Use Equation 4.7.3.

The fixed-free case is statically determinate so there is no internal force P and

$$\sigma_x = 0 \tag{a}$$

For the displacement

$$EA\left(\frac{du}{dx} - \alpha\Delta T\right) = 0 \quad \rightarrow \quad u(x) = \alpha\Delta T x + c \quad \rightarrow \quad u(0) = c = 0 \quad \rightarrow \quad u(x) = \alpha\Delta T x \tag{b}$$

For the fixed-fixed case if $\alpha\Delta T$ is a constant

$$\frac{d}{dx}\left(EA\left(\frac{du}{dx} - \alpha\Delta T\right)\right) = EA\frac{d^2 u}{dx^2} = 0 \quad \rightarrow \quad u(x) = cx + d \tag{c}$$

$$\rightarrow \quad u(0) = d = 0 \quad \rightarrow \quad u(L) = cL = 0 \quad \rightarrow \quad u(x) = 0$$

From Equation 4.7.3

$$\sigma_x = E\varepsilon_x = E\left(\frac{du}{dx} - \alpha\Delta T\right) = -E\alpha\Delta T \tag{d}$$

These solutions can be superimposed with the other solutions for stress and displacement under loading for a bar that is both loaded and heated.

############

4.8 Shearing Stress in an Axially Loaded Bar

From what we have studied so far it would appear that there is one component of normal stress, σ_x, and all other stress components are zero. In terms of internal surfaces on which stress components are defined, that is, planes normal to the rectangular Cartesian coordinates to which we have oriented the bar, this is true. At other orientations of the internal surfaces, however, other stress components can have non zero values.

As we noted in Chapter 1 let us look at an internal surface of the bar which is not normal to the x axis as shown in Figure 4.8.1.

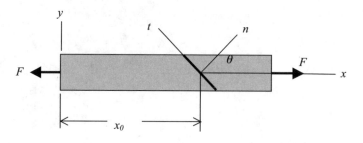

Figure 4.8.1

The normal to the surface is at an angle θ to the x axis. A coordinate system nt is normal and tangent to the inclined surface. The forces and stresses on this surface, that we have labeled A_θ, are shown in Figure 4.8.2.

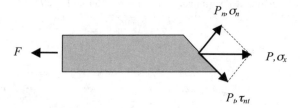

Figure 4.8.2

The normal and shearing stresses are

$$\sigma_n = \frac{P_n}{A_\theta} \qquad \tau_{nt} = -\frac{P_t}{A_\theta} \tag{4.8.1}$$

Since

$$A_\theta = \frac{A}{\cos\theta} \qquad P_n = P\cos\theta \qquad P_t = P\sin\theta \tag{4.8.2}$$

where A is the area normal to the x axis, the stresses are

$$\sigma_n = \frac{P}{A}\cos^2\theta = \sigma_x\cos^2\theta \qquad \tau_{nt} = -\frac{P}{A}\sin\theta\cos\theta = \sigma_x\sin\theta\cos\theta \tag{4.8.3}$$

Note that the normal stress is a maximum at $\theta = 0°$ and a minimum at $\theta = 90°$. The shear stress is zero at $\theta = 0°$ and $90°$ and reaches a maximum value at $\theta = 45°$.

$$(\tau_{nt})_{max} = -\sigma_x\sin 45°\cos 45° = -\frac{\sigma_x}{2} \tag{4.8.4}$$

At that angle the normal stress is

$$\sigma_t = \sigma_x\cos^2 45° = \frac{\sigma_x}{2} \tag{4.8.5}$$

In Figure 4.8.3 we plot the variation in the normal stress and shearing stress as the angle changes. The upper plot is

$$\frac{\sigma_n}{\sigma_x} = \cos^2 \theta \tag{4.8.6}$$

and the lower plot is

$$\frac{\tau_{nt}}{\sigma_x} = -\sin \theta \cos \theta \tag{4.8.7}$$

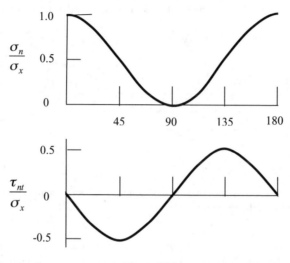

Figure 4.8.3

This emphasizes that the value of the stress component depends not only on the size of the force but also on the orientation of the surface upon which it acts.

A comprehensive discussion of the stress transformation that occurs when axes are rotated is presented in Chapter 9.

4.9 Design of Axially Loaded Bars

In *analysis* we specify the geometry, materials, loads, and restraints and ask for the displacements, strains, and stresses. We wish to **verify** that the structure meets minimum requirements on load carrying ability.

In *design* we are more likely to be given the loads and some of the restraints but are asked what geometry and materials should be used to achieve a certain acceptable value of stress and displacement. We wish to **create** a structure that will meet minimum requirements on load carrying ability.

Design, however, is a much broader subject than that just stated. In addition to the geometric restraints we may have restraints on weight, cost, and ease of manufacture. Safety and margins of safety are ever present. Aesthetics are an additional consideration in many cases, such as the body panels on an automobile and the interior of an airliner. Because of the broad nature of design it is largely a team effort. We cannot begin to study all of these concerns so we shall limit ourselves to geometry, geometric restraints, materials, and weight for a given state of loading. These are the immediate concerns of the structural analyst member of the team.

For an axially loaded slender bar the geometry consists of the length and the cross sectional area. The exact shape of the cross section might be determined by considerations of manufacturing, fasteners and restraints, points or regions of load application, or other considerations. We shall restrict our discussion to the role of modifying the geometry and material to improve the design.

##########

Example 4.9.1

Problem: The bar in Example 4.5.2, repeated here in Figure (a), is made of aluminum for which the yield stress is 255 *MPa*. If the load is applied at $x_0 = \dfrac{L}{2}$, what should the area of the cross section of the bar be to limit the stress to two thirds of the yield stress, that is, limit the stress to 170 *MPa*? Repeat for $x_0 = \dfrac{3L}{4}$.

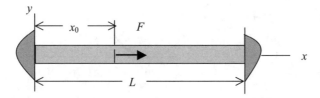

Figure (a)

Solution: Insert values in the equations obtained in Example 4.5.2 and calculate the areas.

In Example 4.5.2 we found the stresses to be

$$\sigma_{1x} = \frac{F}{A}\left(1 - \frac{x_0}{L}\right) \qquad \sigma_{2x} = -\frac{F}{A}\frac{x_0}{L} \tag{a}$$

In this case $x_0 = L/2$. To the left of the load at $x = L/2$ the required area would be

$$\sigma_{1x} = \frac{F}{A}\left(1 - \frac{x_0}{L}\right) = \frac{F}{A}\left(1 - \frac{L}{2L}\right) = \frac{F}{2A} = 170 \; MPa \quad \rightarrow \quad A = \frac{F}{340} \tag{b}$$

To the right of the load the required area would be

$$\sigma_{2x} = -\frac{F}{A}\frac{x_0}{L} = -\frac{F}{A}\frac{L}{2L} = -\frac{F}{2A} = -170 \; MPa \quad \rightarrow \quad A = \frac{F}{340} \tag{c}$$

Both values define the same area.

Now let us ask if the load is located at $x_0 = \dfrac{3L}{4}$ what would be the minimum area. Note that this is a uniform bar throughout its length. To the left of the load the required area would be

$$\sigma_{1x} = \frac{F}{A}\left(1 - \frac{x_0}{L}\right) = \frac{F}{A}\left(1 - \frac{3L}{4L}\right) = \frac{F}{4A} = 170 \; MPa \quad \rightarrow \quad A = \frac{F}{680} \tag{d}$$

To the right of the load the required area would be

$$\sigma_{1x} = -\frac{F}{A}\frac{x_0}{L} = -\frac{F}{A}\frac{3L}{4L} = -\frac{3F}{4A} = -170 \; MPa \quad \rightarrow \quad A = \frac{3F}{680} \tag{e}$$

For a uniform bar the larger of these two areas must prevail.

##########

Consider now that there are conditions on both stress and displacement. In the simplest of cases of a two force member of given length with a known axial force we select a material, stress level, and/or displacement level and calculate the cross section area to satisfy those conditions. Thus, for an internal force $P = F$, a material with Young's modulus, E, a maximum allowable stress, σ_{max}, and a maximum total displacement, δ, we obtain

$$\sigma_x = \frac{P}{A} \quad \rightarrow \quad A = \frac{P}{\sigma_{max}} \quad \text{or} \quad \delta = \frac{FL}{EA} \quad \rightarrow \quad A = \frac{FL}{E\delta} \tag{4.9.1}$$

Then choose the larger area of the two. This would be result in the minimum weight for these conditions and this particular material. A change to a stiffer material would decrease the area based on displacement but might also require a change in the choice of maximum allowable stress.

###########

Example 4.9.2

Problem: A slender bar 1000 *mm* long has an axial load of 8500 *N* as shown in Figure (a). The material choice is between aluminum and steel with the condition that the maximum stress may not exceed two thirds of the yield stress. The minimum weight design is sought.

Case 1: No restriction on displacement
Case 2: The total elongation is restricted to 2 *mm*

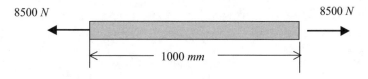

8500 *N*   8500 *N*

1000 *mm*

Figure (a)

Solution: Apply Equations 4.8.1, calculate weights, and compare.
   Materials with the following properties were chosen where ρ is the mass density:

Aluminum: $E = 68950\ Mpa$ Steel: $E = 206800\ MPa$

$\sigma_{yield} = 255\ MPa$ $\sigma_{yield} = 342\ MPa$

$\rho = 2.72 \cdot 10^{-6}\ kg/mm^3$ $\rho = 7.86 \cdot 10^{-6}\ kg/mm^3$

Case 1:
The areas based on stress only and the corresponding weights are

Aluminum: Steel:

$$A = \frac{P}{\frac{2}{3}\sigma_y} = \frac{8500}{255} \cdot \frac{3}{2} = 50\ mm^2 \qquad\qquad A = \frac{P}{\frac{2}{3}\sigma_y} = \frac{8500}{342} \cdot \frac{3}{2} = 37.28\ mm^2 \tag{a}$$

$$\begin{aligned} Wt &= 50 \cdot 1000 \cdot 2.72 \cdot 10^{-6} \\ &= 0.136\ kg \end{aligned} \qquad\qquad \begin{aligned} Wt &= 37.28 \cdot 1000 \cdot 7.86 \cdot 10^{-6} \\ &= 0.293\ kg \end{aligned} \tag{b}$$

The aluminum bar is lighter.

Case 2:

The areas based on the stress and the corresponding weights are the same as Case 1. For the aluminum bar the displacement would be

$$\delta = \frac{PL}{AE} = \frac{8500 \cdot 1000}{50 \cdot 68950} = 2.47 \ mm \tag{c}$$

therefore the area would have to be increased to

$$A = \frac{PL}{\delta E} = \frac{8500 \cdot 1000}{2 \cdot 68950} = 61.64 \ mm^2 \tag{d}$$

to meet both requirements. Its weight would be

$$Wt = 61.64 \cdot 1000 \cdot 2.72 \cdot 10^{-6} = 0.168 \ kg \tag{e}$$

For the steel bar the displacement would be

$$\delta = \frac{PL}{AE} = \frac{8500 \cdot 1000}{37.28 \cdot 206800} = 1.10 \ mm \tag{f}$$

The area needed to meet the stress requirement is more than is needed to meet the displacement criteria, but it cannot be reduced because of the stress criteria. Thus the weight of the steel bar necessary to meet both requirements is

$$Wt = 37.28 \cdot 1000 \cdot 7.86 \cdot 10^{-6} = 0.293 \ kg \tag{g}$$

The aluminum bar wins.

Example 4.9.3

Problem: Redesign the cross sectional area of the beam in Example 4.4.1 for minimum weight. Assume a maximum stress but no displacement criteria; however, knowledge of the displacement is desired.

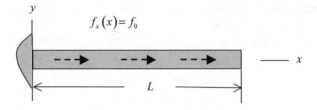

Figure (a)

Solution: Find the area for a constant maximum stress level and the resulting displacement using the equations in Section 4.6.

The internal force is

$$\sum F_x = f_0(L - x) - P(x) = 0 \quad \rightarrow \quad P(x) = f_0(L - x) \tag{a}$$

For a constant stress, σ_x, the area is

$$A = \frac{P}{\sigma_x} = \frac{f_0}{\sigma_x}(L - x) \tag{b}$$

For a bar cross section of constant width the side view would be a triangular shape as shown in Figure (b).

Figure (b)

The displacement is

$$EA\frac{du}{dx} = P \quad \rightarrow \quad u(x) = \frac{1}{E}\int \frac{P}{A}dx = \frac{1}{E}\int \frac{f_0(L-x)}{\dfrac{f_0}{\sigma_x}(L-x)}dx = \frac{\sigma_x}{E}x + c = \frac{\sigma_x}{E}x \qquad (c)$$

where $c = 0$ from the fixed boundary condition at the left end.

For reasons other than stress values the actual choice would probably not be a bar that converges to a sharp point; however, some taper would be desirable in many cases.

############

As the loading gets more complicated the design process becomes more difficult. For statically determinate problems the areas can be calculated directly to meet stress requirements and then displacements can be analyzed to see if a displacement criterion has been met. For indeterminate cases the internal forces depend upon the restraints as well as the loads and so the second order displacement equations must be solved before stress levels can be determined.

############

4.10 Analysis and Design of Pin Jointed Trusses

We have just developed methods of analysis for axially loaded slender bars. Often several bars are joined together to form a structure called a *pin jointed truss*. The joints where bars meet are necessarily pinned to allow rotation so that no moments are applied to the bars, thus all members are axially loaded. It follows also that loads and restraints can only be applied at the joints. Such structures are easy to construct and easy to analyze and have important but limited use in practice. Nevertheless they are an interesting study that provides additional insight into the design, analysis, and construction of structures.

Consider the simplest of trusses – one made of just two members as shown in Figure 4.10.1.

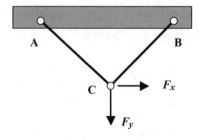

Figure 4.10.1

If all you want is the stress it is easy to find. This is a statically determinate structure and a summation of forces at the point of load application will show you the internal axial load in each member. In fact in Section 2.3 we studied the forces in determinate trusses and found values in Examples 2.3.1 and 2.3.2. Now we must find stresses and displacements.

The stress is easily determined since we know the internal forces in the members.

The deflection requires us to consider joint compatibility in conjunction with knowledge of internal bar stresses. This will be dealt with next, through an example problem.

It is common practice to show only line drawings such as in Figure 4.10.2 with the understanding that the joints are pinned.

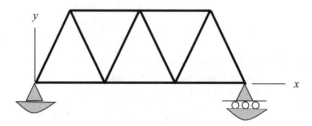

Figure 4.10.2

In this two dimensional structure you can find the support reactions by summation of forces in the x and y directions and the moment about the z axis. Starting at either support you can use summation of forces to find the internal force in each member that meets at that joint. As you work your way along from joint to joint in proper order there are never more than two unknowns at any joint. Thus quite large truss structures can be designed and analyzed in this way. Years ago many highway bridges appeared to be constructed this way and while the joints were not truly pinned the analysis was effective considering the large margins of safety allowed. They still are called truss style bridges. It should be noted that the failure of a single pin or a single member ensures the failure of the structure – so statically determinate pin jointed trusses are not recommended when safety is important.

############

Example 4.10.1

Problem: Consider the simple truss shown in Figure (a). Find the displacements and stresses.

$$\text{Let } A1 = A2 = 400 \ mm^2 \qquad F = 10000 \ N \qquad E = 206840 \ MPa. \tag{a}$$

Solution: Find the force in each member by static equilibrium. From the forces find the stresses. Find the displacements from the elongations of the members.

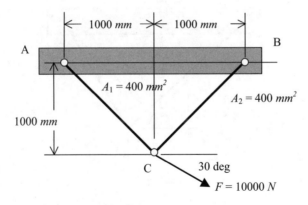

Figure (a)

We draw a free body diagram of the joint at C, where the load is applied as shown in Figure (b).

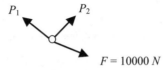

Figure (b)

From summation of forces, where P_1 and P_2 are the internal forces in the left and right truss members respectively, we have

$$\sum F_x = 10000 \cos 30° - P_1 \cos 45° + P_2 \cos 45° = 0$$
$$\sum F_y = -10000 \sin 30° + P_1 \sin 45° + P_2 \sin 45° = 0$$

(b)

We can insert the values and put that in matrix form.

$$0.707 \begin{bmatrix} -1 & 1 \\ 1 & 1 \end{bmatrix} \begin{bmatrix} P_1 \\ P_2 \end{bmatrix} = \begin{bmatrix} -8660 \\ 5000 \end{bmatrix}$$

(c)

Solving we get

$$P_1 = 9660.5 \ N \quad \rightarrow \quad \sigma_1 = \frac{P_1}{A_1} = \frac{9660.5}{400} = 24.15 \ MPa$$

$$P_2 = -2588.4 \ N \quad \rightarrow \quad \sigma_2 = \frac{P_2}{A_2} = \frac{-2588.4}{400} = -6.471 \ MPa$$

(d)

While finding the internal forces and hence the stresses can be easy, finding the displacement is a bit more difficult. We do know that members 1 and 2 will elongate an amount given by (see Example 4.3.1, Equation (l))

$$\delta_1 = \frac{P_1 L_1}{EA_1} = 0.1651 \ mm \qquad \delta_2 = \frac{P_2 L_2}{EA_2} = -0.0442 \ mm$$

(e)

In Figure (c) the dashed lines represent the original unloaded position and the solid lines represent the final loaded position – greatly exaggerated of course.

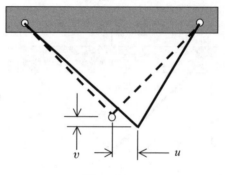

Figure (c)

Let us suppose that the joint C undergoes a displacement of magnitude u in the horizontal direction and a displacement of magnitude v in the vertical direction as indicated. First let us look at member 1 and the possible u and v displacements due to its elongation. Since the actual displacements are very small we note that the change in angle of the position is also very small. Assuming that the angle has not changed we can say that

$$v \sin 45° + u \cos 45° = \delta_1 = 0.1651 \tag{f}$$

and in a similar way that

$$v \sin 45° - u \cos 45° = \delta_2 = -0.0442 \tag{g}$$

Solving for u and v

$$u = \frac{(0.1651 + 0.0442)}{2 \cos 45°} = 0.1480 \ mm$$

$$v = \frac{(0.1651 - 0.0442)}{2 \sin 45°} = 0.0855 \ mm \tag{h}$$

##########

We must pause to note that there are several approximations made in arriving at this final analysis. As we have noted we did not recalculate the stress based on the change in area due to the Poisson's ratio effect. We calculated the forces in the truss members based on the position of the forces in the undeformed structure, that is, we did not account for the position change of the applied forces due to the displacement. In finding the displacements we use the angular position of the truss members in their undeformed positions. Does this make our values invalid or, at least suspect? Not if the displacements are small enough. More detailed analysis confirms that these are valid assumptions for the slender bars of the kind we are considering here.

Many pinned jointed trusses are indeterminate. For example, just add another member to the one in Figure 4.10.1 as shown in Figure 4.10.3.

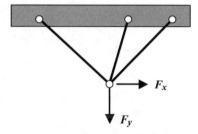

Figure 4.10.3

In two dimensions you now have three unknown internal forces but just two equilibrium equations. The necessary third equation requires information about displacement.

This will be taken up in Chapter 5.

4.11 Work and Energy—Castigliano's Second Theorem

When a force moves through a displacement work is done. Consider the experiment we used to define Young's modulus. The force displacement curve from Figure 3.4.2 is reproduced here as Figure 4.11.1. Within the elastic range the work done as the bar displaces an amount Δ under load F is shown by the shaded triangle. We call this the *external work*, W^e, and note that its value is

$$W^e = \frac{1}{2}F\Delta \qquad (4.11.1)$$

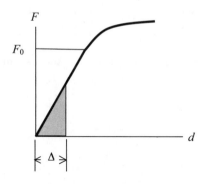

Figure 4.11.1

As the body deforms the external work will produce stresses and strains which store energy in the body equal to the work done. To see this we use the stress-strain curve from Figure 3.4.3 repeated here as Figure 4.11.2. Within the elastic range this energy, called *strain energy*, U, is shown by the shaded triangle.

We note that its value is

$$U = \frac{1}{2}\int_V \sigma_x \varepsilon_x dV \qquad (4.11.2)$$

Assuming the stress is constant across the cross section of the bar we have

$$\varepsilon_x = \frac{\sigma_x}{E} \qquad (4.11.3)$$

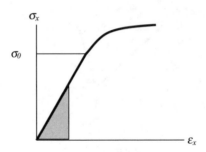

Figure 4.11.2

and so

$$U = \frac{1}{2} \int_V \sigma_x \varepsilon_x dV = \frac{1}{2} \int_V \frac{\sigma_x^2}{E} dV \tag{4.11.4}$$

We define the *internal work*, W^i, done as the negative of the strain energy, that is,

$$W^i = -U \tag{4.11.5}$$

and note the sum of the external and internal work is zero

$$W^e + W^i = W^e - U = 0 \tag{4.11.6}$$

Thus we can say that the external work done is equal to the strain energy stored.

$$W^e = U \tag{4.11.7}$$

This can be used to solve problems, that is, to find the deflection in the direction of the load.

$$\frac{1}{2} F \Delta = \frac{1}{2} \int_V \frac{\sigma_x^2}{E} dV \quad \rightarrow \quad \Delta = \frac{1}{F} \int_V \frac{\sigma_x^2}{E} dV \tag{4.11.8}$$

############

Example 4.11.1

Problem: Consider the axially loaded bar shown in Figure (a). Find the displacement at the end where the load is applied.

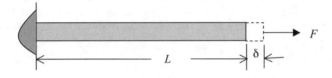

Figure (a)

Solution: Equate external work and internal strain energy and solve for the displacement in terms of the applied force.

The deflection at the right end is $u(L) = \delta$. The external work done as the load is applied and reaches its final value is

$$W^e = \frac{1}{2}F\delta \tag{a}$$

Given that $\sigma_x = P/A$ the strain energy is

$$U = \frac{1}{2}\int_V \sigma_x \varepsilon_x dV = \frac{1}{2}\int_V \frac{\sigma_x^2}{E}dV = \frac{1}{2}\frac{P^2}{EA^2}\int_0^L dx \int_A dA = \frac{1}{2}\frac{P^2 L}{EA} \tag{b}$$

By equating the external work to the internal strain energy we can find the displacement at the point of application of the force and in the direction of the force. Since from static equilibrium $P = F$

$$W^e = U \quad \rightarrow \quad \frac{1}{2}F\delta = \frac{1}{2}\frac{P^2 L}{EA} = \frac{1}{2}\frac{F^2 L}{EA} \quad \rightarrow \quad \delta = \frac{FL}{EA} \tag{c}$$

###########

External work and strain energy can be useful in solving problems where springs are involved. When a spring is displaced by a force work is done and strain energy is stored in the spring.
From Figure 2.3.8 and Equation 2.3.3 we note that

$$W^e = \frac{1}{2}Fd = U = \frac{1}{2}kd^2 \tag{4.11.9}$$

This is used in the next example.

###########

Example 4.11.2

Problem: Find the displacement δ at the point where the load is applied for the rigid bar and spring shown in Figure (a). The bar is 1 meter long.

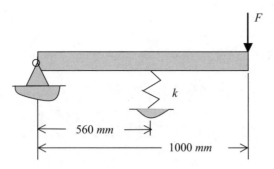

Figure (a)

Solution: Equate external work and strain energy and solve for the displacement.
For small displacements the displacement at the spring, d, is related to the displacement at the end, δ, by

$$d = 0.56\delta \tag{a}$$

Equating external work and strain energy gives us

$$W^e = U \quad \rightarrow \quad \frac{1}{2}F\delta = \frac{1}{2}kd^2 \quad \rightarrow \quad \frac{1}{2}F\delta = \frac{1}{2}k(0.56\delta)^2 \quad \rightarrow \quad \delta = \frac{F}{0.3136 \cdot k} \qquad \text{(b)}$$

Check using static equilibrium by summing moments about the left end.

$$\sum M_z = 560 \cdot kd - 1000 \cdot F = 0 \quad \rightarrow \quad d = \frac{1000 \cdot F}{560 \cdot k}$$

$$\rightarrow \quad \delta = \frac{d}{0.56} = \frac{1000 \cdot F}{0.56 \cdot 560 \cdot k} = \frac{F}{0.3136 \cdot k} \qquad \text{(c)}$$

Static equilibrium and work and energy provided the same answer.

##########

We can also find the displacement in trusses at the location of the force and in the direction of the force using work and energy. We equate the work done by the applied load to the strain energy stored in the truss members. The work done is

$$W^e = \frac{1}{2}F\delta \qquad (4.11.10)$$

The strain energy in a two force member was found in Example 4.11.1 to be

$$U = \frac{1}{2}\frac{P^2 L}{EA} \qquad (4.11.11)$$

and so in a truss we just sum all the members.

$$U = \frac{1}{2}\sum \frac{P_n^2 L_n}{E_n A} \qquad (4.11.12)$$

##########

Example 4.11.3

Problem: Find the displacement of the truss in Example 4.10.1 by work and energy.

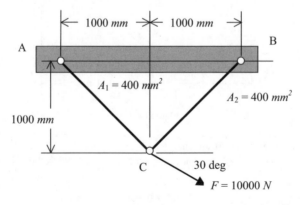

Figure (a)

Solution: Equate external work and strain energy and solve for the displacement.
The strain energy is

$$U = \frac{1}{2}\frac{P_1^2 L_1}{EA} + \frac{1}{2}\frac{P_2^2 L_2}{EA}$$

$$= \frac{1}{2}\frac{(9660.5)^2 \, 1414}{206840 \cdot 400} + \frac{1}{2}\frac{(-2588.4)^2 \, 1414}{206840 \cdot 400} = 854.74 \; N \cdot mm \tag{a}$$

The external work and displacement are

$$\frac{1}{2}F\delta = 854.74 \quad \rightarrow \quad \delta = \frac{2 \cdot 854.46}{F} = \frac{2 \cdot 854.46}{10000} = 0.1709 \; mm \tag{b}$$

Note that this gives the displacement in the direction of the force. This checks with the values in Example 4.10.1, Equation (g), as follows:

$$\delta = \sqrt{u^2 + v^2} = \sqrt{(0.1480)^2 + (0.0855)^2} = 0.1709 \; mm \tag{c}$$

###########

As the trusses get more complicated finding the stresses may be possible with reasonable effort; however, finding the displacements by the geometric resolution of the elongation of each member becomes tedious and prone to error. The simple solution by equating the work done to the strain energy is helpful but limited only to the displacement in the direction of a single load. If there are multiple loads or there is a desire for a displacement at a joint with no load the method does not work. Fortunately there is a work and energy method that can be extended to those cases.

For any statically determinate structure we can find the strain energy in terms of discrete applied forces.

$$U = U(F_1, F_2, \ldots, F_i, \ldots, F_n) \tag{4.11.13}$$

Now let us increase the ith force an amount ΔF_i, do a Taylor's series expansion, and neglect the higher order terms.

$$U = U(F_1, F_2, \ldots, F_i + \Delta F_i, \ldots, F_n)$$

$$= U(F_1, F_2, \ldots, F_i, \ldots, F_n) + \frac{\partial U}{\partial F_i}\Delta F_i + \text{terms of order } (\Delta F_i^2) \text{ and higher} \tag{4.11.14}$$

Now let us look at the same event in another way. If we apply the force ΔF_i before the other forces are applied the external work done by it is

$$\Delta W_e = \frac{1}{2}\Delta F_i \Delta u_i \tag{4.11.15}$$

Since the external work done is equal to the strain energy stored the strain energy would now be

$$U = \frac{1}{2}\Delta F_i \Delta u_i \tag{4.11.16}$$

If now the other forces are applied the force ΔF_i undergoes an additional displacement u_i and does more work in the amount $\Delta F_i u_i$ and the strain energy changes a like amount. Both Δu_i and u_i are displacements in the direction of the force. Since the order of load application does not affect the final value of the strain energy we can say that the total strain energy is now

$$U = (F_1, F_2, \ldots, F_i, \ldots, F_n) + \frac{1}{2}\Delta F_i \Delta u_i + \Delta F_i u_i \qquad (4.11.17)$$

The strain energy represented in Equation 4.11.14 is the same as the strain energy represented in Equation 4.11.17 and so

$$u(F_1, F_2, \ldots, F_i + \Delta F_i, \ldots, F_n) + \frac{1}{2}\Delta F_i \Delta u_i + \Delta F_i u_i$$

$$= u(F_1, F_2, \ldots, F_i, \ldots, F_n) + \frac{\partial U}{\partial F_i}\Delta F_i \qquad (4.11.18)$$

And now we neglect the higher order terms including $\frac{1}{2}\Delta F_i \Delta u_i$ to obtain

$$\Delta F_i u_i + U = U + \frac{\partial U}{\partial F_i}\Delta F_i \quad \rightarrow \quad \Delta F_i u_i = \frac{\partial U}{\partial F_i}\Delta F_i \qquad (4.11.19)$$

From which we can conclude that

$$u_i = \frac{\partial U}{\partial F_i} \qquad (4.11.20)$$

This is known as *Castigliano's second theorem*.

Remember that this works only if the internal strain energy can be presented in terms of external concentrated loads, that is, it is determinate. The resulting displacement is in the direction of the force.

Let us put this to the test with an example.

###########

Example 4.11.4

Problem: Let us reexamine Example 4.10.1 using Castigliano's theorem.

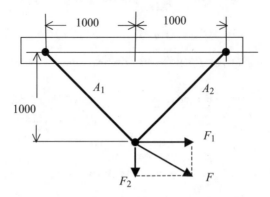

Figure (a)

Solution: Find the strain energy in terms of the applied load and differentiate.

We shall divide the force into components so that we can find the displacement in both the x and y directions.

A free body diagram of the lower joint is shown in Figure (b)

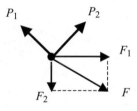

Figure (b)

Summing forces:

$$\sum F_x = F_1 - P_1 \cos 45° + P_2 \cos 45° = 0$$
$$\sum F_y = -F_2 + P_1 \sin 45° + P_2 \sin 45° = 0$$

(a)

Solve for the internal forces in terms of the applied forces.

$$P_1 = \frac{F_1 + F_2}{\sqrt{2}} \qquad P_2 = \frac{F_2 - F_1}{\sqrt{2}}$$

(b)

The strain energy is

$$U = \frac{1}{2}\frac{P_1^2 L_1}{EA} + \frac{1}{2}\frac{P_2^2 L_2}{EA} = \frac{1}{4}\frac{(F_1 + F_2)^2 L_1}{EA} + \frac{1}{4}\frac{(F_2 - F_1)^2 L_2}{EA}$$

(c)

We note that F_1 and F_2 have the values

$$F_1 = F \cos 30° = 10000 \cdot 0.866 = 8660 N$$
$$F_2 = F \sin 30° = 10000 \cdot 0.5 = 5000 N$$

(d)

but we must differentiate first to find the displacements and then insert the values. Since $L_1 = L_2 = L$

$$u = \frac{\partial U}{\partial F_1} = \frac{L}{2EA}\{(F_1 + F_2) - (F_2 - F_1)\} = \frac{L}{EA}F_1 = \frac{1414}{206840 \cdot 400}8660 = 0.1480$$

$$v = \frac{\partial U}{\partial F_2} = \frac{L}{2EA}\{(F_1 + F_2) + (F_2 - F_1)\} = \frac{L}{EA}F_2 = \frac{1414}{206840 \cdot 400}5000 = 0.0855$$

(e)

This agrees with the solution found in Example 4.10.1. Note that the sign difference for the vertical displacement follows from the fact that the displacement here is in the direction of the load.

###########

In cases where we want the displacement where there is no load, place a load there, find the strain energy in terms of all the loads, differentiate with respect to that load, then set the value of that load to zero. Here is a simple example.

############

Example 4.11.5

Problem: Find the displacement of joint B in the following truss in both the x and y directions. The truss members are made of steel with $E = 206,800\ MPa$. The cross section area of member AB is 400 mm^2 and that of member CB is 800 mm^2.

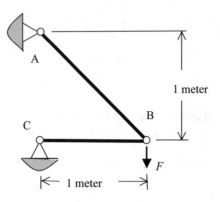

Figure (a)

Solution: Use Castigliano's theorem. To find the horizontal displacement we must add a horizontal force in order to be able to differentiate with respect to it and then let it go to zero.

The horizontal force is added in Figure (b).

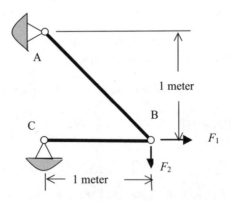

Figure (b)

From summation of forces at point B:

$$\sum F_y = F_{BA} \sin 45° - F_2 = 0 \quad \rightarrow \quad F_{BA} = \frac{F_2}{\sin 45°} = \sqrt{2}F_2$$

$$\sum F_x = F_{BA} \cos 45° + F_{BC} - F_1 = 0 \tag{a}$$

$$\rightarrow \quad F_{BC} = F_1 - F_{BA} \cos 45° = F_1 - \frac{\cos 45°}{\sin 45°}F_2 = F_1 - F_2$$

The strain energy is

$$U = \frac{1}{2} \frac{(F_{BA})^2 L_{BA}}{EA_{BA}} + \frac{1}{2} \frac{(F_{BC})^2 L_{BC}}{EA_{BC}} = \frac{1}{2} \frac{\left(\sqrt{2}F_2\right)^2 L_{BA}}{EA_{BA}} + \frac{1}{2} \frac{(F_1 - F_2)^2 L_{BC}}{EA_{BC}} \qquad (b)$$

Differentiate first to find the displacements and then insert $F_1 = 0$.

$$
\begin{aligned}
v &= \frac{\partial U}{\partial F_2} = \frac{2 F_2 L_{BA}}{EA_{BA}} - \frac{(F_1 - F_2) L_{BC}}{EA_{BC}} = \frac{2 F_2 1414}{206800 \cdot 400} - \frac{(0 - F_2) 1000}{206800 \cdot 800} = 4.032 \cdot 10^{-5} F_2 \\
u &= \frac{\partial U}{\partial F_1} = \frac{(F_1 - F_2) L_{BC}}{EA_{BC}} = \frac{(0 - F_2) 1000}{206800 \cdot 800} = -0.6044 \cdot 10^{-5} F_2
\end{aligned}
\qquad (c)
$$

The displacement in the x direction is positive in the direction of the force F_1 which explains its negative value.

<hr>

Example 4.11.6

Problem: Extend Example 2.5.1 to find displacement of the three dimensional truss shown in Figure (a) in the direction of the applied weight. All three members have the same cross sectional area A and have the same Young's modulus E.

Solution: Given the forces in the members from Example 2.5.1 sum their contributions to the strain energy. Use Castigliano's theorem to find the displacement.

In Example 2.5.1 we found the forces in the members to be

$$F_{AC} = \sqrt{2}W \qquad F_{AD} = F_{AB} = -0.559W \qquad (a)$$

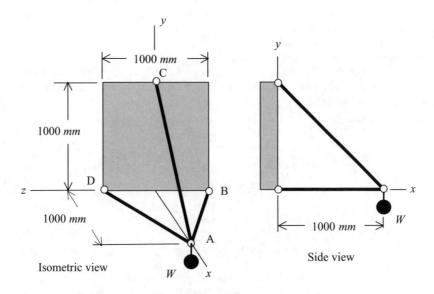

Figure (a)

The strain energy is

$$U = \frac{1}{2EA} \left\{ (F_{AB})^2 L_{AB} + (F_{AC})^2 L_{AC} + (F_{AD})^2 L_{AD} \right\}$$

$$= \frac{1}{2EA} \left\{ 2(-0.559W)^2 L_{AB} + (\sqrt{2}W)^2 L_{AC} \right\} \tag{b}$$

The lengths of the members are

$$L_{AB} = L_{AD} = 1118 \; mm \qquad L_{AC} = 1414 \; mm \tag{c}$$

The displacement is

$$\delta_{yA} = \frac{\partial U}{\partial W} = \frac{1}{2EA} \left\{ 4(-0.559)^2 \, W \cdot 1118 + \left(\sqrt{2} \right)^2 W \cdot 1414 \right\} = 2112.7 \frac{W}{EA} \tag{d}$$

###########

4.12   Summary and Conclusions

The axially loaded bar is studied at some length because the simplified geometry, loadings, and restraints allow us to concentrate on the formulation of the problem and the solution process. In this chapter we have taken a sharp departure from many other traditional introductory texts on the subject of mechanics of solids. We have chosen to derive and then emphasize solving the differential equations in terms of the distributed displacements.

In traditional introductory texts emphasis is placed on the force formulation of statically determinate problems. Stresses are then found from the internal forces determined from the equations of static equilibrium. When statically indeterminate problems are encountered a limited introduction of displacements at the points of restraint application are introduced to provide the additional equations necessary for a solution. Only after extensive use of this method are distributed displacements introduced, if at all.

In our displacement formulation we do consider determinate problems; however, we solve for the distributed displacements, not just the discrete displacement at ends of the bar, using either the first order equations or the second order equations.

We also solve both determinate and indeterminate cases using the second order equation for displacements. In this latter case the internal forces and stresses are found by differentiating the displacements.

The classical analysis equations of the displacement method are summarized in Equations 4.3.8, repeated here as Equation 4.12.1.

$$F_c \;\; \rightarrow \;\; f_x(x) \qquad P(x) \qquad \sigma_x(x) \qquad\qquad \varepsilon_x(x) \qquad u(x) \;\; \leftarrow \;\; \rho$$

$$\frac{dP}{dx} = -f_x(x) \qquad\qquad\qquad \varepsilon_x(x) = \frac{du}{dx}$$

$$\sigma_x(x) = E\varepsilon_x \tag{4.12.1}$$

$$P(x) = \sigma_x A = EA\varepsilon_x = EA\frac{du}{dx}$$

$$\frac{dP}{dx} = \frac{d}{dx} EA\frac{du}{dx} = -f_x(x)$$

Several examples in Sections 4.4 and 4.5 show how these equations are used to find displacements, strains, stresses, internal forces, and restraint forces when EA has a constant value. The first order equation can be used for determinate cases only.

$$EA\frac{du}{dx} = P(x) \quad \rightarrow \quad u(x) = \int \frac{P(x)}{EA}dx + c \qquad (4.12.2)$$

The second order equation can be used for determinate or indeterminate cases.

$$EA\frac{d^2u}{dx^2} = -f_x(x) \quad \rightarrow \quad u(x) = -\frac{1}{EA}\int\int f_n(x)dx + ax + b \qquad (4.12.3)$$

In Section 4.6 we examine the solution for variable cross sections. The first order equation can be used for determinate cases only.

$$EA(x)\frac{du}{dx} = P(x) \quad \rightarrow \quad u = \int \frac{P(x)}{EA(x)}dx + c \qquad (4.12.4)$$

The second order equation can be used for determinate or indeterminate cases.

$$\frac{d}{dx}EA(x)\frac{du}{dx} = -f_x(x) \quad \rightarrow \quad u(x) = -\int \frac{1}{EA(x)}\left(\int f_x(x)dx + a\right)dx + b \qquad (4.12.5)$$

In Section 4.7 we add the thermal strain component to the governing equations. The changes in the equations are summarized here.

$$F_c \quad \rightarrow \quad f_x(x) \qquad P(x) \qquad \sigma_x(x) \qquad \varepsilon_x^{mech}(x) \qquad \Delta T(x) \qquad u(x) \quad \leftarrow \quad \rho$$

$$\frac{dP}{dx} = -f_x(x) \qquad \varepsilon_x^{Total} = \varepsilon_x^{mech} + \varepsilon_{xT} = \frac{\sigma_x}{E} + \alpha\Delta T = \frac{du}{dx}$$

$$\sigma_x(x) = E\frac{du}{dx} - E\alpha\Delta T \qquad (4.12.6)$$

$$P(x) = \sigma_x A = EA\left(\frac{du}{dx} - \alpha\Delta T\right)$$

$$\frac{dP}{dx} = \frac{d}{dx}\left(EA\left(\frac{du}{dx} - \alpha\Delta T\right)\right) = -f_x(x)$$

In Section 4.8 the presence of shearing stress on internal surfaces is examined.

The design of axially loaded bars is discussed briefly in Section 4.9. In analysis the known quantities are geometry, material properties, external forces, and restraints. The unknowns are internal forces, stress, strain, and displacement. In design we often invert the process of selecting some of these quantities. For example, we might ask what geometry and material would be best suited to achieve some known value of stress and displacement.

The analysis and design of pin jointed trusses is examined in Sections 4.10 and 4.11. We equate the external work to the internal strain energy to solve simple problems then extend work and energy methods to include Castigliano's second theorem

$$u_i = \frac{\partial U}{\partial F_i} \qquad (4.12.7)$$

to find displacements in statically determinate axially loaded bars and trusses.

5

A General Method for the Axially Loaded Slender Bar

5.1 Introduction

We have in Equations 4.3.8 the complete set for solving for stress, strain, and displacement in axially loaded bars that are either statically determinate or indeterminate and which are loaded with either one or both of concentrated and distributed loads. As the geometry, loading, and restraint get more complicated the analytical solutions become more time consuming and tedious. In this day of computers there must be a way to organize the solution process more efficiently than is suggested by the example problems considered so far. Note that those examples have the simplest of geometry, restraints, and loads. In practice you rarely find problems that are that simple.

The analytical solution methods examined in Chapter 4 require integrating the differential equations. In the process constants of integration are generated. These constants are found by inserting the solutions into a set of boundary conditions. This results in a set of linear algebraic equations in terms of the unknown constants of integration. The problem is thus reduced to solving sets of linear algebraic equations to find the constants of integration. In the years B.C. (Before Computers) every effort was made to keep the number of simultaneous linear algebraic equations that needed to be solved to a minimum, perhaps not more than a single digit number. Thus, dividing bars into segments and solving the differential equations by matching boundary conditions where segments meet was limited to just a few segments. Today, with modern digital computers, there is virtually no limit to the number of simultaneous linear algebraic equations that can be solved quickly and efficiently.

With the advent of the digital computer a new way of organizing the solution of problems in solid mechanics has evolved and continues to improve with time. For now, we call this the general method because it can be adapted to work with all combinations of material, load, restraint, and geometry. As you shall see distinction between determinate and indeterminate is no longer a consideration in this method.

5.2 Nodes, Elements, Shape Functions, and the Element Stiffness Matrix

Consider an axially loaded slender bar with only concentrated loads. We have seen in Examples 4.4.5 and 4.5.2 that we can express the distributed displacement for each segment of the bar between load locations in terms of the discrete displacements at points of load application and a local coordinate s.

Analysis of Structures: An Introduction Including Numerical Methods, First Edition. Joe G. Eisley and Anthony M. Waas.
© 2011 John Wiley & Sons, Ltd. Published 2011 by John Wiley & Sons, Ltd.

See Example 4.4.5, Equation (i), which is repeated here.

$$u(s_m) = \begin{bmatrix} 1 - \dfrac{s_m}{l_m} & \dfrac{s_m}{l_m} \end{bmatrix} \begin{bmatrix} u_n \\ u_{n+1} \end{bmatrix} \tag{5.2.1}$$

The local coordinate s_m has its origin at one end of the segment which has been numbered segment m.

Let us generalize this a bit by considering a bar divided into segments as shown in Figure 5.2.1. We shall call each segment an *element* and number it. The points at the ends of elements are called *nodes* and also are numbered. Two elements have a common node where they meet, thus node n is common to elements m-1 and m.

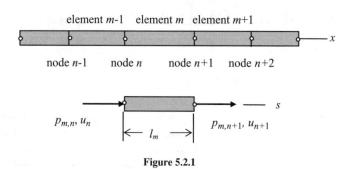

Figure 5.2.1

A particular element, say element m, has a length l_m with the left end labeled node n and the right end labeled node $n + 1$. We define the discrete displacement at node n as u_n and at node $n + 1$ as u_{n+1}. The origin of the coordinate s_m is at node n.

We also define internal forces for element m, also called stress resultants, at node n as $p_{m,n}$ and at node $n + 1$ as $p_{m,n+1}$. These stress resultants are the same as P used in the classical analysis except for the sign convention. Note the direction of these forces in Figure 5.2.1. These forces are chosen to be positive in the same direction as the positive axis direction at both ends because it is easier to keep track of them in the derivation that follows. We used the definition for P (normal away from the face as positive, same as positive normal stress) in the differential equation derivations because it is convenient and is most common in the literature of classical analysis. This new definition is most common in the literature of the numerical analysis of solids.

Applying Equation 5.2.1 to element m gives us the distributed displacement $u(s_m)$.

$$u(s_m) = \begin{bmatrix} 1 - \dfrac{s_m}{l_m} & \dfrac{s_m}{l_m} \end{bmatrix} \begin{bmatrix} u_n \\ u_{n+1} \end{bmatrix} = [n]\{r_m\} \tag{5.2.2}$$

where we have designated

$$[n] = \begin{bmatrix} 1 - \dfrac{s_m}{l_m} & \dfrac{s_m}{l_m} \end{bmatrix} \qquad \{r_m\} = \begin{bmatrix} u_n \\ u_{n+1} \end{bmatrix} \tag{5.2.3}$$

The functions in the matrix $[n]$ are called *shape functions*. We introduce the notation $\{r_m\}$ for the discrete nodal displacement to avoid confusion. To call it $\{u_m\}$ might confuse it with the displacement at node m.

The strain displacement equation is

$$\varepsilon_{xm} = \frac{du_m}{ds_m} = \frac{d}{ds_m}[n]\{r_m\} = \frac{1}{l_m}\begin{bmatrix} -1 & 1 \end{bmatrix} \begin{bmatrix} u_n \\ u_{n+1} \end{bmatrix} \tag{5.2.4}$$

and Hooke's law is

$$\sigma_{xm} = E_m \varepsilon_{xm} = E_m \frac{d}{ds_m} [n] \{r_m\} = \frac{E_m}{l_m} \begin{bmatrix} -1 & 1 \end{bmatrix} \begin{bmatrix} u_n \\ u_{n+1} \end{bmatrix} \tag{5.2.5}$$

At nodes n and $n+1$ we define the internal forces

$$p_{m,n} = -A_m \sigma_{xm} = -E_m A_m \frac{d}{ds_m} [n] \{r_m\} = \frac{E_m A_m}{l_m} \begin{bmatrix} 1 & -1 \end{bmatrix} \begin{bmatrix} u_n \\ u_{n+1} \end{bmatrix}$$

$$\tag{5.2.6}$$

$$p_{m,n+1} = A_m \sigma_{xm} = E_m A_m \frac{d}{ds_m} [n] \{r_m\} = \frac{E_m A_m}{l_m} \begin{bmatrix} -1 & 1 \end{bmatrix} \begin{bmatrix} u_n \\ u_{n+1} \end{bmatrix}$$

The negative value at node n results from defining the positive internal force opposite to the positive direction of the stress, while at $n+1$ they are in the same direction.

The nodal internal forces can be expressed in matrix form.

$$\{p_m\} = \begin{bmatrix} p_{m,n} \\ p_{m,n+1} \end{bmatrix} = \frac{E_m A_m}{l_m} \begin{bmatrix} 1 & -1 \\ -1 & 1 \end{bmatrix} \begin{bmatrix} u_n \\ u_{n+1} \end{bmatrix} = [k_m] \{r_m\} \tag{5.2.7}$$

Now we shall add these new quantities, $\{r_m\}$ and $\{p_m\}$, to the summary of equations shown in Equation 4.3.8 and use the shape functions in Equation 5.2.2 to represent the displacement, strain, stress, and internal forces for an element.

In summary for element m.

$$\{p_m\} \qquad \sigma_{xm} \qquad \varepsilon_{xm} \qquad u(s_m) \qquad \{r_m\}$$

$$u(s_m) = [n]\{r_m\} = \begin{bmatrix} 1 - \dfrac{s_m}{l_m} & \dfrac{s_m}{l_m} \end{bmatrix} \begin{bmatrix} u_n \\ u_{n+1} \end{bmatrix}$$

$$\varepsilon_{xm} = \frac{du_m}{ds_m} = \frac{d}{ds_m} [n]\{r_m\} = \frac{1}{l_m} \begin{bmatrix} -1 & 1 \end{bmatrix} \begin{bmatrix} u_n \\ u_{n+1} \end{bmatrix}$$

$$\sigma_{xm} = E_m \varepsilon_{xm} = E_m \frac{d}{ds_m} [n]\{r_m\} = \frac{E_m}{l_m} \begin{bmatrix} -1 & 1 \end{bmatrix} \begin{bmatrix} u_n \\ u_{n+1} \end{bmatrix}$$

$$\tag{5.2.8}$$

$$p_{m,n} = -A_m \sigma_{xm} = -E_m A_m \frac{d}{ds_m} [n]\{r_m\} = \frac{E_m A_m}{l_m} \begin{bmatrix} 1 & -1 \end{bmatrix} \begin{bmatrix} u_n \\ u_{n+1} \end{bmatrix}$$

$$p_{m,n+1} = A_m \sigma_{xm} = E_m A_m \frac{d}{ds_m} [n]\{r_m\} = \frac{E_m A_m}{l_m} \begin{bmatrix} -1 & 1 \end{bmatrix} \begin{bmatrix} u_n \\ u_{n+1} \end{bmatrix}$$

$$\{p_m\} = \begin{bmatrix} p_{m,n} \\ p_{m,n+1} \end{bmatrix} = \frac{E_m A_m}{l_m} \begin{bmatrix} 1 & -1 \\ -1 & 1 \end{bmatrix} \begin{bmatrix} u_n \\ u_{n+1} \end{bmatrix} = [k_m]\{r_m\}$$

In the first row of Equation 5.2.8 we have the typical unknowns in the formulation of a problem as applied to a single element m.

$\{r_m\}$ - nodal displacements

$u(s_m)$ - distributed displacement

ε_{xm} - normal strain

σ_{xm} - normal stress

$\{p_m\}$ - internal forces

Among the known quantities in any given problem are the geometry, as expressed by A and l_m, and material properties, as expressed by E, of the slender bar.

Starting in the second row where we define the displacement in terms of shape functions and nodal displacements and we move down and to the left to define other quantities in terms of the shape functions. In the third row we have the strain displacement equation, and in the fourth row Hooke's law. Note that since the distributed displacement is a linear function of s_m the distributed strain and stress are constants over the whole length of the element.

In the next two rows we express the internal forces at each node and then combine them in a single matrix.

$$\{p_m\} = \begin{bmatrix} p_{m,n} \\ p_{m,n+1} \end{bmatrix} = \frac{E_m A_m}{l_m} \begin{bmatrix} 1 & -1 \\ -1 & 1 \end{bmatrix} \begin{bmatrix} u_n \\ u_{n+1} \end{bmatrix} = [k_m]\{r_m\} \tag{5.2.9}$$

where the matrix

$$[k_m] = \frac{E_m A_m}{l_m} \begin{bmatrix} 1 & -1 \\ -1 & 1 \end{bmatrix} \tag{5.2.10}$$

is called the *element stiffness matrix*.

In the next example we shall show that the element stiffness matrix contains exactly the same information as the differential equations in Chapter 4.

############

Example 5.2.1

Problem: Demonstrate that the element stiffness matrix contains the same information as the equations in Chapter 4 by applying it to the bar in Example 4.3.1, Figure (d). The figure is repeated here as Figure (a).

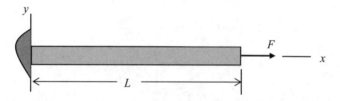

Figure (a)

Solution: Take the entire bar as a single element and adapt the element matrix equations to this configuration.

Call the bar element 1 and number the nodes as shown in Figure (b).

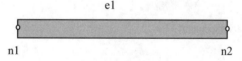

Figure (b)

The element matrix equations, Equation 5.2.11, are then

$$\begin{bmatrix} p_{1,1} \\ p_{1,2} \end{bmatrix} = \frac{E_1 A_1}{L_1} \begin{bmatrix} 1 & -1 \\ -1 & 1 \end{bmatrix} \begin{bmatrix} u_1 \\ u_2 \end{bmatrix} \tag{a}$$

From Figure (a) we note that the boundary conditions are

$$u_1 = 0 \qquad p_{1,2} = F \tag{b}$$

Also we note that u_2 and $p_{1,1}$ are unknown. Since $p_{1,1}$ is an unknown restraint force let us assign it the symbol

$$p_{1,1} = R_1 \tag{c}$$

Note that $p_{1,1}$ and R_1 have the same sign convention.
The matrix equations now become

$$\begin{bmatrix} R_1 \\ F \end{bmatrix} = \frac{EA}{L} \begin{bmatrix} 1 & -1 \\ -1 & 1 \end{bmatrix} \begin{bmatrix} 0 \\ u_2 \end{bmatrix} \tag{d}$$

If we expand the matrix equations and solve for the unknowns we get

$$F = \frac{EA}{L}(-1 \cdot 0 + 1 \cdot u_2) = \frac{EA}{L}u_2 \quad \rightarrow \quad u_2 = \frac{FL}{EA}$$

$$R_1 = \frac{EA}{L}(1 \cdot 0 - 1 \cdot u_2) = -\frac{EA}{L}u_2 = -\frac{EA}{L}\frac{FL}{EA} = -F \tag{e}$$

To find the stress we use Equation 5.2.5.

$$\sigma_x = \frac{E}{L}\begin{bmatrix} -1 & 1 \end{bmatrix}\begin{bmatrix} 0 \\ u_2 \end{bmatrix} = \frac{E}{L}\begin{bmatrix} -1 & 1 \end{bmatrix}\begin{bmatrix} 0 \\ \dfrac{FL}{EA} \end{bmatrix} = \frac{E}{L}\frac{FL}{EA} = \frac{F}{A} \tag{f}$$

To find the distributed displacement use Equation 5.2.2. Since s_1 and x are the same coordinate we have

$$u(x) = \begin{bmatrix} 1 - \dfrac{x}{L} & \dfrac{x}{L} \end{bmatrix}\begin{bmatrix} 0 \\ u_2 \end{bmatrix} = \frac{x}{L}u_2 = \frac{x}{L}\frac{FL}{EA} = \frac{F}{EA}x \tag{g}$$

This is the same as the solution obtained in Example 4.3.1.

This almost trivial example is presented to emphasize that the element matrix equations contain exactly the same information as the classical equations presented in Chapter 4 for bars with concentrated loads.

##########

5.3 The Assembled Global Equations and Their Solution

A bar with several concentrated loads must be divided into several elements. The shape functions of Equation 5.2.2 apply to any and all segments between the concentrated loads. Consider that a bar with multiple concentrated loads has been divided into a number of segments and the nodes and elements are numbered as indicated in Figure 5.3.1. We make sure that there is a node at each location of an applied force; however, we may have nodes at other locations where the loads are zero. The elements may be different lengths.

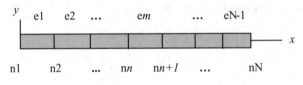

Figure 5.3.1

To formulate the problem we must write a stiffness matrix for each element and then combine all the elements to represent the whole bar. Then we must add the applied loads and restraints to define the complete set of equations governing the behavior of the bar. The process of combining all the elements is called the *assembly* of the various element stiffness matrices to form the *global stiffness matrix*.

For element 1 we have just nodes 1 and 2. We can say

$$\{p_1\} = \begin{bmatrix} p_{1,1} \\ p_{1,2} \end{bmatrix} = \frac{E_1 A_1}{l_1} \begin{bmatrix} 1 & -1 \\ -1 & 1 \end{bmatrix} \begin{bmatrix} u_1 \\ u_2 \end{bmatrix} = [k_1]\{r_1\} \tag{5.3.1}$$

For any intermediate element m we have nodes n and $n+1$. We have

$$\{p_m\} = \begin{bmatrix} p_{m,n} \\ p_{m,n+1} \end{bmatrix} = \frac{E_m A_m}{l_m} \begin{bmatrix} 1 & -1 \\ -1 & 1 \end{bmatrix} \begin{bmatrix} u_n \\ u_{n+1} \end{bmatrix} = [k_m]\{r_m\} \tag{5.3.2}$$

For the last element N-1 we have nodes N-1 and N. We have

$$\{p_{N-1}\} = \begin{bmatrix} p_{N-1,N-1} \\ p_{N-1,N} \end{bmatrix} = \frac{E_{N-1} A_{N-1}}{l_{N-1}} \begin{bmatrix} 1 & -1 \\ -1 & 1 \end{bmatrix} \begin{bmatrix} u_{N-1} \\ u_N \end{bmatrix} = [k_{N-1}]\{r_{N-1}\} \tag{5.3.3}$$

Note that we always have one more node than we have elements.

We immediately encounter a problem when we try to assemble these equations. The assembly process requires a summation of the element stiffness matrices but by the rules of matrix addition. Two sets of equations in matrix form

$$\{X\} = [A]\{x\} \qquad \{Y\} = [B]\{y\} \tag{5.3.4}$$

can be assembled, or summed, only if $\{x\}$ is exactly equal to $\{y\}$ and $[A]$ and $[B]$ have the same number of rows and columns, or the same *order*. If so, we obtain

$$\{Z\} = \{X\} + \{Y\} = [A]\{x\} + [B]\{y\} = ([A] + [B])\{x\} \tag{5.3.5}$$

Element matrices in Equations 5.3.1-3 cannot be added directly because each one refers to a different pair of nodes. To overcome this problem we imbed each element matrix in a global matrix format. For example, for element 1 the element stiffness matrix is reformatted as follows.

$$\{p_1\} = \begin{bmatrix} p_{1,1} \\ p_{1,2} \\ 0 \\ 0 \\ \cdots \\ 0 \end{bmatrix} = [k_1]\{u\} = \begin{bmatrix} \dfrac{E_1 A_1}{l_1} & -\dfrac{E_1 A_1}{l_1} & 0 & 0 & \cdots & 0 \\ -\dfrac{E_1 A_1}{l_1} & \dfrac{E_1 A_1}{l_1} & 0 & 0 & \cdots & 0 \\ 0 & 0 & 0 & 0 & \cdots & 0 \\ 0 & 0 & 0 & 0 & \cdots & 0 \\ \cdots & \cdots & \cdots & \cdots & \cdots & \cdots \\ 0 & 0 & 0 & 0 & \cdots & 0 \end{bmatrix} \begin{bmatrix} u_1 \\ u_2 \\ u_3 \\ u_4 \\ \cdots \\ u_N \end{bmatrix} \tag{5.3.6}$$

This has exactly the same information as the shorter form. The stiffness matrix for element 2 becomes

$$
\{p_2\} =
\begin{bmatrix}
0 \\
p_{2,2} \\
p_{2,3} \\
0 \\
\cdots \\
0
\end{bmatrix}
= [k_2]\{u\} =
\begin{bmatrix}
0 & 0 & 0 & 0 & \cdots & 0 \\
0 & \dfrac{E_2 A_2}{l_2} & -\dfrac{E_2 A_2}{l_2} & 0 & \cdots & 0 \\
0 & -\dfrac{E_2 A_2}{l_2} & \dfrac{E_2 A_2}{l_2} & 0 & \cdots & 0 \\
0 & 0 & 0 & 0 & \cdots & 0 \\
\cdots & \cdots & \cdots & \cdots & \cdots & \cdots \\
0 & 0 & 0 & 0 & \cdots & 0
\end{bmatrix}
\begin{bmatrix}
u_1 \\
u_2 \\
u_3 \\
u_4 \\
\cdots \\
u_N
\end{bmatrix}
\tag{5.3.7}
$$

and so on until all the elements are reformatted. Now when all the element matrices are added together we get the global equation set shown in Equation 5.3.8.

$$
\begin{bmatrix}
p_{1,1} \\
p_{1,2} + p_{2,2} \\
p_{2,3} + p_{3,3} \\
p_{3,4} + p_{4,4} \\
\cdots \\
p_{N-1,N-1} + p_{N-1,N}
\end{bmatrix}
$$

$$
=
\begin{bmatrix}
\dfrac{E_1 A_1}{l_1} & -\dfrac{E_1 A_1}{l_1} & 0 & 0 & \cdots & 0 \\
-\dfrac{E_1 A_1}{l_1} & \dfrac{E_1 A_1}{l_1} + \dfrac{E_2 A_2}{l_2} & -\dfrac{E_2 A_2}{l_2} & 0 & \cdots & 0 \\
0 & -\dfrac{E_2 A_2}{l_2} & \dfrac{E_2 A_2}{l_2} + \dfrac{E_3 A_3}{l_3} & -\dfrac{E_3 A_3}{l_3} & \cdots & 0 \\
0 & 0 & -\dfrac{E_3 A_3}{l_3} & \dfrac{E_3 A_3}{l_3} + \dfrac{E_4 A_4}{l_4} & \cdots & 0 \\
\cdots & \cdots & \cdots & \cdots & \cdots & \cdots \\
0 & 0 & 0 & 0 & \cdots & \dfrac{E_{N-1} A_{N-1}}{l_{N-1}}
\end{bmatrix}
\begin{bmatrix}
u_1 \\
u_2 \\
u_3 \\
u_4 \\
\cdots \\
u_N
\end{bmatrix}
$$

$$
\tag{5.3.8}
$$

Now we must define the forces acting at the nodes. The forces acting at node $n+1$ are shown in Figure 5.3.2. The applied load F_{n+1} is shown above the node but is acting on the node.

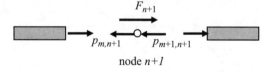

node $n+1$

Figure 5.3.2

From equilibrium of the forces at the node the external load F_{n+1} is equal to the sum of the internal nodal forces, so

$$
F_{n+1} - p_{m,n+1} - p_{m+1,n+1} = 0 \quad \rightarrow \quad p_{m,n+1} + p_{m+1,n+1} = F_{n+1}
\tag{5.3.9}
$$

It follows that

$$
\begin{bmatrix}
p_{1,1} \\
p_{1,2} + p_{2,2} \\
p_{2,3} + p_{3,3} \\
p_{3,4} + p_{4,4} \\
\cdots \\
p_{N-1,N-1} + p_{N-1,N}
\end{bmatrix}
=
\begin{bmatrix}
F_1 \\
F_2 \\
F_3 \\
F_4 \\
\cdots \\
F_N
\end{bmatrix}
= \{F\}
\tag{5.3.10}
$$

We now represent Equation 5.3.8 as

$$
\{F\} = [K]\{r\}
\tag{5.3.11}
$$

where $[K]$ is called the *global stiffness matrix*. Since the applied forces $\{F\}$ are known it would appear that $\{F\} = [K]\{r\}$ is an acceptable set of equations to solve. If we try we run into a problem. It so happens that $[K]$ is a singular matrix, that is, the determinant of $[K]$ is equal to zero.

$$
|K| = 0
\tag{5.3.12}
$$

This means that the equations have no unique solution. Why is that? We have not yet applied boundary conditions. To complete the problem we must add boundary restraints and note that restraint forces exist at nodes where displacements are restrained. At least one of the nodal displacements must have a known value to satisfy static equilibrium. At any restrained node we must add a restraint force to the force vector $\{F\}$. For example, if the displacement at node 1 is zero then $u_1 = 0$ and there is a restraint force R_1 added as follows:

$$
\begin{bmatrix}
R_1 \\
F_2 \\
F_3 \\
F_4 \\
\cdots \\
F_N
\end{bmatrix}
=
\begin{bmatrix}
\dfrac{E_1 A_1}{l_1} & -\dfrac{E_1 A_1}{l_1} & 0 & 0 & \cdots & 0 \\[2ex]
-\dfrac{E_1 A_1}{l_1} & \dfrac{E_1 A_1}{l_1} + \dfrac{E_2 A_2}{l_2} & -\dfrac{E_2 A_2}{l_2} & 0 & \cdots & 0 \\[2ex]
0 & -\dfrac{E_2 A_2}{l_2} & \dfrac{E_2 A_2}{l_2} + \dfrac{E_3 A_3}{l_3} & -\dfrac{E_3 A_3}{l_3} & \cdots & 0 \\[2ex]
0 & 0 & -\dfrac{E_3 A_3}{l_3} & \dfrac{E_3 A_3}{l_3} + \dfrac{E_4 A_4}{l_4} & \cdots & 0 \\[2ex]
\cdots & \cdots & \cdots & \cdots & \cdots & \cdots \\[1ex]
0 & 0 & 0 & 0 & \cdots & \dfrac{E_{N-1} A_{N-1}}{l_{N-1}}
\end{bmatrix}
\begin{bmatrix}
0 \\
u_2 \\
u_3 \\
u_4 \\
\cdots \\
u_N
\end{bmatrix}
\tag{5.3.13}
$$

Of course $[K]$ is still a singular matrix; however, we now have one less unknown nodal displacement and so there is one less equation to solve as a result of applying the restraint. We now do what is called *partitioning* the matrix equations. Let us separate the first equation from the rest as indicated by the dashed lines in Equation 5.3.14 and write the remaining equations separately.

$$
\begin{bmatrix} R_1 \\ \hline F_2 \\ F_3 \\ F_4 \\ \cdots \\ F_N \end{bmatrix} = \begin{bmatrix} \dfrac{E_1 A_1}{l_1} & -\dfrac{E_1 A_1}{l_1} & 0 & 0 & \cdots & 0 \\ \hline -\dfrac{E_1 A_1}{l_1} & \dfrac{E_1 A_1}{l_1} + \dfrac{E_2 A_2}{l_2} & -\dfrac{E_2 A_2}{l_2} & 0 & \cdots & 0 \\ 0 & -\dfrac{E_2 A_2}{l_2} & \dfrac{E_2 A_2}{l_2} + \dfrac{E_3 A_3}{l_3} & -\dfrac{E_3 A_3}{l_3} & \cdots & 0 \\ 0 & 0 & -\dfrac{E_3 A_3}{l_3} & \dfrac{E_3 A_3}{l_3} + \dfrac{E_4 A_4}{l_4} & \cdots & 0 \\ \cdots & \cdots & \cdots & \cdots & \cdots & \cdots \\ 0 & 0 & 0 & 0 & \cdots & \dfrac{E_{N-1} A_{N-1}}{l_{N-1}} \end{bmatrix} \begin{bmatrix} 0 \\ \hline u_2 \\ u_3 \\ u_4 \\ \cdots \\ u_N \end{bmatrix}
$$

$$(5.3.14)$$

We get the following set of equations with known forces and unknown displacements. This set is not singular.

$$
\begin{bmatrix} F_2 \\ F_3 \\ F_4 \\ \cdots \\ F_N \end{bmatrix} = \begin{bmatrix} \dfrac{E_1 A_1}{l_1} + \dfrac{E_2 A_2}{l_2} & -\dfrac{E_2 A_2}{l_2} & 0 & \cdots & 0 \\ -\dfrac{E_2 A_2}{l_2} & \dfrac{E_2 A_2}{l_2} + \dfrac{E_3 A_3}{l_3} & -\dfrac{E_3 A_3}{l_3} & \cdots & 0 \\ 0 & -\dfrac{E_3 A_3}{l_3} & \dfrac{E_3 A_3}{l_3} + \dfrac{E_4 A_4}{l_4} & \cdots & 0 \\ \cdots & \cdots & \cdots & \cdots & \cdots \\ 0 & 0 & 0 & \cdots & \dfrac{E_{N-1} A_{N-1}}{l_{N-1}} \end{bmatrix} \begin{bmatrix} u_2 \\ u_3 \\ u_4 \\ \cdots \\ u_N \end{bmatrix}
$$

$$(5.3.15)$$

These equations can be solved for the nodal displacements. And once they are solved we can find the unknown restraint force with the other partitioned equation.

$$
[R_1] = \begin{bmatrix} -\dfrac{E_1 A_1}{l_1} & 0 & 0 & \cdots & 0 \end{bmatrix} \begin{bmatrix} u_2 \\ u_3 \\ u_4 \\ \cdots \\ u_N \end{bmatrix}
$$

$$(5.3.16)$$

To find the stresses we apply Equation 5.2.5 for each element.

$$
\sigma_{xm} = E_m \varepsilon_{xm} = E_m \frac{d}{ds} [n] \{r_m\} = \frac{E_m}{l_m} \begin{bmatrix} -1 & 1 \end{bmatrix} \begin{bmatrix} u_n \\ u_{n+1} \end{bmatrix}
$$

$$(5.3.17)$$

This relatively simple act of defining shape functions turns the problem from one of solving differential equations to that of directly solving linear algebraic equations. The differential equations are solved when creating the shape functions once and for all for each class of elements we examine.

Note that the strain displacement, stress-strain, and equilibrium equations are satisfied without compromise. For a bar with truly concentrated loads this is not an approximation. All these relations are found directly from integration of the governing equation for an axially loaded slender bar from classical theory. This is exactly what you have been studying in Chapter 4. It is just presented in a different way.

The restraints can be applied at either end or at both ends, thus possible formulations of a problem can include

$$
\begin{bmatrix} F_1 \\ F_2 \\ F_3 \\ F_4 \\ \cdots \\ \text{----} \\ R_N \end{bmatrix} =
\begin{bmatrix}
\dfrac{E_1 A_1}{l_1} & -\dfrac{E_1 A_1}{l_1} & 0 & 0 & \cdots & 0 \\[2ex]
-\dfrac{E_1 A_1}{l_1} & \dfrac{E_1 A_1}{l_1}+\dfrac{E_2 A_2}{l_2} & -\dfrac{E_2 A_2}{l_2} & 0 & \cdots & 0 \\[2ex]
0 & -\dfrac{E_2 A_2}{l_2} & \dfrac{E_2 A_2}{l_2}+\dfrac{E_3 A_3}{l_3} & -\dfrac{E_3 A_3}{l_3} & \cdots & 0 \\[2ex]
0 & 0 & -\dfrac{E_3 A_3}{l_3} & \dfrac{E_3 A_3}{l_3}+\dfrac{E_4 A_4}{l_4} & \cdots & 0 \\[1ex]
\cdots & \cdots & \cdots & \cdots & \cdots & \cdots \\[1ex]
0 & 0 & 0 & 0 & \cdots & \dfrac{E_{N-1} A_{N-1}}{l_{N-1}}
\end{bmatrix}
\begin{bmatrix} u_1 \\ u_2 \\ u_3 \\ u_4 \\ \cdots \\ \text{----} \\ 0 \end{bmatrix}
$$

$$(5.3.18)$$

and

$$
\begin{bmatrix} R_1 \\ \text{----} \\ F_2 \\ F_3 \\ F_4 \\ \cdots \\ \text{----} \\ R_N \end{bmatrix} =
\begin{bmatrix}
\dfrac{E_1 A_1}{l_1} & -\dfrac{E_1 A_1}{l_1} & 0 & 0 & \cdots & 0 \\[2ex]
-\dfrac{E_1 A_1}{l_1} & \dfrac{E_1 A_1}{l_1}+\dfrac{E_2 A_2}{l_2} & -\dfrac{E_2 A_2}{l_2} & 0 & \cdots & 0 \\[2ex]
0 & -\dfrac{E_2 A_2}{l_2} & \dfrac{E_2 A_2}{l_2}+\dfrac{E_3 A_3}{l_3} & -\dfrac{E_3 A_3}{l_3} & \cdots & 0 \\[2ex]
0 & 0 & -\dfrac{E_3 A_3}{l_3} & \dfrac{E_3 A_3}{l_3}+\dfrac{E_4 A_4}{l_4} & \cdots & 0 \\[1ex]
\cdots & \cdots & \cdots & \cdots & \cdots & \cdots \\[1ex]
0 & 0 & 0 & 0 & \cdots & \dfrac{E_{N-1} A_{N-1}}{l_{N-1}}
\end{bmatrix}
\begin{bmatrix} 0 \\ \text{----} \\ u_2 \\ u_3 \\ u_4 \\ \cdots \\ \text{----} \\ 0 \end{bmatrix}
$$

$$(5.3.19)$$

We would normally call a problem defined by the constraints in Equations 5.3.13 and 5.3.18 statically determinate and a problem defined by Equations 5.3.19 statically indeterminate and use different approaches in solving them. By this method we need make no such distinctions.

In summary, we assign nodes where known concentrated loads and where known concentrated restraints are applied and at other points if we wish. We then divide the bar into elements and number the nodes and elements.

Next we find the element matrix for each element

$$
\{p_m\} = \begin{bmatrix} p_{m,n} \\ p_{m,n+1} \end{bmatrix} = \begin{bmatrix} -A_m \sigma_{m,n} \\ A_m \sigma_{m,n+1} \end{bmatrix} = \frac{E_m A_m}{l_m} \begin{bmatrix} 1 & -1 \\ -1 & 1 \end{bmatrix} \begin{bmatrix} u_n \\ u_{n+1} \end{bmatrix} = [k_m]\{r_m\} \tag{5.3.20}
$$

and imbed each element stiffness matrix in the global format. The element matrices can then be assembled (summed) to form the global matrix equations.

$$\{F\} = \sum \{p_m\} = \sum [k_m]\{r\} = [K]\{r\} \tag{5.3.21}$$

Next the known nodal restraints are inserted in $\{r\}$; the corresponding unknown restraint forces in $\{F\}$; and the known applied loads in $\{F\}$. This leaves us with unknown nodal displacements in $\{r\}$ and unknown restraint forces in $\{F\}$ to be determined. We then partition the global equations into on set containing the unknown nodal displacements and the known applied forces that we solve for the nodal displacements. In partitioning another set is found for solving for the restraint forces in terms of the now known nodal displacements

Once we know the nodal displacements we can find the distributed displacement, the strain, the stresses, and the internal forces in each element from the equations presented in Equation 5.2.6.

This whole process is best understood by examples.

###########

Example 5.3.1

Problem: A bar with both ends fixed has a concentrated force at the midpoint as shown in Figure (a). Find the displacements, internal forces, and stresses. Use the general method with two elements.

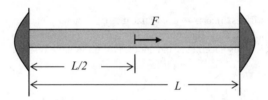

Figure (a)

Solution: Divide into elements, number the nodes and elements, assemble the element matrices, add the applied forces and restraints, partition, and solve.

With nodes at the two ends and at the location of the applied load we choose two elements of equal length. The nodes and elements are numbered in Figure (b).

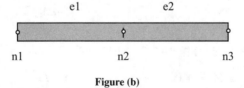

Figure (b)

In order to assemble the element matrices we rewrite the stiffness matrix for each element by imbedding it in the global matrix format.

For element 1

$$\begin{bmatrix} p_{1,1} \\ p_{1,2} \end{bmatrix} = \frac{2EA}{L} \begin{bmatrix} 1 & -1 \\ -1 & 1 \end{bmatrix} \begin{bmatrix} u_1 \\ u_2 \end{bmatrix} \quad \rightarrow \quad \begin{bmatrix} p_{1,1} \\ p_{1,2} \\ 0 \end{bmatrix} = \frac{2EA}{L} \begin{bmatrix} 1 & -1 & 0 \\ -1 & 1 & 0 \\ 0 & 0 & 0 \end{bmatrix} \begin{bmatrix} u_1 \\ u_2 \\ u_3 \end{bmatrix} \tag{a}$$

For element 2

$$
\begin{bmatrix} p_{2,2} \\ p_{2,3} \end{bmatrix} = \frac{2EA}{L} \begin{bmatrix} 1 & -1 \\ -1 & 1 \end{bmatrix} \begin{bmatrix} u_2 \\ u_3 \end{bmatrix} \quad \rightarrow \quad \begin{bmatrix} 0 \\ p_{2,2} \\ p_{2,3} \end{bmatrix} = \frac{2EA}{L} \begin{bmatrix} 0 & 0 & 0 \\ 0 & 1 & -1 \\ 0 & -1 & 1 \end{bmatrix} \begin{bmatrix} u_1 \\ u_2 \\ u_3 \end{bmatrix} \tag{b}
$$

Now assemble these two.

$$
\begin{bmatrix} p_{1,1} \\ p_{1,2} + p_{2,2} \\ p_{2,3} \end{bmatrix} = \frac{2EA}{L} \begin{bmatrix} 1 & -1 & 0 \\ -1 & 2 & -1 \\ 0 & -1 & 1 \end{bmatrix} \begin{bmatrix} u_1 \\ u_2 \\ u_3 \end{bmatrix} \tag{c}
$$

Looking closely we note that the nodal displacements are known at the restraints and the rest are unknown. Likewise the restraint forces are unknown at the restraints and applied forces are known at all other nodes. The applied loads and restraints for this case are quite simple and have been inserted in Equation (d). Symbols for the unknown restraint forces have been added.

$$
\begin{bmatrix} R_1 \\ F \\ R_3 \end{bmatrix} = \frac{2EA}{L} \begin{bmatrix} 1 & -1 & 0 \\ -1 & 2 & -1 \\ 0 & -1 & 1 \end{bmatrix} \begin{bmatrix} 0 \\ u_2 \\ 0 \end{bmatrix} \tag{d}
$$

As noted, the global stiffness matrix is a singular matrix, meaning its determinant is zero. Thus this set of equations does not have a unique solution. The next step is to partition the matrix equation into one set for finding the nodal displacements and another for finding the restraint forces. The dashed lines show the partitions.

$$
\begin{bmatrix} R_1 \\ \hline F \\ \hline R3 \end{bmatrix} = \frac{2EA}{L} \left[\begin{array}{c:c:c} 1 & -1 & 0 \\ \hdashline -1 & 2 & -1 \\ \hdashline 0 & -1 & 1 \end{array} \right] \begin{bmatrix} 0 \\ \hline u_2 \\ \hline 0 \end{bmatrix} \tag{e}
$$

The equation for the displacement and its solution is

$$
F = \frac{2EA}{L} \cdot 2u_2 \quad \rightarrow \quad u_2 = \frac{FL}{4EA} \tag{f}
$$

The equations for the restraint forces and their solution are

$$
\begin{bmatrix} R_1 \\ R_3 \end{bmatrix} = \frac{2EA}{L} \begin{bmatrix} -1 \\ -1 \end{bmatrix} u_2 \quad \rightarrow \quad R_1 = -\frac{2EA}{L} \cdot \frac{FL}{4EA} = -\frac{F}{2} = R_3 \tag{g}
$$

The internal stresses from Equation 5.2.5 for element 1 are

$$
\sigma_{x1} = \frac{E_1}{l_1} \begin{bmatrix} -1 & 1 \end{bmatrix} \begin{bmatrix} u_1 \\ u_2 \end{bmatrix} = \frac{2E}{L} \begin{bmatrix} -1 & 1 \end{bmatrix} \begin{bmatrix} 0 \\ \frac{FL}{4EA} \end{bmatrix} = \frac{F}{2A} \tag{h}
$$

and for element 2

$$\sigma_{x2} = \frac{E_2}{l_2}\begin{bmatrix} -1 & 1 \end{bmatrix}\begin{bmatrix} u_2 \\ u_3 \end{bmatrix} = \frac{2E}{L}\begin{bmatrix} -1 & 1 \end{bmatrix}\begin{bmatrix} \dfrac{FL}{4EA} \\ 0 \end{bmatrix} = -\frac{F}{2A} \tag{i}$$

To find the distributed displacements we note that $s_1 = x$ and $s_2 = x - L/2$. Then for each element from Equation 5.2.2 we have

$$u_1(s_1) = \begin{bmatrix} 1 - \dfrac{2s_1}{L} & \dfrac{2s_1}{L} \end{bmatrix}\begin{bmatrix} u_1 \\ u_2 \end{bmatrix} = \begin{bmatrix} 1 - \dfrac{2s_1}{L} & \dfrac{2s_1}{L} \end{bmatrix}\begin{bmatrix} 0 \\ \dfrac{FL}{4EA} \end{bmatrix} = \frac{F}{2EA}s_1$$

$$\rightarrow u_1(x) = \frac{F}{2EA}x \tag{j}$$

$$u_2(s_2) = \begin{bmatrix} 1 - \dfrac{2s_2}{L} & \dfrac{2s_2}{L} \end{bmatrix}\begin{bmatrix} u_2 \\ u_3 \end{bmatrix} = \begin{bmatrix} 1 - \dfrac{2s_2}{L} & \dfrac{2s_2}{L} \end{bmatrix}\begin{bmatrix} \dfrac{FL}{4EA} \\ 0 \end{bmatrix} = \frac{FL}{4EA}\left(1 - \frac{2s_2}{L}\right)$$

$$\rightarrow u_2(x) = \frac{F}{2EA}\left(1 - \frac{x}{L}\right) \tag{k}$$

Plots of displacements and stresses are given in Figure (c). The displacements between nodes are always linear so we may simply draw straight lines between nodal displacements. The internal forces and stresses are always constant between nodes.

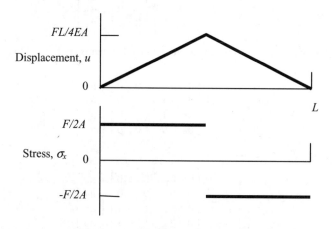

Figure (c)

Compare with Example 4.5.2.

Example 5.3.2

Problem: Repeat Example 4.4.5 using the general method. The bar is shown in Figure (a). Find the displacements and stresses.

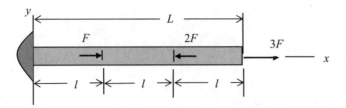

Figure (a)

Solution: Assemble the global stiffness matrix equations, add the forces and restraints, partition the equations, and solve.

Divide the bar into three elements and number the nodes and elements as shown in Figure (b).

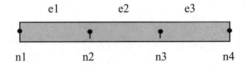

Figure (b)

Putting all the elements together into a set of equations describing the entire bar requires the assembly of the element matrices. We note that E, A, and l are the same in all three elements where $l = L/3$.

$$\{p_1\} = \begin{bmatrix} p_{1,1} \\ p_{1,2} \\ 0 \\ 0 \end{bmatrix} = \frac{EA}{l} \begin{bmatrix} 1 & -1 & 0 & 0 \\ -1 & 1 & 0 & 0 \\ 0 & 0 & 0 & 0 \\ 0 & 0 & 0 & 0 \end{bmatrix} \begin{bmatrix} u_1 \\ u_2 \\ u_3 \\ u_4 \end{bmatrix} = [k_1]\{r\} \qquad (a)$$

This conveys exactly the same information as Equation 5.2.10 but is specialized to element 1.

For element 2

$$\{p_2\} = \begin{bmatrix} 0 \\ p_{2,2} \\ p_{2,3} \\ 0 \end{bmatrix} = \frac{EA}{l} \begin{bmatrix} 0 & 0 & 0 & 0 \\ 0 & 1 & -1 & 0 \\ 0 & -1 & 1 & 0 \\ 0 & 0 & 0 & 0 \end{bmatrix} \begin{bmatrix} u_1 \\ u_2 \\ u_3 \\ u_4 \end{bmatrix} = [k_2]\{r\} \qquad (b)$$

For element 3

$$\{p_3\} = \begin{bmatrix} 0 \\ 0 \\ p_{3,3} \\ p_{3,4} \end{bmatrix} = \frac{EA}{l} \begin{bmatrix} 0 & 0 & 0 & 0 \\ 0 & 0 & 0 & 0 \\ 0 & 0 & 1 & -1 \\ 0 & 0 & -1 & 1 \end{bmatrix} \begin{bmatrix} u_1 \\ u_2 \\ u_3 \\ u_4 \end{bmatrix} = [k_3]\{r\} \qquad (c)$$

The assembled global matrix equations for the entire bar are found by adding Equations (a), (b), and (c).

$$\{F\} = \begin{bmatrix} F_1 \\ F_2 \\ F_3 \\ F_4 \end{bmatrix} = \begin{bmatrix} p_{1,1} \\ p_{1,2} + p_{2,2} \\ p_{2,3} + p_{3,3} \\ p_{3,4} \end{bmatrix} = \frac{EA}{l} \begin{bmatrix} 1 & -1 & 0 & 0 \\ -1 & 2 & -1 & 0 \\ 0 & -1 & 2 & -1 \\ 0 & 0 & -1 & 1 \end{bmatrix} \begin{bmatrix} u_1 \\ u_2 \\ u_3 \\ u_4 \end{bmatrix} = [K]\{r\} \qquad (d)$$

where $[K]$ is the global stiffness matrix for the entire structure.

Let us now introduce the applied loads, restraints, and restraint forces. The restraint is zero displacement at node 1 and applied loads are concentrated forces at nodes 2, 3, and 4.

The global equations become

$$
\begin{bmatrix} R_1 \\ \hline F \\ -2F \\ 3F \end{bmatrix} = \frac{EA}{l} \left[\begin{array}{c:ccc} 1 & -1 & 0 & 0 \\ \hdashline -1 & 2 & -1 & 0 \\ 0 & -1 & 2 & -1 \\ 0 & 0 & -1 & 1 \end{array} \right] \begin{bmatrix} 0 \\ \hline u_2 \\ u_3 \\ u_4 \end{bmatrix}
\tag{e}
$$

Partitioning leaves us with three equations for finding the three unknown displacements.

$$
\begin{bmatrix} F \\ -2F \\ 3F \end{bmatrix} = \frac{EA}{l} \begin{bmatrix} 2 & -1 & 0 \\ -1 & 2 & -1 \\ 0 & -1 & 1 \end{bmatrix} \begin{bmatrix} u_2 \\ u_3 \\ u_4 \end{bmatrix}
\tag{f}
$$

Once this set of equation is solved we can then find the restraint force from the other partitioned equation which can now be written as

$$
[R_1] = \frac{EA}{l} \begin{bmatrix} -1 & 0 & 0 \end{bmatrix} \begin{bmatrix} u_2 \\ u_3 \\ u_4 \end{bmatrix}
\tag{g}
$$

Solving Equation (f) by the usual substitution methods of elementary algebra is not difficult but this might be a good time to get familiar with one or more of the numerical and symbolic equation solving software packages. As our matrices grow larger through the use of many more elements it will be necessary to turn to these tools. Instructions for using Mathematica are given in Appendix C.

By factoring the F from the loading matrix we can rewrite Equation (f) as

$$
\frac{Fl}{EA} \begin{bmatrix} 1 \\ -2 \\ 3 \end{bmatrix} = \begin{bmatrix} 2 & -1 & 0 \\ -1 & 2 & -1 \\ 0 & -1 & 1 \end{bmatrix} \begin{bmatrix} u_2 \\ u_3 \\ u_4 \end{bmatrix}
\tag{h}
$$

Using Mathematica we obtain the solutions

$$
\begin{bmatrix} u_2 \\ u_3 \\ u_4 \end{bmatrix} = \frac{Fl}{EA} \begin{bmatrix} 2 \\ 3 \\ 6 \end{bmatrix} \qquad R_1 = \frac{EA}{l} \begin{bmatrix} -1 & 0 & 0 \end{bmatrix} \begin{bmatrix} u_2 \\ u_3 \\ u_4 \end{bmatrix} = -2F
\tag{i}
$$

Notice that a free body diagram of the entire bar would have provided $R_1 = -2F$. So this checks out. Given the nodal displacements the distributed displacements are linear functions found from

$$
u_1(s_1) = \begin{bmatrix} 1 - \dfrac{s_1}{l} & \dfrac{s_1}{l} \end{bmatrix} \begin{bmatrix} u_1 \\ u_2 \end{bmatrix}
$$

$$
u_2(s_2) = \begin{bmatrix} 1 - \dfrac{s_2}{l} & \dfrac{s_2}{l} \end{bmatrix} \begin{bmatrix} u_2 \\ u_3 \end{bmatrix}
\tag{j}
$$

$$
u_3(s_3) = \begin{bmatrix} 1 - \dfrac{s_3}{l} & \dfrac{s_3}{l} \end{bmatrix} \begin{bmatrix} u_3 \\ u_4 \end{bmatrix}
$$

As noted before we can simply draw straight lines between nodal displacements in a plot. This is shown in Figure (c).

Given the nodal displacements the stresses can be found for each element from Equations 5.2.5. These have the same sign convention as the classical analysis.

$$\sigma_{x1} = \frac{E}{l}\begin{bmatrix} -1 & 1 \end{bmatrix}\begin{bmatrix} u_1 \\ u_2 \end{bmatrix} = \frac{E}{l}\begin{bmatrix} -1 & 1 \end{bmatrix}\begin{bmatrix} 0 \\ \dfrac{2Fl}{EA} \end{bmatrix} = \frac{2F}{A}$$

$$\sigma_{x2} = \frac{E}{l}\begin{bmatrix} -1 & 1 \end{bmatrix}\begin{bmatrix} u_2 \\ u_3 \end{bmatrix} = \frac{E}{l}\begin{bmatrix} -1 & 1 \end{bmatrix}\begin{bmatrix} \dfrac{2Fl}{EA} \\ \dfrac{3Fl}{EA} \end{bmatrix} = \frac{F}{A} \qquad \text{(k)}$$

$$\sigma_{x3} = \frac{E}{l}\begin{bmatrix} -1 & 1 \end{bmatrix}\begin{bmatrix} u_3 \\ u_4 \end{bmatrix} = \frac{E}{l}\begin{bmatrix} -1 & 1 \end{bmatrix}\begin{bmatrix} \dfrac{3Fl}{EA} \\ \dfrac{6Fl}{EA} \end{bmatrix} = \frac{3F}{A}$$

And if you want the internal forces they are simply the stresses times the area.
These are exactly the same values obtained in Example 4.4.5. The stresses are plotted in Figure (c).

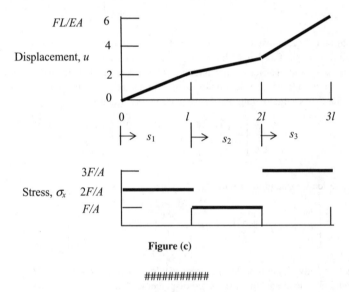

Figure (c)

##########

As we noted three equations in three unknowns can be solved analytically; however, as the number of concentrated loads and hence the number of elements increases an analytical solution becomes tedious, although there are symbolic solver programs available that will ease the burden. It is normal to convert to numerical values even before assembling the equations and to use numerical linear equation solving software to provide answers. In fact, numerical solutions are common even for small problems.

The global stiffness matrix depends only on the geometry and material properties of the bar. It is independent of a particular loading or restraints. The next example defines different loads and restraints but uses the same global stiffness matrix [K] that was used in Example 5.3.2.

############

Example 5.3.3

Problem: A uniform slender bar has loading and restraints as shown in Figure (a). Find the displacements and stresses.

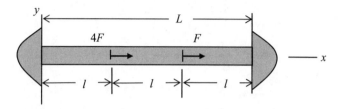

Figure (a)

Solution: Modify the global set of matrix equations in Example 5.3.2 and solve.

The assembled global stiffness matrix is the same as in Example 5.3.2. When you add the loads, restraints and restraint forces for this problem the global set of equations becomes

$$
\begin{bmatrix} R_1 \\ 4F \\ F \\ R_4 \end{bmatrix} = \frac{EA}{l} \begin{bmatrix} 1 & -1 & 0 & 0 \\ -1 & 2 & -1 & 0 \\ 0 & -1 & 2 & -1 \\ 0 & 0 & -1 & 1 \end{bmatrix} \begin{bmatrix} 0 \\ u_2 \\ u_3 \\ 0 \end{bmatrix}
\tag{a}
$$

The partitioned equations are

$$
\begin{bmatrix} 4F \\ F \end{bmatrix} = \frac{EA}{l} \begin{bmatrix} 2 & -1 \\ -1 & 2 \end{bmatrix} \begin{bmatrix} u_2 \\ u_3 \end{bmatrix}
\qquad
\begin{bmatrix} R_1 \\ R_4 \end{bmatrix} = \frac{EA}{l} \begin{bmatrix} -1 & 0 \\ 0 & -1 \end{bmatrix} \begin{bmatrix} u_2 \\ u_3 \end{bmatrix}
\tag{b}
$$

and the solutions are

$$
\begin{bmatrix} u_2 \\ u_3 \end{bmatrix} = \frac{Fl}{EA} \begin{bmatrix} 3 \\ 2 \end{bmatrix}
\qquad
\begin{bmatrix} R_1 \\ R_2 \end{bmatrix} = F \begin{bmatrix} -3 \\ -2 \end{bmatrix}
\tag{c}
$$

The stresses are

$$
\sigma_{x,1} = \frac{E}{l} \begin{bmatrix} -1 & 1 \end{bmatrix} \begin{bmatrix} 0 \\ u_2 \end{bmatrix} = \frac{E}{l} \begin{bmatrix} -1 & 1 \end{bmatrix} \begin{bmatrix} 0 \\ \dfrac{3Fl}{EA} \end{bmatrix} = \frac{3F}{A}
$$

$$
\sigma_{x,2} = \frac{E}{l} \begin{bmatrix} -1 & 1 \end{bmatrix} \begin{bmatrix} u_2 \\ u_3 \end{bmatrix} = \frac{E}{l} \begin{bmatrix} -1 & 1 \end{bmatrix} \begin{bmatrix} \dfrac{3Fl}{EA} \\ \dfrac{2Fl}{EA} \end{bmatrix} = \frac{-F}{A}
\tag{d}
$$

$$
\sigma_{x,3} = \frac{E}{l} \begin{bmatrix} -1 & 1 \end{bmatrix} \begin{bmatrix} u_3 \\ 0 \end{bmatrix} = \frac{E}{l} \begin{bmatrix} -1 & 1 \end{bmatrix} \begin{bmatrix} \dfrac{2Fl}{EA} \\ 0 \end{bmatrix} = \frac{-2F}{A}
$$

Given the nodal displacements the distributed displacements are linear functions between nodes. The stresses are constants between nodes. These results are plotted in Figure (b)

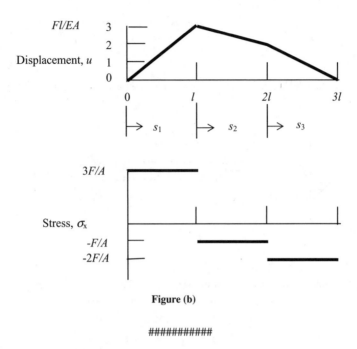

Figure (b)

###########

5.4   A General Method—Distributed Applied Loads

To allow the method just described to work with distributed applied loads we must replace them with a set of concentrated loads that are in some sense statically equivalent. Replacing distributed loads with equivalent concentrated loads has been used in the past with the analytical methods discussed in Chapter 4 but has been restricted to a small number of concentrated loads because of the number of linear algebraic equations generated in the process of matching boundary conditions. With the advent of the digital computer this restriction is lifted. If a sufficiently large number of equivalent concentrated loads are used the answer closely approximates the exact solution based on distributed loads.

One method of creating equivalent concentrated loads can be explained by the process shown in the next three figures. Consider the axially loaded bar in Figure 5.4.1 with a plot of the distributed axial load shown in Figure 5.4.2. We divide the bar into elements and the distributed load into segments as shown in Figure 5.4.2.

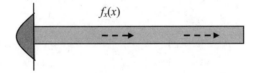

Figure 5.4.1

The area of each rectangular block can be the actual area under the corresponding segment of the curve or can be approximated by taking the value of the distributed force at the mid point of the segment times the length of the segment. We take that area under each block, which is the total force for that segment, and put half of it at each side of the segment as shown in Figure 5.4.3. These form a set of concentrated loads that is closely statically equivalent to the distributed load. The double arrows indicate

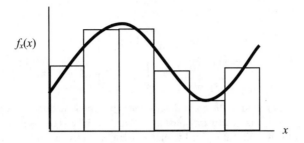

Figure 5.4.2

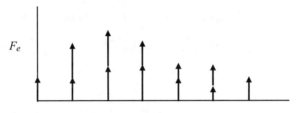

Figure 5.4.3

that the particular concentrated load comes from half the block on either side. The blocks need not be of equal width.

If we designate the equivalent concentrated forces for each element as

$$\{F_{em}\} = \begin{bmatrix} F_{em,n} \\ F_{em,n+1} \end{bmatrix} \tag{5.4.1}$$

Then the total equivalent forces are

$$\{F_e\} = \sum \{F_{em}\} \tag{5.4.2}$$

The element equivalent forces must be embedded in a global format for summing.

In the early days the nodal loads were calculated by the method described in the above figures. Later a formal method of calculating them was used that is presented in Equation 5.4.3 for, say, element m.

$$\{F_{em}\} = \int_0^{l_m} [n]^T f_x(s_m)\, ds_m \tag{5.4.3}$$

In Equation 5.4.3 the matrix $[n]^T$ is the transpose of $[n]$. Since $[n]$ is a row matrix, $[n]^T$ is a column matrix. The coordinate s_m is a local coordinate with origin at node n. These are called *kinematically equivalent nodal loads* and are identified with the subscript e.

We shall derive Equation 5.4.3 in Chapter 11 using the principle of virtual work. The kinematically equivalent nodal loads are based on the principal that there is work equivalence between real applied loads acting over the real displacements and the nodal equivalent loads acting over the nodal displacements. Virtual work is an alternative statement of equilibrium that is equivalent to Newton's laws in a "weak" or distributed sense for continuous structures.

For the axial case for element m

$$\{F_{em}\} = \int_0^{l_m} [n]^T f_x(s_m)\, ds_m = \int_0^{l_m} \begin{bmatrix} 1 - \dfrac{s_m}{l_m} \\[2mm] \dfrac{s_m}{l_m} \end{bmatrix} f_x(s_m)\, ds_m = \begin{bmatrix} F_{em,n} \\ F_{em,n+1} \end{bmatrix} \tag{5.4.4}$$

Our gallery of quantities used in the analysis of slender bars now reads as shown in Equation 5.4.5. Everything from $\{p_m\}$ to $\{r_m\}$ is exactly the same as in Equation 5.2.8. The distributed load is converted into equivalent nodal loads which are added to any concentrated loads to obtain the total loading. The restraints ρ are also added.

In summary

$$\{F\} \qquad \{F_c\} \qquad f_x(s) \;\rightarrow\; \{p_m\} \qquad \sigma_{xm} \qquad \varepsilon_{xm} \qquad u(s_m) \qquad \{r_m\} \;\leftarrow\; \{\rho\}$$

$$\{F_{em}\} = \int_0^{l_m} [n]^T f_x(s_m)\, ds_m \qquad \{F_e\} = \sum \{F_{em}\} \qquad u(s_m) = [n]\{r_m\}$$

$$\{F\} = \{F_c\} + \{F_e\} \qquad\qquad\qquad \varepsilon_{xm} = \frac{du_m}{ds_m} = \frac{d}{ds_m}[n]\{r_m\}$$

$$\sigma_{xm} = E_m \varepsilon_{xm} = E_m \frac{d}{ds_m}[n]\{r_m\} \tag{5.4.5}$$

$$p_{m,n} = -A_m \sigma_{xm} = -E_m A_m \frac{d}{ds_m}[n]\{r_m\}$$

$$\rightarrow \quad \{p_m\} = \begin{bmatrix} p_{m,n} \\ p_{m,n+1} \end{bmatrix} = [k_m]\{r_m\}$$

$$p_{m,n+1} = A_m \sigma_{xm} = E_m A_m \frac{d}{ds_m}[n]\{r_m\}$$

$$\{F\} = \sum \{p_m\} = \sum [k_m]\{r\} = [K]\{r\}$$

We shall illustrate the two methods for obtaining equivalent nodal loads in Example 5.4.1.

##########

Example 5.4.1

Problem: An axially loaded bar has a uniform distributed applied load f_0 with units of force per unit length (N/mm). The bar is divided into six equal elements as shown in Figure (a). Find the equivalent nodal loads.

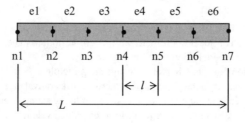

Figure (a)

Solution: Find the equivalent nodal loads by both methods.

Using the first method the load on each element is

$$f_0 l_m = \frac{f_0 L}{6} \tag{a}$$

and so the equivalent concentrated load on each element is found by putting half of this load on each node.

$$\{F_{em}\} = \begin{bmatrix} \dfrac{f_0 L}{12} \\[2ex] \dfrac{f_0 L}{12} \end{bmatrix} \tag{b}$$

The total equivalent load in this case is

$$\{F_e\} = \begin{bmatrix} \dfrac{f_0 L}{12} \\[2ex] \dfrac{f_0 L}{12} + \dfrac{f_0 L}{12} \\[2ex] \dfrac{f_0 L}{12} + \dfrac{f_0 L}{12} \\[2ex] \dfrac{f_0 L}{12} + \dfrac{f_0 L}{12} \\[2ex] \dfrac{f_0 L}{12} + \dfrac{f_0 L}{12} \\[2ex] \dfrac{f_0 L}{12} + \dfrac{f_0 L}{12} \\[2ex] \dfrac{f_0 L}{12} \end{bmatrix} = \begin{bmatrix} \dfrac{f_0 L}{12} \\[2ex] \dfrac{f_0 L}{6} \\[2ex] \dfrac{f_0 L}{6} \\[2ex] \dfrac{f_0 L}{6} \\[2ex] \dfrac{f_0 L}{6} \\[2ex] \dfrac{f_0 L}{6} \\[2ex] \dfrac{f_0 L}{12} \end{bmatrix} \tag{c}$$

Using the second method for each element

$$\{F_{em}\} = \int_0^{l_m} \begin{bmatrix} 1 - \dfrac{s_m}{l_m} \\[2ex] \dfrac{s_m}{l_m} \end{bmatrix} f_0 \, ds_m = f_0 \begin{bmatrix} \int_0^{l_m} \left(1 - \dfrac{s_m}{l_m}\right) ds_m \\[2ex] \int_0^{l_m} \left(\dfrac{s_m}{l_m}\right) ds_m \end{bmatrix} = \begin{bmatrix} \dfrac{f_0 l_m}{2} \\[2ex] \dfrac{f_0 l_m}{2} \end{bmatrix} = \begin{bmatrix} \dfrac{f_0 L}{12} \\[2ex] \dfrac{f_0 L}{12} \end{bmatrix} \tag{d}$$

and so the total is the same as Equation (c).

Note that this set of equivalent nodal loads applies to all boundary conditions.

For more complex distributions the two methods may differ a bit with the second one the more accurate. As the elements increase in number and are, therefore, shorter in length the accuracy of the displacements and stresses improves. We shall demonstrate just how accurate in examples coming up.

###########

Now let us test the accuracy of using equivalent nodal loads with a numerical example that we can compare to an exact analytical solution using a distributed load.

############

Example 5.4.2

Problem: Consider the bar in Example 4.4.1 as shown in Figure (a). Find the displacement and stress. Compare with the traditional analytical method.

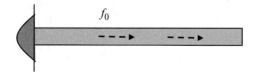

Figure (a)

Solution: Find the equivalent nodal loads; assemble the global matrix equations; introduce the loads, restraints, and restraint forces and solve. Compare with the solution in Example 4.4.1.

Suppose we start with a very crude approximation of only two elements as in Figure (b).

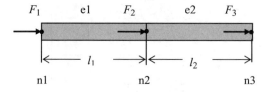

Figure (b)

Let us assume the $l_1 = l_2 = L/2$ and the E and A are the same for both elements. The element stiffness matrix for element 1 is

$$\begin{bmatrix} p_{1,1} \\ p_{1,2} \end{bmatrix} = \frac{2EA}{L} \begin{bmatrix} 1 & -1 \\ -1 & 1 \end{bmatrix} \begin{bmatrix} u_1 \\ u_2 \end{bmatrix} = [k_1] \{r_1\} \tag{a}$$

and for element 2

$$\begin{bmatrix} p_{2,2} \\ p_{2,3} \end{bmatrix} = \frac{2EA}{L} \begin{bmatrix} 1 & -1 \\ -1 & 1 \end{bmatrix} \begin{bmatrix} u_2 \\ u_3 \end{bmatrix} = [k_2] \{r_2\} \tag{b}$$

When these are imbedded in the global format the assembled global matrix equations are

$$\{F\} = \frac{2EA}{L} \begin{bmatrix} 1 & -1 & 0 \\ -1 & 2 & -1 \\ 0 & -1 & 1 \end{bmatrix} \begin{bmatrix} u_1 \\ u_2 \\ u_3 \end{bmatrix} = [K] \{r\} \tag{c}$$

Let us assign values for the equivalent nodal loads. By either method the total load on each element is $f_0 L/2$. Place half of this at each end, that is, $f_0 L/4$. At node 1 there is also a reaction force R_1 that we must account for. Also at node 1 the displacement is zero. Our matrix equations become

$$\left\{ \begin{array}{c} R_1 + \dfrac{f_0 L}{4} \\[2mm] \dfrac{f_0 L}{4} + \dfrac{f_0 L}{4} \\[2mm] \dfrac{f_0 L}{4} \end{array} \right\} = \dfrac{2EA}{L} \begin{bmatrix} 1 & -1 & 0 \\ -1 & 2 & -1 \\ 0 & -1 & 1 \end{bmatrix} \begin{bmatrix} 0 \\ u_2 \\ u_3 \end{bmatrix} = [K]\{r\} \tag{d}$$

Note that at node 1 there is both an applied load and a restraint force.
We partition the equations as shown in Equation (e).

$$\left\{ \begin{array}{c} R_1 + \dfrac{f_0 L}{4} \\[1mm] \hdashline \\[-2mm] \dfrac{f_0 L}{4} + \dfrac{f_0 L}{4} \\[2mm] \dfrac{f_0 L}{4} \end{array} \right\} = \dfrac{2EA}{L} \left[\begin{array}{c:cc} 1 & -1 & 0 \\ \hdashline -1 & 2 & -1 \\ 0 & -1 & 1 \end{array} \right] \begin{bmatrix} 0 \\ \hdashline u_2 \\ u_3 \end{bmatrix} = [K]\{r\} \tag{e}$$

We obtain two matrix equations

$$\begin{bmatrix} \dfrac{f_0 L}{2} \\[2mm] \dfrac{f_0 L}{4} \end{bmatrix} = \dfrac{2EA}{L} \begin{bmatrix} 2 & -1 \\ -1 & 1 \end{bmatrix} \begin{bmatrix} u_2 \\ u_3 \end{bmatrix} \tag{f}$$

and

$$R_1 + \dfrac{f_0 L}{4} = \dfrac{2EA}{L} \begin{bmatrix} -1 & 0 \end{bmatrix} \begin{bmatrix} u_2 \\ u_3 \end{bmatrix} = -\dfrac{2EA}{L} u_2 \tag{g}$$

The solutions are

$$\begin{bmatrix} u_2 \\ u_3 \end{bmatrix} = \begin{bmatrix} \dfrac{3 f_0 L^2}{8EA} \\[3mm] \dfrac{f_0 L^2}{2EA} \end{bmatrix} \qquad R_1 = -f_0 L \tag{h}$$

How well does this compare with the exact answer? From Example 4.4.1, Equation (f) we get

$$u(x) = \dfrac{f_0}{EA}\left(Lx - \dfrac{x^2}{2} \right) \quad \rightarrow \quad u\left(\dfrac{L}{2}\right) = \dfrac{3 f_0 L^2}{8EA} \quad u(L) = \dfrac{f_0 L^2}{2EA} \tag{i}$$

At 0, $L/2$, and L the answers are the same. From a plot of the displacements in Figure (c) we see the difference.

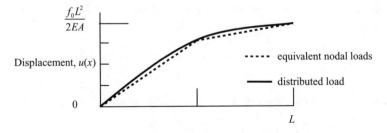

Figure (c)

The displacement between points of concentrated load application is always a linear function. It is a quadratic function in the exact case; however, the approximate curve is not far off.

For the stresses from Example 4.4.1, Equation (b)

$$\sigma_x(x) = \frac{f_0}{A}(L - x) \quad \rightarrow \quad \sigma_x(0) = \frac{f_0 L}{A} \quad \sigma_x\left(\frac{L}{2}\right) = \frac{f_0 L}{2A} \quad \sigma_x(L) = 0 \tag{j}$$

The stresses from equivalent nodal loads are

$$\sigma_{x1} = \frac{E}{l}[-1 \quad 1]\begin{bmatrix} 0 \\ u_2 \end{bmatrix} = \frac{2E}{L}[-1 \quad 1]\begin{bmatrix} 0 \\ \dfrac{3f_0 L^2}{8EA} \end{bmatrix} = \frac{3f_0 L}{4A}$$

$$\sigma_{x2} = \frac{E}{l}[-1 \quad 1]\begin{bmatrix} u_2 \\ u_3 \end{bmatrix} = \frac{2E}{L}[-1 \quad 1]\begin{bmatrix} \dfrac{3f_0 L^2}{8EA} \\ \dfrac{f_0 L^2}{2EA} \end{bmatrix} = \frac{f_0 L}{4A} \tag{k}$$

When using equivalent nodal loads the stress is constant in each element as represented by the dotted lines in Figure (d) while the exact answer for a distributed load shows a linearly varying stress represented by the solid line.

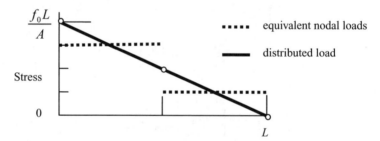

Figure (d)

The first reaction is that this is not a very good way to solve for the stresses. But note that the average stress at node 2 using equivalent nodal loads is surprisingly accurate. Compare that average to the exact value

$$\sigma_{ave,2} = \frac{1}{2}\left(\frac{3}{4}\frac{f_0 L}{A} + \frac{1}{4}\frac{f_0 L}{A}\right) = \frac{f_0 L}{2A} \quad \sigma_x\left(\frac{L}{2}\right) = \frac{f_0}{A}\left(L - \frac{L}{2}\right) = \frac{f_0 L}{2A} \tag{l}$$

and they agree exactly. Also let us use the stress at the support based on the support reaction (see Equation (h)) to get

$$P(0) = -R_1 = f_0 L \quad \rightarrow \quad \sigma_x(0) = \frac{P(0)}{A} = \frac{f_0 L}{A} \tag{m}$$

Likewise use the known value at the right end where there is no load to obtain

$$P(L) = 0 \quad \rightarrow \quad \sigma_x(L) = \frac{P(L)}{A} = \frac{0}{A} = 0 \tag{n}$$

and place these values on the graph repesented by small circles. A curve fitted through these values at the ends and the nodal average value at node 2 is the same as the exact answer.

Clearly the stress between nodes is not accurate. But the stress is accurate at the nodes if you use the loads at the boundary to find the stress there and average the stress at interior nodes. A curve fitted to these values coincides with the analytical answer in this case and is very close in more complicated cases of distributed applied loads. This is common practice.

Now let us double the number of elements to four and compare the answers. The process of assembling the matrix equations is just the same. You get

$$
\{F\} =
\begin{bmatrix}
R_1 + \dfrac{f_0 L}{8} \\[2mm]
\dfrac{f_0 L}{4} \\[2mm]
\dfrac{f_0 L}{4} \\[2mm]
\dfrac{f_0 L}{4} \\[2mm]
\dfrac{f_0 L}{8}
\end{bmatrix}
=
\dfrac{4EA}{L}
\begin{bmatrix}
1 & -1 & 0 & 0 & 0 \\
-1 & 2 & -1 & 0 & 0 \\
0 & -1 & 2 & -1 & 0 \\
0 & 0 & -1 & 2 & -1 \\
0 & 0 & 0 & -1 & 1
\end{bmatrix}
\begin{bmatrix}
u_1 \\ u_2 \\ u_3 \\ u_4 \\ u_5
\end{bmatrix}
\tag{o}
$$

Let us assign some numerical values. Let the material be aluminum for which $E = 68950 \, N/mm^2$. The length is $L = 200$ and the cross section area is $4 \, mm^2$. The load is $f_0 = 2.5 \, N/mm$. Numerical values are inserted and the partitions are selected.

$$
\begin{bmatrix}
R_1 + 62.5 \\ \hline
125 \\
125 \\
125 \\
62.5
\end{bmatrix}
= 5516
\begin{bmatrix}
1 & -1 & 0 & 0 & 0 \\ \hline
-1 & 2 & -1 & 0 & 0 \\
0 & -1 & 2 & -1 & 0 \\
0 & 0 & -1 & 2 & -1 \\
0 & 0 & 0 & -1 & 1
\end{bmatrix}
\begin{bmatrix}
0 \\ \hline
u_2 \\ u_3 \\ u_4 \\ u_5
\end{bmatrix}
\tag{p}
$$

The partitoned matrix equations are

$$
\begin{bmatrix}
125 \\ 125 \\ 125 \\ 62.5
\end{bmatrix}
= 5516
\begin{bmatrix}
2 & -1 & 0 & 0 \\
-1 & 2 & -1 & 0 \\
0 & -1 & 2 & -1 \\
0 & 0 & -1 & 1
\end{bmatrix}
\begin{bmatrix}
u_2 \\ u_3 \\ u_4 \\ u_5
\end{bmatrix}
\qquad R_1 + 62.5 = -5516u_2
\tag{q}
$$

This is easily solved for the unknown nodal displacements and the restraint force using one of the software solvers or with your calculator.

$$
\{r\} =
\begin{bmatrix}
u_2 \\ u_3 \\ u_4 \\ u_5
\end{bmatrix}
=
\begin{bmatrix}
0.0793 \\ 0.1360 \\ 0.1700 \\ 0.1813
\end{bmatrix}
\qquad R_1 = -499.92
\tag{r}
$$

This compares to the actual restraint force of -500.

Once again use the restraint force at node 1 and the actual applied load at node 5 to find the stresses there.

$$\sigma_{x1} = \frac{500}{4} = 125 \, N/mm^2 \qquad \sigma_{x5} = 0 \tag{s}$$

The stresses at node 2 are

$$(\sigma_x)_{1,2} = \frac{4 \bullet 68950}{200} \begin{bmatrix} -1 & 1 \end{bmatrix} \begin{bmatrix} 0.0 \\ 0.0793 \end{bmatrix} = 109.35 \, N/mm^2$$

$$(\sigma_x)_{2,3} = \frac{4 \bullet 68950}{200} \begin{bmatrix} -1 & 1 \end{bmatrix} \begin{bmatrix} 0.0793 \\ 0.1360 \end{bmatrix} = 78.19 \, N/mm^2 \tag{t}$$

The average stress at node 2 is

$$\sigma_{ave,2} = \frac{109.35 + 78.19}{2} = 93.77 \, N/mm^2 \tag{u}$$

The analytical value is 93.75. The displacement and stress values at all the nodes are given in Table 5.4.1. Just for comparison we added the results from an 8 element equivalent nodal load solution. The displacement values are given in Table 5.4.1 for 2, 4, and 8 elements and for the exact analytical solution at the node locations.

Table 5.4.1

Value of x	0	25	50	75	100	125	150	175	200
Displ. 2 elem.	0.0000				0.1360				0.1813
Displ. 4 elem.	0.0000		0.0793		0.1360		0.1700		0.1813
Displ. 8 elem.	0.0000	0.0425	0.0793	0.1105	0.1360	0.1558	0.1700	0.1785	0.1813
Displ. analytic	0.0000	0.0425	0.0793	0.1105	0.1360	0.1558	0.1700	0.1785	0.1813
Stress 2 elem.	125.022				62.503				0.000
Stress 4 elem.	124.980		93.770		62.538		31.234		0.000
Stress 8 elem.	125.000	109.355	93.772	78.189	62.469	46.886	31.303	15.583	0.000
Stress analytic	125.000	109.375	93.750	78.125	62.500	46.875	31.250	15.625	0.000

It is customary to draw a smooth curve through the nodal values for displacement and for average stress in presenting this data. Within plotting accuracy the curves fall right on top of each other when several elements are used.

Example 5.4.3

Problem: Repeat Example 4.5.3 by the general method.

$$u(L) = -\delta \tag{a}$$

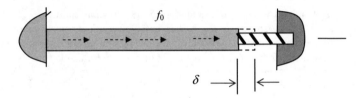

Figure (a)

Solution: Borrow the equivalent nodal loads and global stiffness matrix from Example 5.4.2. Insert the new displacement constraints and restraint forces, partition, and solve.

Using just two elements the partitioned global equations are

$$
\begin{Bmatrix} R_1 + \dfrac{f_0 L}{4} \\ \hdashline \dfrac{f_0 L}{4} + \dfrac{f_0 L}{4} \\ \hdashline R_3 + \dfrac{f_0 L}{4} \end{Bmatrix} = \frac{2EA}{L} \left[\begin{array}{c:cc} 1 & -1 & 0 \\ \hdashline -1 & 2 & -1 \\ \hdashline 0 & -1 & 1 \end{array} \right] \begin{bmatrix} 0 \\ \hdashline u_2 \\ \hdashline -\delta \end{bmatrix}
\tag{a}
$$

The partitioned equation for displacement and the solution is given in Equation (b).

$$
\frac{f_0 L}{2} = \frac{2EA}{L}(2u_2 + \delta) \quad \rightarrow \quad u_2 = \frac{f_0 L^2}{8EA} - \frac{\delta}{2}
\tag{b}
$$

This agrees with the analytical solution in Example 4.5.3, Equation (e) at $x = L/2$.
The partitioned equations for the restraints are

$$
\begin{bmatrix} R_1 + \dfrac{f_0 L}{4} \\ R_3 + \dfrac{f_0 L}{4} \end{bmatrix} = \frac{2EA}{L} \begin{bmatrix} -1 & 0 \\ -1 & 1 \end{bmatrix} \begin{bmatrix} u_2 \\ -\delta \end{bmatrix}
\tag{c}
$$

and the values of the restraint forces are

$$
R_1 = -\frac{f_0 L}{2} + EA\frac{\delta}{L} \qquad R_3 = -\frac{f_0 L}{2} - EA\frac{\delta}{L}
\tag{d}
$$

These also agree with the values found in Example 4.5.3. The stresses are

$$
\sigma_{x1} = \frac{2E}{L} \begin{bmatrix} -1 & 1 \end{bmatrix} \begin{bmatrix} 0 \\ u_2 \end{bmatrix} = \frac{2E}{L} \begin{bmatrix} -1 & 1 \end{bmatrix} \begin{bmatrix} 0 \\ \dfrac{f_0 L^2}{8EA} - \dfrac{\delta}{2} \end{bmatrix} = \frac{f_0 L}{4A} - E\frac{\delta}{L}
$$

$$
\sigma_{x2} = \frac{2E}{L} \begin{bmatrix} -1 & 1 \end{bmatrix} \begin{bmatrix} u_2 \\ -\delta \end{bmatrix} = \frac{2E}{L} \begin{bmatrix} -1 & 1 \end{bmatrix} \begin{bmatrix} \dfrac{f_0 L^2}{8EA} - \dfrac{\delta}{2} \\ -\delta \end{bmatrix} = -\frac{f_0 L}{4A} - E\frac{\delta}{L}
\tag{e}
$$

In the following plots the answers from Example 4.5.3 are shown as solid lines. The values found here are dotted lines. Circles are shown for the stresses based on restraint forces and the average value at node 2. Once again the curve fitted to the nodal values for stresses agrees closely with the analytical solution. We should use more elements to get a better approximation for displacements.

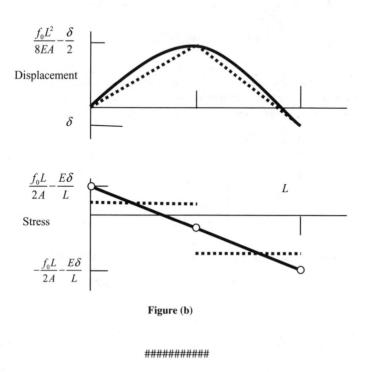

Figure (b)

############

Example 5.4.4

Problem: Find the displacements, internal axial forces, and stresses by the general method using equivalent nodal loads for the bar shown in Figure (a). Use two elements. Compare with the exact analytical solution.

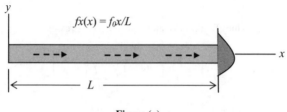

Figure (a)

Solution: The bar is divided into two elements and the nodes and elements are numbered exactly as in the previous three examples. The global stiffness matrix will be the same. The equivalent nodal loads and restraints will differ. Find the equivalent nodal loads, complete the global equations, partition, and solve.

First find the equivalent nodal loads for element 1.

$$
\begin{bmatrix} F_{e1,1} \\ F_{e1,2} \end{bmatrix} = \int_0^l [n]^T f_x(s_1)\,ds_1 = \int_0^{\frac{L}{2}} \begin{bmatrix} 1 - \dfrac{2s_1}{L} \\[2ex] \dfrac{2s_1}{L} \end{bmatrix} f_0 \dfrac{s_1}{L}\,ds_1
$$

$$
= f_0 \begin{bmatrix} \int_0^{\frac{L}{2}} \left(\dfrac{s_1}{L} - \dfrac{2s_1^2}{L^2} \right) ds_1 \\[3ex] \int_0^{\frac{L}{2}} \left(\dfrac{2s_1^3}{L^2} \right) ds_1 \end{bmatrix} = f_0 \begin{bmatrix} \left(\dfrac{s_1^2}{2L} - \dfrac{2s_1^3}{3L^2} \right) \\[3ex] \dfrac{2s_1^3}{3L^2} \end{bmatrix} \Bigg|_0^{\frac{L}{2}} = \begin{bmatrix} \dfrac{f_0 L}{24} \\[3ex] \dfrac{f_0 L}{12} \end{bmatrix} \qquad \text{(a)}
$$

For element 2 you must make a coordinate transformation. If you use the new coordinate s_2 you must note that $x = s_2 + L/2$ and replace x in the loading as shown in Equation (b).

$$
\begin{bmatrix} F_{e2,2} \\ F_{e2,3} \end{bmatrix} = \int_0^l [n]^T f_x(s_2)\,ds_2 = \int_0^{\frac{L}{2}} \begin{bmatrix} 1 - \dfrac{2s_2}{L} \\[2ex] \dfrac{2s_2}{L} \end{bmatrix} f_0 \dfrac{\left(s_2 + \frac{L}{2}\right)}{L}\,ds_2
$$

$$
= \int_0^{\frac{L}{2}} \begin{bmatrix} 1 - \dfrac{2s_2}{L} \\[2ex] \dfrac{2s_2}{L} \end{bmatrix} f_0 \dfrac{s_2}{L}\,ds_2 + \int_0^{\frac{L}{2}} \begin{bmatrix} 1 - \dfrac{2s_2}{L} \\[2ex] \dfrac{2s_2}{L} \end{bmatrix} \dfrac{f_0}{2}\,ds_2 \qquad \text{(b)}
$$

Note that the first of these integrals is the same as the one in Equation (a). The second is one half that found in Example 5.4.2 since the value of the load is $f_0/2$.

$$
\begin{bmatrix} F_{e2,2} \\ F_{e2,3} \end{bmatrix} = \begin{bmatrix} \dfrac{f_0 L}{24} \\[3ex] \dfrac{f_0 L}{12} \end{bmatrix} + \begin{bmatrix} \dfrac{f_0 L}{8} \\[3ex] \dfrac{f_0 L}{8} \end{bmatrix} = \begin{bmatrix} \dfrac{f_0 L}{6} \\[3ex] \dfrac{5 f_0 L}{24} \end{bmatrix} \qquad \text{(c)}
$$

You could have replaced s_2 with $s_2 = x - L/2$ in the integral in Equation (b) and integrated from $L/2$ to L. That would be a bit more difficult.

Or you can arrive at the values for element 2 another way. The loading is triangular as plotted here.

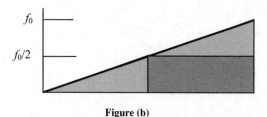

Figure (b)

We have just found the equivalent nodal loads for the triangular load in element 1. Element 2 has an equal triangular load and a uniform load as shown by the darker shaded area. The uniform portion of the load has two equivalent nodal loads each equal to the one half the total load represented by that area. The triangular load contribution to element 2 is the same as for element 1.

The total load from the uniform part is

$$F = \frac{f_0}{2} \cdot \frac{L}{2} \tag{d}$$

So the equivalent nodal load contribution of the uniform part is

$$\begin{bmatrix} \Delta F_{e2,2} \\ \Delta F_{e2,3} \end{bmatrix} = \begin{bmatrix} \dfrac{f_0 L}{8} \\ \dfrac{f_0 L}{8} \end{bmatrix} \tag{e}$$

Add to it the triangular contribution.

$$\begin{bmatrix} F_{e2,2} \\ F_{e2,3} \end{bmatrix} = \begin{bmatrix} \dfrac{f_0 L}{8} \\ \dfrac{f_0 L}{8} \end{bmatrix} + \begin{bmatrix} \dfrac{f_0 L}{24} \\ \dfrac{f_0 L}{12} \end{bmatrix} = \begin{bmatrix} \dfrac{f_0 L}{6} \\ \dfrac{5 f_0 L}{24} \end{bmatrix} \tag{f}$$

The combination of all the equivalent nodal loads is in the assembled equations, Equations (g). The assembled equations are

$$\left\{ \begin{array}{c} \dfrac{f_0 L}{24} \\[2mm] \dfrac{f_0 L}{12} + \dfrac{f_0 L}{6} \\[2mm] R_3 + \dfrac{5 f_0 L}{24} \end{array} \right\} = \frac{2EA}{L} \begin{bmatrix} 1 & -1 & 0 \\ -1 & 2 & -1 \\ 0 & -1 & 1 \end{bmatrix} \begin{bmatrix} u_1 \\ u_2 \\ 0 \end{bmatrix} \tag{g}$$

The partitioned equations and solutions are

$$\begin{bmatrix} \dfrac{f_0 L}{24} \\[2mm] \dfrac{f_0 L}{4} \end{bmatrix} = \frac{2EA}{L} \begin{bmatrix} 1 & -1 \\ -1 & 2 \end{bmatrix} \begin{bmatrix} u_1 \\ u_2 \end{bmatrix} \quad \rightarrow \quad \begin{bmatrix} u_1 \\ u_2 \end{bmatrix} = \begin{bmatrix} \dfrac{f_0 L^2}{6EA} \\[2mm] \dfrac{7 f_0 L^2}{48EA} \end{bmatrix} \tag{h}$$

$$R_3 + \frac{5 f_0 L}{24} = \frac{2EA}{L} \begin{bmatrix} 0 & -1 \end{bmatrix} \begin{bmatrix} u_1 \\ u_2 \end{bmatrix} = \frac{2EA}{L} \begin{bmatrix} 0 & -1 \end{bmatrix} \begin{bmatrix} \dfrac{f_0 L^2}{6EA} \\[2mm] \dfrac{7 f_0 L^2}{48EA} \end{bmatrix} = -\frac{7 f_0 L}{24} \tag{i}$$

$$\rightarrow R_3 = -\frac{7 f_0 L}{24} - \frac{5 f_0 L}{24} = -\frac{12 f_0 L}{24} = -\frac{f_0 L}{2}$$

The internal forces and stresses are

$$P_{1,2} = \frac{2EA}{L}\begin{bmatrix} -1 & 1 \end{bmatrix}\begin{bmatrix} u_1 \\ u_2 \end{bmatrix} = \frac{2EA}{L}\begin{bmatrix} -1 & 1 \end{bmatrix}\begin{bmatrix} \frac{f_0 L^2}{6EA} \\ \frac{7f_0 L^2}{48EA} \end{bmatrix} = -\frac{f_0 L}{24} \quad \rightarrow \quad \sigma_{x1} = -\frac{f_0 L}{24A}$$

(j)

$$P_{2,3} = \frac{2EA}{L}\begin{bmatrix} -1 & 1 \end{bmatrix}\begin{bmatrix} u_2 \\ 0 \end{bmatrix} = \frac{2EA}{L}\begin{bmatrix} -1 & 1 \end{bmatrix}\begin{bmatrix} \frac{7f_0 L^2}{48EA} \\ 0 \end{bmatrix} = -\frac{7f_0 L}{24} \quad \rightarrow \quad \sigma_{x2} = -\frac{7f_0 L}{24A}$$

These displacements and internal forces are plotted in Figure (c) as dotted lines.
The average internal force at node 2 is

$$P_{2ave} = \frac{1}{2}\left(-\frac{f_0 L}{24} - \frac{7f_0 L}{24} \right) = -\frac{f_0 L}{6}$$

(k)

and the forces at each end are

$$P(0) = 0 \qquad P(L) = R_3 = -\frac{f_0 L}{2}$$

(l)

These values of internal forces at the nodes are plotted in Figure (c) as circles and a dashed curve is fitted through them.

The analytical expressions for the internal force and the stress can be found from static equilibrium.

$$\frac{dP}{dx} = -f_x(x) = -f_0\frac{x}{L} \quad \rightarrow \quad P(x) = -\frac{f_0 x^2}{2L} \quad \rightarrow \quad \sigma_x(x) = \frac{P(x)}{A} = -\frac{f_0 x^2}{2LA}$$

(m)

Particular values of the analytical solution for the internal forces at the nodal points are

$$P(0) = 0 \qquad P\left(\frac{L}{2}\right) = -\frac{f_0 L}{8} \qquad P(L) = -\frac{f_0 L}{2}$$

(n)

The displacements are found using the first order differential equation.

$$EA\frac{du}{dx} = P(x) = -\frac{f_0 x^2}{2L} \quad \rightarrow \quad u(x) = -\frac{f_0 x^3}{6EAL} + c$$

$$\rightarrow \quad u(L) = -\frac{f_0 L^2}{6EA} + c = 0 \quad \rightarrow \quad c = \frac{f_0 L^2}{6EA}$$

(o)

$$\rightarrow \quad u(x) = \frac{f_0}{6EAL}\left(L^3 - x^3 \right)$$

Particular values at the nodes are

$$u(0) = \frac{f_0 L^2}{6EA} \qquad u\left(\frac{L}{2}\right) = \frac{7f_0 L^2}{48} \qquad u(L) = 0$$

(p)

The analytical solutions are plotted in Figure (c) as solid lines.

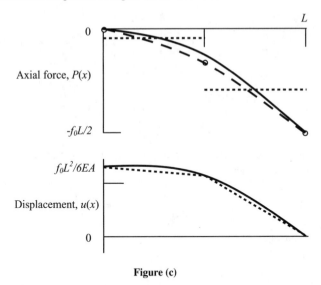

Figure (c)

Using a few more elements would greatly improve the solution.

###########

We have called this a general method because it works for all loadings and all restraints. There need be no distinction between determinate and indeterminate bars. Now we must admit that it has another much better known name. It is called the *finite element method* (FEM) and the process is *finite element analysis* (FEA).

###########

5.5   Variable Cross Sections

To use the elements stiffness matrices that we have developped we must have a uniform *EA* in each element. We shall approximate each element with a variable cross section with one that has a constant average value. An example will demonstrate the validity of this approach.

###########

Example 5.5.1

Problem: Do both Examples 4.6.1 and 4.6.2 using the general method (FEM). The figures for these two problems are repeated here as Figure (a) and Figure (b).

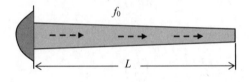

Figure (a)

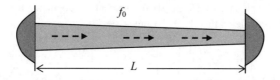

Figure (b)

Solution: Divide the bar into segments of constant cross section. Assemble the global stiffness matrix, find the equivalent nodal loads, assign boundary restraints, partition the equations, and solve.

First we shall show the bar divided into only two elements, as in Figure (c), to show the process. In reality more elements should be used to more closely approximate the exact answer; however, even this crude approximation does quite well for most purposes.

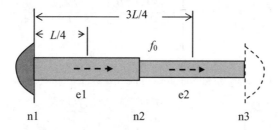

Figure (c)

The support at the right end is free in Example 4.6.1 and fixed in 4.6.2, hence the ghostly image at the right end. Remember that we use the same global stiffness matrix in each case.

The cross sectional area of the tapered beam is

$$A(x) = A_0 \left(1 - \frac{x}{2L}\right) \tag{a}$$

Each element has a constant cross section taking on the value of A at the midpoint of the variable cross section.

$$A_1 = A_0 \left(1 - \frac{\frac{L}{4}}{2L}\right) = \frac{7}{8} A_0 \qquad A_2 = A_0 \left(1 - \frac{\frac{3L}{4}}{2L}\right) = \frac{5}{8} A_0 \tag{b}$$

The loading is divided into concentrated loads for each element at each node. The total load at each node is

$$\text{Node 1:} \frac{f_0 L}{4} \qquad \text{Node 2:} \frac{f_0 L}{4} + \frac{f_0 L}{4} \qquad \text{Node 3:} \frac{f_0 L}{4} \tag{c}$$

We must pause here and point out that two approximations have been introduced: (1) the distributed load is replaced with equivalent nodal loads, and (2) the tapered cross section is replaced with uniform cross sections in each element.

The element stiffness matrix in general is

$$[k_n] = \frac{E_n A_n}{l_n} \begin{bmatrix} 1 & -1 \\ -1 & 1 \end{bmatrix} \tag{d}$$

Let us imbed this in the global stiffness matrix for element 1 recognizing that $A_1 = \frac{7}{8}A_0$ and $l_1 = \frac{L}{2}$ and also for element 2 recognizing that $A_2 = \frac{5}{8}A_0$ and $l_2 = \frac{L}{2}$.

$$[k_1] = \begin{bmatrix} \dfrac{7EA_0}{4L} & -\dfrac{7EA_0}{4L} & 0 \\[2mm] -\dfrac{7EA_0}{4L} & \dfrac{7EA_0}{4L} & 0 \\[2mm] 0 & 0 & 0 \end{bmatrix} \qquad [k_2] = \begin{bmatrix} 0 & 0 & 0 \\[2mm] 0 & \dfrac{5EA_0}{4L} & -\dfrac{5EA_0}{4L} \\[2mm] 0 & -\dfrac{5EA_0}{4L} & \dfrac{5EA_0}{4L} \end{bmatrix} \qquad (e)$$

The complete assembled global stiffness matrix is

$$\{F\} = \begin{bmatrix} F_1 \\ F_2 \\ F_3 \end{bmatrix} = [K]\{r\} = \begin{bmatrix} \dfrac{7EA_0}{4L} & -\dfrac{7EA_0}{4L} & 0 \\[2mm] -\dfrac{7EA_0}{4L} & \dfrac{7EA_0}{4L} + \dfrac{5EA_0}{4L} & -\dfrac{5EA_0}{4L} \\[2mm] 0 & -\dfrac{5EA_0}{4L} & \dfrac{5EA_0}{4L} \end{bmatrix} \begin{bmatrix} u_1 \\ u_2 \\ u_3 \end{bmatrix} \qquad (f)$$

First let us apply the boundary conditions for Figure (b), where both ends are fixed.

$$\{F\} = \begin{bmatrix} R_1 + \dfrac{f_0L}{4} \\[2mm] \hline \dfrac{f_0L}{2} \\[2mm] \hline R_3 + \dfrac{f_0L}{4} \end{bmatrix} = [k]\{r\} = \begin{bmatrix} \dfrac{7EA_0}{4L} & \vdots & -\dfrac{7EA_0}{4L} & \vdots & 0 \\[2mm] \hline -\dfrac{7EA_0}{4L} & \vdots & \dfrac{7EA_0}{4L} + \dfrac{5EA_0}{4L} & \vdots & -\dfrac{5EA_0}{4L} \\[2mm] \hline 0 & \vdots & -\dfrac{5EA_0}{4L} & \vdots & \dfrac{5EA_0}{4L} \end{bmatrix} \begin{bmatrix} 0 \\ \hline u_2 \\ \hline 0 \end{bmatrix} \qquad (g)$$

The partitioned equation to obtain the displacement is

$$\left(\frac{7EA_0}{4L} + \frac{5EA_0}{4L} \right) u_2 = \frac{f_0L}{2} \quad \rightarrow \quad u_2 = \frac{1}{6}\frac{f_0L^2}{EA_0} \qquad (h)$$

and the equations for the restraint forces are

$$R_1 + \frac{f_0L}{4} = -\frac{7EA_0}{4L}u_2 \quad \rightarrow \quad R_1 = -0.5417 f_0L$$

$$R_3 + \frac{f_0L}{4} = -\frac{5EA_0}{4L}u_2 \quad \rightarrow \quad R_3 = -0.4583 f_0L \qquad (i)$$

We can compare these to the exact answers found in Example 4.6.2. The deflection at $x = \dfrac{L}{2}$ is

$$u\left(\frac{L}{2} \right) = 0.1698 \frac{f_0L^2}{EA_0} \qquad (j)$$

The approximate value from the FE method is

$$u_2 = 0.1667 \frac{f_0L^2}{EA_0} \qquad (k)$$

The support reactions are

$$\text{Exact: } P(0) = 0.5572 f_0 L \qquad P(L) = -0.4428 f_0 L$$
$$\text{FEM: } R_1 = -0.5417 f_0 L \qquad R_3 = -0.4583 f_0 L \tag{l}$$

The differences in sign result from the different definition of positive forces between internal forces and support restraints.

The high accuracy for the displacement at the midpoint and the forces at the support when using such a crude model is amazing. The axial forces in the two elements are

$$\begin{bmatrix} p_{1,1} \\ p_{1,2} \end{bmatrix} = \frac{7A_0}{8} \frac{2E}{L} \begin{bmatrix} 1 & -1 \\ -1 & 1 \end{bmatrix} \begin{bmatrix} 0 \\ u_2 \end{bmatrix} = \frac{7A_0}{8} \frac{2E}{L} \begin{bmatrix} 1 & -1 \\ -1 & 1 \end{bmatrix} \begin{bmatrix} 0 \\ \frac{1}{6} \frac{f_0 L^2}{EA_0} \end{bmatrix} = \frac{7}{24} f_0 L \begin{bmatrix} -1 \\ 1 \end{bmatrix}$$

$$\begin{bmatrix} p_{2,2} \\ p_{2,3} \end{bmatrix} = \frac{5A_0}{8} \frac{2E}{L} \begin{bmatrix} 1 & -1 \\ -1 & 1 \end{bmatrix} \begin{bmatrix} u_2 \\ 0 \end{bmatrix} = \frac{5A_0}{8} \frac{2E}{L} \begin{bmatrix} 1 & -1 \\ -1 & 1 \end{bmatrix} \begin{bmatrix} \frac{1}{6} \frac{f_0 L^2}{EA_0} \\ 0 \end{bmatrix} = \frac{5}{24} f_0 L \begin{bmatrix} 1 \\ -1 \end{bmatrix} \tag{m}$$

The average value of P at node 2 is

$$P\left(\frac{L}{2}\right) = \frac{1}{2}(p_{1,2} + p_{2,3}) = \frac{1}{2}\frac{7-5}{24}f_0 L = \frac{1}{24}f_0 L = 0.0417 f_0 L \tag{n}$$

The stresses in each element are based on the area of the element and are uniform throughout the element. Of course the stresses will not be so accurate; however, with a little creative interpretation they are not so bad. If you use the actual area at the supports then the stresses there are reasonably accurate.

$$\text{Exact: } \sigma_x(0) = 0.5572 \frac{f_0 L}{A_0} \qquad \sigma_x(L) = -0.8856 \frac{f_0 L}{A_0}$$

$$\text{FEM: } \sigma_x(0) = 0.5417 \frac{f_0 L}{A_0} \qquad \sigma_x(L) = -0.9166 \frac{f_0 L}{A_0} \tag{o}$$

By the FEM at $x = \frac{L}{2}$ we have two different stress components, one for element one and one for element two. These values are

$$\sigma_{x1} = \frac{8}{7}0.2917\frac{f_0 L}{A_0} = 0.3333\frac{f_0 L}{A_0} \qquad \sigma_{x2} = -\frac{8}{5}0.2083\frac{f_0 L}{A_0} = -0.3333\frac{f_0 L}{A_0} \tag{p}$$

This is terrible; we can't have both a positive and negative stress at the same point with a distributed load. Of course, it is the result of concentrating forces at that point. The exact stress is

$$\sigma_x\left(\frac{L}{2}\right) = 0.0572\frac{f_0 L}{A_0} \tag{q}$$

Once again we find the average between the stresses in elements one and two.

$$\sigma_{12ave} = 0.0 \tag{r}$$

That is not so bad. Since this is a region of low stress the inaccuracy is not much of a problem. Remember that this is a very crude model. Using a few more elements improves accuracy tremendously. Plots are presented in Figure (d) for comparison. The solid lines are from the exact solutions obtained in

Section 4.6, Example 4.6.2. The dotted lines are from the values obtained here. The dashed curves are fitted through the nodal values for axial force and stress.

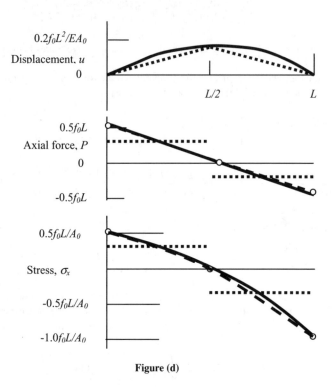

Figure (d)

Because we use equivalent nodal loads the displacements are linear functions for each element. The internal forces and stresses are constant in each element. As we learned in previous examples we can improve the approximations by using the restraint forces to find the internal forces and stresses at the two ends and average the values at interior nodes. Both the constant values and the curves connecting the interpretive values are shown as dotted lines in the plots.

Now let us also redo Example 4.6.1. Our matrix equations are found quickly from Equations (f).

$$\{F\} = \begin{bmatrix} R_1 + \dfrac{f_0 L}{4} \\ \hdashline \dfrac{f_0 L}{2} \\ \dfrac{f_0 L}{4} \end{bmatrix} = [k]\{r\} = \begin{bmatrix} \dfrac{7EA_0}{4L} & -\dfrac{7EA_0}{4L} & 0 \\ \hdashline -\dfrac{7EA_0}{4L} & \dfrac{7EA_0}{4L} + \dfrac{5EA_0}{4L} & -\dfrac{5EA_0}{4L} \\ 0 & -\dfrac{5EA_0}{4L} & \dfrac{5EA_0}{4L} \end{bmatrix} \begin{bmatrix} 0 \\ \hdashline u_2 \\ u_3 \end{bmatrix} \quad (s)$$

The partitioned equations are

$$\begin{bmatrix} \dfrac{f_0 L}{2} \\ \dfrac{f_0 L}{4} \end{bmatrix} = \begin{bmatrix} \dfrac{7EA_0}{4L} + \dfrac{5EA_0}{4L} & -\dfrac{5EA_0}{4L} \\ -\dfrac{5EA_0}{4L} & \dfrac{5EA_0}{4L} \end{bmatrix} \begin{bmatrix} u_2 \\ u_3 \end{bmatrix} \quad (t)$$

and

$$R_1 = -\frac{7EA_0}{4L}u_2 - \frac{f_0L}{4}$$

(u)

Solving for the displacements we get

$$u_2 = \frac{3}{7}\frac{f_0L^2}{EA_0} = 0.4286\frac{f_0L^2}{EA_0} \qquad u_3 = \frac{22}{35}\frac{f_0L^2}{EA_0} = 0.6286\frac{f_0L^2}{EA_0}$$

(v)

The exact values are

$$u\left(\frac{L}{2}\right) = 0.4246\frac{f_0L^2}{EA_0} \qquad u(L) = 0.6137\frac{f_0L^2}{EA_0}$$

(w)

The approximate restraint force is

$$R_1 = -\frac{7EA_0}{4L}u_2 - \frac{f_0L}{4} = -\frac{7EA_0}{4L}\frac{3}{7}\frac{f_0L^2}{EA_0} - \frac{f_0L}{4} = -f_0L$$

(x)

which, of course, is the exact answer.
The internal forces are

$$\begin{bmatrix} p_{1,1} \\ p_{1,2} \end{bmatrix} = \frac{7A_0}{8}\frac{2E}{L}\begin{bmatrix} 1 & -1 \\ -1 & 1 \end{bmatrix}\begin{bmatrix} 0 \\ u_2 \end{bmatrix} = \frac{7A_0}{8}\frac{2E}{L}\begin{bmatrix} 1 & -1 \\ -1 & 1 \end{bmatrix}\begin{bmatrix} 0 \\ \frac{3}{7}\frac{f_0L^2}{EA_0} \end{bmatrix} = \frac{3}{4}f_0L\begin{bmatrix} -1 \\ 1 \end{bmatrix}$$

$$\begin{bmatrix} p_{2,2} \\ p_{2,3} \end{bmatrix} = \frac{5A_0}{8}\frac{2E}{L}\begin{bmatrix} 1 & -1 \\ -1 & 1 \end{bmatrix}\begin{bmatrix} u_2 \\ u_3 \end{bmatrix} = \frac{5A_0}{8}\frac{2E}{L}\begin{bmatrix} 1 & -1 \\ -1 & 1 \end{bmatrix}\begin{bmatrix} \frac{3}{7}\frac{f_0L^2}{EA_0} \\ \frac{22}{35}\frac{f_0L^2}{EA_0} \end{bmatrix} = \frac{1}{4}f_0L\begin{bmatrix} -1 \\ 1 \end{bmatrix}$$

(y)

The internal forces and stresses must be interpreted once again. Using the correct area and the restraint force you get the correct stress at $x = 0$. The stress at $x = L$ is known to be zero.
Using the constant values of stress in each element we get

$$\sigma_1 = \left(\frac{3}{4}f_0L\right) \div \left(\frac{7}{8}A_0\right) = 0.8571\frac{f_0L}{A_0} \qquad \sigma_2 = \left(\frac{1}{4}f_0L\right) \div \left(\frac{5}{8}A_0\right) = 0.4\frac{f_0L}{A_0}$$

(z)

The average value at node 2 is

$$\sigma_{12ave} = 0.6286\frac{f_0L}{A_0}$$

(α)

The exact value is

$$\sigma\left(\frac{L}{2}\right) = 0.6667\frac{f_0L}{A_0}$$

(β)

The approximate values are shown in Figure (e) as dotted lines. The solid lines are the exact values from Section 4.6, Example 4.6.1.
The moral of our story is that the values of displacement and stress at the nodes are quite accurate even for crude models such as just two elements. Let us assure you that with a few more nodes and elements

the accuracy is excellent. By fitting a curve through the nodal values you can get a sufficiently accurate value between nodes.

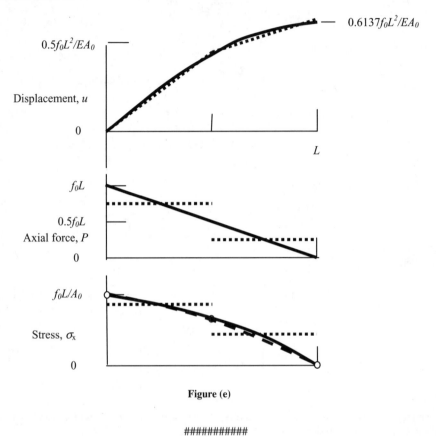

Figure (e)

###########

As you might have expected by now the general method is quite ripe for being programmed for computer analysis. In such cases the number of elements can be increased with little additional effort and answers can more closely approximate the exact answers.

5.6   Analysis and Design of Pin-jointed Trusses

We shall now apply FEA to pin jointed trusses. The first thing that we must recognize is that the element stiffness matrix that we derived earlier for an axially loaded member is aligned with the x axis while those in a truss structure can be at various angles to the x axis. Since all external forces and displacements are measured by their components in the x and y directions (for 2D trusses) we must define each element stiffness matrix in terms of the displacement components in the x and y directions. Consider an element m at an angle α to the x axis as shown in Figure 5.6.1.

What we need to do is find the displacement vectors u_α and v_α at each node in terms of its components in the xy axes directions. From Figure 5.6.1 we see that

$$u_{\alpha n} = u_n \cos \alpha + v_n \sin \alpha \qquad v_{\alpha n} = -u_n \sin \alpha + v_n \cos \alpha$$
$$u_{\alpha n+1} = u_{n+1} \cos \alpha + v_{n+1} \sin \alpha \qquad v_{\alpha n+1} = -u_{n+1} \sin \alpha + v_{n+1} \cos \alpha$$

$$(5.6.1)$$

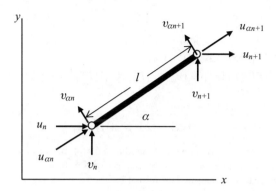

Figure 5.6.1

This can be put in matrix form

$$\begin{bmatrix} u_{\alpha n} \\ v_{\alpha n} \end{bmatrix} = \begin{bmatrix} \cos\alpha & \sin\alpha \\ -\sin\alpha & \cos\alpha \end{bmatrix} \begin{bmatrix} u_n \\ v_n \end{bmatrix} \qquad \begin{bmatrix} u_{\alpha n+1} \\ v_{\alpha n+1} \end{bmatrix} = \begin{bmatrix} \cos\alpha & \sin\alpha \\ -\sin\alpha & \cos\alpha \end{bmatrix} \begin{bmatrix} u_{n+1} \\ v_{n+1} \end{bmatrix} \qquad (5.6.2)$$

For the two noded axial element there is no v_α component, that is, the loads and displacements can only lie along the axis of the member, so we can set

$$v_{\alpha n} = v_{\alpha n+1} = 0 \qquad (5.6.3)$$

Let us define the nodal displacements as

$$\{r_m\} = \begin{bmatrix} u_n \\ v_n \\ u_{n+1} \\ v_{n+1} \end{bmatrix} \qquad \{r_{\alpha m}\} = \begin{bmatrix} u_{\alpha n} \\ 0 \\ u_{\alpha n+1} \\ 0 \end{bmatrix} \qquad (5.6.4)$$

We can combine Equations 5.6.2 into one matrix set of equations.

$$\{r_{\alpha m}\} = \begin{bmatrix} u_{\alpha n} \\ 0 \\ u_{\alpha n+1} \\ 0 \end{bmatrix} = \begin{bmatrix} \cos\alpha & \sin\alpha & 0 & 0 \\ -\sin\alpha & \cos\alpha & 0 & 0 \\ 0 & 0 & \cos\alpha & \sin\alpha \\ 0 & 0 & -\sin\alpha & \cos\alpha \end{bmatrix} \begin{bmatrix} u_n \\ v_n \\ u_{n+1} \\ v_{n+1} \end{bmatrix} = [T]\{r_m\} \qquad (5.6.5)$$

It just happens that this transformation matrix $[T]$ has the interesting property that its inverse is equal to its transpose, or $[T]^{-1} = [T]^T$. You can verify that $[T]^T [T] = [1]$ where $[1]$ is the unit matrix

$$[1] = \begin{bmatrix} 1 & 0 & 0 & 0 \\ 0 & 1 & 0 & 0 \\ 0 & 0 & 1 & 0 \\ 0 & 0 & 0 & 1 \end{bmatrix} \qquad (5.6.6)$$

This allows us to state that

$$\{r_{\alpha m}\} = [T]\{r_m\} \qquad \{r_m\} = [T]^T \{r_{\alpha m}\} \qquad (5.6.7)$$

and also that

$$\{p_{\alpha m}\} = [T]\{p_m\} \qquad \{p_m\} = [T]^T \{p_{\alpha m}\} \qquad (5.6.8)$$

Since

$$\{p_{\alpha m}\} = [k_{\alpha m}]\{r_{\alpha m}\}$$
$$[T]\{p_m\} = [k_{\alpha m}][T]\{r_m\} \tag{5.6.9}$$
$$[T]^T[T]\{p_m\} = \{p_m\} = [T]^T[k_{\alpha m}][T]\{r_m\} = [k_m]\{r_m\}$$

It follows that

$$[k_m] = [T]^T[k_{\alpha m}][T] \tag{5.6.10}$$

Since $[T]$ is a 4×4 matrix $[k_{\alpha m}]$ must be formatted as a 4×4 matrix as well. Note that $[k_{\alpha m}]$ is the element stiffness matrix for the two force member, that is, for $\alpha = 0$.

This element stiffness matrix in the standard format is

$$[k_{\alpha m}]\{r_{\alpha m}\} = \frac{EA}{l}\begin{bmatrix} 1 & -1 \\ -1 & 1 \end{bmatrix}\begin{bmatrix} u_{\alpha n} \\ u_{\alpha n+1} \end{bmatrix} \tag{5.6.11}$$

can be presented in the format

$$[k_{\alpha m}]\{r_{\alpha m}\} = \frac{EA}{l}\begin{bmatrix} 1 & 0 & -1 & 0 \\ 0 & 0 & 0 & 0 \\ -1 & 0 & 1 & 0 \\ 0 & 0 & 0 & 0 \end{bmatrix}\begin{bmatrix} u_{\alpha n} \\ 0 \\ u_{\alpha n+1} \\ 0 \end{bmatrix} \tag{5.6.12}$$

so that the operation in Equation 5.6.10 can be carried out.

This is best understood in an example.

##########

Example 5.6.1

Problem: Find the element stiffness matrix for typical elements at 45° and 135°, 60° and 120°, and 90° and 180°. These will be used in the following examples.

Solution: Find the matrix $[T]$ and then find $[k_m]$ for each case.

For 45° the matrix $[T]$ is

$$[T_{45}] = \begin{bmatrix} \cos\alpha & \sin\alpha & 0 & 0 \\ -\sin\alpha & \cos\alpha & 0 & 0 \\ 0 & 0 & \cos\alpha & \sin\alpha \\ 0 & 0 & -\sin\alpha & \cos\alpha \end{bmatrix} = \frac{\sqrt{2}}{2}\begin{bmatrix} 1 & 1 & 0 & 0 \\ -1 & 1 & 0 & 0 \\ 0 & 0 & 1 & 1 \\ 0 & 0 & -1 & 1 \end{bmatrix} \tag{a}$$

The stiffness matrix for $\alpha = 45°$ is

$$[k_{45}] = [T_{45}]^T[k_0][T_{45}] = \frac{\sqrt{2}}{2}\begin{bmatrix} 1 & -1 & 0 & 0 \\ 1 & 1 & 0 & 0 \\ 0 & 0 & 1 & -1 \\ 0 & 0 & 1 & 1 \end{bmatrix}\frac{EA}{l}\begin{bmatrix} 1 & 0 & -1 & 0 \\ 0 & 0 & 0 & 0 \\ -1 & 0 & 1 & 0 \\ 0 & 0 & 0 & 0 \end{bmatrix}\frac{\sqrt{2}}{2}\begin{bmatrix} 1 & 1 & 0 & 0 \\ -1 & 1 & 0 & 0 \\ 0 & 0 & 1 & 1 \\ 0 & 0 & -1 & 1 \end{bmatrix}$$

$$= \frac{EA}{2l}\begin{bmatrix} 1 & 1 & -1 & -1 \\ 1 & 1 & -1 & -1 \\ -1 & -1 & 1 & 1 \\ -1 & -1 & 1 & 1 \end{bmatrix} \tag{b}$$

For $\alpha = 135°$ the matrix $[T]$ is

$$[T_{135}] = \begin{bmatrix} \cos\alpha & \sin\alpha & 0 & 0 \\ -\sin\alpha & \cos\alpha & 0 & 0 \\ 0 & 0 & \cos\alpha & \sin\alpha \\ 0 & 0 & -\sin\alpha & \cos\alpha \end{bmatrix} = \frac{\sqrt{2}}{2}\begin{bmatrix} -1 & 1 & 0 & 0 \\ -1 & -1 & 0 & 0 \\ 0 & 0 & -1 & 1 \\ 0 & 0 & -1 & -1 \end{bmatrix} \tag{c}$$

The stiffness matrix for $\alpha = 135°$ is

$$[k_{135}] = [T_{135}]^T [k_0][T_{135}]$$

$$= \frac{\sqrt{2}}{2}\begin{bmatrix} -1 & -1 & 0 & 0 \\ 1 & -1 & 0 & 0 \\ 0 & 0 & -1 & -1 \\ 0 & 0 & 1 & -1 \end{bmatrix}\frac{EA}{l}\begin{bmatrix} 1 & 0 & -1 & 0 \\ 0 & 0 & 0 & 0 \\ -1 & 0 & 1 & 0 \\ 0 & 0 & 0 & 0 \end{bmatrix}\frac{\sqrt{2}}{2}\begin{bmatrix} -1 & 1 & 0 & 0 \\ -1 & -1 & 0 & 0 \\ 0 & 0 & -1 & 1 \\ 0 & 0 & -1 & -1 \end{bmatrix}$$

$$= \frac{EA}{2l}\begin{bmatrix} 1 & -1 & -1 & 1 \\ -1 & 1 & 1 & -1 \\ -1 & 1 & 1 & -1 \\ 1 & -1 & -1 & 1 \end{bmatrix} \tag{d}$$

For $\alpha = 60°$ the matrix $[T]$ is

$$[T_{60}] = \begin{bmatrix} \cos\alpha & \sin\alpha & 0 & 0 \\ -\sin\alpha & \cos\alpha & 0 & 0 \\ 0 & 0 & \cos\alpha & \sin\alpha \\ 0 & 0 & -\sin\alpha & \cos\alpha \end{bmatrix} = \frac{1}{2}\begin{bmatrix} 1 & \sqrt{3} & 0 & 0 \\ -\sqrt{3} & 1 & 0 & 0 \\ 0 & 0 & 1 & \sqrt{3} \\ 0 & 0 & -\sqrt{3} & 1 \end{bmatrix} \tag{e}$$

The stiffness matrix for $\alpha = 60°$ is

$$[k_{60}] = [T_{60}]^T [k_0][T_{60}]$$

$$= \frac{1}{2}\begin{bmatrix} 1 & -\sqrt{3} & 0 & 0 \\ \sqrt{3} & 1 & 0 & 0 \\ 0 & 0 & 1 & -\sqrt{3} \\ 0 & 0 & \sqrt{3} & 1 \end{bmatrix}\frac{EA}{l}\begin{bmatrix} 1 & 0 & -1 & 0 \\ 0 & 0 & 0 & 0 \\ -1 & 0 & 1 & 0 \\ 0 & 0 & 0 & 0 \end{bmatrix}\frac{1}{2}\begin{bmatrix} 1 & \sqrt{3} & 0 & 0 \\ -\sqrt{3} & 1 & 0 & 0 \\ 0 & 0 & 1 & \sqrt{3} \\ 0 & 0 & -\sqrt{3} & 1 \end{bmatrix}$$

$$= \frac{EA}{4l}\begin{bmatrix} 1 & \sqrt{3} & -1 & -\sqrt{3} \\ \sqrt{3} & 3 & -\sqrt{3} & -3 \\ -1 & -\sqrt{3} & 1 & \sqrt{3} \\ -\sqrt{3} & -3 & \sqrt{3} & 3 \end{bmatrix} \tag{f}$$

For $\alpha = 120°$ the matrix $[T]$ is

$$[T_{120}] = \begin{bmatrix} \cos\alpha & \sin\alpha & 0 & 0 \\ -\sin\alpha & \cos\alpha & 0 & 0 \\ 0 & 0 & \cos\alpha & \sin\alpha \\ 0 & 0 & -\sin\alpha & \cos\alpha \end{bmatrix} = \frac{1}{2}\begin{bmatrix} -1 & \sqrt{3} & 0 & 0 \\ -\sqrt{3} & -1 & 0 & 0 \\ 0 & 0 & -1 & \sqrt{3} \\ 0 & 0 & -\sqrt{3} & -1 \end{bmatrix} \tag{g}$$

The stiffness matrix for $\alpha = 120°$ is

$$[k_{120}] = [T_{120}]^T [k_0] [T_{120}]$$

$$= \frac{1}{2}\begin{bmatrix} -1 & -\sqrt{3} & 0 & 0 \\ \sqrt{3} & -1 & 0 & 0 \\ 0 & 0 & -1 & -\sqrt{3} \\ 0 & 0 & \sqrt{3} & -1 \end{bmatrix} \frac{EA}{l} \begin{bmatrix} 1 & 0 & -1 & 0 \\ 0 & 0 & 0 & 0 \\ -1 & 0 & 1 & 0 \\ 0 & 0 & 0 & 0 \end{bmatrix} \frac{1}{2}\begin{bmatrix} -1 & \sqrt{3} & 0 & 0 \\ -\sqrt{3} & -1 & 0 & 0 \\ 0 & 0 & -1 & \sqrt{3} \\ 0 & 0 & -\sqrt{3} & -1 \end{bmatrix}$$

$$= \frac{EA}{4l}\begin{bmatrix} 1 & -\sqrt{3} & -1 & \sqrt{3} \\ -\sqrt{3} & 3 & \sqrt{3} & -3 \\ -1 & \sqrt{3} & 1 & -\sqrt{3} \\ \sqrt{3} & -3 & -\sqrt{3} & 3 \end{bmatrix} \tag{h}$$

For $\alpha = 90°$ the matrix $[T]$ is

$$[T_{90}] = \begin{bmatrix} \cos\alpha & \sin\alpha & 0 & 0 \\ -\sin\alpha & \cos\alpha & 0 & 0 \\ 0 & 0 & \cos\alpha & \sin\alpha \\ 0 & 0 & -\sin\alpha & \cos\alpha \end{bmatrix} = \begin{bmatrix} 0 & 1 & 0 & 0 \\ -1 & 0 & 0 & 0 \\ 0 & 0 & 0 & 1 \\ 0 & 0 & -1 & 0 \end{bmatrix} \tag{i}$$

The stiffness matrix for $\alpha = 90°$ is

$$[k_{90}] = [T_{90}]^T [k_0] [T_{90}] = \begin{bmatrix} 0 & -1 & 0 & 0 \\ 1 & 0 & 0 & 0 \\ 0 & 0 & 0 & -1 \\ 0 & 0 & 1 & 0 \end{bmatrix} \frac{EA}{l} \begin{bmatrix} 1 & 0 & -1 & 0 \\ 0 & 0 & 0 & 0 \\ -1 & 0 & 1 & 0 \\ 0 & 0 & 0 & 0 \end{bmatrix} \begin{bmatrix} 0 & 1 & 0 & 0 \\ -1 & 0 & 0 & 0 \\ 0 & 0 & 0 & 1 \\ 0 & 0 & -1 & 0 \end{bmatrix}$$

$$= \frac{EA}{l}\begin{bmatrix} 0 & 0 & 0 & 0 \\ 0 & 1 & 0 & -1 \\ 0 & 0 & 0 & 0 \\ 0 & -1 & 0 & 1 \end{bmatrix} \tag{j}$$

For $\alpha = 180°$ the matrix $[T]$ is

$$[T_{180}] = \begin{bmatrix} \cos\alpha & \sin\alpha & 0 & 0 \\ -\sin\alpha & \cos\alpha & 0 & 0 \\ 0 & 0 & \cos\alpha & \sin\alpha \\ 0 & 0 & -\sin\alpha & \cos\alpha \end{bmatrix} = \begin{bmatrix} -1 & 0 & 0 & 0 \\ 0 & -1 & 0 & 0 \\ 0 & 0 & -1 & 0 \\ 0 & 0 & 0 & -1 \end{bmatrix} \tag{k}$$

The stiffness matrix for $\alpha = 180°$ is

$$[k_{180}] = [T_{180}]^T [k_0] [T_{180}] = \begin{bmatrix} -1 & 0 & 0 & 0 \\ 0 & -1 & 0 & 0 \\ 0 & 0 & -1 & 0 \\ 0 & 0 & 0 & -1 \end{bmatrix} \frac{EA}{l} \begin{bmatrix} 1 & 0 & -1 & 0 \\ 0 & 0 & 0 & 0 \\ -1 & 0 & 1 & 0 \\ 0 & 0 & 0 & 0 \end{bmatrix} \begin{bmatrix} -1 & 0 & 0 & 0 \\ 0 & -1 & 0 & 0 \\ 0 & 0 & -1 & 0 \\ 0 & 0 & 0 & -1 \end{bmatrix}$$

$$= \frac{EA}{l} \begin{bmatrix} 1 & 0 & -1 & 0 \\ 0 & 0 & 0 & 0 \\ -1 & 0 & 1 & 0 \\ 0 & 0 & 0 & 0 \end{bmatrix} \tag{1}$$

Example 5.6.2

Problem: Find the displacements and stresses for the pin jointed truss in Figure (a). All members are made of the same material and have the same cross section area. We have numbered the nodes and elements on the figure. Let

$$F_H = 400\,N \qquad F_V = 1000\,N \tag{a}$$

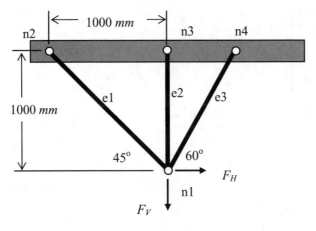

Figure (a)

Solution: Assemble the global equations, partition, and solve. The method for matrix multiplication and the equation solving which follow are presented for Mathematica in Appendix C. Note that this is an indeterminate truss.

The length of the members are

$$l_1 = \sqrt{2} \cdot 1000\,mm \qquad l_2 = 1000\,mm \qquad l_3 = \frac{2}{\sqrt{3}} \cdot 1000\,mm \tag{b}$$

Now the assembly process is a bit more complicated because the element matrices do not just descend down the diagonal as in the case of a single axially loaded member. In this case nodes 1 and 2 are associated with element 1, nodes 1 and 3 are associated with element 2, and nodes 1 and 4 are associated with element 3. Each element matrix must be imbeded in the larger global matrix as shown here.

The stiffness matrix for element 1 ($\alpha = 135°$)is given in Equation (c).

$$
\begin{bmatrix} p_{1,1x} \\ p_{1,1y} \\ p_{1,2x} \\ p_{1,2y} \end{bmatrix} = \frac{EA}{2l_1} \begin{bmatrix} 1 & -1 & -1 & 1 \\ -1 & 1 & 1 & -1 \\ -1 & 1 & 1 & -1 \\ 1 & -1 & -1 & 1 \end{bmatrix} \begin{bmatrix} u_1 \\ v_1 \\ u_2 \\ v_2 \end{bmatrix} = \frac{EA}{1000} \begin{bmatrix} \dfrac{\sqrt{2}}{4} & -\dfrac{\sqrt{2}}{4} & -\dfrac{\sqrt{2}}{4} & \dfrac{\sqrt{2}}{4} \\ -\dfrac{\sqrt{2}}{4} & \dfrac{\sqrt{2}}{4} & \dfrac{\sqrt{2}}{4} & -\dfrac{\sqrt{2}}{4} \\ -\dfrac{\sqrt{2}}{4} & \dfrac{\sqrt{2}}{4} & \dfrac{\sqrt{2}}{4} & -\dfrac{\sqrt{2}}{4} \\ \dfrac{\sqrt{2}}{4} & -\dfrac{\sqrt{2}}{4} & -\dfrac{\sqrt{2}}{4} & \dfrac{\sqrt{2}}{4} \end{bmatrix} \begin{bmatrix} u_1 \\ v_1 \\ u_2 \\ v_2 \end{bmatrix} \quad (c)
$$

It is embedded in the global matrix format in Equation (d).

$$
\begin{bmatrix} p_{1,1x} \\ p_{1,1y} \\ p_{1,2x} \\ p_{1,2y} \\ 0 \\ 0 \\ 0 \\ 0 \end{bmatrix} = \frac{EA}{1000} \begin{bmatrix} \dfrac{\sqrt{2}}{4} & -\dfrac{\sqrt{2}}{4} & -\dfrac{\sqrt{2}}{4} & \dfrac{\sqrt{2}}{4} & 0 & 0 & 0 & 0 \\ -\dfrac{\sqrt{2}}{4} & \dfrac{\sqrt{2}}{4} & \dfrac{\sqrt{2}}{4} & -\dfrac{\sqrt{2}}{4} & 0 & 0 & 0 & 0 \\ -\dfrac{\sqrt{2}}{4} & \dfrac{\sqrt{2}}{4} & \dfrac{\sqrt{2}}{4} & -\dfrac{\sqrt{2}}{4} & 0 & 0 & 0 & 0 \\ \dfrac{\sqrt{2}}{4} & -\dfrac{\sqrt{2}}{4} & -\dfrac{\sqrt{2}}{4} & \dfrac{\sqrt{2}}{4} & 0 & 0 & 0 & 0 \\ 0 & 0 & 0 & 0 & 0 & 0 & 0 & 0 \\ 0 & 0 & 0 & 0 & 0 & 0 & 0 & 0 \\ 0 & 0 & 0 & 0 & 0 & 0 & 0 & 0 \\ 0 & 0 & 0 & 0 & 0 & 0 & 0 & 0 \end{bmatrix} \begin{bmatrix} u_1 \\ v_1 \\ u_2 \\ v_2 \\ u_3 \\ v_3 \\ u_4 \\ v_4 \end{bmatrix} \quad (d)
$$

Element 2 ($\alpha = 90°$) imbedded in the global matrix format is given in Equation (e).

$$
\begin{bmatrix} p_{2,1x} \\ p_{2,1y} \\ p_{2,3x} \\ p_{2,3y} \end{bmatrix} = \frac{EA}{1000} \begin{bmatrix} 0 & 0 & 0 & 0 \\ 0 & 1 & 0 & -1 \\ 0 & 0 & 0 & 0 \\ 0 & -1 & 0 & 1 \end{bmatrix} \begin{bmatrix} u_1 \\ v_1 \\ u_3 \\ v_3 \end{bmatrix}
$$

$$
\begin{bmatrix} p_{2,1x} \\ p_{2,y} \\ 0 \\ 0 \\ p_{2,3x} \\ p_{2,3y} \\ 0 \\ 0 \end{bmatrix} = \frac{EA}{1000} \begin{bmatrix} 0 & 0 & 0 & 0 & 0 & 0 & 0 & 0 \\ 0 & 1 & 0 & 0 & 0 & -1 & 0 & 0 \\ 0 & 0 & 0 & 0 & 0 & 0 & 0 & 0 \\ 0 & 0 & 0 & 0 & 0 & 0 & 0 & 0 \\ 0 & 0 & 0 & 0 & 0 & 0 & 0 & 0 \\ 0 & -1 & 0 & 0 & 0 & 1 & 0 & 0 \\ 0 & 0 & 0 & 0 & 0 & 0 & 0 & 0 \\ 0 & 0 & 0 & 0 & 0 & 0 & 0 & 0 \end{bmatrix} \begin{bmatrix} u_1 \\ v_1 \\ u_2 \\ v_2 \\ u_3 \\ v_3 \\ u_4 \\ v_4 \end{bmatrix} \quad (e)
$$

Element 3 ($\alpha = 60°$) imbedded in the global matrix format is given in Equation (f).

$$
\begin{bmatrix} p_{3,1x} \\ p_{3,1y} \\ p_{3,4x} \\ p_{3,4y} \end{bmatrix} = \frac{EA}{4l} \begin{bmatrix} 1 & \sqrt{3} & -1 & -\sqrt{3} \\ \sqrt{3} & 3 & -\sqrt{3} & -3 \\ -1 & -\sqrt{3} & 1 & \sqrt{3} \\ -\sqrt{3} & -3 & \sqrt{3} & 3 \end{bmatrix} \begin{bmatrix} u_1 \\ v_1 \\ u_4 \\ v_4 \end{bmatrix}
$$

$$
= \frac{EA}{1000} \begin{bmatrix} \dfrac{\sqrt{3}}{8} & \dfrac{3}{8} & -\dfrac{\sqrt{3}}{8} & -\dfrac{3}{8} \\[2mm] \dfrac{3}{8} & \dfrac{3}{8}\sqrt{3} & -\dfrac{3}{8} & -\dfrac{3}{8}\sqrt{3} \\[2mm] -\dfrac{\sqrt{3}}{8} & -\dfrac{3}{8} & \dfrac{\sqrt{3}}{8} & \dfrac{3}{8} \\[2mm] -\dfrac{3}{8} & -\dfrac{3}{8}\sqrt{3} & \dfrac{3}{8} & \dfrac{3}{8}\sqrt{3} \end{bmatrix} \begin{bmatrix} u_1 \\ v_1 \\ u_4 \\ v_4 \end{bmatrix} \tag{f}
$$

$$
\begin{bmatrix} p_{3,1x} \\ p_{3,1y} \\ 0 \\ 0 \\ 0 \\ 0 \\ p_{3,4x} \\ p_{3,4y} \end{bmatrix} = \frac{EA}{1000} \begin{bmatrix} \dfrac{\sqrt{3}}{8} & \dfrac{3}{8} & 0 & 0 & 0 & 0 & -\dfrac{\sqrt{3}}{8} & -\dfrac{3}{8} \\[2mm] \dfrac{3}{8} & \dfrac{3}{8}\sqrt{3} & 0 & 0 & 0 & 0 & -\dfrac{3}{8} & -\dfrac{3}{8}\sqrt{3} \\[2mm] 0 & 0 & 0 & 0 & 0 & 0 & 0 & 0 \\[2mm] 0 & 0 & 0 & 0 & 0 & 0 & 0 & 0 \\[2mm] 0 & 0 & 0 & 0 & 0 & 0 & 0 & 0 \\[2mm] 0 & 0 & 0 & 0 & 0 & 0 & 0 & 0 \\[2mm] -\dfrac{\sqrt{3}}{8} & -\dfrac{3}{8} & 0 & 0 & 0 & 0 & \dfrac{\sqrt{3}}{8} & \dfrac{3}{8} \\[2mm] -\dfrac{3}{8} & -\dfrac{3}{8}\sqrt{3} & 0 & 0 & 0 & 0 & \dfrac{3}{8} & \dfrac{3}{8}\sqrt{3} \end{bmatrix} \begin{bmatrix} u_1 \\ v_1 \\ u_2 \\ v_2 \\ u_3 \\ v_3 \\ u_4 \\ v_4 \end{bmatrix}
$$

The only nonzero displacements are at node one. To find the displacements we need only those terms in the assembled matrices that are part of the partitioned matrix equations.

The assembled matrices are given in Equation (g) with just the necessary terms for finding the displacements. The symbol c is just to identify the location of the other terms in the global stiffness matrix.

$$
\begin{bmatrix} F_H \\ -F_V \\ \hline R_{2x} \\ R_{2y} \\ R_{3x} \\ R_{3y} \\ R_{4x} \\ R_{4y} \end{bmatrix} = \begin{bmatrix} p_{1,1x} + p_{2,1x} + p_{3,1x} \\ p_{1,1y} + p_{2,1y} + p_{3,1y} \\ \hline c \\ c \\ c \\ c \\ c \\ c \end{bmatrix} = \frac{EA}{1000} \begin{bmatrix} 0.57006 & 0.02145 & c & c & c & c & c & c \\ 0.02145 & 2.00307 & c & c & c & c & c & c \\ \hline c & c & c & c & c & c & c & c \\ c & c & c & c & c & c & c & c \\ c & c & c & c & c & c & c & c \\ c & c & c & c & c & c & c & c \\ c & c & c & c & c & c & c & c \\ c & c & c & c & c & c & c & c \end{bmatrix} \begin{bmatrix} u_1 \\ v_1 \\ \hline 0 \\ 0 \\ 0 \\ 0 \\ 0 \\ 0 \end{bmatrix} \tag{g}
$$

Partition and solve for the displacements.

$$\begin{bmatrix} 400 \\ -1000 \end{bmatrix} = \frac{EA}{1000} \begin{bmatrix} 0.57006 & 0.02145 \\ 0.02145 & 2.00307 \end{bmatrix} \begin{bmatrix} u_1 \\ v_1 \end{bmatrix}$$

$$\rightarrow \begin{bmatrix} u_1 \\ v_1 \end{bmatrix} = \frac{1000}{EA} \begin{bmatrix} 720.76 \\ -506.95 \end{bmatrix}$$

(h)

Now you could complete the global stiffness matrix and partition and solve for the restraint forces and then resolve those forces to find the internal axial forces in each member. It is easier to solve for the internal forces directly since we already have those matrix equations formed. The internal force in element 1 can be found from the following equations.

$$\begin{bmatrix} p_{1,1x} \\ p_{1,1y} \\ p_{1,2x} \\ p_{1,2y} \end{bmatrix} = \frac{EA}{1000} \begin{bmatrix} \dfrac{\sqrt{2}}{4} & -\dfrac{\sqrt{2}}{4} & -\dfrac{\sqrt{2}}{4} & \dfrac{\sqrt{2}}{4} \\ -\dfrac{\sqrt{2}}{4} & \dfrac{\sqrt{2}}{4} & \dfrac{\sqrt{2}}{4} & -\dfrac{\sqrt{2}}{4} \\ -\dfrac{\sqrt{2}}{4} & \dfrac{\sqrt{2}}{4} & \dfrac{\sqrt{2}}{4} & -\dfrac{\sqrt{2}}{4} \\ \dfrac{\sqrt{2}}{4} & -\dfrac{\sqrt{2}}{4} & -\dfrac{\sqrt{2}}{4} & \dfrac{\sqrt{2}}{4} \end{bmatrix} \begin{bmatrix} u_1 \\ v_1 \\ 0 \\ 0 \end{bmatrix}$$

(i)

$$\rightarrow \begin{bmatrix} p_{1,2x} \\ p_{1,2y} \end{bmatrix} = \frac{EA}{1000} \begin{bmatrix} -\dfrac{\sqrt{2}}{4} & \dfrac{\sqrt{2}}{4} \\ \dfrac{\sqrt{2}}{4} & -\dfrac{\sqrt{2}}{4} \end{bmatrix} \begin{bmatrix} u_1 \\ v_1 \end{bmatrix}$$

$$= \begin{bmatrix} -0.35355 & 0.35355 \\ 0.35355 & -0.35355 \end{bmatrix} \begin{bmatrix} 720.76 \\ -506.95 \end{bmatrix} = \begin{bmatrix} -434.06 \\ 434.06 \end{bmatrix} N$$

The internal axial force in member 1 is

$$P_1 = \sqrt{(-434.06)^2 + (434.06)^2} = 613.85 \, N$$

(j)

The internal force is element 2 is found from

$$\begin{bmatrix} p_{2,1x} \\ p_{2,1y} \\ p_{2,3x} \\ p_{2,3y} \end{bmatrix} = \frac{EA}{1000} \begin{bmatrix} 0 & 0 & 0 & 0 \\ 0 & 1 & 0 & -1 \\ 0 & 0 & 0 & 0 \\ 0 & -1 & 0 & 1 \end{bmatrix} \begin{bmatrix} u_1 \\ v_1 \\ 0 \\ 0 \end{bmatrix}$$

$$\rightarrow \begin{bmatrix} p_{2,3x} \\ p_{2,3y} \end{bmatrix} = \frac{EA}{1000} \begin{bmatrix} 0 & 0 \\ 0 & -1 \end{bmatrix} \begin{bmatrix} u_1 \\ v_1 \end{bmatrix} = \begin{bmatrix} 0 & 0 \\ 0 & -1 \end{bmatrix} \begin{bmatrix} 720.76 \\ -506.95 \end{bmatrix} = \begin{bmatrix} 0 \\ 506.95 \end{bmatrix} N$$

(k)

The internal axial force in member 2 is

$$P_2 = 506.95 \, N$$

(l)

The internal force is element 3 is found from

$$
\begin{bmatrix} P_{3,1x} \\ P_{3,1y} \\ P_{3,4x} \\ P_{3,4y} \end{bmatrix} = \frac{EA}{1000} \begin{bmatrix} \dfrac{\sqrt{3}}{8} & \dfrac{3}{8} & -\dfrac{\sqrt{3}}{8} & -\dfrac{3}{8} \\ \dfrac{3}{8} & \dfrac{3}{8}\sqrt{3} & -\dfrac{3}{8} & -\dfrac{3}{8}\sqrt{3} \\ -\dfrac{\sqrt{3}}{8} & -\dfrac{3}{8} & \dfrac{\sqrt{3}}{8} & \dfrac{3}{8} \\ -\dfrac{3}{8} & -\dfrac{3}{8}\sqrt{3} & \dfrac{3}{8} & \dfrac{3}{8}\sqrt{3} \end{bmatrix} \begin{bmatrix} u_1 \\ v_1 \\ 0 \\ 0 \end{bmatrix}
$$

$$
\rightarrow \begin{bmatrix} P_{3,4x} \\ P_{3,4y} \end{bmatrix} = \frac{EA}{1000} \begin{bmatrix} -\dfrac{\sqrt{3}}{8} & -\dfrac{3}{8} \\ -\dfrac{3}{8} & -\dfrac{3}{8}\sqrt{3} \end{bmatrix} \begin{bmatrix} u_1 \\ v_1 \end{bmatrix} \tag{m}
$$

$$
= \frac{EA}{1000} \begin{bmatrix} -0.21651 & -0.375 \\ -0.375 & -0.64952 \end{bmatrix} \begin{bmatrix} 720.76 \\ -506.95 \end{bmatrix} = \begin{bmatrix} 34.06 \\ 58.99 \end{bmatrix} N \tag{n}
$$

The internal axial force in member 3 is

$$
P_3 = \sqrt{(34.06)^2 + (58.99)^2} = 68.12 \, N \tag{o}
$$

As a check see if the truss member force components and the applied loads are in equilibrium at node 1.

$$
\sum F_y = (434.06 + 506.95 + 58.99 - 1000) = 0
$$
$$
\sum F_x = (-434.06 + 0 + 34.06 + 400) = 0 \tag{p}
$$

They check.

######### #

The solution of even such a simple truss problem like this by either the classical differential equation method or the modern finite element method is quite a chore. Fortunately, the finite element method has been programmed for graphical input of the geometry, automatic creation of nodes and elements, assembly of the matrices, and direct application of loads and restraints to the assembled equations. Partitioning of the equation and their solution is also automated. And additional bonus is the graphical as well as numerical presentation of the results.

5.7 Summary and Conclusions

The finite element method is especially useful for formulating problems for numerical analysis. The equations for this method are summarized in Equations 5.4.5 and repeated here in

Equation 5.7.1.

$$\{F\} \qquad \{F_c\} \qquad f_x(s) \qquad \rightarrow \qquad \{p_m\} \qquad \sigma_{xm} \qquad \varepsilon_{xm} \qquad u(s_m) \qquad \{r_m\} \leftarrow \{\rho\}$$

$$\{F_{em}\} = \int_0^{l_m} [n]^T f_x(s_m) \, ds_m \qquad \{F_e\} = \sum \{F_{em}\} \qquad u(s_m) = [n]\{r_m\}$$

$$\{F\} = \{F_c\} + \{F_e\} \qquad\qquad\qquad \varepsilon_{xm} = \frac{du_m}{ds_m} = \frac{d}{ds_m}[n]\{r_m\}$$

$$\sigma_{xm} = E_m \varepsilon_{xm} = E_m \frac{d}{ds_m}[n]\{r_m\}$$

$$(5.7.1)$$

$$p_{m,n} = -A_m \sigma_{xm} = -E_m A_m \frac{d}{ds_m}[n]\{r_m\}$$

$$\rightarrow \quad \{p_m\} = \begin{bmatrix} p_{m,n} \\ p_{m,n+1} \end{bmatrix} = [k_m]\{r_m\}$$

$$p_{m,n+1} = A_m \sigma_{xm} = E_m A_m \frac{d}{ds_m}[n]\{r_m\}$$

$$\{F\} = \sum \{p_m\} = \sum [k_m]\{r\} = [K]\{r\}$$

The critical factor here is the discovery and recognition of shape functions $[n]$.

$$u(s_m) = [n]\{r_m\} = \begin{bmatrix} 1 - \dfrac{s_m}{l_m} & \dfrac{s_m}{l_m} \end{bmatrix} \begin{bmatrix} u_n \\ u_{n+1} \end{bmatrix} \qquad (5.7.2)$$

Note that for axial bars with concentrated loads the equations for the classical method are exactly satisfied. To use this method distributed loads are replaced with equivalent concentrated nodal loads. This is done using Equation 5.4.4 repeated here as Equation 5.7.3. The derivation of this relation is given in a later chapter.

$$\{F_{em}\} = \int_0^{l_m} [n]^T f_x(s_m) \, ds_m = \int_0^{l_m} \begin{bmatrix} 1 - \dfrac{s_m}{l_m} \\ \dfrac{s_m}{l_m} \end{bmatrix} f_x(s_m) \, ds_m = \begin{bmatrix} F_{em,n} \\ F_{em,n+1} \end{bmatrix} \qquad (5.7.3)$$

In Section 5.6 we extend the analysis to pin jointed trusses and introduce the transformation for axial element stiffness matrices oriented at positions other than along the x axis. The transformation matrix is

$$[T] = \begin{bmatrix} \cos\alpha & \sin\alpha & 0 & 0 \\ -\sin\alpha & \cos\alpha & 0 & 0 \\ 0 & 0 & \cos\alpha & \sin\alpha \\ 0 & 0 & -\sin\alpha & \cos\alpha \end{bmatrix} \qquad (5.7.4)$$

6

Torsion

6.1 Introduction

In this chapter we consider the special case of a slender bar or *shaft* that is loaded in pure torque about its centroidal axis and is restrained from any rigid body motion. This has come to be known as the subject of *torsion*. We shall assume that any local effects of load application and restraint are neglected. There are, of course, local effects but we depend upon St. Venant's principle to allow us to gain useful information about the behavior if the shaft is long and slender.

The equations and solution methods found here for torsion are very similar in form to those in Chapters 4 and 5 that govern axial deformation. In fact, we will depend heavily on what we learned there to understand what is happening here.

6.2 Torsional Displacement, Strain, and Stress

To gain a preliminary understanding of torsion we conduct an experiment. Consider a slender bar with a circular cross section restrained at $x = 0$ and with a concentrated applied moment or torque M_L at $x = L$ as shown in Figure 6.2.1. A double arrowhead is used to indicate a moment. The moment is positive by the right hand rule. Just how the moment is applied is not yet defined. The body deforms in rotation and the total angle of twist at $x = L$ as a result of the load is designated as ϕ.

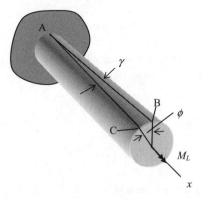

Figure 6.2.1

Analysis of Structures: An Introduction Including Numerical Methods, First Edition. Joe G. Eisley and Anthony M. Waas.
© 2011 John Wiley & Sons, Ltd. Published 2011 by John Wiley & Sons, Ltd.

Experimental evidence confirms that the distributed angle of twist $\beta(x)$ is proportional to the distance along the bar. Then

$$\beta(x) = \frac{\phi}{L}x \tag{6.2.1}$$

Let us look at a free body portion of the bar as shown in Figure 6.2.2 and invoke static equilibrium.

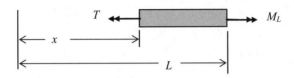

Figure 6.2.2

From summation of moments the internal torque T is a constant equal to the applied moment M_L.

$$\sum M_x = M_L - T = 0 \quad \rightarrow \quad T = M_L \tag{6.2.2}$$

We now adopt cylindrical coordinates with the origin at the center of the circular cross section and with r any point along the radius of the cross section. In Section 3.2 we define the shearing strain as a change in angular displacement. Such an angular change is noted in Figure 6.2.1 and is the angle between lines AB and AC at the outer radius. For small displacements the sine of the angle is equal to the angle and shearing strain at the outer surface is seen to be

$$\gamma_{max} = \frac{\phi}{L}r_{max} \tag{6.2.3}$$

It varies linearly to zero at the origin of the radius; therefore, the shearing strain at any point in the body is

$$\gamma(x,r) = \frac{\phi}{L}r \tag{6.2.4}$$

Given that Hooke's law applies we have from Equation 3.4.13

$$\tau = G\gamma \tag{6.2.5}$$

where G is the *shear modulus*. Thus

$$\tau(x,r) = \frac{G\phi}{L}r \tag{6.2.6}$$

We note that the torque on any cross section is

$$T = \int_A \tau r dA = G \int_A \gamma r dA = G\frac{\phi}{L} \int_A r^2 dA \tag{6.2.7}$$

We define

$$\int_A r^2 dA = J \tag{6.2.8}$$

and we note that this is the polar moment of inertia of the circular cross section and is known as the *torsional constant*. Thus we obtain

$$T = \frac{GJ\phi}{L} \quad \rightarrow \quad \phi = \frac{TL}{GJ} = \frac{M_L L}{GJ} \tag{6.2.9}$$

The stress in terms of the applied load and the properties of the bar is found by combining Equations 6.2.6 and 6.2.9.

$$\tau = \frac{G\phi}{L}r = \frac{G}{L}\left(\frac{TL}{GJ}\right)r = \frac{Tr}{J} \tag{6.2.10}$$

Note, this is true only for a circular cross section. Non circular cross sections require a more advanced application of the theory of elasticity to define the stress distribution and the torsional constant. That is beyond the scope of this work for the most part; however, in Sections 6.5–6.8 we shall extend the analysis to consider the torsional stress in a class of cross sections of the so called thin walled type.

The shear stress distribution given in Equation 6.2.10 is a linear function in the radial direction as shown in Figure 6.2.3.

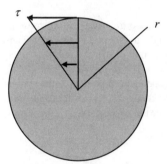

Figure 6.2.3

The maximum shear stress is found at the outer radius.

$$\tau_{max} = \frac{T}{J}r_{max} \tag{6.2.11}$$

It may be noted that the polar moment of inertia of a solid circular section is (see Appendix B)

$$J = \frac{\pi r_{max}^4}{2} \tag{6.2.12}$$

An actual experiment can be performed to find the value of the shear modulus for a particular material. We can take a circular shaft of that particular material, fasten it at one end, and apply a moment at the other, as in Figure 6.2.1, and gradually increase the moment M_L as we measure the rotation ϕ and plot the results. We will get a plot very like the one in Figure 3.4.2, shown here with appropriate symbols for torsion in Figure 6.2.4.

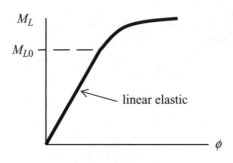

Figure 6.2.4

The slope of the strait portion we shall call k_t then

$$k_t = \frac{M_L}{\phi} \qquad \phi = \frac{M_L}{k_t} \qquad\qquad (6.2.13)$$

where k_t is called the *torsional stiffness* or *torsional spring constant*. The proportional limit is reached at M_{L0}. This is the point at which the curve in no longer linear. These values apply to the particular bar tested and are dependent upon the length of the bar and the radius of the cross section as well as the material.

To obtain the shear modulus independent of particular dimensions of the bar and thus related only to the material of the bar we replot the data in terms of stress and strain as shown is Figure 6.2.5. We then find G as the slope of the straight portion of the curve.

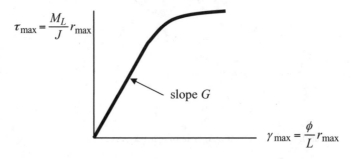

Figure 6.2.5

6.3   Derivation and Solution of the Governing Equations

Now consider the more general case of a circular shaft loaded with some combination of distributed torsional loads, $t(x)$, and concentrate torsional loads, M_c, as shown, for example, in Figure 6.3.1.

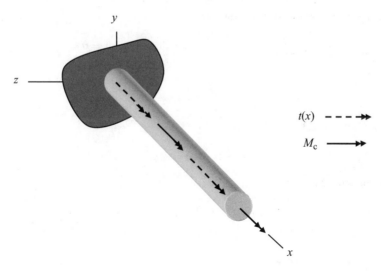

Figure 6.3.1

We shall denote the distributed displacement as $\beta(x)$. We may surmise as a generalization of Equation 6.2.4 that the local value of strain is proportional to the local rate of change of angular displacement. The strain displacement equation is then

$$\gamma(x,r) = \frac{d\beta}{dx}r \tag{6.3.1}$$

Tests verify that this is a valid assumption for long slender bars with circular cross sections. Combining Equations 6.2.5 and 6.3.1 we get for Hooke's law

$$\tau(x,r) = G\gamma = G\frac{d\beta}{dx}r \tag{6.3.2}$$

The internal torque $T(x)$ is

$$T(x) = \int_A \tau r dA = G\int_A \gamma r dA = G\frac{d\beta}{dx}\int_A r^2 dA = GJ\frac{d\beta}{dx} \tag{6.3.3}$$

Consider the equilibrium of a short segment of the bar of length dx with a distributed moment $t_x(x)$ applied as shown in Figure 6.3.2. We include the values of the internal torque at x and $x + dx$ from a Taylor series expansion neglecting higher order terms.

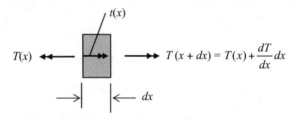

Figure 6.3.2

Summing the torques we satisfy static equilibrium.

$$\sum M_x = T(x) + \frac{dT}{dx}dx - T(x) + t(x)dx = 0 \qquad (6.3.4)$$

which reduces to

$$\frac{dT}{dx} = -t(x) \qquad (6.3.5)$$

Combining Equations 6.3.3 and 6.3.5 we obtain

$$\frac{d}{dx}GJ\frac{d\beta}{dx} = -t(x) \qquad (6.3.6)$$

From Equations 6.3.2 and 6.3.3 we get

$$\tau(x,r) = G\gamma(x,r) = G\frac{d\beta}{dx}r = \frac{T(x)}{J}r \qquad (6.3.7)$$

In summary

$$
\begin{array}{ccccc}
M_c & \rightarrow & t_x(x) & T(x) & \tau(x,r) & \gamma(x,r) & \beta(x) & \leftarrow \rho
\end{array}
$$

$$
\frac{dT}{dx} = -t_x(x) \qquad\qquad \gamma(x,r) = \frac{d\beta}{dx}r
$$

$$
\tau(x,r) = G\gamma = G\frac{d\beta}{dx}r
$$

$$(6.3.8)$$

$$
T(x) = \int_A \tau r dA = G\int_A \gamma r dA = G\frac{d\beta}{dx}\int_A r^2 dA = GJ\frac{d\beta}{dx} \quad\rightarrow\quad \tau(x,r) = \frac{T(x)}{J}r
$$

$$
\frac{dT}{dx} = \frac{d}{dx}GJ\frac{d\beta}{dx} = -t(x)
$$

In the first row of Equation 6.3.8 we have the typical unknowns in the formulation of a torsion problem.

$\beta(x)$ — distributed rotational displacement
$\gamma(x,r)$ — shearing strain in cylindrical coordinates
$\tau(x,r)$ — shearing stress in cylindrical coordinates
$T(x)$ — internal torque about the x axis

and the known quantities

$t_x(x)$ — distributed applied torque
M_c — concentrated applied torques
ρ — displacement restraints

The arrows are there to indicate that both M_c and ρ are boundary conditions to be satisfied in the process of solving the problem.

The equations connecting these quantities are in rows two and three through four.

$$\frac{dT}{dx} = -t_x(x) \quad - \text{ static equilibrium}$$

$$\gamma = \frac{d\beta}{dx}r \quad - \text{ strain displacement}$$

$$\tau = G\gamma = G\frac{d\beta}{dx}r \quad - \text{ Hooke's law for shear}$$

$$T = \int_A \tau r dA = GJ\frac{d\beta}{dx} \quad - \text{ internal torque definition}$$

and in the fifth row equilibrium is given in terms of the single unknown displacement.

$$\frac{d}{dx}GJ\frac{d\beta}{dx} = -t(x) \tag{6.3.9}$$

And, of course, when GJ is constant

$$GJ\frac{d^2\beta}{dx^2} = -t(x) \tag{6.3.10}$$

Now everything we learned in Chapter 4 about solving a second order differential equation by direct integration applies here. From Equation 6.3.10 we get

$$\frac{d\beta}{dx} = -\frac{1}{GJ}\int t_x(x)dx + a \quad \rightarrow \quad \beta(x) = -\frac{1}{GJ}\int \left(\int t_x(x)dx\right) dx + ax + b \tag{6.3.11}$$

where the constants of integration, a and b, are found by assigning boundary conditions.

Having found the rotational displacement $\beta(x)$ we then find the internal torque T from Equation 6.3.3 and then the stress τ from Equation 6.3.7. That is, in order we find

$$\beta(x) \quad \rightarrow \quad T = GJ\frac{d\beta}{dx} \quad \rightarrow \quad \tau = \frac{T}{J}r \tag{6.3.12}$$

This solution works for both statically determinate and indeterminate cases.

In the determinate case the internal torque, T, can be found directly from the equations of static equilibrium and so we can go directly to the stress in Equation 6.3.7.

If the displacement is desired we use

$$\beta(x) = \frac{1}{GJ}\int T(x)dx + c \tag{6.3.13}$$

where c is the constant of integration to be determined from a displacement boundary condition. We shall illustrate both approaches to a solution in the next simple example.

Here we pause to remind you that the set of equations and flow chart for the torsion problem is identical to the set of equations and flow chart for axial deformation if we make the following substitutions shown in Table 6.3.1.

Table 6.3.1

Axial	Torsional
$u(x)$	$\beta(x)$
EA	GJ
$f(x)$	$t(x)$
$dP/dx = -f(x)$	$dT/dx = -t(x)$
.	.
.	.
.	.

############

Example 6.3.1

Problem: A shaft with a circular cross section is fixed at one end and has a concentrated applied torque at the other end as shown in Figure (a). Find the stresses and displacements of the shaft.

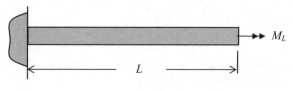

Figure (a)

Note: This is the problem we used as an experiment in Section 6.2 to aid in the derivation of the torsional equations. We shall now use it to illustrate how the differential equations and boundary conditions are used to solve the displacements and stresses.

Solution: First find the internal torque, the stress, and then solve the first order equation for the displacement. Then repeat the solution using the second order displacement equation.

As a first step, use equilibrium to find the internal torque in terms of the applied moment. A simple free body diagram of a portion of the bar is shown in Figure 6.2.2, and repeated here as Figure (b)

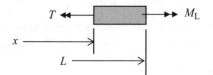

Figure (b)

From summation of moments the internal torque is

$$\sum M_x = M_L - T = 0 \quad \rightarrow \quad T = M_L \tag{a}$$

From Equation 6.3.7 the stress is

$$\tau = \frac{T}{J}r = \frac{M_L}{J}r \tag{b}$$

The displacement is obtained by integrating the first order equation.

$$GJ\frac{d\beta}{dx} = M_L \quad \rightarrow \quad \beta(x) = \frac{M_L}{GJ}\int dx + c = \frac{M_L}{GJ}x + c \tag{c}$$

From the boundary condition on the displacement at the left end

$$\beta(0) = \frac{M_L}{GJ} \cdot 0 + c = 0 \quad \rightarrow \quad c = 0 \quad \rightarrow \quad \beta(x) = \frac{M_L}{GJ}x \tag{d}$$

Now let us repeat the solution by starting with the second order equation for the displacement, Equation 6.3.10, and integrating twice. We get

$$GJ\frac{d^2\beta}{dx^2} = 0 \quad \rightarrow \quad \frac{d\beta}{dx} = a \quad \rightarrow \quad \beta(x) = ax + b \tag{e}$$

Now apply the boundary conditions for displacement on the left end and for applied load on the right end to find the constants of integration.

$$\beta(0) = a \cdot 0 + b = 0 \quad \rightarrow \quad b = 0$$

$$GJ\frac{d\beta(L)}{dx} = GJa = M_L \quad \rightarrow \quad a = \frac{M_L}{GJ} \quad \rightarrow \quad \beta(x) = \frac{M_L}{GJ}x \tag{f}$$

Now that we know $\beta(x)$ we can find T

$$T = GJ\frac{d\beta}{dx} = GJ\frac{M_L}{GJ} = M_L \tag{g}$$

and then find τ

$$\tau = \frac{T}{J}r = \frac{M_L}{J}r \tag{h}$$

Take a quick look at Example 4.3.1 for the axially loaded bar. This is exactly the same problem but with different symbols. Mathematically the torsional and axial cases for finding the displacements and internal forces are identical. This will apply to many other problems in this chapter.

Example 6.3.2

Problem: A shaft with a circular cross section is fixed at one end and has two concentrated applied torques as shown in Figure (a). Find the stresses and displacements of the shaft.

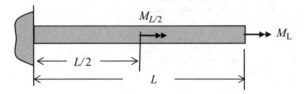

Figure (a)

Solution: Divide the shaft into two regions and integrate each equation. Apply boundary conditions to determine the values of the constants of integration.

The stresses are easy to find because this is a statically determinate case. The internal torque is

$$\text{Region 1:} \sum M_z = M_{L/2} + M_L - T_1 = 0 \quad \rightarrow \quad T_1 = M_{L/2} + M_L \quad 0 \le x \le \frac{L}{2}$$

$$\text{Region 2:} \sum M_z = M_L - T_2 = 0 \quad \rightarrow \quad T_2 = M_L \quad \frac{L}{2} < x \le L \tag{a}$$

The stresses are found by inserting the internal torque values in Equation 6.3.7.

$$\tau_1 = \frac{M_{L/2} + M_L}{J}r \quad 0 \le x \le \frac{L}{2}$$

$$\tau_2 = \frac{M_L}{J}r \quad \frac{L}{2} \le x \le L \tag{b}$$

The displacements require a bit more work. Let us call the displacements

$$\beta_1 \text{ from } 0 \le x \le \frac{L}{2} \quad \text{and} \quad \beta_2 \text{ from } \frac{L}{2} \le x \le L \tag{c}$$

then

$$GJ\frac{d\beta_1}{dx} = M_{L/2} + M_L \quad \rightarrow \quad \beta_1 = \frac{M_{L/2} + M_L}{GJ}x + a_1$$

$$GJ\frac{d\beta_2}{dx} = M_L \quad \rightarrow \quad \beta_2 = \frac{M_L}{GJ}x + a_2 \tag{d}$$

The boundary conditions are on the displacement at the left end and continuity of displacement at the point where the two displacement functions meet.

$$\beta_1(0) = 0 \qquad \beta_1\left(\frac{L}{2}\right) = \beta_2\left(\frac{L}{2}\right) \tag{e}$$

Therefore

$$\beta_1(0) = \frac{M_{L/2} + M_L}{GJ}\cdot 0 + a_1 = 0 \quad \rightarrow \quad a_1 = 0$$

$$\beta_1\left(\frac{L}{2}\right) = \beta_2\left(\frac{L}{2}\right) = \frac{M_{L/2} + M_L}{GJ}\frac{L}{2} = \frac{M_L}{GJ}\frac{L}{2} + a_2 \quad \rightarrow \quad a_2 = \frac{M_{L/2}L}{2GJ} \tag{f}$$

The displacements are

$$\beta_1 = \frac{M_{L/2} + M_L}{GJ}x \qquad \beta_2 = \frac{M_L}{GJ}x + \frac{M_{L/2}L}{2GJ} \tag{g}$$

The same results can be obtained by integrating the second order equation twice for each region and applying a boundary condition at each end and continuity of displacement and torque at $x = L/2$. The equations and their solution are

$$GJ\frac{d^2\beta_1}{dx^2} = 0 \quad \rightarrow \quad \beta_1 = a_1 x + b_1$$

$$GJ\frac{d^2\beta_2}{dx^2} = 0 \quad \rightarrow \quad \beta_2 = a_2 x + b_2 \tag{h}$$

The boundary conditions are the displacement at the left end, the applied moment at the right end, equilibrium of external and internal moments and continuity of displacement where the two regions meet.

$$\beta_1(0) = 0$$

$$GJ\frac{d\beta_2(L)}{dx} = M_L$$

$$GJ\frac{d\beta_2\left(\frac{L}{2}\right)}{dx} - GJ\frac{d\beta_1\left(\frac{L}{2}\right)}{dx} + M_{L/2} = 0 \tag{i}$$

$$\beta_2\left(\frac{L}{2}\right) = \beta_1\left(\frac{L}{2}\right)$$

Apply the boundary condition at the left end.

$$\beta(0)_1 = a_1 \cdot 0 + b_1 = 0 \quad \rightarrow \quad b_1 = 0 \tag{j}$$

Next satisfy the applied torque at the right end.

$$GJ\frac{d\beta_2(L)}{dx} = GJa_2 = M_L \quad \rightarrow \quad a_2 = \frac{M_L}{GJ} \tag{k}$$

Now satisfy equilibrium at the point where the two regions meet.

$$GJ\frac{d\beta_2\left(\frac{L}{2}\right)}{dx} - GJ\frac{d\beta_1\left(\frac{L}{2}\right)}{dx} + M_{L/2} = GJa_2 - GJa_1 + M_{L/2} = M_L - GJa_1 + M_{L/2} = 0 \tag{1}$$

$$\rightarrow \quad a_1 = \frac{M_L + M_{L/2}}{GJ}$$

Finally, satisfy continuity of displacement at the midpoint.

$$\beta_2\left(\frac{L}{2}\right) = \beta_1\left(\frac{L}{2}\right) = a_2\cdot\frac{L}{2} + b_2 = a_1\cdot\frac{L}{2} + b_1 = \frac{M_L}{GJ}\frac{L}{2} + b_1 = \frac{M_L + M_{L/2}}{GJ}\frac{L}{2} + b_2 \tag{m}$$

$$\rightarrow \quad b_2 = \frac{M_{L/2}}{GJ}\frac{L}{2}$$

The displacements are

$$\beta_1 = \frac{M_{L/2} + M_L}{GJ}x \qquad \beta_2 = \frac{M_L}{GJ}x + \frac{M_{L/2}L}{2GJ} \tag{n}$$

Now that the displacements are known you can find the internal torques and the stresses as shown in Equation 6.3.12, repeated here as Equation (o).

$$\beta(x) \quad \rightarrow \quad T = GJ\frac{d\beta}{dx} \quad \rightarrow \quad \tau = \frac{T}{J}r \tag{o}$$

Example 6.3.3

Problem: Consider other loadings and restraints, for example, a uniform distributed applied moment t_0 with the left end fixed, as shown in Figure (a), or a shaft with both ends fixed and a concentrated applied moment at x_0 as shown in Figure (b).

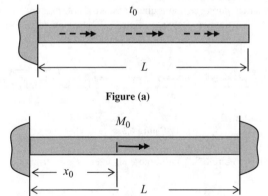

Figure (a)

Figure (b)

Solution: Integrate the second order equation and apply boundary conditions to determine the values of the constants of integration.

Since the second order equation and restraints for torsion are the same form as the second order equation and restraints for an axially loaded bar the solution methods are identical. Just the symbols and the quantities they represent are different. Therefore the solution procedure is exactly the same for these two problems as found in Examples 4.4.1 and 4.4.2.

We shall just summarize the results here.

For the shaft in Figure (a)

$$\beta(x) = \frac{t_0}{GJ}\left(Lx - \frac{x^2}{2}\right)$$

$$T = GJ\frac{d\beta}{dx} = t_0(L - x) \qquad \tau = G\frac{d\beta}{dx}r = \frac{t_0}{J}(L - x)r$$

(a)

For the shaft in Figure (b)

$$\beta_1(x) = \frac{M_{x_0}}{GJ}\left(1 - \frac{x_0}{L}\right)x \qquad\qquad \beta_2(x) = \frac{M_{x_0}}{GJ}\left(1 - \frac{x}{L}\right)x_0$$

$$T_1 = GJ\frac{d\beta}{dx} = M_{x_0}\left(1 - \frac{x_0}{L}\right) \qquad \tau_1 = G\frac{d\beta}{dx}r = \frac{M_{x_0}}{J}\left(1 - \frac{x_0}{L}\right)r$$

$$T_2 = GJ\frac{d\beta}{dx} = -M_{x_0}\frac{x_0}{L} \qquad\qquad \tau_2 = G\frac{d\beta}{dx}r = -\frac{M_{x_0}}{J}\frac{x_0}{L}r$$

(b)

###########

In summary there are four possible sets of boundary restraint. For one the shaft can be in static equilibrium under a set of applied loads as shown in Figure 6.3.3.

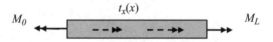

$$M_0 \qquad\qquad t_x(x) \qquad\qquad M_L$$

Figure 6.3.3

In such cases we assign the origin of the coordinate system to be the point from which the displacements are measured. For all practical purposes it amounts to the kinds of restraints in the next figure.

The one in Figure 6.3.3 and the two shown in Figure 6.3.4 are statically determinate.

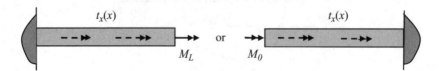

Figure 6.3.4

The fourth case shown in Figure 6.3.5 with both ends fixed is indeterminate.

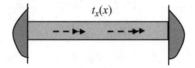

Figure 6.3.5

Either the first order or the second order equation for displacement may be used in the first three cases. The second order equation must be used in the fourth case.

If $t(x)$ is discontinuous, or E is discontinuous, or A is discontinuous, or midpoint concentrated loads are present the shaft must be divided into regions with separate equations for each region. Continuity boundary conditions between regions are then imposed. This process presented in detail in Chapter 4 for the axial case applies here as well.

6.4 Solutions from the Theory of Elasticity

How do the solutions in Section 6.3 hold up as solutions to the three dimensional equations of elasticity? We note that the torsional stress component, τ, in Equation 6.2.10 can be resolved into two components in rectangular Cartesian coordinates as shown in Figure 6.4.1.

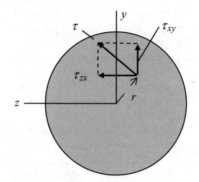

Figure 6.4.1

It follows that

$$\tau = \frac{T}{J}r \quad \rightarrow \quad \tau_{xy} = \frac{T}{J}z \qquad \tau_{zx} = \frac{T}{J}y \tag{6.4.1}$$

In the absence of body forces these together with the other stresses

$$\sigma_x = \sigma_y = \sigma_z = \tau_{yz} = 0 \tag{6.4.2}$$

exactly satisfy the three dimensional equations of equilibrium, Equations 2.6.13.

$$
\begin{bmatrix}
\dfrac{\partial}{\partial x} & 0 & 0 & \dfrac{\partial}{\partial y} & 0 & \dfrac{\partial}{\partial z} \\[2ex]
0 & \dfrac{\partial}{\partial y} & 0 & \dfrac{\partial}{\partial x} & \dfrac{\partial}{\partial z} & 0 \\[2ex]
0 & 0 & \dfrac{\partial}{\partial z} & 0 & \dfrac{\partial}{\partial y} & \dfrac{\partial}{\partial x}
\end{bmatrix}
\begin{bmatrix}
0 \\ 0 \\ 0 \\ \dfrac{T}{J}z \\ 0 \\ \dfrac{T}{J}y
\end{bmatrix}
=
\begin{bmatrix}
0 \\ 0 \\ 0
\end{bmatrix}
\tag{6.4.3}
$$

They also satisfy the conditions of applied load on the boundaries provided that the cross section is circular with the end loads distributed the same as the stresses in the interior.

Compatibility is also satisfied. In Section 3.3 it is noted that compatibility requires that certain equations involving the second derivatives of the strain components must be satisfied. In this case the strain components are

$$\gamma_{xy} = \frac{\tau_{xy}}{G} = \frac{T}{GJ}z \qquad \gamma_{yz} = 0 \qquad \gamma_{zx} = \frac{\tau_{zx}}{G} = -\frac{T}{GJ}y \tag{6.4.4}$$

The second derivatives of linear functions are zero and therefore the conditions of compatibility are satisfied. For the one equation

$$\frac{\partial^2 \varepsilon_x}{\partial y^2} + \frac{\partial^2 \varepsilon_y}{\partial x^2} = \frac{\partial^2 \gamma_{xy}}{\partial x \partial y} \quad \rightarrow \quad \frac{\partial^2}{\partial y^2}(0) + \frac{\partial^2}{\partial x^2}(0) = \frac{\partial^2}{\partial x \partial y}\left(\frac{T}{GJ}z\right) = 0 \tag{6.4.5}$$

Similar results are obtained for the other compatibility equations in three dimensions.

If you change the geometry, load distribution, or restraints you no longer have an exact solution with the torsional theory presented here. Non circular sections, except as noted in Section 6.5 for thin wall cross sections, require a more advanced theory. Since the circular section usually is the most efficient it is frequently used in structures where torsion is dominant if other requirements allow.

Additional stress components are generated for a circular section if the end loads are anything other than linear with radius, or if the end is restrained in a particular way. In such cases we depend upon St. Venant's principle to save us from our sins. It can be shown that the effects are localized near the restraint or load application.

This leads us to ask just how loads are applied to a shaft and precisely what effect do they have both locally and globally? In Figure 6.4.2 we represent a shaft with the left end fixed and with a distributed load applied tangentially to a portion of the surface at the right end. This provides a pure torque at the right end but with a different distribution of local stress from the linear with radius values of Equation 6.2.10.

Figure 6.4.2

A computer based numerical solution using the full three dimensional equations of elasticity was obtained for a specific case. The Unigraphics I-DEAS software is used. The basis for these equations and the validity of this solution are presented in Chapters 11 and 12.

Contour plots of the shear stress values are shown in Figure 6.4.3.

The stress over most of the length is equal to that obtained by Equation 6.2.10. What is shown on the cylindrical surface is the stress on that surface or τ_{max}. For this particular geometry and loading the maximum shear stress at the surface obtained from Equation 6.2.11 is 0.8 MPa. At the loaded end there are higher stresses near the edge between the loaded surface and the free surface. They reach a maximum of 1.5 MPa in the region shown in red (darkest in gray scale) in Figure 6.4.3.

This is another example of stress concentration that occurs at regions of abrupt change in geometry, restraint, or loading. Note that the region of transition is very short, thus confirming St. Venant once

RESULTS: 2- B. C. 1, STRESS_2, LOAD SET 1
STRESS - MAX SHEAR M N: 5. 21E- 03 MAX: 1. 50E+00
FRAME OF REF: PART

VALUE OPTI ON: ACTUAL

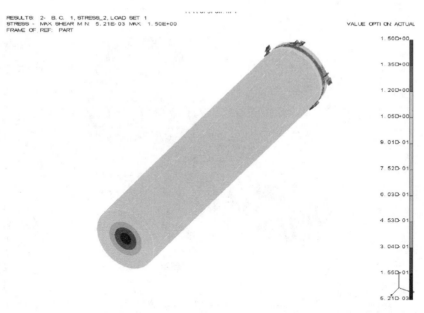

1. 50D+00
1. 35D+00
1. 20D+00
1. 05D+00
9. 01D- 01
7. 52D- 01
6. 03D- 01
4. 53D- 01
3. 04D- 01
1. 55D- 01
5. 21D- 03

Figure 6.4.3

again. At the fixed end it is essentially the same as for the rest of the shaft, that is, we can confirm a linear distribution with radius. Thus we conclude that in this case the restraint does not impose conditions that would modify the results from the simplified theory any significant amount.

In practice torsional loadings are often provided by pulleys or gears attached to the shaft so that the shaft acts as an axle. We add a disk to the shaft in Figure 6.4.4 and apply a tangential load as before. The load magnitude is chosen to provide the same total torque as in the previous example. The left end is fixed as before.

Figure 6.4.4

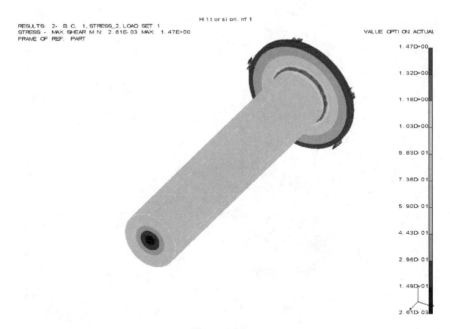

RESULTS: 2- B.C. 1, STRESS_2, LOAD SET 1
STRESS - MAX SHEAR M N: 2.61E-03 MAX: 1.47E+00
FRAME OF REF: PART

H.\t or si on .nf 1

VALUE OPTION: ACTUAL

1.47D+00
1.32D+00
1.18D+00
1.03D+00
8.83D-01
7.36D-01
5.90D-01
4.43D-01
2.96D-01
1.49D-01
2.61D-03

Figure 6.4.5

Contour plots of the shear stress values are shown in Figure 6.4.5.

The solution over most of the shaft is the same. Once again we get a stress concentration as the diameter changes abruptly where the shaft and disk join. The highest value of shear stress is 1.47 *MPa* which is only slightly different from the previous case. The color over the main portion is slightly different due to the way the contours are divided between the highest and lowest stress value. The stress values on the shaft are the same.

One way to lessen the maximum stresses is to avoid abrupt changes in geometry, loading, or restraint. In Figure 6.4.6 we have added a fillet between the shaft and the disk.

Figure 6.4.6

The contour plot is shown in Figure 6.4.7.

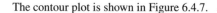

Figure 6.4.7

The maximum stress is now just 0.904 *MPa*. This is just a bit over the 0.8 *MPa* for the shaft. Once again the color changes due to the way the contours are divided. Stress concentrations are discussed in more detail in later chapters. In any case the simplified theory works over most of the length of the shaft and therefore is an efficient way to get a first approximation to the behavior of the shaft.

6.5   Torsional Stress in Thin Walled Cross Sections

The equations derived and solved in Section 6.2 apply for hollow circular cross sections as shown in Figure 6.5.1 as well. The shear stress obtained from Equation 6.2.10 has the distribution as shown. The inner radius is r_i and the outer radius is r_o.

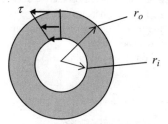

Figure 6.5.1

In this case the polar moment of inertia or torsional constant is (See Appendix B)

$$J = \frac{\pi}{2}\left(r_o^4 - r_i^4\right) \tag{6.5.1}$$

This torsional constant may be used in all the torsional equations for hollow circular sections.

In cases where the inner radius is only slightly smaller than the outer radius the values of the shear stress are only slightly different along the inner and outer radii. It is common practice to find the shear stress at the mid radius between the inner and outer radii and multiply it by the difference in two radii, or the wall thickness of the cylinder, which we shall designate as b. This is called the *shear flow* and has the units of force per unit length along the circumference. This is illustrated in Figure 6.5.2 where the mid radius is designated r_{mid} and the shear flow q where

$$q = \tau_{mid} b \tag{6.5.2}$$

The shear stress is assumed to be constant across the wall thickness.

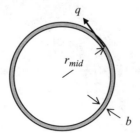

Figure 6.5.2

We can write the torsional constant in terms of the mid radius and the wall thickness.

$$J = \frac{\pi}{2}\left(r_o^4 - r_i^4\right) = \frac{\pi}{2}\left[\left(r_{mid} + \frac{b}{2}\right)^4 - \left(r_{mid} - \frac{b}{2}\right)^4\right] \tag{6.5.3}$$

If we expand the right side and neglect all terms of order b^2 and higher, we obtain the following approximation for the torsional constant when b is small compared to r_{mid}.

$$J = 2\pi r_{mid}^3 b \tag{6.5.4}$$

Noting that

$$q = \tau b = \frac{T r_{mid} b}{J} = \frac{T r_{mid} b}{2\pi r_{mid}^3 b} = \frac{T}{2\pi r_{mid}^2} = \frac{T}{2 A_T} \tag{6.5.5}$$

where $A_T = \pi r_{mid}^2$ is the area enclosed by the circle of radius r_{mid}. The subscript is added so that this area will not be confused with the area of the cross section material.

This can be extended to non circular thin walled closed cross sections. Because there is no shear stress on the inner and outer surfaces of the tube and because the wall is so thin no significant shear stress can exist except tangent to the wall of the tube. Thus we assume a constant shear flow all around the cross section. This assumption is verified by experiment.

Consider a tube with the cross section shown in Figure 6.5.3 and a constant wall thickness.

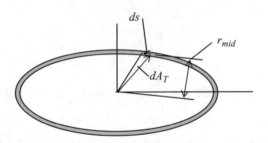

Figure 6.5.3

Assuming a constant shear flow the torque is

$$T = \int q r_{mid} ds = q \int r_{mid} ds \tag{6.5.6}$$

where the integral is around the closed boundary. This integral can be replaced by an area integral when we note that the triangular area is

$$dA_T = \frac{1}{2} r_{mid} ds \tag{6.5.7}$$

Thus

$$T = 2q \int dA_T = 2q A_T \tag{6.5.8}$$

where A_T is the area enclosed by the thin walled tube. This applies to any closed thin walled section.
Given the torque at a cross section the shear flow and shear stress are

$$q = \frac{T}{2A_T} \qquad \tau = \frac{T}{2A_T b} \tag{6.5.9}$$

###########

Example 6.5.1

Problem: Find the stress in a tube with a rectangular thin walled cross section with dimensions shown in Figure (a) in terms of the torque T. The wall thickness is a constant value b.

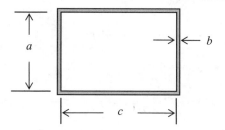

Figure (a)

Solution: Apply Equation 6.5.9.
If the wall is quite thin it makes little difference if the outside or if the inside or if the median dimensions are used in defining the enclosed area.

$$q = \frac{T}{2A_T} = \frac{T}{2ac} \quad \rightarrow \quad \tau = \frac{T}{2acb} \tag{a}$$

###########

6.6 Work and Energy—Torsional Stiffness in a Thin Walled Tube

Work and strain energy can be used to find displacements at points where concentrated moments are applied. The applied moments move through rotations to provide the external work. The strain energy in torsion for a circular shaft of length L is

$$U_T = \frac{1}{2}\int \gamma\tau\,dV = \frac{1}{2}\int \frac{\tau^2}{G}\,dV = \frac{1}{2}\int \frac{1}{G}\left(\frac{Tr}{J}\right)^2 dV = \frac{1}{2}\int_0^L \frac{T^2}{GJ^2}\,dx \int_A r^2\,dA = \frac{1}{2}\int_0^L \frac{T^2}{GJ}\,dx$$

(6.6.1)

In Example 6.6.1 we revisit Example 6.2.1.

###########

Example 6.6.1

Problem: Find the displacement at the right end. The cross section is circular.

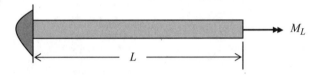

Figure (a)

Solution: Equate the external work to the internal strain energy and solve for the displacement. The external work done by the applied moment is

$$W^e = \frac{1}{2}M_L\beta(L) \tag{a}$$

Equate this to the strain energy and solve.

$$\frac{1}{2}M_L\beta(L) = \frac{1}{2}\int_0^L \frac{T^2}{GJ}\,dx = \frac{1}{2}\frac{T^2L}{GJ} \quad\rightarrow\quad \beta(L) = \frac{TL}{GJ} \tag{b}$$

This agrees with the result we obtained in Example 6.3.1, Equation (f).

###########

When there are multiple loads on a determinate shaft, including distributed and concentrated loads, for example, as in Figure 6.6.1, we can use Castigliano's second theorem to find the displacements at the points of concentrated load application. Castigliano's theorem was first presented in Chapter 4, Section 4.11.

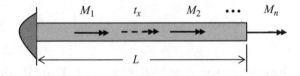

Figure 6.6.1

First find the internal torques in terms of the applied moments, that is, find $T_i(M_1, M_2, \ldots, M_n, t_x)$ for each region. The total strain energy is then the sum of the strain energies for each region.

$$U_T(M_1, M_2, \ldots, M_n, t_x) = \sum_{i=1}^{n} \frac{1}{2} \frac{T_i^2 L}{GJ} \qquad (6.6.2)$$

The displacements at the points of concentrated load application are

$$\frac{\partial U_T}{\partial M_i} = \beta_i \qquad (6.6.3)$$

###########

Example 6.6.2

Problem: A shaft is fixed at the left end. It is loaded with a concentrated applied moment at $x = 400$ *mm* and a distributed moment t over half the length as shown in Figure (a). Find the displacement at the right end where the concentrated moment is applied. The distributed moment is

$$t(x) = -210 \frac{N \cdot mm}{mm} \qquad 0 \le x \le 200$$

$$t(x) = 0 \qquad 200 \le x \le 400 \qquad (a)$$

The material is steel with a shear modulus of 80,000 *MPa*. The polar moment of inertia is $J = 20000$ *mm*⁴.

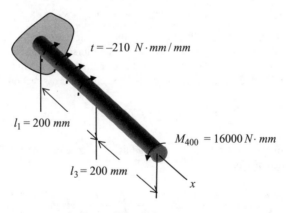

Figure (a)

Solution: Use Castigliano's second theorem.

We must carry the concentrated moment at the tip as a symbolic M_{400} in order to be able to take the derivative. We can enter the numerical value after that.

$$T_1 = M_{400} - 210 \cdot x \quad N \cdot mm \qquad 0 \le x \le 200$$
$$T_2 = M_{400} \quad N \cdot mm \qquad 200 \le x \le 400 \qquad (b)$$

The strain energy is

$$
\begin{aligned}
U_T &= \frac{1}{2GJ}\left(\int_0^{200} T_1^2 dx + \int_{200}^{400} T_2^2 dx+\right)\\
&= \frac{1}{2GJ}\left(\int_0^{200}(M_{400}-210\cdot x)^2 dx + \int_{200}^{400}(M_{400})^2 dx\right)\\
&= \frac{1}{2GJ}\left\{\int_0^{200}(M_{400})^2 dx - \int_0^{200}(2M_{400}\cdot 210\cdot x)dx + \int_0^{200}(210\cdot x)^2 dx + \int_{200}^{400}(M_{400})^2 dx\right\}\\
&= \frac{1}{2GJ}\left\{(M_{400})^2\cdot 200 - M_{400}\cdot 210\cdot 200^2 + 70\cdot 210\cdot 200^3 + (M_{400})^2\cdot 200\right\}\\
&= \frac{1}{2GJ}\left\{(M_{400})^2\cdot 400 - M_{400}\cdot 210\cdot 200^2 + 70\cdot 210\cdot 200^3\right\} \tag{c}
\end{aligned}
$$

The displacement is

$$
\begin{aligned}
\beta(400) &= \frac{\partial U_T}{\partial M_{400}} = \frac{1}{2GJ}\left\{2\cdot 400\cdot M_{400} - 210\cdot 200^2\right\}\\
&= \frac{2\cdot 400\cdot 16000 - 210\cdot 200^2}{2\cdot 80000\cdot 20000} = 0.001375 \text{ rad} = 0.0788° \tag{d}
\end{aligned}
$$

If the displacement is desired at some other point place a symbolic concentrated applied load at that point, find the strain energy, differentiate with respect to that load, and then set the value of the concentrated load to zero.

###########

The torsional constant for the thin walled tube of a non circular cross section is not the polar moment of inertia. To use the displacement equations requires a valid torsional constant. We can use the work and energy principles to obtain and effective torsional constant which we shall label J_{eff}.

Suppose the shaft in Example 6.6.1 has a thin walled non circular cross section. The internal torque is

$$
\sum M_x = M_L - T = 0 \quad \rightarrow \quad T = M_L \tag{6.6.4}
$$

If we define a coordinate s that follows the mid radius of the tube as shown in Figure 6.5.3 the strain energy in the thin walled tube is

$$
U = \frac{1}{2}\int_V \tau\gamma\,dV = \frac{1}{2}\int_V \frac{\tau^2}{G}dV = \frac{1}{2G}\int_V \frac{T^2}{(2A_T b)^2}dxdA = \frac{T^2 L}{8A_T^2 G}\int_s \frac{ds}{b} \tag{6.6.5}
$$

The integral is along the path of the local mid radius.

The external work done by the applied moment is

$$
W^e = \frac{1}{2}M_L\phi = \frac{1}{2}M_L\beta(L) \tag{6.6.6}
$$

where ϕ is the total angle of twist at the loaded end (see Figure 6.2.1).

Equating the external work to the strain energy gives us

$$W^e = \frac{1}{2} M_L \beta(L) = U = \frac{T^2 L}{8 A_T^2 G} \int_s \frac{ds}{b} \quad \rightarrow \quad \beta(L) = \frac{TL}{4 A_T^2 G} \int_s \frac{ds}{b} \tag{6.6.7}$$

For a circular section we note that

$$\beta(L) = \frac{M_L L}{GJ} \tag{6.6.8}$$

So let us define an effective torsion constant for the non circular section

$$\beta(L) = \frac{TL}{4 A_T^2 G} \int_s \frac{ds}{b} = \frac{M_L L}{GJ_{eff}} \quad \rightarrow \quad J_{eff} = \frac{1}{\frac{1}{4 A_T^2} \int_s \frac{ds}{b}} = \frac{4 A_T^2}{\int_s \frac{ds}{b}} \tag{6.6.9}$$

If you use this formula to calculate the J_{eff} of a thin walled circular section you get

$$J_{eff} = \frac{4 A_T^2}{\int_s \frac{ds}{b}} = \frac{4 \left(\pi r_{mid}^2 \right)^2 b}{2 \pi r_{mid}} = 2 \pi r_{mid}^3 b = J \tag{6.6.10}$$

This new formula agrees with the value obtained before for the circular section in Equation 6.5.4. This effective torsional stiffness can be used to find the displacement using Equation 6.3.10

$$GJ_{eff} \frac{d^2 \beta}{dx^2} = -t(x) \tag{6.6.11}$$

and to find the internal torque using Equation 6.3.3

$$GJ_{eff} \frac{d\beta}{dx} = T \tag{6.6.12}$$

###########

Example 6.6.3

Problem: The shaft in Example 6.6.1 has the thin walled cross section in Example 6.5.1. What is its displacement?

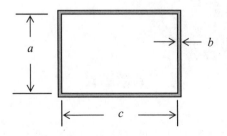

Figure (a)

Solution: Calculate J_{eff} and insert it in the displacement formula.

For this cross section

$$J_{eff} = \frac{4A_T^2}{\int_s \frac{ds}{b}} = \frac{4(ac)^2 b}{2a + 2c} = \frac{2(ac)^2 b}{a + c} \tag{a}$$

From Example 6.3.1

$$\beta(x) = \frac{M_L}{GJ_{eff}} x = \frac{M_L}{G} \frac{a + c}{2(ac)^2 b} x \tag{b}$$

and the total twist angle is

$$\beta(L) = \frac{M_L}{GJ_{eff}} L = \frac{M_L}{G} \frac{a + c}{2(ac)^2 b} L \tag{c}$$

###########

Let us confirm the assumption that the stress is constant in a non circular thin walled tube by a numerical solution that accounts for the local effects of loading and restraint. This is the same analysis method used for the illustrations in Section 6.4 that will be explained in more detail in Chapters 11 and 12.

The first example is one with an oval cross section as shown in Figure 6.6.2.

Figure 6.6.2

Once again the left end is fixed and a torque is applied to the right end by tangent surface loads. The shear stress distribution is shown in Figure 6.6.3.

RESULTS: 2- B.C. 1, STRESS_2, LOAD SET 1
STRESS - MAX SHEAR M N: 2.22E+00 MAX: 9.93E+00
FRAME OF REF: PART

VALUE OPTION: ACTUAL
SHELL SURFACE: TOP

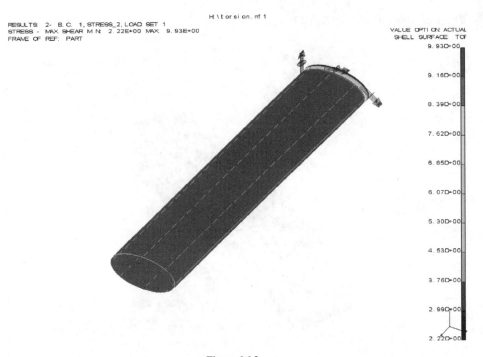

9.93D+00

9.16D+00

8.39D+00

7.62D+00

6.85D+00

6.07D+00

5.30D+00

4.53D+00

3.76D+00

2.99D+00

2.22D+00

Figure 6.6.3

The dark color does not show well in gray scale. The point is that the whole surface is one color indicating a constant stress except for a very short region near the load application.

Next we solve for a square cross section as shown in Figure 6.6.4.

Figure 6.6.4

The stress distribution is shown in Figure 6.5.5.

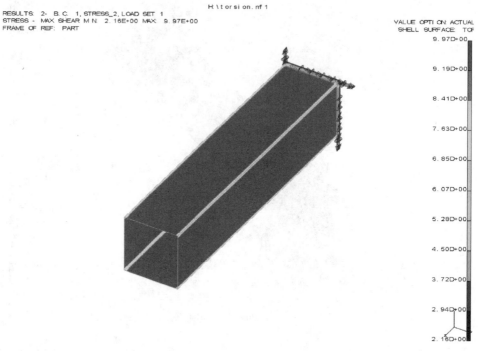

Figure 6.6.5

This shows a small difference at the sharp corners where the stress components make a 90° degree change in direction. On the flat surfaces the stress is constant.

Let us round off the corners as shown in Figure 6.6.6 and try again.

Figure 6.6.6

The stress contours are shown in Figure 6.6.7.

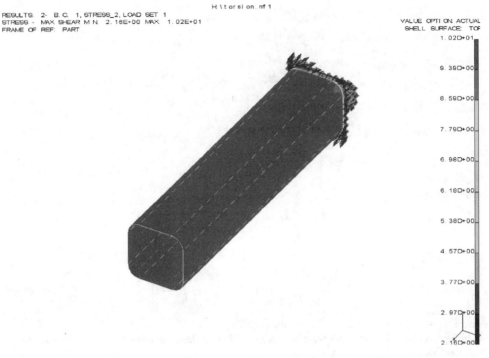

H:\torsion.mf 1
RESULTS: 2- B.C. 1, STRESS_2, LOAD SET 1
STRESS - MAX SHEAR MN: 2.16E+00 MAX: 1.02E+01
FRAME OF REF: PART

VALUE OPTION: ACTUAL
SHELL SURFACE: TOP

1.02D+01
9.39D+00
8.59D+00
7.79D+00
6.98D+00
6.18D+00
5.38D+00
4.57D+00
3.77D+00
2.97D+00
2.16D+00

Figure 6.6.7

The stresses are constant except for the locale of the applied forces.

6.7 Torsional Stress and Stiffness in Multicell Sections

We can extend our results to the torsion in slender bars with multi celled thin walled cross sections. The main assumption is that all cells of the cross section experience the same amount of twist. Experimental evidence confirms that this is a reasonable assumption.

Consider the multi celled section in Figure 6.7.1.

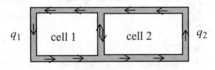

q_1 cell 1   cell 2 q_2

Figure 6.7.1

From Equation 6.5.8 we know that

$$T = 2q A_T \tag{6.7.1}$$

therefore, for two cells

$$T = 2q_1 A_{T1} + 2q_2 A_{T2} \tag{6.7.2}$$

From Equations 6.6.12 and 6.6.9 we have

$$GJ_{eff}\frac{d\beta}{dx} = T \qquad J_{eff} = \frac{4A_T^2}{\displaystyle\int_s \frac{ds}{b}} \tag{6.7.3}$$

Combining we obtain

$$\frac{d\beta}{dx} = \frac{1}{GJ_{eff}}(2q_1 A_{T1} + 2q_2 A_{T2}) \tag{6.7.4}$$

We also know that

$$\frac{d\beta}{dx} = \frac{1}{2A_T}\oint \frac{q}{Gb}ds \tag{6.7.5}$$

and so for cell 1

$$\frac{d\beta}{dx} = \frac{1}{2GA_{T1}}\left(q_1\oint_{s_1}\frac{ds}{b} - q_2\oint_{s_{12}}\frac{ds}{b}\right) \tag{6.7.6}$$

and for cell 2

$$\frac{d\beta}{dx} = \frac{1}{2GA_{T2}}\left(q_2\oint_{s_{12}}\frac{ds}{b} - q_1\oint_{s_{12}}\frac{ds}{b}\right) \tag{6.7.7}$$

where s_{12} is the length common to both cells.

In Equations 6.7.4 and 6.7.6-7 we have three equations in three unknowns, q_1, q_2, and $d\beta/dx$. Let us do an example.

###########

Example 6.7.1

Problem: A two celled cross section has the dimensions shown in Figure (a). The wall thickness is a constant value of b. Find the shear flows and the rate of twist for a pure internal torque T.

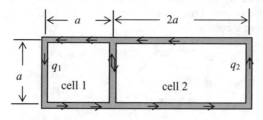

Figure (a)

Solution: Solve for the three unknowns using the equations noted above.

We can calculate the following:

$$A_{T1} = a^2 \qquad A_{T2} = 2a^2 \qquad \oint_{s_{12}}\frac{ds}{b} = \frac{a}{b} \qquad \oint_{s_1}\frac{ds}{b} = \frac{4a}{b} \qquad \oint_{s_2}\frac{ds}{b} = \frac{6a}{b} \tag{a}$$

It follows that

$$T = 2q_1 A_{T1} + 2q_2 A_{T2} = 2a^2 q_1 + 4a^2 q_2 \tag{b}$$

$$\frac{d\beta}{dx} = \frac{1}{2GA_{T1}} \left(q_1 \oint_{s_1} \frac{ds}{b} - q_2 \oint_{s_{12}} \frac{ds}{b} \right) = \frac{1}{2Ga^2} \left(q_1 \frac{4a}{b} - q_2 \frac{a}{b} \right) \tag{c}$$

$$\frac{d\beta}{dx} = \frac{1}{2GA_{T2}} \left(q_2 \oint_{s_{12}} \frac{ds}{b} - q_1 \oint_{s_{12}} \frac{ds}{b} \right) = \frac{1}{4Ga^2} \left(q_2 \frac{6a}{b} - q_1 \frac{a}{b} \right) \tag{d}$$

Equations (c) and (d) may be rewritten and solved for the shear flows in terms of the rate of twist.

$$4q_1 - q_2 = 2Gab\frac{d\beta}{dx} \qquad \qquad q_1 = \frac{16}{23} Gab \frac{d\beta}{dx}$$
$$\longrightarrow \tag{e}$$
$$-q_1 + 6q_2 = 4Gab\frac{d\beta}{dx} \qquad \qquad q_2 = \frac{18}{23} Gab \frac{d\beta}{dx}$$

The torque can now be written in terms of the rate of twist.

$$T = 2a^2 q_1 + 4a^2 q_2 = 2a^2 \left(\frac{16}{23} Gab \frac{d\beta}{dx} \right) + 4a^2 \left(\frac{18}{23} Gab \frac{d\beta}{dx} \right) = \frac{104}{23} Ga^3 b \frac{d\beta}{dx} \tag{f}$$

Since

$$T = GJ_{eff} \frac{d\beta}{dx} = \frac{104}{23} Ga^3 b \frac{d\beta}{dx} \qquad \longrightarrow \qquad J_{eff} = \frac{104}{23} a^3 b = 4.52 a^3 b \tag{g}$$

and

$$q_1 = 0.154 \frac{T}{a^2} \qquad q_2 = 0.173 \frac{T}{a^2} \tag{h}$$

The shear flow in the interior web is

$$q_2 - q_1 = (0.173 - 0.154) \frac{T}{a^2} = 0.019 \frac{T}{a^2} \tag{i}$$

Let us check to see how the interior web affects the result by removing it and comparing the answers. If the interior web is not present, then

$$\frac{d\beta}{dx} = \frac{1}{2GA_T} \oint_s \frac{q}{b} ds = \frac{1}{2G(3a^2)} \frac{q}{b} 8a = \frac{4}{3} \frac{q}{abG} \tag{j}$$
$$T = 2A_T q = 6a^2 q$$

Therefore

$$T = 6a^2 \frac{3}{4} abG \frac{d\beta}{dx} = 4.50a^3 bG \frac{d\beta}{dx} = GJ_{eff} \frac{d\beta}{dx} \tag{k}$$

Thus

$$J_{eff} = 4.50a^3 b \qquad q = 0.17 \frac{T}{a^2} \tag{l}$$

The internal web has a small effect. While the interior web has a small effect on the torsional stress we shall find out later that the effect on shear stress in bending is much greater.

############

6.8   Torsional Stress and Displacement in Thin Walled Open Sections

It has been observed that the torsional stress in a narrow rectangular section has a stress distribution that is essentially parallel to the outer surface as illustrated in Figure 6.8.1.

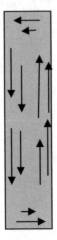

Figure 6.8.1

When the rectangle is very narrow and the section is subjected to pure torque the shear stress can be approximated as a linear distribution through the thickness as shown in Figure 6.8.2.

The stress does have to stay parallel to the surface at the top and bottom; however, over most of the section the stress can be approximated. The stress at the narrow end is found to be small and can be neglected as far as finding maximum values of stress is concerned.

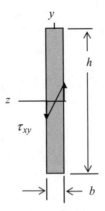

Figure 6.8.2

The stress is found to be

$$\tau_{xy} = \frac{T}{J_{eff}} z \qquad\qquad (6.8.1)$$

where the maximum stress is found at the two surfaces, that is, at $z = \pm b/2$, Without proof the torsional constant for the thin walled section is found to be

$$J_{eff} = \frac{hb^3}{3} \tag{6.8.2}$$

The displacement is found from

$$GJ_{eff}\frac{d\beta}{dx} = T \tag{6.8.3}$$

This can be extended to thin walled cross sections of other shapes. For the cross section in Figure 6.8.3a the effective torsional constant is given by Equation 6.8.2 where h is the length of the arc midway between the inner and outer edges. It can also be applied to multiple sections as, for example, in Figure 6.8.3b.

$$J_{eff} = \frac{h_1 b_1^3}{3} + \frac{h_2 b_2^3}{3} \tag{6.8.4}$$

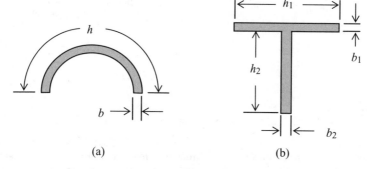

(a)   (b)

Figure 6.8.3

###########

Example 6.8.1

Problem: A cantilever bar with a cross section as defined in Figure (a) has a pure torque applied at the free end. Find the stress and displacement.

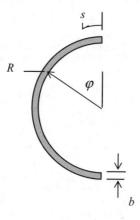

Figure (a)

Solution: Use Equations 6.8.1–6.8.3.

The torsional constant is

$$J_{eff} = \frac{\pi R b^3}{3} \tag{a}$$

The torsional stress is

$$\tau = \frac{T}{J_{eff}} \delta = \frac{3T}{\pi R b^3} \delta \tag{b}$$

where δ is measured normal to the mid radius and has the maximum values of $\pm b/2$.

The torsional displacement is

$$GJ_{eff} \frac{d\beta}{dx} = T \quad \rightarrow \quad \frac{d\beta}{dx} = \frac{T}{GJ_{eff}} \quad \rightarrow \quad \beta(x) = \frac{T}{GJ_{eff}} x + a \tag{c}$$

This is a cantilever bar so

$$\beta(0) = a = 0 \quad \rightarrow \quad \beta(x) = \frac{T}{GJ_{eff}} x \tag{d}$$

###########

This is a good time to compare the torsional stiffness and maximum stresses in open and closed thin walled sections. In Section 6.5, Equation 6.5.9 we note that the shear stress in a closed section under a pure toque load is

$$q = \frac{T}{2A_T} \quad \rightarrow \quad \tau = \frac{T}{2A_T b} \tag{6.8.5}$$

The torsion stiffness is given in Section 6.6, Equation 6.6.9, to be

$$J_{eff} = \frac{1}{\dfrac{1}{4A_T^2} \displaystyle\int_s \frac{ds}{b}} = \frac{4A_T^2}{\displaystyle\int_s \frac{ds}{b}} \tag{6.8.6}$$

For a circular thin walled section this reduces to the polar moment of inertia as shown in Equation 6.6.10.

$$J_{eff} = \frac{4A_T^2}{\displaystyle\int_s \frac{ds}{b}} = 2\pi r_{mid}^3 b = J \tag{6.8.7}$$

To compare the stresses in an open section use Equations 6.8.1 and 6.8.2 and note that

$$(\tau_{xy})_{max} = \frac{T}{J_{eff}} z = \frac{3T}{hb^3} \frac{b}{2} = \frac{3}{2} \frac{T}{hb^2} \tag{6.8.8}$$

The torsional rigidity of the open section is given by Equation 6.8.2.

Let us compare a specific case of a circular cross section.

############

Example 6.8.2

Problem: A circular thin walled shaft in pure torsion with a closed section is compare to one with a slit running the full length producing an open section.

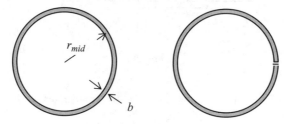

Figure (a)

Solution: Compare the stresses and effective torsional stiffness.
The maximum stress in the closed section is

$$\tau_c = \frac{T}{2A_T b} = \frac{T}{2\pi r_{mid}^2 b} \tag{a}$$

The maximum stress in the open section is

$$\tau_o = \frac{3}{2}\frac{T}{hb^2} = \frac{3}{4}\frac{T}{\pi r_{mid} b^2} \tag{b}$$

The ratio of the stresses of the open to closed is

$$\frac{\tau_o}{\tau_c} = \frac{3}{4}\frac{T}{\pi r_{mid} b^2} \bigg/ \frac{T}{2\pi r_{mid}^2 b} = \frac{3}{2}\frac{r}{b}$$

For example if the radius is 10 *mm* and the wall thickness is 1 *mm* the stress would be 15 times higher in the open section.
The ratio of torsional rigidities is

$$\frac{J_o}{J_c} = \frac{2\pi r_{mid} b^3}{3} \bigg/ 2\pi r_{mid}^3 b = \frac{1}{3}\frac{b^2}{r^2} \tag{c}$$

For an open section with the same dimensions the ratio would be 1/300, that is the open section would have a twist displacement of 300 times that of the closed section for the same applied torque.

############

6.9 A General (Finite Element) Method

Let us consider a shaft loaded only with concentrated moments. We divide the shaft into segments and number the nodes and elements just as we did in the axial case. Then label the nodal rotations and internal torques as in Figure 6.9.1.

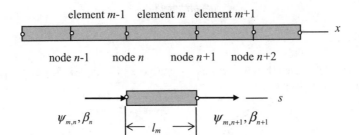

Figure 6.9.1

We shall use $\{\phi_m\}$ to represent the nodal rotations and $\{\psi_m\}$ to represent the internal nodal torques.

$$\{\phi_m\} = \begin{bmatrix} \beta_n \\ \beta_{n+1} \end{bmatrix} \qquad \{\psi_m\} = \begin{bmatrix} \psi_{m,n} \\ \psi_{m,n+1} \end{bmatrix} \tag{6.9.1}$$

We define shape functions for element m that present the distributed rotational displacement in terms of the nodal values of rotation.

$$\beta(s_m) = [n]\{\phi_m\} \tag{6.9.2}$$

By analogy with the axial case the shape functions for element m are

$$\beta(s_m) = [n]\{\phi_m\} = \begin{bmatrix} 1 - \dfrac{s_m}{l_m} & \dfrac{s_m}{l_m} \end{bmatrix} \begin{bmatrix} \beta_n \\ \beta_{n+1} \end{bmatrix} \tag{6.9.3}$$

Using Equation 6.3.1 the strain in terms of the nodal displacements is

$$\gamma_m(r) = \frac{d\beta}{ds_m} r = \frac{1}{l_m} \begin{bmatrix} -1 & 1 \end{bmatrix} \begin{bmatrix} \beta_n \\ \beta_{n+1} \end{bmatrix} r \tag{6.9.4}$$

and from Hooke's law the stress is

$$\tau_m(r) = G\gamma_m = G\frac{d\beta}{ds_m} r = \frac{G}{l_m} \begin{bmatrix} -1 & 1 \end{bmatrix} \begin{bmatrix} \beta_n \\ \beta_{n+1} \end{bmatrix} r \tag{6.9.5}$$

For element m, the internal moments at the ends of the elements are

$$\psi_{m,n} = -\int_A \tau_m r dA = GJ\frac{d\beta}{ds_m} = \frac{GJ}{l_m} \begin{bmatrix} 1 & -1 \end{bmatrix} \begin{bmatrix} \beta_n \\ \beta_{n+1} \end{bmatrix}$$

$$\psi_{m,n+1} = \int_A \tau_m r dA = GJ\frac{d\beta}{ds_m} = \frac{GJ}{l_m} \begin{bmatrix} -1 & 1 \end{bmatrix} \begin{bmatrix} \beta_n \\ \beta_{n+1} \end{bmatrix} \tag{6.9.6}$$

Equations 6.9.6 can be put in matrix form

$$\{\psi_m\} = \begin{bmatrix} \psi_{m,n} \\ \psi_{m,n+1} \end{bmatrix} = \frac{G_m J_m}{l_m} \begin{bmatrix} 1 & -1 \\ -1 & 1 \end{bmatrix} \begin{bmatrix} \beta_n \\ \beta_{n+1} \end{bmatrix} = [k_m]\{\phi_m\} \tag{6.9.7}$$

and so the element stiffness matrix is

$$[k_m] = \frac{G_m J_m}{l_m} \begin{bmatrix} 1 & -1 \\ -1 & 1 \end{bmatrix} \tag{6.9.8}$$

The element stiffness matrices are summed to provide the global stiffness matrix.

$$\{M\} = \sum \{\psi_m\} = \sum [k_m]\{\phi_m\} = [K]\{\phi\} \tag{6.9.9}$$

Does this look familiar? What we learned in Chapter 5 applies with only a change in symbols and coefficients.

The equivalent nodal loads for element m are found from

$$\{M_{em}\} = \int_0^{l_m} [n]^T t_x(s)\,ds \tag{6.9.10}$$

and the total equivalent nodal load matrix is the sum of the element matrices.

$$\{M_e\} = \sum \{M_{em}\} \tag{6.9.11}$$

These are added to any actual concentrated applied moments.

$$\{M\} = \{M_c\} + \{M_{enl}\} \tag{6.9.12}$$

By analogy with the axial case (compare with Equation 5.4.5) we can summarize the torsional case as follows:

$$\{M\} \quad \{M_c\} \quad t_x(s_m) \quad \rightarrow \quad \{\psi_m\} \quad \tau_m(r) \quad \gamma_m(r) \quad \beta(s_m) \quad \{\phi_m\} \quad \leftarrow \quad \{\rho\}$$

$$\{M_{em}\} = \int_0^{l_m} [n]^T t_x(s_m)\,ds_m \qquad \{M_e\} = \sum \{M_{em}\} \qquad \beta(s_m) = [n]\{\phi_m\}$$

$$\{M\} = \{M_c\} + \{M_{enl}\} \qquad \gamma_m(r) = \frac{d\beta}{ds_m}r = \frac{d}{ds_m}[n]\{\phi_m\}r$$

$$\tau_m(r) = G\frac{d\beta}{ds_m}r = G\frac{d}{ds_m}[n]\{\phi_m\}r \tag{6.9.13}$$

$$\psi_{m,n} = -GJ\frac{d\beta}{ds_m} = -GJ\frac{d}{ds_m}[n]\{\phi_m\}$$

$$\rightarrow \quad \{\psi_m\} = \begin{bmatrix} \psi_{m,n} \\ \psi_{m,n+1} \end{bmatrix} = [k_m]\{\phi_m\}$$

$$\psi_{m,n+1} = GJ\frac{d\beta}{ds_m} = GJ\frac{d}{ds_m}[n]\{\phi_m\}$$

$$\{M\} = \sum \{\psi_m\} = \sum [k_m]\{\phi_m\} = [K]\{\phi\}$$

In the first row we have the typical unknowns in the formulation of the equations for a single element

$\{\phi_m\}$ — nodal rotational displacements
$\beta(s_m)$ — distributed rotational displacement
$\gamma_m(r)$ — shear strain
$\tau_m(r)$ — shear stress
$\{\psi_m\}$ — internal moments

and the knowns

$\{\rho\}$ — nodal displacement restraints
$\{M_c\}$ — concentrated moments at nodes
$\{M_e\}$ — equivalent nodal moments

The assembly process is exactly the same as the one we just learned in Chapter 5. The result is a set of global equations.

$$\{M\} = \sum \{\psi_m\} = \sum [k_m]\{\phi_m\} = [K]\{\phi\} \tag{6.9.14}$$

The boundary restraints are added to the global nodal matrix and the restrain moments to the global moment matrix, the equations are partitioned and solved.

###########

Example 6.9.1

Problem: Find the FEM equations for the shaft in Example 6.2.2.

Solution: Divide into elements, find the element stiffness matrices, assemble the global matrix equations, add the boundary conditions, and partition.

Let us use just two elements as shown in Figure (a).

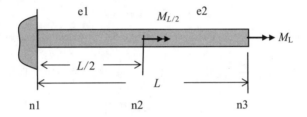

Figure (a)

By analogy with the axial case the assembled global equations are

$$\{M\} = [K]\{\phi\} = \begin{bmatrix} Q_1 \\ M_{L/2} \\ M_L \end{bmatrix} = \frac{2GJ}{L} \begin{bmatrix} 1 & -1 & 0 \\ -1 & 2 & -1 \\ 0 & -1 & 1 \end{bmatrix} \begin{bmatrix} 0 \\ \beta_2 \\ \beta_3 \end{bmatrix} \tag{a}$$

This may be partitioned to find the nodal displacements

$$\begin{bmatrix} M_{L/2} \\ M_L \end{bmatrix} = \frac{2GJ}{L} \begin{bmatrix} 2 & -1 \\ -1 & 1 \end{bmatrix} \begin{bmatrix} \beta_2 \\ \beta_3 \end{bmatrix} \tag{b}$$

Generally numerical values are introduced and the equations solved by a linear algebraic equation solver rather than finding an analytical solution. Use Equation 6.9.5 to find the stresses and Equation 6.9.3 to find the distributed displacements. The restraint moment is found from the other partitioned equation.

$$Q_1 = \frac{2GJ}{L} \begin{bmatrix} -1 & 0 \end{bmatrix} \begin{bmatrix} \beta_2 \\ \beta_3 \end{bmatrix} \tag{c}$$

###########

At this point it is not so clear why the finite element method is better than the classical one we used. On simple problems it is not. As problems get more complicated it comes into its own because it lends itself to numerical solutions using existing software. The next example problem is complicated enough that it would be more difficult to solve the traditional analytical way.

Again, we remind you of the similarity of the "general method" for the axial problem and the "general method" for the torsional problem just presented.

############

Example 6.9.2

Problem: A shaft is fixed on both ends, has changes in cross section, as shown in Figure (a) and has three applied concentrated moments, say, at $x = L/4$, $L/2$, and $3L/4$. Find the displacements and stresses.

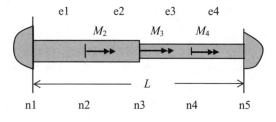

Figure (a)

Solution: Divide into elements, find the element stiffness matrices, assemble the global matrix equations, add the boundary conditions, and solve. The element and nodal numbering is shown on Figure (a).

To do this analytically by the traditional method would require solving the second order equation four times and satisfying eight boundary conditions. Let us set up some conditions. For example, assume that the shaft is divided into four elements of equal length, the value of GJ for elements e1 and e2 is twice that for elements e3 and e4, that is,

$$G_3 J_3 = G_4 J_4 = \frac{G_1 J_1}{2} \quad \text{and} \quad G_2 J_2 = G_1 J_1 \tag{a}$$

The matrices for the four elements are

$$\begin{bmatrix} \psi_{1,1} \\ \psi_{1,2} \end{bmatrix} = \frac{4G_1 J_1}{L} \begin{bmatrix} 1 & -1 \\ -1 & 1 \end{bmatrix} \begin{bmatrix} \beta_1 \\ \beta_2 \end{bmatrix}$$

$$\begin{bmatrix} \psi_{2,2} \\ \psi_{2,3} \end{bmatrix} = \frac{4G_2 J_2}{L} \begin{bmatrix} 1 & -1 \\ -1 & 1 \end{bmatrix} \begin{bmatrix} \beta_2 \\ \beta_3 \end{bmatrix} = \frac{4G_1 J_1}{L} \begin{bmatrix} 1 & -1 \\ -1 & 1 \end{bmatrix} \begin{bmatrix} \beta_2 \\ \beta_3 \end{bmatrix}$$

$$\begin{bmatrix} \psi_{3,3} \\ \psi_{3,4} \end{bmatrix} = \frac{4G_3 J_3}{L} \begin{bmatrix} 1 & -1 \\ -1 & 1 \end{bmatrix} \begin{bmatrix} \beta_3 \\ \beta_4 \end{bmatrix} = \frac{2G_1 J_1}{L} \begin{bmatrix} 1 & -1 \\ -1 & 1 \end{bmatrix} \begin{bmatrix} \beta_3 \\ \beta_4 \end{bmatrix} \tag{b}$$

$$\begin{bmatrix} \psi_{4,4} \\ \psi_{4,5} \end{bmatrix} = \frac{2G_4 J_4}{L} \begin{bmatrix} 1 & -1 \\ -1 & 1 \end{bmatrix} \begin{bmatrix} \beta_4 \\ \beta_5 \end{bmatrix} = \frac{2G_1 J_1}{L} \begin{bmatrix} 1 & -1 \\ -1 & 1 \end{bmatrix} \begin{bmatrix} \beta_4 \\ \beta_5 \end{bmatrix}$$

These must be imbedded in the global format and summed. The assembled equations are

$$\begin{bmatrix} Q_1 \\ M_2 \\ M_3 \\ M_4 \\ Q_5 \end{bmatrix} = \frac{2G_1 J_1}{L} \begin{bmatrix} 2 & -2 & 0 & 0 & 0 \\ -2 & 4 & -2 & 0 & 0 \\ 0 & -2 & 3 & -1 & 0 \\ 0 & 0 & -1 & 2 & -1 \\ 0 & 0 & 0 & -1 & 1 \end{bmatrix} \begin{bmatrix} 0 \\ \beta_2 \\ \beta_3 \\ \beta_4 \\ 0 \end{bmatrix} \tag{c}$$

After partitioning

$$\begin{bmatrix} M_2 \\ M_3 \\ M_4 \end{bmatrix} = \frac{2G_1 J_1}{L} \begin{bmatrix} 4 & -2 & 0 \\ -2 & 3 & -1 \\ 0 & -1 & 2 \end{bmatrix} \begin{bmatrix} \beta_2 \\ \beta_3 \\ \beta_4 \end{bmatrix} \tag{d}$$

and

$$\begin{bmatrix} Q_1 \\ Q_5 \end{bmatrix} = \frac{2G_1 J_1}{L} \begin{bmatrix} -2 & 0 & 0 \\ 0 & 0 & -1 \end{bmatrix} \begin{bmatrix} \beta_2 \\ \beta_3 \\ \beta_4 \end{bmatrix} \tag{e}$$

Let us introduce some numerical values. Consider a steel shaft with $G = 75,000\ MPa$. Its diameter at the fixed end is 20 mm; its length is 500 mm. It follows that

$$J_1 = \frac{\pi}{2} r_1^4 = \frac{\pi}{2} 10^4 = 15708\ mm^4 = J_2$$
$$J_3 = \frac{J_1}{2} = 7854\ mm^4 = \frac{\pi}{2} r_3^4 = J_4 \quad \rightarrow \quad r_3 = 8.409\ mm = r_4 \tag{f}$$

Consider two cases of applied moments ($N \cdot mm$).

$$\begin{array}{llll} \text{Case 1} & M_2 = 50000 & M_3 = 34000 & M_4 = 25679 \\ \text{Case 2} & M_2 = 10000 & M_3 = -20000 & M_4 = 30000 \end{array} \tag{g}$$

Case 1:
Substituting the values for Case 1 into Equation (c) we have

$$\begin{bmatrix} 50000 \\ 34000 \\ 25679 \end{bmatrix} = \frac{2 \cdot 75000 \cdot 15708}{500} \begin{bmatrix} 4 & -2 & 0 \\ -2 & 3 & -1 \\ 0 & -1 & 2 \end{bmatrix} \begin{bmatrix} \beta_2 \\ \beta_3 \\ \beta_4 \end{bmatrix} \tag{h}$$

Using an equation solver for Case 1 we get

$$\begin{bmatrix} \beta_2 \\ \beta_3 \\ \beta_4 \end{bmatrix} = \begin{bmatrix} 7.734 \cdot 10^{-3} \\ 10.136 \cdot 10^{-3} \\ 7.806 \cdot 10^{-3} \end{bmatrix} rad = \begin{bmatrix} 0.443 \\ 0.582 \\ 0.447 \end{bmatrix} deg \tag{i}$$

For Case 1 the maximum stress in each element, that is, the stress at the surface, is

$$\tau_1 = \frac{4G}{L} \begin{bmatrix} -1 & 1 \end{bmatrix} \begin{bmatrix} \beta_1 \\ \beta_2 \end{bmatrix} = \frac{4 \cdot 75000}{500} \begin{bmatrix} -1 & 1 \end{bmatrix} \begin{bmatrix} 0 \\ 0.007734 \end{bmatrix} 10 = 46.404\ MPa$$

$$\tau_2 = \frac{4G}{L} \begin{bmatrix} -1 & 1 \end{bmatrix} \begin{bmatrix} \beta_2 \\ \beta_3 \end{bmatrix} = \frac{4 \cdot 75000}{500} \begin{bmatrix} -1 & 1 \end{bmatrix} \begin{bmatrix} 0.007734 \\ 0.010163 \end{bmatrix} 10 = 14.574\ MPa$$

$$\tau_3 = \frac{4G}{L} \begin{bmatrix} -1 & 1 \end{bmatrix} \begin{bmatrix} \beta_3 \\ \beta_4 \end{bmatrix} = \frac{4 \cdot 75000}{500} \begin{bmatrix} -1 & 1 \end{bmatrix} \begin{bmatrix} 0.010163 \\ 0.007806 \end{bmatrix} 8.409 = -11.892\ MPa$$

$$\tau_4 = \frac{4G}{L} \begin{bmatrix} -1 & 1 \end{bmatrix} \begin{bmatrix} \beta_4 \\ \beta_5 \end{bmatrix} = \frac{4 \cdot 75000}{500} \begin{bmatrix} -1 & 1 \end{bmatrix} \begin{bmatrix} 0.007806 \\ 0 \end{bmatrix} 8.409 = -39.3844\ MPa$$

$$\tag{j}$$

Case 2:
Substituting the values for Case 2 into Equation (c) we have

$$\begin{bmatrix} 10000 \\ -20000 \\ 30000 \end{bmatrix} = \frac{2 \cdot 75000 \cdot 15708}{500} \begin{bmatrix} 4 & -2 & 0 \\ -2 & 3 & -1 \\ 0 & -1 & 2 \end{bmatrix} \begin{bmatrix} \beta_2 \\ \beta_3 \\ \beta_4 \end{bmatrix} \tag{k}$$

Using an equation solver for Case 2 we get

$$\begin{bmatrix} \beta_2 \\ \beta_3 \\ \beta_4 \end{bmatrix} = \begin{bmatrix} 5.30515 \cdot 10^{-4} \\ 0 \\ 3.18309 \cdot 10^{-3} \end{bmatrix} rad = \begin{bmatrix} 0.0304 \\ 0 \\ 0.1824 \end{bmatrix} deg \qquad (1)$$

The stresses may be found using the same equations as in Equation (i). Note that once the main stiffness matrix is assembled a variety of loading conditions can be solved using the one entry for the stiffness matrix. Other restraints can also be used but the global matrix must then be repartitioned.

Example 6.9.3

Problem: A shaft is fixed at both ends as shown in Figure (a). The three sections each have a different diameter. The center section has a uniform distributed applied moment applied over all its length. The material is the same in all parts. Write down the equations for finding the displacements and stresses.

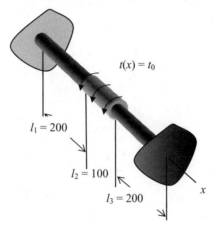

$t(x) = t_0$

$l_1 = 200$

$l_2 = 100$

$l_3 = 200$

x

Figure (a)

Solution: Divide into elements, find the element stiffness matrices, assemble the global matrix equations, add the boundary conditions, and partition.
  Three elements will do.

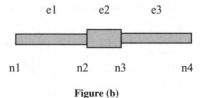

e1   e2   e3

n1   n2   n3   n4

Figure (b)

The equivalent nodal loads on element 2 are

$$\begin{bmatrix} M_2 \\ M_3 \end{bmatrix} = \begin{bmatrix} \dfrac{t_0 100}{2} \\ \dfrac{t_0 100}{2} \end{bmatrix} = \begin{bmatrix} 50 t_0 \\ 50 t_0 \end{bmatrix} \qquad (a)$$

The assembled equations are

$$\{F\} = [K\{r\}] = \begin{bmatrix} F_1 \\ F_2 \\ F_3 \\ F_4 \end{bmatrix} = \begin{bmatrix} \dfrac{GJ_1}{200} & -\dfrac{GJ_1}{200} & 0 & 0 \\[2mm] -\dfrac{GJ_1}{200} & \dfrac{GJ_1}{200} + \dfrac{GJ_2}{100} & -\dfrac{GJ_2}{100} & 0 \\[2mm] 0 & -\dfrac{GJ_2}{100} & \dfrac{GJ_2}{100} + \dfrac{GJ_3}{200} & -\dfrac{GJ_3}{200} \\[2mm] 0 & 0 & -\dfrac{GJ_3}{200} & \dfrac{GJ_3}{200} \end{bmatrix} \begin{bmatrix} \beta_1 \\ \beta_2 \\ \beta_3 \\ \beta_4 \end{bmatrix} \quad (b)$$

When boundary conditions, restraints, and loads are inserted

$$\begin{bmatrix} Q_1 \\ 50t_0 \\ 50t_0 \\ Q_4 \end{bmatrix} = \begin{bmatrix} \dfrac{GJ_1}{200} & -\dfrac{GJ_1}{200} & 0 & 0 \\[2mm] -\dfrac{GJ_1}{200} & \dfrac{GJ_1}{200} + \dfrac{GJ_2}{100} & -\dfrac{GJ_2}{100} & 0 \\[2mm] 0 & -\dfrac{GJ_2}{100} & \dfrac{GJ_2}{100} + \dfrac{GJ_3}{200} & -\dfrac{GJ_3}{200} \\[2mm] 0 & 0 & -\dfrac{GJ_3}{200} & \dfrac{GJ_3}{200} \end{bmatrix} \begin{bmatrix} 0 \\ \beta_2 \\ \beta_3 \\ 0 \end{bmatrix} \quad (c)$$

Partition the equations for finding the displacements.

$$\begin{bmatrix} 50t_0 \\ 50t_0 \end{bmatrix} = \begin{bmatrix} \dfrac{GJ_1}{200} + \dfrac{GJ_2}{100} & -\dfrac{GJ_2}{100} \\[2mm] -\dfrac{GJ_2}{100} & \dfrac{GJ_2}{100} + \dfrac{GJ_3}{200} \end{bmatrix} \begin{bmatrix} \beta_2 \\ \beta_3 \end{bmatrix} \quad (d)$$

The equations for the restraint forces are

$$\begin{bmatrix} Q_1 \\ Q_4 \end{bmatrix} = \begin{bmatrix} -\dfrac{GJ_1}{200} & 0 \\[2mm] 0 & -\dfrac{GJ_3}{200} \end{bmatrix} \begin{bmatrix} \beta_2 \\ \beta_3 \end{bmatrix} \quad (e)$$

Finally the internal torques are found from these equations

$$\begin{bmatrix} \psi_{1,1} \\ \psi_{1,2} \end{bmatrix} = GJ_1 \begin{bmatrix} 1 & -1 \\ -1 & 1 \end{bmatrix} \begin{bmatrix} 0 \\ \beta_2 \end{bmatrix}$$

$$\begin{bmatrix} \psi_{2,2} \\ \psi_{2,3} \end{bmatrix} = GJ_2 \begin{bmatrix} 1 & -1 \\ -1 & 1 \end{bmatrix} \begin{bmatrix} \beta_2 \\ \beta_3 \end{bmatrix} \quad (f)$$

$$\begin{bmatrix} \psi_{3,3} \\ \psi_{3,4} \end{bmatrix} = GJ_3 \begin{bmatrix} 1 & -1 \\ -1 & 1 \end{bmatrix} \begin{bmatrix} \beta_3 \\ 0 \end{bmatrix}$$

And the stresses are found from

$$\tau_1(x) = \frac{\psi_1}{J_1} r \quad \tau_2(x) = \frac{\psi_2}{J_2} r \quad \tau_3(x) = \frac{\psi_3}{J_3} r \quad (g)$$

Example 6.9.4

Problem: A shaft is fixed at the left end. It is loaded with concentrated applied moments at $x = 200$ *mm* and 300 *mm* as shown in Figure (a). In addition it is subjected to an applied rotation, $\beta_L = 0.002$ *rad*, at the right end. Write down the appropriate equations and boundary conditions for finding the displacements, internal torques, and stresses.

The material is steel with a shear modulus of 80,000 *MPa*. The polar moments of inertia are $J_1 = 20000$ *mm*$^4 = J_3$ and $J_2 = 40000$ *mm*^{4}.

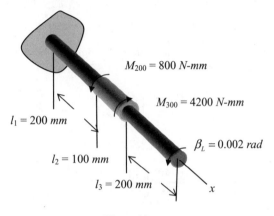

$M_{200} = 800$ *N-mm*

$M_{300} = 4200$ *N-mm*

$l_1 = 200$ *mm*

$\beta_L = 0.002$ *rad*

$l_2 = 100$ *mm*

$l_3 = 200$ *mm*

x

Figure (a)

Solution: Divide into elements, find the element stiffness matrices, assemble the global matrix equations, partition, and solve.

The values for GJ are

$$GJ_1 = 16 \cdot 10^8 = GJ_3 \quad GJ_2 = 32 \cdot 10^8 \tag{a}$$

Three elements will do.

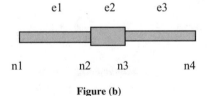

e1   e2   e3

n1   n2   n3   n4

Figure (b)

The assembled equations are

$$\{M\} = [K\{\phi\}] = \begin{bmatrix} M_1 \\ M_2 \\ M_3 \\ M_4 \end{bmatrix} = \begin{bmatrix} \dfrac{GJ_1}{200} & -\dfrac{GJ_1}{200} & 0 & 0 \\[2mm] -\dfrac{GJ_1}{200} & \dfrac{GJ_1}{200} + \dfrac{GJ_2}{100} & -\dfrac{GJ_2}{100} & 0 \\[2mm] 0 & -\dfrac{GJ_2}{100} & \dfrac{GJ_2}{100} + \dfrac{GJ_3}{200} & -\dfrac{GJ_3}{200} \\[2mm] 0 & 0 & -\dfrac{GJ_3}{200} & \dfrac{GJ_3}{200} \end{bmatrix} \begin{bmatrix} \beta_1 \\ \beta_2 \\ \beta_3 \\ \beta_4 \end{bmatrix} \tag{b}$$

When displacement boundary conditions, restraint forces, GJ values, and loads are inserted

$$
\begin{bmatrix} Q_1 \\ \hdashline 800 \\ -4200 \\ \hdashline Q_4 \end{bmatrix} = 16 \cdot 10^8
\begin{bmatrix}
\dfrac{1}{200} & -\dfrac{1}{200} & 0 & 0 \\[2mm]
\hdashline
-\dfrac{1}{200} & \dfrac{1}{200}+\dfrac{2}{100} & -\dfrac{2}{100} & 0 \\[2mm]
0 & -\dfrac{2}{100} & \dfrac{2}{100}+\dfrac{1}{200} & -\dfrac{1}{200} \\[2mm]
\hdashline
0 & 0 & -\dfrac{1}{200} & \dfrac{1}{200}
\end{bmatrix}
\begin{bmatrix} 0 \\ \hdashline \beta_2 \\ \beta_3 \\ \hdashline 0.02 \end{bmatrix}
\tag{c}
$$

Partition the equations for find the displacements. This time the fourth column is multiplied by a known value and so must be included in the partition.

$$
\begin{bmatrix} 800 \\ -4200 \end{bmatrix} = \frac{16 \cdot 10^8}{200}
\begin{bmatrix} 5 & -4 & 0 \\ -4 & 5 & -1 \end{bmatrix}
\begin{bmatrix} \beta_2 \\ \beta_3 \\ 0.02 \end{bmatrix}
\tag{d}
$$

The equations for the restraint forces are

$$
\begin{bmatrix} Q_1 \\ Q_4 \end{bmatrix} = \frac{16 \cdot 10^8}{200}
\begin{bmatrix} -1 & 0 & 0 \\ 0 & -1 & 1 \end{bmatrix}
\begin{bmatrix} \beta_2 \\ \beta_3 \\ 0.02 \end{bmatrix}
\tag{e}
$$

Finally the internal torques are found from these equations

$$
\begin{bmatrix} \psi_{1,1} \\ \psi_{1,2} \end{bmatrix} = GJ_1 \begin{bmatrix} 1 & -1 \\ -1 & 1 \end{bmatrix} \begin{bmatrix} 0 \\ \beta_2 \end{bmatrix}
$$

$$
\begin{bmatrix} \psi_{2,2} \\ \psi_{2,3} \end{bmatrix} = GJ_2 \begin{bmatrix} 1 & -1 \\ -1 & 1 \end{bmatrix} \begin{bmatrix} \beta_2 \\ \beta_3 \end{bmatrix}
\tag{f}
$$

$$
\begin{bmatrix} \psi_{3,3} \\ \psi_{3,4} \end{bmatrix} = GJ_3 \begin{bmatrix} 1 & -1 \\ -1 & 1 \end{bmatrix} \begin{bmatrix} \beta_3 \\ 0.02 \end{bmatrix}
$$

And the stresses are found from

$$
\tau_1(x) = \frac{\psi_1}{J_1} r \quad \tau_2(x) = \frac{\psi_2}{J_2} r \quad \tau_3(x) = \frac{\psi_3}{J_3} r
\tag{g}
$$

###########

6.10 Continuously Variable Cross Sections

If J varies with x we can adapt the equations as we demonstrated for the axial case. If the shaft is statically determinate we find the internal torque and from it the stress.

$$
\tau(x,r) = \frac{T(x)}{J(x)} r
\tag{6.10.1}
$$

Finding the displacement requires integrating

$$
\frac{d\beta}{dx} = \frac{T(x)}{GJ(x)} \quad \rightarrow \quad \beta(x) = \frac{1}{G} \int \frac{T(x)}{J(x)} dx + c
\tag{6.10.2}
$$

or

$$\frac{d}{dx}GJ(x)\frac{d\beta}{dx} = -t_x(x) \quad \rightarrow \quad \beta(x) = -\int \frac{1}{GJ(x)}\left(\int t_x(x)dx + a\right)dx + b \tag{6.10.3}$$

If the shaft is statically indeterminate we must use the second order equations. Thus

$$\frac{d}{dx}GJ(x)\frac{d\beta}{dx} = -t_x(x) \tag{6.10.4}$$

Once we find the displacement $\beta(x)$ by integrating twice and applying constraints to find the constants of integration we can find the internal torque.

$$T(x) = GJ(x)\frac{d\beta}{dx} \tag{6.10.5}$$

and the stress is

$$\tau(x,r) = \frac{T(x)}{J(x)}r \tag{6.10.6}$$

We learned in the axial case that except for the simplest functions representing $J(x)$ direct integration is an unwieldy process. It is more common to approximate the variation by a series of constant cross sections, such as is represented in Figure 6.10.1. In practice several sections might be used, not just the three shown. The distributed load would be replaced by equivalent nodal loads and the general method used.

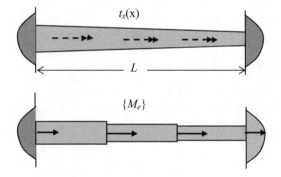

Figure 6.10.1

In Section 5.5 we learned how careful interpretation of the results can achieve a high degree of accuracy. We would use the restraint torques at the ends to find the stresses there and use average values at the nodes.

6.11 Summary and Conclusions

Since the equations for torsion differ from the axial case only in the symbols used for the coefficients everything we learned in the axial case applies here. For this reason we do not do a lot of examples.

The equations for the classical analytical solution of torsional problems for slender shafts with circular or thin walled closed cross sections are summarized in Equations 6.3.8 and are repeated here as Equation 6.11.1.

$$M_c \quad \rightarrow \quad t_x(x) \qquad T(x) \qquad \tau(x,r) \qquad \gamma(x,r) \qquad \beta(x) \quad \leftarrow \quad \rho$$

$$\frac{dT}{dx} = -t_x(x) \qquad\qquad\qquad \gamma(x,r) = \frac{d\beta}{dx}r$$

$$\tau(x,r) = G\gamma = G\frac{d\beta}{dx}r$$

$$T(x) = \int_A \tau r dA = G\int_A \gamma r dA = G\frac{d\beta}{dx}\int_A r^2 dA = GJ\frac{d\beta}{dx} \quad \rightarrow \quad \tau = \frac{T}{J}r \qquad (6.11.1)$$

$$\frac{dT}{dx} = \frac{d}{dx}GJ\frac{d\beta}{dx} = -t(x)$$

The equations for the general method (FEM) are summarized in Equations 6.9.2 and are repeated here as Equation 6.11.2.

$$\{M\} \quad \{M_c\} \quad t_x(s_m) \quad \rightarrow \quad \{\psi_m\} \quad \tau_m(r) \quad \gamma_m(r) \quad \beta(s_m) \quad \{\phi_m\} \quad \leftarrow \quad \{\rho\}$$

$$\{M_{em}\} = \int_0^{l_m} [n]^T t_x(s_m)ds_m \qquad\qquad \{M_e\} = \sum \{M_{em}\} \qquad\qquad \beta(s_m) = [n]\{\phi_m\}$$

$$\{M\} = \{M_c\} + \{M_{enl}\} \qquad\qquad\qquad \gamma_m(r) = \frac{d\beta}{ds_m}r = \frac{d}{ds_m}[n]\{\phi_m\}r$$

$$\tau_m(r) = G\frac{d\beta}{ds_m}r = G\frac{d}{ds_m}[n]\{\phi_m\}r \qquad\qquad (6.11.2)$$

$$\psi_{m,n} = -GJ\frac{d\beta}{ds_m} = -GJ\frac{d}{ds_m}[n]\{\phi_m\}$$

$$\psi_{m,n+1} = GJ\frac{d\beta}{ds_m} = GJ\frac{d}{ds_m}[n]\{\phi_m\} \qquad \rightarrow \quad \{\psi_m\} = \begin{bmatrix} \psi_{m,n} \\ \psi_{m,n+1} \end{bmatrix} = [k_m]\{\phi_m\}$$

$$\{M\} = \sum \{\psi_m\} = \sum [k_m]\{\phi_m\} = [K]\{\phi\}$$

We must emphasize that while the axial findings are true for any cross section shape the findings for torsion apply only to circular or thin walled cross sections.

7

Classical Analysis of the Bending of Beams

7.1 Introduction

We now consider slender bars with transverse loads and moments. A slender member that can carry such loads is referred to as a beam. If in addition the member also has axial loads that are compressive we call it a *beam column*. The effect of axial loading on a beam column is treated in detail in Chapter 13.

Just as in the previous cases we have included in earlier chapters we place the x axis at the centroid of the cross section but we must be careful with the orientation of the y and z axes. To simplify the problem we consider that the y and z axes are *principal axes of inertia* of the cross section area. Just what this means is explained in Section 7.2. We must also deal with sign conventions that change according to the type of analysis. The so-called traditional sign convention, or the "elasticity" sign convention, that is used throughout this chapter is defined in the next section.

7.2 Area Properties—Sign Conventions

7.2.1 Area Properties

Consider a cross section with a general shape such as is shown in Figure 7.2.1 with the x axis normal to the cross section.

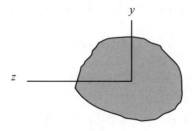

Figure 7.2.1

Analysis of Structures: An Introduction Including Numerical Methods, First Edition. Joe G. Eisley and Anthony M. Waas.
© 2011 John Wiley & Sons, Ltd. Published 2011 by John Wiley & Sons, Ltd.

The x axis is a *centroidal* axis if

$$\int_A ydA = 0 \qquad \int_A zdA = 0 \qquad\qquad (7.2.1)$$

The *area moments of inertia* are

$$I_{zz} = \int_A y^2 dA \qquad I_{yy} = \int_A z^2 dA \qquad\qquad (7.2.2)$$

and the *area product of inertia* is

$$I_{yz} = \int_A yzdA \qquad\qquad (7.2.3)$$

In many cases $I_{yz} = 0$. This occurs when either the xy or the xz axes plane is a plane of symmetry. For cross sections without symmetry it is possible to orient the yz axes so that $I_{yz} = 0$. When $I_{yz} = 0$ the axes are called *principal axes of inertia*. While it is possible to derive the beam equations for any orientation of the axes, that is, when I_{yz} may not be zero, we shall limit discussion in this chapter to beams with the x axis centroidal and the yz axes aligned as principal axes. In such cases the equations governing displacement in the two principal axis directions are uncoupled. For the most part we shall derive equations and give examples only for applied forces in the y direction and applied moments about the z axis. Very similar equations are used in the xz plane. Since the equations in the two planes are uncoupled they can be solved independently.

Information on finding the centroids and the moments and product of inertia of cross section areas, including finding the orientation of the yz axes so that $I_{yz} = 0$, may be found in Appendix B.

In many of our examples we shall use a rectangular cross section, such as that illustrated in Figure 7.2.2, for illustrating the distribution of the shear and normal stresses.

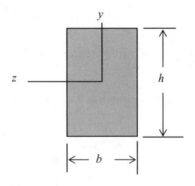

Figure 7.2.2

The centroid of this section is easily determined by symmetry and its moments and product of inertia are

$$I_{zz} = \int_A y^2 dA = b \int_{-h/2}^{h/2} y^2 dy = \frac{by^3}{3}\bigg|_{-h/2}^{h/2} = \frac{bh^3}{12} \qquad I_{yy} = \frac{b^3 h}{12} \qquad I_{yz} = 0 \qquad (7.2.4)$$

This and several other cross sections and their properties are found and illustrated in Appendix B.

The extension to more general cross section shapes, including cases where $I_{yz} \neq 0$, is presented in Chapter 10.

7.2.2 Sign Conventions

Sign conventions can be confusing and the different standards used in different books and publications do not help; however, nearly everyone uses the same conventions for stress and strain. For a reminder, positive values for stress are shown in Figure 7.2.3 in two dimensions. (See Chapter 1, Section 1.7.)

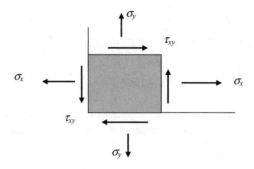

Figure 7.2.3

The sign convention for positive internal forces and moments in beams is usually as shown in Figure 7.2.4 in two dimensions in the xy plane when explaining traditional analytical methods. $M(x)$ is the internal moment about the z axis and $V(x)$ is the internal shear force in the y direction. The direction of the normal to the face of interest determines the positive direction.

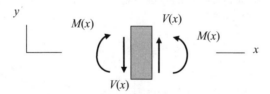

Figure 7.2.4

The shear force $V(x)$ and the moment $M(x)$ are stress resultants, that is,

$$V(x) = \int_A \tau_{xy} dA \qquad M(x) = -\int_A y\sigma_x dA \qquad (7.2.5)$$

The minus sign in the moment definition is needed because of the convention that a positive stress at a positive y produces a negative moment. At this time we do not yet know just how the stress is distributed on the cross section. That will be coming right up.

The sign convention for applied loads usually is to have all applied forces as positive in the positive direction of the axes and all applied moments as positive according to the right hand rule when a right hand rectangular Cartesian coordinate system is used. The same convention applies for forces and moments at the restraints. Note that while the right hand rules apply to internal moments on a face whose normal is in the positive x direction it does not apply to internal moments on a face whose normal is in the negative x direction.

7.3   Derivation and Solution of the Governing Equations

First consider a slender bar acted upon by pure moments at each end as shown in Figure 7.3.1. For convenience the origin of the coordinates is placed at the mid point or centroid of the volume of the beam.

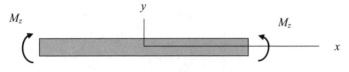

Figure 7.3.1

It is observed in experiments that, as the beam deforms laterally in the y direction an amount $v(x)$, plane cross sections normal to the initially straight centroidal axis remain essentially plane and normal to the deformed shape of the curve passing through the centroids of the cross sections. This is illustrated in Figure 7.3.2.

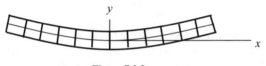

Figure 7.3.2

As the plane cross section rotates any point on it may also undergo a displacement u in the x direction. Consider a segment of the beam as shown in Figure 7.3.3 in both the undeformed and deformed position. The displacement is greatly exaggerated.

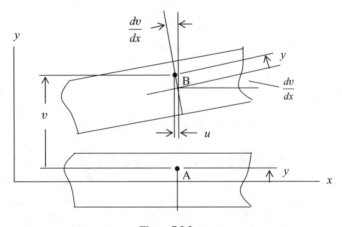

Figure 7.3.3

We are particularly interested in the movement of a point on the beam from its undeformed position A to its deformed position B. For small deflections it can be assumed that the plane cross section moves up an amount v and rotates an amount dv/dx. For a slender beam the Poisson's ratio effect can be ignored and so the point remains a distance y from the centroid of the cross section. This assumes that all points on a cross section at a given value of x displace the same amount in the y direction.

For small displacements there is a relation between the u and v displacements as expressed in Equation 7.3.1

$$u(x, y) = -y \frac{dv(x)}{dx} \tag{7.3.1}$$

While u is a function of both x and y, the displacement v is a function of x only.
From the strain displacement equation the normal strain on the cross section is then

$$\varepsilon_x(x, y) = \frac{du(x, y)}{dx} = \frac{d}{dx}\left(-y\frac{dv}{dx}\right) = -y\frac{d^2v(x)}{dx^2} \tag{7.3.2}$$

and from Hooke's law the stress is

$$\sigma_x(x, y) = E\varepsilon_x(x, y) = -yE\frac{d^2v(x)}{dx^2} \tag{7.3.3}$$

The moment on an internal surface is

$$M(x) = -\int_A \sigma_x y \, dA = E\frac{d^2v}{dx^2}\int_A y^2 \, dA = EI_{zz}\frac{d^2v(x)}{dx^2} \tag{7.3.4}$$

In cases where the moment can be found from static equilibrium, that is, the beam is statically determinate, you can find the normal stress immediately by combining Equations 7.3.3 and 7.3.4.

$$\sigma_x(x, y) = -\frac{M(x)y}{I_{zz}} \tag{7.3.5}$$

You integrate Equation 7.3.4 twice to find the displacement.

When the beam is not statically determinate we must find some other way to satisfy equilibrium. Consider a slender bar that is restrained to remove rigid body motion and is acted upon by a lateral distributed load as shown in Figure 7.3.4. We shall restrict the case to actions taking place in the xy plane and assume that the yz axes are principal axes of inertia.

In Figure 7.3.4 it appears that we have applied the lateral load to the upper surface. Actually it is assumed to be applied along the centroidal axes. Eventually, we shall show that it has little effect whether applied to the upper or lower surface, the centroid axis, or as a body force.

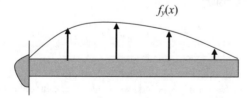

$f_y(x)$

Figure 7.3.4

If lateral loads are present there must be both internal shear forces and stresses and internal moments and normal stresses. Consider a thin slice of length dx of this beam, as shown in Figure 7.3.5, in the initial undeformed configuration. Later when we treat buckling problems due to destabilizing axial loads we need to consider the forces on the thin slice in the deformed configuration. For now, since the deformations are small writing the equilibrium equations in the undeformed configuration incurs no significant error.

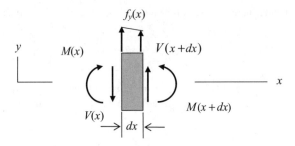

Figure 7.3.5

Using a Taylor's series expansion and neglecting higher order terms we recognize that

$$M(x + dx) = M(x) + \frac{dM}{dx}dx \quad + \text{higher order terms}$$
$$V(x + dx) = V(x) + \frac{dV}{dx}dx \quad + \text{higher order terms}$$
(7.3.6)

Summation of forces in the y direction and summation of moments about the z axis produce these two equilibrium equations.

$$\sum F_y = V(x) + \frac{dV}{dx}dx - V(x) + f_y(x)dx = 0 \quad \rightarrow \quad \frac{dV}{dx} = -f_y(x)$$
$$\sum M_z = M(x) + \frac{dM}{dx}dx - M(x) + Vdx - f_y dx\frac{dx}{2} = 0 \quad \rightarrow \quad \frac{dM}{dx} = -V(x)$$
(7.3.7)

Since we are still assuming plane sections remain plane we can continue to use Equation 7.3.4 and combine it with Equations 7.3.7.

$$\frac{dM}{dx} = \frac{d}{dx}EI_{zz}\frac{d^2v}{dx^2} = -V(x)$$
(7.3.8)

$$-\frac{dV}{dx} = \frac{d^2M}{dx^2} = \frac{d^2}{dx^2}EI_{zz}\frac{d^2v}{dx^2} = f_y(x)$$
(7.3.9)

And, of course, if EI_{zz} is constant

$$EI_{zz}\frac{d^3v}{dx^3} = -V(x) \qquad EI_{zz}\frac{d^4v}{dx^4} = f_y(x)$$
(7.3.10)

Can we still accurately assume plane sections remain plane? Yes and no. The shear force does cause a shear deformation that warps the cross section; however, if the beam is slender enough that effect is negligible. At some value, as the beam is less and less slender, the shear deformation becomes important and the assumption of plane sections no longer holds. For now we shall proceed with slender beams for which we can neglect shear deformation, but that does not mean we can neglect the shear stress.

So far we have equations for finding the displacement, internal moment and normal stress, and internal shear force, but not the shear stress. To find the shear stress consider a slice of the beam and the normal stresses acting on it as shown in Figure 7.3.6a. For the time being consider this to be a rectangular cross section.

In Figure 7.3.6b we show a portion of the section above the coordinate value of y. The unbalanced normal force caused by the normal stresses in the x direction must be balanced by the horizontal shear force resulting in the shear stress τ_{xy}.

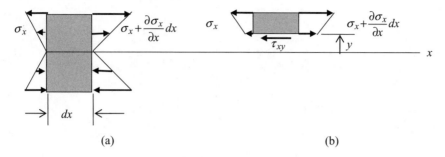

Figure 7.3.6

If we assume that the shear stress is constant across the width, b, of the cross section we get

$$\sum F_x = \int_y^{\frac{h}{2}} \left(\sigma_x + \frac{\partial \sigma_x}{\partial x} dx \right) b dy - \int_y^{\frac{h}{2}} \sigma_x b dy - \tau_{xy} b dx = 0 \qquad (7.3.11)$$

or

$$\int_y^{\frac{h}{2}} \frac{\partial \sigma_x}{\partial x} dx b dy - \tau_{xy} b dx = 0 \quad \rightarrow \quad \tau_{xy} = \int_y^{\frac{h}{2}} \frac{\partial \sigma_x}{\partial x} dy \qquad (7.3.12)$$

The integration is carried out over the cross section area above the coordinate y. This area is defined as

$$dA_y = b dy \qquad (7.3.13)$$

and is given the subscript y to identify it from the total cross sectional area.

When Equations 7.3.5 and 7.3.7 are noted we get

$$\tau_{xy}(x, y) = \frac{1}{b} \int_{A_y} \frac{\partial \sigma_x}{\partial x} dA_y = -\frac{1}{b} \int_{A_y} \frac{dM}{dx} \frac{y}{I_{zz}} dA_y$$

$$= -\frac{1}{I_{zz} b} \frac{dM}{dx} \int_{A_y} y dA_y = \frac{V(x)}{I_{zz} b} \int_{A_y} y dA_y \qquad (7.3.14)$$

For a rectangular cross section of height h and width b we get

$$\tau_{xy}(x, y) = \frac{V(x)}{I_{zz} b} \int_{A_y} y dA_y = \frac{V(x)}{I_{zz} b} \int_y^{\frac{h}{2}} y b dy = \frac{V(x)}{2 I_{zz}} \left(\frac{h^2}{4} - y^2 \right) \qquad (7.3.15)$$

For a rectangular section the magnitude of the shear stress is parabolic as illustrated in Figure 7.3.7. The maximum value is at $y = 0$.

$$\tau_{max}(x, 0) = \frac{V(x) h^2}{8 I_{zz}} = \frac{3 V(x)}{2 b h} \qquad (7.3.16)$$

The direction of the shear stress is parallel to the surface.

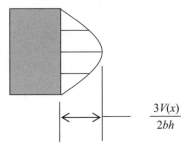

$$\frac{3V(x)}{2bh}$$

Figure 7.3.7

This formula is reasonably accurate for beams with rectangular cross sections; however, is it not adequate for cross sections that have abrupt changes in width. In such cases the more advanced theory of elasticity may necessary. We shall consider shear in other cross section shapes in Section 7.9 and in Chapter 10. In particular we can adapt beam theory to thin walled cross sections.

To summarize, we have listed the quantities and equations used to solve the beam bending problem.

$$M_c, F_c \rightarrow f_y(x) \qquad V(x) \qquad M(x) \qquad \sigma_x(x, y) \qquad \varepsilon_x(x, y) \qquad u(x, y) \qquad v(x) \leftarrow \rho$$

$$\frac{dV}{dx} = -f_y \qquad \frac{dM}{dx} = -V(x) \qquad\qquad\qquad\qquad u(x, y) = -y\frac{dv}{dx}$$

$$\varepsilon_x(x, y) = \frac{du}{dx} = -y\frac{d^2v}{dx^2}$$

$$\sigma_x(x, y) = E\varepsilon_x = -Ey\frac{d^2v}{dx^2}$$

$$M(x) = -\int_A \sigma_x y\, dA = E\frac{d^2v}{dx^2}\int_A y^2\, dA = EI_{zz}\frac{d^2v}{dx^2} \quad \rightarrow \quad \sigma_x = -\frac{M(x)\, y}{I_{zz}}$$

$$\frac{dM}{dx} = \frac{d}{dx}EI_{zz}\frac{d^2v}{dx^2} = -V(x) \quad \rightarrow \quad \tau_{xy}(x, y) = \frac{V(x)}{I_{zz}b}\int_{A_y} y\, dA_y$$

$$\frac{d^2M}{dx^2} = \frac{d^2}{dx^2}EI_{zz}\frac{d^2v}{dx^2} = f_y(x)$$

$$(7.3.17)$$

Listed in the first row are the unknown quantities in a typical problem

    $v(x)$ – distributed displacement in the y direction

    $u(x, y)$ – displacement in the x direction

    $\varepsilon_x(x, y)$ – normal strain

    $\sigma_x(x, y)$ – normal stress

    $M(x)$ – internal moment

    $V(x)$ – internal shear force

and also the known quantities

    $f_y(x)$ – distributed applied loads – force per unit length

    M_c, F_c – concentrated applied moments and forces

    ρ – displacement restraints

The arrows are there to indicate that M_c, F_c, and ρ are boundary conditions to be satisfied in the process of solving for constants of integration. Note that the shear displacement, strain, and stress are not used to derive the displacement equations and therefore do not appear in the first row. Shear displacement is neglected and the shear stress is derived from the normal stress by an equilibrium argument as noted above.

In the second row we have the equilibrium equations on the left and the plane sections remain plane relation on the right.

$$\frac{dM}{dx} = -V(x) \quad - \text{ static equilibrium of moments}$$

$$\frac{dV}{dx} = -f_y(x) \quad - \text{ static equilibrium of forces}$$

$$u(x, y) = -y\frac{dv}{dx} \quad - \text{ plane sections remain plane assumption}$$

In the third row we have the strain displacement relation and in the fourth row Hooke's law.

$$\varepsilon_x(x, y) = \frac{du}{dx} \quad - \text{ strain displacement}$$

$$\sigma_x(x, y) = E\varepsilon_x(x, y) - \text{ Hooke's law for material properties}$$

In row five we define the moment in terms of displacement by combining the above terms and then show the stress in terms of the moment.

$$M(x) = -\int_A \sigma_x\, y dA = E\frac{d^2v}{dx^2}\int_A y^2 dA = EI_{zz}\frac{d^2v}{dx^2} \quad \rightarrow \quad \sigma_x(x, y) = -\frac{M(x)\,y}{I_{zz}} \qquad (7.3.18)$$

In row six we define the shear force in terms of the displacement and the shear stress in terms of the shear force.

$$\frac{d}{dx}EI_{zz}\frac{d^2v}{dx^2} = -V(x) \quad \rightarrow \quad Z\tau_{xy}(x, y) = \frac{V(x)}{I_{zz}b}\int_{A_y} y dA \qquad (7.3.19)$$

Finally, in the last row we have the equilibrium equation relating the displacement and the applied load.

$$\frac{d^2}{dx^2}EI_{zz}\frac{d^2v}{dx^2} = f_y(x) \qquad (7.3.20)$$

Of course, when EI_{zz} is constant we get

$$EI_{zz}\frac{d^3v}{dx^3} = -V(x) \qquad EI_{zz}\frac{d^4v}{dx^4} = f_y(x) \qquad (7.3.21)$$

The solution to Equations 7.3.20 and 7.3.21 involve straight forward integration. For Equations 7.3.21 we have

$$\frac{d^4v}{dx^4} = \frac{f_y(x)}{EI_{zz}}$$

$$\rightarrow \frac{d^3v}{dx^3} = \frac{1}{EI_{zz}} \int f_y(x)dx + a$$

$$\rightarrow \frac{d^2v}{dx^2} = \frac{1}{EI_{zz}} \int\int f_y(x)dx + ax + b \qquad (7.3.22)$$

$$\rightarrow \frac{dv}{dx} = \frac{1}{EI_{zz}} \int\int\int f_y(x)dx + \frac{a}{2}x^2 + bx + c$$

$$\rightarrow v(x) = \frac{1}{EI_{zz}} \int\int\int\int f_x(x)dx + \frac{a}{6}x^3 + \frac{b}{2}x^2 + cx + d$$

where a, b, c, and d are the constants of integration. The four boundary conditions needed to find the constants of integration are from the restraints and/or applied loads at the ends of the beam segments, that is, some combination of the values of the displacement, slope, moment, or shear force at the ends of the beam segments.

Equations 7.3.22 apply to both statically determinate and indeterminate beams. In the determinate case you can find the internal moment directly from the equations of static equilibrium and use the second order equation from Equation 7.3.4 to find the displacement.

$$EI_{zz}\frac{d^2v}{dx^2} = M(x) \qquad (7.3.23)$$

In this case the solution is

$$\frac{d^2v}{dx^2} = \frac{M}{EI_{zz}}$$

$$\rightarrow \frac{dv}{dx} = \frac{1}{EI_{zz}} \int M(x)dx + e \qquad (7.3.24)$$

$$\rightarrow v(x) = \frac{1}{EI_{zz}} \int\int M(x)dx + ex + f$$

where e and f are the constants of integration. The constants of integration are found by applying the boundary conditions on displacement and slope.

With all these equations swirling before your eyes it is time for some examples to help clarify what is going on.

############

Example 7.3.1

Problem: Find the stresses and displacements in a beam loaded by pure concentrated moments of equal value at the ends as shown in Figure 7.3.1 and reproduced here in Figure (a). The coordinate system origin is placed at the midpoint as shown.

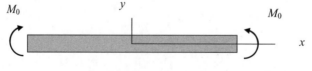

M_0 y M_0 x

Figure (a)

Solution: Since this beam is statically determinate we can find the internal moment and shear force, find the stresses, and then use the second order equation for the displacement.

From static equilibrium the internal moment has a constant value

$$\sum M_z = M_0 - M(x) = 0 \quad \rightarrow \quad M(x) = M_0 \tag{a}$$

and so the stress is a constant in x and linear in y, that is

$$\sigma_x(x, y) = -\frac{M(x)y}{I_{zz}} = -\frac{M_0 y}{I_{zz}} \tag{b}$$

There is no shear force and therefore no shear stresses and so

$$EI_{zz}\frac{d^3v}{dx^3} = \frac{dM}{dx} = -V(x) = 0 \tag{c}$$

We can use the second order equation to find the displacement.

$$EI_{zz}\frac{d^2v}{dx^2} = M_0 \quad \rightarrow \quad \frac{dv}{dx} = \frac{M_0 x}{EI_{zz}} + e \quad \rightarrow \quad v(x) = \frac{M_0 x^2}{2EI_{zz}} + ex + f \tag{d}$$

There are no physical restraints; however, the beam is in static equilibrium so we can assign displacements to orient the beam with respect to the coordinate system.

Let us attach the coordinate system at the origin so that

$$v(0) = 0 \qquad \frac{dv(0)}{dx} = 0 \tag{e}$$

By inserting the displacement solution into these boundary conditions we obtain

$$v(0) = f = 0 \quad \rightarrow \quad \frac{dv(0)}{dx} = e = 0 \tag{f}$$

The deflection is

$$v(x) = \frac{M_0 x^2}{2EI_{zz}} \tag{g}$$

The plot of the displacement is given in Figure (b).

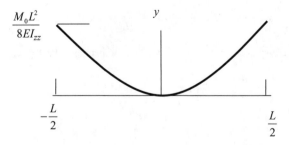

Figure (b)

Example 7.3.2

Problem: Find the stresses and the displacement of a cantilever beam fixed at the left end and with a single concentrated load at the right end as shown in Figure (a). The load is applied on the centroidal x axis. The cross section is rectangular with height h and width b.

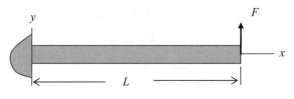

Figure (a)

Solution: Let us first solve this using the fourth order equation. Then we shall repeat the solution using the second order equation.

Using the fourth order equation:
Since there is no distributed applied load we have

$$EI_{zz}\frac{d^4v}{dx^4} = f_y(x) = 0 \quad \rightarrow \quad \frac{d^4v}{dx^4} = 0 \quad \rightarrow \quad \frac{d^3v}{dx^3} = a \quad \rightarrow \quad \frac{d^2v}{dx^2} = ax + b$$

$$\rightarrow \quad \frac{dv}{dx} = \frac{a}{2}x^2 + bx + c \quad \rightarrow \quad v(x) = \frac{a}{6}x^3 + \frac{b}{2}x^2 + cx + d \tag{a}$$

The boundary conditions are

$$v(0) = 0 \qquad \frac{dv}{dx}(0) = 0 \qquad EI_{zz}\frac{d^3v}{dx^3}(L) = -F \qquad EI_{zz}\frac{d^2v}{dx^2}(L) = 0 \tag{b}$$

Thus

$$v(0) = d = 0 \qquad \frac{dv}{dx}(0) = c = 0 \qquad EI_{zz}\frac{d^3v}{dx^3}(L) = EI_{zz}a = -F \quad \rightarrow \quad a = -\frac{F}{EI_{zz}}$$

$$EI_{zz}\frac{d^2v}{dx^2}(L) = EI_{zz}(aL + b) = 0 \quad \rightarrow \quad b = \frac{FL}{EI_{zz}} \tag{c}$$

The displacement is

$$v(x) = \frac{a}{6}x^3 + \frac{b}{2}x^2 + cx + d = -\frac{F}{6EI_{zz}}x^3 + \frac{FL}{2EI_{zz}}x^2 + 0 \cdot x + 0 = \frac{F}{EI_{zz}}\left(\frac{Lx^2}{2} - \frac{x^3}{6}\right) \tag{d}$$

We differentiate the displacement to find the bending moment and shear force.

$$EI_{zz}\frac{dv}{dx} = F\left(Lx - \frac{x^2}{2}\right)$$

$$EI_{zz}\frac{d^2v}{dx^2} = F(L - x) = M \quad \rightarrow \quad M = F(L - x) \tag{e}$$

$$EI_{zz}\frac{d^3v}{dx^3} = -F = -V \quad \rightarrow \quad V = F$$

From this we can obtain the stresses.

$$\sigma_x(x, y) = -\frac{M}{I_{zz}}y = -\frac{F(L-x)}{I_{zz}}y \qquad \tau_{xy}(x, y) = \frac{F}{I_{zz}b}\int_{A_y} ydA_y \qquad \text{(f)}$$

If the cross section is rectangular with a height h and a width b the shear stress is

$$\tau_{xy}(x, y) = \frac{F}{I_{zz}b}\int_{A_y} ydA_y = \frac{F}{I_{zz}b}\int_y^{\frac{h}{2}} ybdy = \frac{F}{2I_{zz}}\left(\frac{h^2}{4} - y^2\right) \qquad \text{(g)}$$

It is a common practice to plot the shear force and bending moment to have a visual picture of their distribution. This makes it easy to spot where these quantities are the largest and hence where the stresses are the maximum. In this case it is very simple as shown in Figure (b).

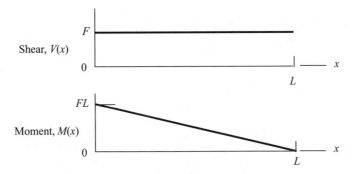

Figure (b)

Clearly the maximum normal stress is at $x = 0$ and at the largest value of y on the cross section. If the cross section is rectangular of height h and width b then the maximum value is

$$\sigma_{max} = \sigma_x\left(0, \pm\frac{h}{2}\right) = \pm\frac{FL}{I_{zz}}\frac{h}{2} \qquad \text{(h)}$$

The shear force is a constant for all values of x. The maximum shear stress occurs at $y = 0$.

$$\tau_{max} = \tau_{xy}(x, 0) = \frac{F}{2I_{zz}}\frac{h^2}{4} = \frac{3F}{2bh} \qquad \text{(i)}$$

Using the second order equation:

For an alternative solution first recognize that this is a statically determinate beam. The internal moment and the shear force can be obtained directly from a free body diagram.

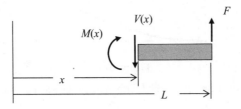

Figure (c)

From static equilibrium

$$\sum M_z = F(L - x) - M(x) = 0 \quad \rightarrow \quad M(x) = F(L - x)$$
$$\sum F_y = F - V(x) = 0 \quad \rightarrow \quad V(x) = F$$

(j)

and the stresses are

$$\sigma_x(x, y) = -\frac{M}{I_{zz}}y = -\frac{F(L - x)}{I_{zz}}y \qquad \tau_{xy}(x, y) = \frac{F}{I_{zz}b}\int_{A_y} y dA_y$$

(k)

The normal stress and shear stress are found directly as in Equation (f). Since we can determine the internal moment directly from static equilibrium the second order equation suffices for finding the displacement.

$$\frac{d^2v}{dx^2} = \frac{M(x)}{EI_{zz}} = \frac{F}{EI_{zz}}(L - x)$$

(l)

Integrate twice

$$\frac{dv}{dx} = \frac{F}{EI_{zz}}\left(Lx - \frac{x^2}{2}\right) + e \quad \rightarrow \quad v(x) = \frac{F}{EI_{zz}}\left(\frac{Lx^2}{2} - \frac{x^3}{6}\right) + ex + f$$

(m)

where e and f are constants of integration.

Apply the boundary conditions and obtain the solution. When the second order equation is used the two boundary conditions are both displacement restraints. That is because a minimum of two displacement restraints are needed on a beam to prevent rigid body motion.

$$v(0) = f = 0 \qquad \frac{dv}{dx}(0) = e = 0 \quad \rightarrow \quad v(x) = \frac{F}{EI_{zz}}\left(\frac{Lx^2}{2} - \frac{x^3}{6}\right)$$

(n)

Note that the displacement is a cubic polynomial, the same as Equation (d). The maximum displacement is at $x = L$.

$$v_{\text{max}} = v(l) = \frac{FL^3}{3EI_{zz}}$$

(o)

A plot of the displacement is given in Figure (c).

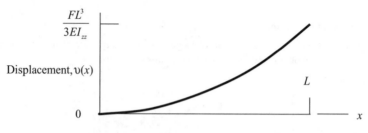

Figure (d)

For a statically determinate beam either the fourth order or the second order equation is suitable.

############

7.4   The Statically Determinate Case

With EI_{zz} and $f_y(x)$ continuous functions of x and concentrated loads at ends the possible statically determinate cases include a beam with all external forces in equilibrium for which Example 7.3.1 is a special case and beams with the restraints as shown in Figure 7.4.1.

All loads including concentrated forces and moments at a boundary are satisfied by using static equilibrium to find both $V(x)$ and $M(x)$.

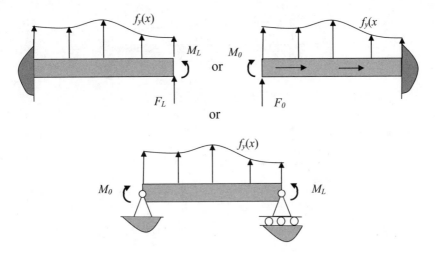

Figure 7.4.1

Once the internal shear forces and bending moments are found you can find the stresses.

$$\sigma_x(x, y) = -\frac{M(x)}{I_{zz}}y \qquad \tau_{xy}(x, y) = \frac{V(x)}{I_{zz}b}\int_{A_y} y\,dA \qquad (7.4.1)$$

To find displacements

$$EI_{zz}\frac{d^2v}{dx^2} = M(x) \quad \rightarrow \quad \frac{dv(x)}{dx} = \int \frac{M(x)}{EI_{zz}}dx + e \rightarrow v(x) = \int\int \frac{M(x)}{EI_{zz}}dx + ex + f \qquad (7.4.2)$$

The constants of integration are found by imposing two displacement restraints, either

$$v(0) = 0 \quad \text{and} \quad \frac{dv(0)}{dx} = 0$$

$$\text{or} \quad v(L) = 0 \quad \text{and} \quad \frac{dv(L)}{dx} = 0 \qquad (7.4.3)$$

$$\text{or} \quad v(0) = 0 \quad \text{and} \quad v(L) = 0$$

When distributed loading is discontinuous, and/or concentrated loads are at midpoints, or EI_{zz} is discontinuous, the equations must be written for each segment. Several possibilities are represented in Figure 7.4.2. There are, of course, an infinite number of variations.

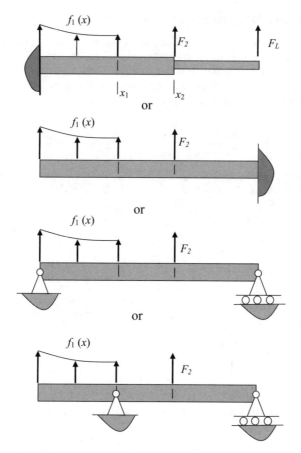

Figure 7.4.2

All loads including concentrated loads at a boundary are satisfied by using static equilibrium to find $M_n(x)$ and $V_n(x)$ in each region.

To find the stresses in each region

$$\sigma_{xn}(x, y) = -\frac{M_n(x)}{I_n} y \qquad \tau_{xyn}(x, y) = \frac{V_n(x)}{I_n b_n} \int_{A_y} y dA \qquad (7.4.4)$$

To find displacements:

$$E_1 I_1 \frac{d^2 v_1}{dx^2} = M_1(x) \rightarrow \frac{dv_1}{dx} = \int \frac{M_1(x)}{E_1 I_1} dx + e_1 \rightarrow v_1(x) = \iint \frac{M_1(x)}{E_1 I_1} dx dx + e_1 x + f_1$$

$$E_2 I_2 \frac{d^2 v_2}{dx^2} = M_2(x) \quad \rightarrow \quad \frac{dv_2}{dx} = \int \frac{M_2(x)}{E_2 I_2} dx + e_2 \rightarrow \quad v_2(x) = \iint \frac{M_2(x)}{E_2 I_2} dx dx + e_2 x + f_2$$

$$\cdots$$

$$E_n I_n \frac{d^2 v_n}{dx^2} = M_n(x) \quad \rightarrow \quad \frac{dv_n}{dx} = \int \frac{M_n(x)}{E_n I_n} dx + e_n \rightarrow \quad v_n(x) = \iint \frac{M_n(x)}{E_n I_n} dx dx + e_n x + f_n$$

$$(7.4.5)$$

The constants of integration are found in part by imposing displacement restraints as given in Equation 7.4.3 in the first three cases respectively. The last example in Figure 7.4.2 is a special case of a simple support at a location other than the ends. As long as there are only two simple supports they may be any place along the beam and it will still be statically determinate.

In addition we must enforce continuity of displacement and slope between regions of integration.

$$v_1(x_1) = v_2(x_1) \qquad v_2(x_2) = v_3(x_2) \quad \cdots \quad v_{n-1}(x_{n-1}) = v_n(x_{n-1})$$

$$\frac{dv_1(x_1)}{dx} = \frac{dv_2(x_1)}{dx} \qquad \frac{dv_2(x_2)}{dx} = \frac{dv_3(x_2)}{dx} \quad \cdots \quad \frac{dv_{n-1}(x_{n-1})}{dx} = \frac{dv_n(x_n)}{dx} \qquad (7.4.6)$$

When the second order equations are used the shear forces, V, and the internal moments, M, are found first and the stresses are calculated. Then the displacement is found by integrating twice. The constants of integration are evaluated by using two displacement boundary conditions.

Here are some more examples of statically determinate cases.

###########

Example 7.4.1

Problem: Find the stress and displacement in a cantilever beam with a uniform distributed load as shown in Figure (a).

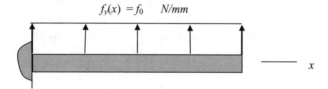

$$f_y(x) = f_0 \quad N/mm$$

Figure (a)

Solution: First find the internal shear force and moment to find the stresses and then solve the second order equation for the displacement.

A free body diagram of a portion of the beam showing the internal force and moment is shown in Figure (b).

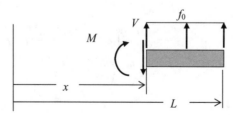

Figure (b)

From Figure (b) the internal shear and moment are

$$\sum F_y = f_0(L - x) - V(x) = 0 \quad \rightarrow \quad V(x) = f_0(L - x)$$

$$\sum M_z = f_0(L - x)\frac{(L - x)}{2} - M(x) = 0 \quad \rightarrow \quad M(x) = \frac{f_0}{2}(L - x)^2 \tag{a}$$

The shear and bending moment diagrams are shown in Figure (c).

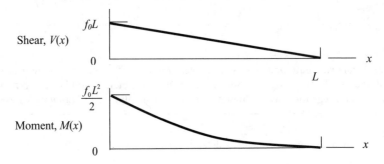

Figure (c)

The normal stress is

$$\sigma_x(x, y) = -\frac{M(x)y}{I_{zz}} = -\frac{f_0}{2I_{zz}}(L - x)^2 y \tag{b}$$

The shear stress is

$$\tau_{xy}(x, y) = \frac{V}{I_{zz}b}\int_{A_y} y dA_y = \frac{f_0(L - x)}{I_{zz}b}\int_{A_y} y dA_y \tag{c}$$

If the cross section is rectangular with height h and width b the shear stress is

$$\tau_{xy}(x, y) = \frac{f_0(L - x)}{2I_{zz}}\left(\frac{h^2}{4} - y^2\right) \tag{d}$$

In such a case the maximum stresses are at $x = 0$ and at $y = \pm h/2$ for normal stress and $y = 0$ for shear stress.

$$\sigma_{max} = \sigma_x\left(0, \pm\frac{h}{2}\right) = \pm\frac{f_0 L^2}{4I_{zz}}h \qquad \tau_{max} = \tau_{xy}(0, 0) = \frac{f_0 L}{8I_{zz}}h^2 \tag{e}$$

Integrate twice to find the displacement.

$$\frac{d^2v}{dx^2} = \frac{M}{EI_{zz}} = \frac{f_0}{2EI_{zz}}(L - x)^2$$

$$\rightarrow \quad \frac{dv}{dx} = \frac{f_0}{2EI_{zz}}\left(L^2 x - Lx^2 + \frac{x^3}{3}\right) + e \tag{f}$$

$$\rightarrow \quad v(x) = \frac{f_0}{2EI_{zz}}\left(\frac{L^2 x^2}{2} - \frac{Lx^3}{3} + \frac{x^4}{12}\right) + ex + f$$

The boundary conditions are

$$v(0) = f = 0 \qquad \frac{dv}{dx}(0) = e = 0 \tag{g}$$

The displacement is

$$v(x) = \frac{f_0}{2EI_{zz}} \left(\frac{L^2 x^2}{2} - \frac{Lx^3}{3} + \frac{x^4}{12} \right) \tag{h}$$

Note that the displacement is a fourth order polynomial. The maximum displacement is at $x = L$.

$$v_{max} = v(L) = \frac{f_0 L^4}{8EI_{zz}} \tag{i}$$

Example 7.4.2

Problem: Find the stresses and displacements in a beam with two transverse forces as shown in Figure (a). The cross section is a rectangle of height h and width b.

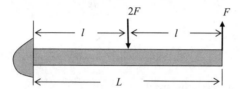

Figure (a)

Solution: We can divide the beam into two regions with separate equations covering each region. This is statically determinate and so we can find the moments and shear forces in each region to be

$$\sum M_z = F(2l - x) - 2F(l - x) - M_1 = 0$$

Region 1:
$$\rightarrow M_1 = F(2l - x) - 2F(l - x) = Fx \quad 0 \le x \le l$$
$$\sum F_y = F - 2F - V_1 = 0 \quad \rightarrow \quad V_1 = -F$$
$$\sum M_z = F(2l - x) - M_2 = 0 \quad \rightarrow \quad M_2 = F(2l - x) \tag{a}$$

Region 2:
$$\qquad\qquad\qquad\qquad\qquad\qquad l \le x \le 2l$$
$$\sum F_y = F - V_2 = 0 \quad \rightarrow \quad V_2 = F$$

The shear and bending moment diagrams are shown in Figure (b).

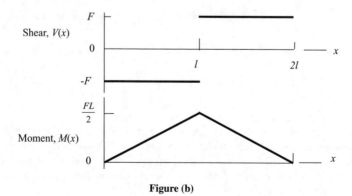

Figure (b)

One can quickly see where the moments and shear forces are a maximum and consequently where the largest stresses occur. The maximum shear stress occurs at $y = 0$ all along the beam; however its value is negative along the left half and positive along the right half.

$$\tau_{xy}(x, y) = \frac{V}{2I_{zz}} \left(\frac{h^2}{4} - y^2 \right) \quad \rightarrow \quad \tau_{max} = \pm \frac{F}{8I_{zz}} h^2 \tag{b}$$

The maximum bending stress occurs at $x = \dfrac{L}{2}$ and at $y = \pm \frac{h}{2}$.

$$\sigma_x(x, y) = -\frac{FL}{2I_{zz}} y \quad \rightarrow \quad \sigma_{max} = \pm \frac{FL}{4I_{zz}} h \tag{c}$$

The equations governing the displacement in each region are

$$EI_{zz} \frac{d^2 v_1}{dx^2} = Fx \qquad EI_{zz} \frac{d^2 v_2}{dx^2} = F(2l - x) \tag{d}$$

The boundary conditions are

$$v_1(0) = 0 \qquad \frac{dv_1}{dx}(0) = 0 \qquad v_1(l) = v_2(l) \qquad \frac{dv_1}{dx}(l) = \frac{dv_2}{dx}(l) \tag{e}$$

The solutions are found by integrating twice in each region.

$$\frac{dv_1}{dx}(x) = \frac{F}{EI_{zz}} \left(\frac{x^2}{2} \right) + e_1 \qquad v_1(x) = \frac{F}{EI_{zz}} \left(\frac{x^3}{6} \right) + e_1 x + f_1$$

$$\frac{dv_2}{dx}(x) = \frac{F}{EI_{zz}} \left(2lx - \frac{x^2}{2} + e_2 \right) \qquad v_2(x) = \frac{F}{EI_{zz}} \left(lx^2 - \frac{x^3}{6} + e_2 x + f_2 \right) \tag{f}$$

Applying boundary conditions, the displacement in the first region is

$$v_1(0) = f_1 = 0 \qquad \frac{dv_1}{dx}(0) = e_1 = 0 \quad \rightarrow \quad v_1(x) = \frac{F}{EI_{zz}} \frac{x^3}{6} \qquad 0 \le x \le \frac{L}{2} \tag{g}$$

The boundary conditions for the second region are the continuity of slope and displacement where the two regions meet.

$$\frac{dv_1}{dx}(l) = \frac{F}{EI_{zz}} \frac{l^2}{2} = \frac{dv_2}{dx}(l) = \frac{F}{EI_{zz}} \left(2l^2 - \frac{l^2}{2} + e_2 \right) \quad \rightarrow \quad e_2 = -l^2$$

$$v_1(l) = \frac{F}{EI_{zz}} \frac{l^3}{6} = v_2(l) = \frac{F}{EI_{zz}} \left(l^3 - \frac{l^3}{6} - l^3 + f_2 \right) \quad \rightarrow \quad f_2 = \frac{l^3}{3} \tag{h}$$

The displacement in the second region is

$$v_2(x) = \frac{F}{EI_{zz}} \left(lx^2 - \frac{x^3}{6} - l^2 x + \frac{l^3}{3} \right) \qquad l \le x \le 2l \tag{i}$$

Note that in each region the displacement is a cubic polynomial.

It is sometimes useful to define the displacements in term of the discrete values of displacement at the points of load application and restraint and in terms of local coordinates. See Example 4.4.5 in the axial case. In the bending case the discrete displacements must include both the lateral displacement and the slope in order to represent the cubic polynomial, that is, we need four quantities to define the four coefficients of the cubic polynomial.

We use the following notation for the slopes

$$\theta = \frac{dv}{dx} \tag{j}$$

The plot of the displacements based on Equations (g) and (i) is shown in the Figure (c).

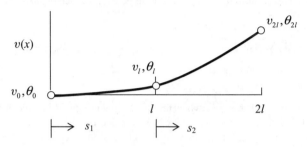

Figure (c)

To this plot we have added two local coordinates, s_1 and s_2 and the lateral displacements and slopes at the discrete points of load application and restraint. Let us define the cubic polynomial for region 2 where the displacements are v_l, θ_l at $s_2 = 0$ and v_{2l}, θ_{2l} at $s_2 = l$.

The distributed displacement and slope are

$$v(s_2) = as_2^3 + bs_2^2 + cs_2 + d$$

$$\theta(s_2) = \frac{dv}{dx}(s_2) = 3as_2^2 + 2bs_2 + c \tag{k}$$

Evaluate the distributed displacements at the discrete points $s_2 = 0, l$.

$$\begin{aligned} v(0) = d = v_l && \theta(0) = c = \theta_l \\ v(l) = al^3 + bl^2 + \theta_l l + v_l = v_{2l} && \theta(l) = 3al^2 + 2bl + \theta_l = \theta_{2l} \end{aligned} \tag{l}$$

Solve for a and b in terms of the discrete values of displacement and slope.

$$a = \frac{1}{l^3}(2v_l + \theta_l l - 2v_{2l} + \theta_{2l}l) \qquad b = \frac{1}{l^2}(-3v_l - 2\theta_l l + 3v_{2l} - \theta_{2l}l) \tag{m}$$

The displacement in region 2 may be written

$$v(s_2) = \frac{1}{l^3}(2v_l + \theta_l l - 2v_{2l} + \theta_{2l}l)s_2^3 + \frac{1}{l^2}(-3v_l - 2\theta_l l + 3v_{2l} - \theta_{2l}l)s_2^2 + \theta_l s_2 + v_l \tag{n}$$

This may be rearranged into the following matrix form

$$v(s_2) = \frac{1}{l^3}\begin{bmatrix} l^3 - 3ls_2^2 + 2s_2^3 & l^3 s_2 - 2l^2 s_2^2 + ls_2^3 & 3ls_2^2 - 2s_2^3 & -l^2 s_2^2 + ls_2^3 \end{bmatrix} \begin{bmatrix} v_l \\ \theta_l \\ v_{2l} \\ \theta_{2l} \end{bmatrix} \tag{o}$$

In a similar fashion we can write the displacement in region one to be

$$v(s_1) = \frac{1}{l^3}\begin{bmatrix} l^3 - 3ls_1^2 + 2s_1^3 & l^3 s_1 - 2l^2 s_1^2 + ls_1^3 & 3ls_1^2 - 2s_1^3 & -l^2 s_1^2 + ls_1^3 \end{bmatrix} \begin{bmatrix} v_0 \\ \theta_0 \\ v_l \\ \theta_l \end{bmatrix} \tag{p}$$

These are the shape functions for a beam element which we shall utilize in Chapter 8 when we introduce the general (finite element) method for the beam.

##########

In fact this format given in Equations (o) and (p) in Example 7.4.2 applies to all regions between concentrated loads in any uniform beam segment, or, in general for element m

$$v(s_m) = \frac{1}{l_m^3} \left[l_m^3 - 3l_m s_m + 2s_m^3 \quad l_m^3 s_m - 2l_m^2 s_m^2 + l_m s_m^3 \quad 3l_m s_m^2 - 2s_m^3 \quad -l_m^2 s_m^2 + l_m s_m^3 \right] \begin{bmatrix} v_n \\ \theta_n \\ v_{n+1} \\ \theta_{n+1} \end{bmatrix}$$

$$(7.4.7)$$

Now why did we go to all this fuss when the displacements given in Equation (g) look simpler and easier to use? It is because Equation 7.4.7 can be applied to all beams with concentrated loads – determinate or indeterminate. And it will make the solution of more complicated cases of loading and restraint much easier. We shall explore the use of these functions in great detail in Chapter 8.

Much of what has been written on the subject of beams, especially in the introductory texts, dwells at great length on determinate beams. Because of the importance of bending moments and shear forces a visual representation is sought. So these texts spend much time on bending moment and shear diagrams. From this visual representation you quickly see where the bending moments and shear forces are a maximum and thus where the normal stresses and shear stresses are a maximum. However, in real life many beams are statically indeterminate and the bending moments and shear forces cannot be found from static equilibrium alone. In such cases you must integrate the fourth order equations to find the displacements, apply boundary conditions to find the constants of integration and then differentiate twice to find the moments and three times to find the shear forces. We shall demonstrate this in Section 7.6.

First let us take a look at the use of work and energy in the statically determinate case.

7.5 Work and Energy—Castigliano's Second Theorem

The strain energy in bending for a beam of length L consists of a contribution from the normal stress and another from the shear stress.

$$U_B = U_n + U_s = \frac{1}{2} \int \sigma_x \varepsilon_x dV + \frac{1}{2} \int \gamma_{xy} \tau_{xy} dV = \frac{1}{2} \int \frac{\sigma_x^2}{E} dV + \frac{1}{2} \int \frac{\tau_{xy}^2}{G} dV \qquad (7.5.1)$$

where

$$\sigma_x(x, y) = -\frac{M(x)y}{I_{zz}} \qquad \tau_{xy}(x, y) = \frac{V(x)}{I_{zz}b} \int_{A_y} y dA_y \qquad (7.5.2)$$

The contribution to the strain energy of the normal bending stress is

$$U_n = \frac{1}{2} \int \sigma_x \varepsilon_x dV = \frac{1}{2} \int \frac{\sigma_x^2}{E} dV = \frac{1}{2} \int \frac{1}{E} \left(-\frac{My}{I_{zz}} \right)^2 dV$$

$$= \frac{1}{2} \int_0^L \frac{M^2}{EI_{zz}^2} dx \int_A y^2 dA = \frac{1}{2} \int_0^L \frac{M^2}{EI_{zz}} dx \qquad (7.5.3)$$

The contribution to the strain energy of the shear stress in bending is

$$U_s = \frac{1}{2} \int \gamma_{xy} \tau_{xy} dV = \frac{1}{2} \int \frac{\tau_{xy}^2}{G} dV = \frac{1}{2G} \int \left(\frac{V}{I_{zz}b} \int_{A_y} y dA_y \right)^2 dV \qquad (7.5.4)$$

With Castigliano's second theorem (see Chapter 4, Section 4.11) we can find the deflection at the point of application of a concentrated load in the direction of that load.

$$v_i = \frac{\partial U}{\partial F_i} \qquad (7.5.5)$$

In the next two examples we shall learn that the contribution of the shear strain energy is very small and usually can be neglected.

###########

Example 7.5.1

Problem: Consider the common case of a beam of uniform material and cross section and with one end fixed and a concentrated force at the other end as shown in Figure (a). The cross section is rectangular with a height h and width b. Find the displacement at the right end. See Example 7.3.2 for comparison.

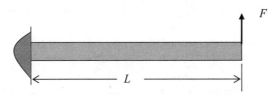

Figure (a)

Solution: Use Castigliano's second theorem.
From equilibrium the internal moment is

$$\sum M_z = F(L-x) - M(x) = 0 \quad \rightarrow \quad M(x) = F(L-x) \qquad (a)$$

The contribution to the strain energy of the normal bending stress in this case is

$$U_n = \frac{1}{2}\int_0^L \frac{M^2}{EI_{zz}} dx = \frac{1}{2}\int_0^L \frac{F^2(L-x)^2}{EI_{zz}} dx = \frac{F^2}{2EI_{zz}}\left(L^2 x - Lx^2 + \frac{x^3}{3}\right)\Big|_0^L = \frac{F^2 L^3}{6EI_{zz}} \qquad (b)$$

The shear force is constant, that is,

$$\sum V_y = F - V(x) = 0 \quad \rightarrow \quad V(x) = F \qquad (c)$$

For the rectangular cross section the shear stress is from Equation 7.3.14

$$\tau_{xy}(x,y) = \frac{V(x)}{2I_{zz}}\left(\frac{h^2}{4} - y^2\right) = \frac{F}{2I_{zz}}\left(\frac{h^2}{4} - y^2\right) \qquad (d)$$

The strain energy of shear is

$$U_s = \frac{1}{2}\int \gamma_{xy}\tau_{xy}dV = \frac{1}{2}\int \frac{\tau_{xy}^2}{G} dV = \frac{F^2 b}{8GI_{zz}^2}\int_{-\frac{h}{2}}^{\frac{h}{2}}\int_0^L \left(\frac{h^2}{4} - y^2\right)^2 dxdy = \frac{F^2 bh^5 L}{240GI_{zz}^2} \qquad (e)$$

Using Castigliano's second theorem we have the displacement at the point of load application.

$$v(L) = \frac{\partial U_B}{\partial F} = \frac{\partial U_n}{\partial F} + \frac{\partial U_s}{\partial F} = \frac{\partial}{\partial F}\left(\frac{F^2 L^3}{6EI_{zz}}\right) + \frac{\partial}{\partial F}\left(\frac{F^2 bh^5 L}{96GI_{zz}^2}\right) = \frac{FL^3}{3EI_{zz}} + \frac{Fbh^5 L}{120GI_{zz}^2} \qquad (f)$$

The value found in Example 7.3.2, Equation (m), by solving the differential equation is

$$v(L) = \frac{FL^3}{3EI_{zz}} \tag{g}$$

###########

This very suspiciously looks as if we found only the normal stress contribution to the displacement when we solved the differential equation. In fact, that is what happened because our assumption in developing the beam theory neglects shear deformation. That the shear contribution can be neglected is examined in the next example.

###########

Example 7.5.2

Problem: Examine the contribution of shear energy to the displacement of the beam in Example 7.5.1.

Solution: Compare the displacements with and without the strain energy of shear stress.

Let us take a specific case to compare. Use an aluminum beam with $E = 68950\ MPa$ and $G = 26520\ MPa$. It has the dimensions

$$L = 1000\ mm \qquad h = 100\ mm \qquad b = 40\ mm \tag{a}$$

Its moment of inertia is

$$I_{zz} = \frac{bh^3}{12} = \frac{40 \cdot 100^3}{12} \tag{b}$$

The strain energy of normal stresses in bending is

$$U_n = \frac{1}{2} \cdot \frac{F^2 L^3}{3EI_{zz}} = \frac{1}{2} \cdot \frac{F^2 \cdot 1000^3 \cdot 12}{3 \cdot 68950 \cdot 40 \cdot 100^3} = F^2 \cdot 0.0007252 \tag{c}$$

The strain energy of shear stresses in bending is

$$U_s = \frac{F^2 bh^5 L}{240 G I_{zz}^2} = \frac{F^2 bh^5 L}{240 G} \cdot \frac{12^2}{b^2 h^6} = \frac{3F^2 L}{5bhG} = \frac{3 \cdot 1000}{5 \cdot 40 \cdot 100 \cdot 26520} F^2 = 5.656 \times 10^{-6} F^2 \tag{d}$$

The ratio of the strain energy of normal stresses to the strain energy of shear stresses is

$$\frac{U_n}{U_s} = \frac{7.252 \times 10^{-4}}{5.656 \times 10^{-6}} = 128.22 \tag{e}$$

The displacement under the load including shear energy is

$$v(L) = F \cdot 2 \cdot \left(7.252 \times 10^{-4} + 5.626 \times 10^{-6}\right) = 0.0014617 F \tag{f}$$

From bending energy alone the displacement is

$$v(L) = F \cdot 2 \cdot (0.0007252) = F \cdot 0.0014504 \tag{g}$$

This is a 1.94 % change. The contribution of shear to displacement is usually neglected for slender beams.

###########

When there are multiple loads on a determinate beam, including distributed and concentrated forces and concentrated moments, for example, as shown in Figure 7.5.1, we can use Castigliano's theorem to find the displacements at the points of concentrated load application.

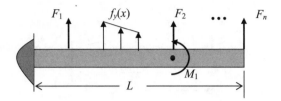

Figure 7.5.1

First find the strain energy as a function of all the loading.

$$U_B\left(F_1, F_2, \ldots, F_n, f_{y1}, \ldots, f_{yn}, M_1, \ldots M_n\right)$$ (7.5.5)

Do not insert numerical values for the loads until after the differentiation has been done.
The displacements at the points of concentrated load application are

$$v_i = \frac{\partial U_B}{\partial F_i} \qquad \theta_i = \frac{\partial U_B}{\partial M_i}$$ (7.5.6)

where

$$\theta_i = \frac{dv_i}{dx}$$ (7.5.7)

Note that differentiation with respect to a force finds the lateral displacement and differentiation with respect to a moment finds the slope.

If you wish to find the displacement or slope at a point where no load is applied you introduce a force or moment at that point, differentiate to obtain the expression for that displacement and then set the load to zero.

###########

7.6   The Statically Indeterminate Case

With EI_{zz} and $f_x(x)$ continuous functions of x and three or more displacement restraints the fourth order equation must be used.

$$\frac{d^2}{dx^2}EI_{zz}\frac{d^2v}{dx^2} = f_y(x)$$ (7.6.1)

and, of course, when EI_{zz} is constant

$$EI_{zz}\frac{d^4v}{dx^4} = f_y(x)$$ (7.6.2)

When EI_{zz} is constant we integrate to find the displacement as explained in Equation 7.3.22.

$$v(x) = \frac{1}{EI_{zz}} \int \int \int \int f_y(x)dx + \frac{a}{6}x^3 + \frac{b}{2}x^2 + cx + d$$ (7.6.3)

Distributed loads are accounted for but concentrated loads must enter through the boundary conditions. When the beam is indeterminate the constants of integration are found by three or more boundary displacements and one boundary force or four boundary displacements. Some of the possible configurations of restraints are shown in Figure 7.6.1.

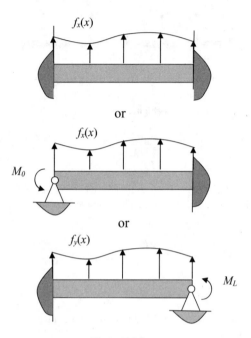

Figure 7.6.1

For the beam with both ends fixed the boundary conditions are

$$v(0) = 0 \qquad \frac{dv(0)}{dx} = 0 \qquad v(l) = 0 \qquad \frac{dv(L)}{dx} = 0 \tag{7.6.4}$$

For the two beams with a simple support at one end and a fixed support at the other we have

$$v(0) = 0 \quad \text{and} \quad EI\frac{d^2v(0)}{dx^2} = M_0 \quad \text{and} \quad v(l) = \frac{dv(L)}{dx} = 0 \tag{7.6.5}$$

or

$$v(0) = \frac{dv(0)}{dx} = 0 \quad \text{and} \quad v(l) = 0 \quad \text{and} \quad EI\frac{d^2v(L)}{dx^2} = M_L \tag{7.6.6}$$

To find the stresses differentiate the displacement.

$$EI\frac{d^2v}{dx^2} = M(x) \quad \rightarrow \quad \sigma_x(x, y) = -\frac{M(x)}{I}y$$

$$EI\frac{d^3v}{dx^3} = -V(x) \quad \rightarrow \quad \tau_{xy}(x, y) = \frac{V(x)}{Ib}\int_{A_y} ydA \tag{7.6.7}$$

When distributed loading is discontinuous, and/or concentrated loads are at midpoints, or EI is discontinuous but constant for each segment the equations.

$$\frac{d^4v_1}{dx^4} = \frac{f_{y1}(x)}{E_1I_1} \quad \rightarrow \quad v_2(x) = \frac{1}{E_1I_2}\int\int\int\int f_{y1}(x)dx + \frac{a_1}{6}x^3 + \frac{b_1}{2}x^2 + c_1x + d_1$$

$$\frac{d^4v_2}{dx^4} = \frac{f_{y2}(x)}{E_2I_2} \quad \rightarrow \quad v_2(x) = \frac{1}{E_2I_2}\int\int\int\int f_{y2}(x)dx + \frac{a_2}{6}x^3 + \frac{b_2}{2}x^2 + c_2x + d_2 \qquad (7.6.8)$$

$$\cdots$$

$$\frac{d^4v_i}{dx^4} = \frac{f_{yi}(x)}{E_iI_i} \quad \rightarrow \quad v_i(x) = \frac{1}{E_iI_i}\int\int\int\int f_{yi}(x)dx + \frac{a_i}{6}x^3 + \frac{b_i}{2}x^2 + c_ix + d_i$$

The constants of integration are found by imposing boundary conditions. Here we can have a great variety of restraints because we can have intermediate supports as well as end supports. There must be a minimum of two displacement restraints but there may be many more. For example the beam shown in Figure 7.6.2 has five displacement restraints. It would have to be divided into at least seven segments in order to solve.

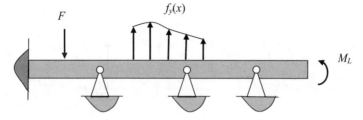

Figure 7.6.2

Apply whatever displacement restraints are necessary at the ends of each segment of the beam. Then apply continuity of displacement and slope where segments meet. Let two adjacent segments be identified by the subscripts i and $i + 1$ and the coordinate where they meet by x_j.

$$v_i(x_j) = v_{i+1}(x_j) \qquad \frac{dv_i(x_j)}{dx} = \frac{dv_{i+1}(x_j)}{dx} \qquad (7.6.9)$$

and where necessary apply equilibrium of internal and external forces between segments. When concentrated moments M_j or forces F_j are present this means

$$E_iI_i\frac{d^2v_i(x_j)}{dx^2} - M_j - E_{i+1}I_{i+1}\frac{d^2v_{i+1}(x_j)}{dx^2} = 0$$

$$-E_iI_i\frac{d^3v_i(x_j)}{dx^3} + F_j + E_{i+1}I_{i+1}\frac{d^3v_i(x_j)}{dx^3} = 0 \qquad (7.6.10)$$

Once the displacement are known differentiate to find the stresses.

$$E_iI_i\frac{d^2v_i(x)}{dx^2} = M_i(x) \quad \rightarrow \quad \sigma_{xi} = -\frac{M_i(x)}{I_i}y$$

$$E_iI_i\frac{d^3v_i(x)}{dx^3} = -V_i(x) \quad \rightarrow \quad \tau_{xyi} = \frac{V_i(x)}{Ib}\int_{A_y}ydA_y \qquad (7.6.11)$$

There are too many possibilities to list them all.

When the fourth order equation is used the displacements are found first by integration. The constants of integration are found from a combination of displacement boundary conditions and force boundary conditions. The stresses are then found by differentiating the displacements to find the internal shear forces and moments and inserting these values in the appropriate formulas. If the problem is indeterminate you must use the fourth order equations. If the problem is determinate you may use either the second order or the fourth order equations

###########

Example 7.6.1

Problem: Find the stresses and displacement of a uniform beam with both ends fixed and loaded with a uniform distributed applied load as shown in Figure (a). The cross section is rectangular with a height of *h* and width *b*.

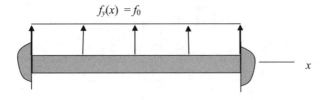

$f_y(x) = f_0$

x

Figure (a)

Solution: Solve the fourth order beam equation for the displacements and then differentiate to find the internal shear force and moment and hence the stresses.

A free body diagram of a portion of the beam in Figure (b) shows that there are four unknowns – the internal force and moment and the restraint force and moment. We have just two equations of static equilibrium; therefore, this is a statically indeterminate problem.

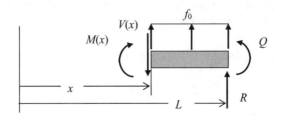

Figure (b)

We turn to the fourth order equation in such cases. For this problem we have

$$EI_{zz}\frac{d^4v}{dx^4} = f_y(x) = f_0 \qquad\qquad (a)$$

Integrating four times we obtain four constants of integration.

$$\frac{d^4v}{dx^4} = \frac{f_0}{EI_{zz}}$$

$$\rightarrow \quad \frac{d^3v}{dx^3} = \frac{f_0}{EI_{zz}}x + a$$

$$\rightarrow \quad \frac{d^2v}{dx^2} = \frac{f_0}{2EI_{zz}}x^2 + ax + b \tag{b}$$

$$\rightarrow \quad \frac{dv}{dx} = \frac{f_0}{6EI_{zz}}x^3 + \frac{a}{2}x^2 + bx + c$$

$$\rightarrow \quad v(x) = \frac{f_0}{24EI_{zz}}x^4 + \frac{ax^3}{6} + \frac{bx^2}{2} + cx + d$$

Boundary conditions are

$$v(0) = d = 0$$

$$\frac{dv}{dx}(0) = c = 0$$

$$v(L) = \frac{f_0}{24EI_{zz}}L^4 + \frac{aL^3}{6} + \frac{bL^2}{2} = 0 \tag{c}$$

$$\frac{dv}{dx}(L) = \frac{f_0}{6EI_{zz}}L^3 + \frac{a}{2}L^2 + bL = 0$$

and so

$$a = -\frac{f_0 L}{2EI_{zz}} \quad\quad b = \frac{f_0 L^2}{12EI_{zz}} \quad\quad c = 0 \quad\quad d = 0 \tag{d}$$

Our final answer for the displacement is

$$v(x) = \frac{f_0}{24EI_{zz}}x^4 - \frac{f_0 L}{12EI_{zz}}x^3 + \frac{f_0 L^2}{24EI_{zz}}x^2 = \frac{f_0}{24EI_{zz}}(x^4 - 2Lx^3 + L^2x^2) \tag{e}$$

The shear force is

$$V(x) = -EI_{zz}\frac{d^3v}{dx^3} = -f_0 x + \frac{f_0 L}{2} = f_0\left(\frac{L}{2} - x\right) \tag{f}$$

and the bending moment is

$$M(x) = EI_{zz}\frac{d^2v}{dx^2} = f_0\left(\frac{x^2}{2} - \frac{Lx}{2} + \frac{L^2}{12}\right) \tag{g}$$

The shear and bending moment diagrams are shown in Figure (c).

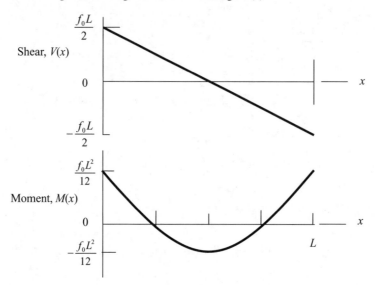

Figure (c)

The shear and normal stresses are

$$\tau_{xy}(x, y) = \frac{V}{I_{zz}b} \int_{A_y} y dA_y = \frac{f_0}{I_{zz}b} \left(\frac{L}{2} - x\right) \int_{A_y} y dA_y = \frac{f_0}{I_{zz}} \left(\frac{L}{2} - x\right) \left(\frac{h^2}{8} - \frac{y^2}{2}\right)$$

$$\sigma_x(x, y) = -\frac{M(x)}{I_{zz}} y = -\frac{f_0}{I_{zz}} \left(\frac{x^2}{2} - \frac{Lx}{2} + \frac{L^2}{12}\right) y$$

(h)

The maximum shear stresses and the maximum bending stresses are at the fixed ends.

$$\tau_{max} = \pm\frac{f_0 L}{16 I_{zz}} h^2 \qquad \sigma_{max} = \pm\frac{f_0 L^2}{24 I_{zz}} h$$

(i)

###########

Example 7.6.2

Problem: Find and plot the displacements, shear forces, and bending moments for the beam shown in Figure (a).

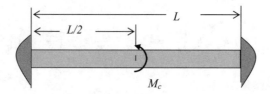

Figure (a)

Solution: Use the fourth order equations.

The beam is indeterminate so we must divide it into two regions and use the fourth order equations and find the displacements first.

$$EI_{zz}\frac{d^4v_1}{dx^4} = 0 \quad \rightarrow \quad \frac{d^3v_1}{dx^3} = a_1 \quad \rightarrow \quad \frac{d^2v_1}{dx^2} = a_1x + b_1$$

$$\rightarrow \quad \frac{dv_1}{dx} = a_1\frac{x^2}{2} + b_1x + c_1 \quad \rightarrow \quad v_1 = a_1\frac{x^3}{6} + b_1\frac{x^2}{2} + c_1x + d_1$$

$$EI_{zz}\frac{d^4v_2}{dx^4} = 0 \quad \rightarrow \quad \frac{d^3v_2}{dx^3} = a_2 \quad \rightarrow \quad \frac{d^2v_2}{dx^2} = a_2x + b_2$$ (a)

$$\rightarrow \quad \frac{dv_2}{dx} = a_2\frac{x^2}{2} + b_2x + c_2 \quad \rightarrow \quad v_2 = a_2\frac{x^3}{6} + b_2\frac{x^2}{2} + c_2x + d_2$$

The boundary conditions at the ends are

$$v_1(0) = 0 \qquad \rightarrow \qquad d_1 = 0$$

$$\frac{dv_1(0)}{dx} = 0 \qquad \rightarrow \qquad c_1 = 0$$ (b)

$$\frac{dv_2(L)}{dx} = 0 \qquad \rightarrow \qquad a_2\frac{L^2}{2} + b_2L + c_2 = 0$$

$$v_2(l) = 0 \qquad \rightarrow \qquad a_2\frac{L^3}{6} + b_2\frac{L^2}{2} + c_2L + d_2 = 0$$

The boundary conditions for continuity between the two regions at $L/2$ are

$$v_1\left(\frac{L}{2}\right) = v_2\left(\frac{L}{2}\right) \quad \rightarrow \quad a_1\frac{L^3}{48} + b_1\frac{L^2}{8} = a_2\frac{L^3}{48} + b_2\frac{L^2}{8} + c_2\frac{L}{2} + d_2$$

$$\frac{dv_1\left(\frac{L}{2}\right)}{dx} = \frac{dv_2\left(\frac{L}{2}\right)}{dx} \quad \rightarrow \quad a_1\frac{L^2}{8} + b_1\frac{L}{2} = a_2\frac{L^2}{8} + b_2\frac{L}{2} + c_2$$

$$EI_{zz}\frac{d^2v_1\left(\frac{L}{2}\right)}{dx^2} - M_c - EI_{zz}\frac{d^2v_2\left(\frac{L}{2}\right)}{dx^2} = 0 \quad \rightarrow \quad a_1\frac{L}{2} + b_1 - \frac{M_c}{EI_{zz}} - a_2\frac{L}{2} - b_2 = 0$$ (c)

$$\frac{d^3v_1\left(\frac{L}{2}\right)}{dx^3} - \frac{d^3v_2\left(\frac{L}{2}\right)}{dx^3} = 0 \quad \rightarrow \quad a_1 = a_2$$

We can use $a_1 = a_2$ to simplify this set of linear algebraic equations.

$$a_2\frac{L^2}{2} + b_2L + c_2 = 0$$

$$a_2\frac{L^3}{6} + b_2\frac{L^2}{2} + c_2L + d_2 = 0$$

$$b_1\frac{L^2}{8} - b_2\frac{L^2}{8} - c_2\frac{L}{2} - d_2 = 0$$ (d)

$$b_1\frac{L}{2} - b_2\frac{L}{2} - c_2 = 0$$

$$b_1 - b_2 = \frac{M_c}{EI_{zz}}$$

The five algebraic equations for the constants can be put in matrix form.

$$
\begin{bmatrix}
0 & \dfrac{L^2}{2} & L & 1 & 0 \\[2mm]
0 & \dfrac{L^3}{6} & \dfrac{L^2}{2} & L & 1 \\[2mm]
\dfrac{L^2}{8} & 0 & -\dfrac{L^2}{8} & -\dfrac{L}{2} & -1 \\[2mm]
\dfrac{L}{2} & 0 & -\dfrac{L}{2} & -1 & 0 \\[2mm]
1 & 0 & -1 & 0 & 0
\end{bmatrix}
\begin{bmatrix}
b_1 \\ a_2 \\ b_2 \\ c_2 \\ d_2
\end{bmatrix}
=
\begin{bmatrix}
0 \\ 0 \\ 0 \\ 0 \\ \dfrac{M_c}{EI_{zz}}
\end{bmatrix}
\tag{e}
$$

Using a symbolic solver (see Appendix C) the constants are found to be

$$
\begin{bmatrix}
a_1 \\ b_1 \\ a_2 \\ b_2 \\ c_2 \\ d_2
\end{bmatrix}
=
\frac{M_c}{EI_{zz}}
\begin{bmatrix}
\dfrac{3}{2L} \\[2mm]
-\dfrac{1}{4} \\[2mm]
\dfrac{3}{2L} \\[2mm]
-\dfrac{5}{4} \\[2mm]
\dfrac{L}{2} \\[2mm]
-\dfrac{L^2}{8}
\end{bmatrix}
\tag{f}
$$

The displacements are

$$
\begin{aligned}
v_1 &= a_1 \frac{x^3}{6} + b_1 \frac{x^2}{2} = \frac{M_c}{4EI_{zz}L} x^3 - \frac{M_c}{8EI_{zz}} x^2 \\[2mm]
v_2 &= a_2 \frac{x^3}{6} + b_2 \frac{x^2}{2} + c_2 x + d_2 = \frac{M_c}{4EI_{zz}L} x^3 - \frac{5M_c}{8EI_{zz}L} x^2 + \frac{M_c L}{2EI_{zz}} x - \frac{M_c L^2}{8EI_{zz}}
\end{aligned}
\tag{g}
$$

The bending moments are

$$
\begin{aligned}
M_1 &= EI_{zz} \frac{d^2 v_1}{dx^2} = EI_{zz}(a_1 x + b_1) = \frac{3M_c}{2L} x - \frac{M_c}{4} \\[2mm]
M_2 &= EI_{zz} \frac{d^2 v_2}{dx^2} = EI_{zz}(a_2 x + b_2) = \frac{3M_c}{2L} x - \frac{5M_c}{4}
\end{aligned}
\tag{h}
$$

The shear forces are

$$
\begin{aligned}
V_1 &= -EI_{zz} \frac{d^3 v_1}{dx^3} = -EI_{zz} a_1 = -\frac{3M_c}{2L} \\[2mm]
V_2 &= -EI_{zz} \frac{d^3 v_2}{dx^3} = -EI_{zz} a_2 = -\frac{3M_c}{2L}
\end{aligned}
\tag{i}
$$

Plots of shear and bending moment are given in Figure (b)

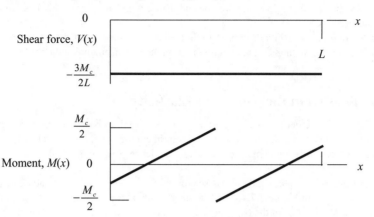

Figure (b)

The stresses are

$$\sigma_1 = -\frac{M_1}{I_{zz}}y = -\left(\frac{3M_c}{2L}x - \frac{M_c}{4}\right)\frac{y}{I_{zz}} \qquad \sigma_2 = -\frac{M_2}{I_{zz}}y = -\left(\frac{3M_c}{2L}x - \frac{5M_c}{4}\right)\frac{y}{I_{zz}} \tag{j}$$

$$\tau_1 = \tau_2 = \frac{V}{I_{zz}b}\int_{A_y}ydA_y = -\frac{3M_c}{4LI_{zz}}\left(\frac{h^2}{4} - y^2\right)$$

The bending stress is maximum at $x = L/2$ and $y = \pm\frac{h}{2}$. The shear stress has a constant value throughout the length and is maximum at $y = 0$.

Plots of the displacements are given in Figure (c).

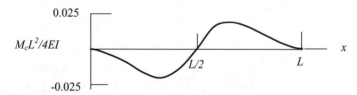

Figure (c)

In the indeterminate case we solve the fourth order equation for the displacement and then find the bending moments and shear forces in order to find the stresses.

############

In the determinate case the bending moments and shear forces are inputs to the solution for the stresses and the displacements. In the indeterminate case the bending moments and shear forces are output from the displacement solution.

It is important to note that the fourth order equation is equally successful in solving determinate and indeterminate cases. In the past, because of the limitations in analysis capability, many structures were designed to be determinate. Now with more powerful methods this is no longer the norm. In fact, the concept of fail safe structures demands indeterminacy, that is, you must have multiple internal load paths in the structure so that a local failure does not cause a global failure. This inherently implies an indeterminate structure.

So we shall not dwell on methods created to find bending moments and shear forces as inputs to the problem. In our favored methods these are outputs from the displacement solution.

7.7   Solutions from the Theory of Elasticity

Let us pause and ask whether the beam equations and their solutions based on plane sections remaining plane are, if fact, valid. We have, so far, invoked experimental evidence. The fact that these equations have been used successfully for so many decades adds credence. Still we must ask under what conditions they might fail.

Consider a beam with a rectangular cross section with moments applied at each end via a linearly varying stress as shown in Figure 7.7.1. The load distribution at the ends of the beam exactly duplicates the internal stress obtained from the beam equations. Let the cross section have a height h and width b. The length of the beam is L and the origin of the coordinates is at the centroid of the cross section as shown.

Let the end loads be expressed by a force per unit area of value

$$f_x \left(\pm \frac{L}{2}, y, z \right) = \mp \frac{2f_0}{h} y \qquad (7.7.1)$$

The loads on all other surfaces are zero and there are no body forces.

Let us assume that the state of stress throughout the beam is

$$\sigma_x = -\frac{2f_0}{h} y \qquad \sigma_y = \sigma_z = \tau_{xy} = \tau_{yz} = \tau_{zx} = 0 \qquad (7.7.2)$$

and ask if this stress distribution satisfies the complete equations of elasticity.

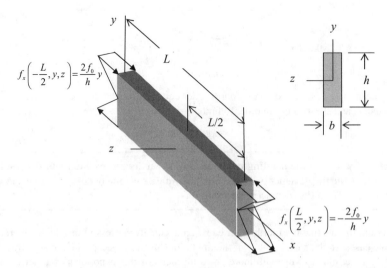

Figure 7.7.1

Upon substituting these values into the three dimensional equations of equilibrium (see Chapter 2, Section 2.6, Equations 2.6.13) we find that they are satisfied identically if the body forces are zero.

$$
\begin{bmatrix}
\dfrac{\partial}{\partial x} & 0 & 0 & \dfrac{\partial}{\partial y} & 0 & \dfrac{\partial}{\partial z} \\[2ex]
0 & \dfrac{\partial}{\partial y} & 0 & \dfrac{\partial}{\partial x} & \dfrac{\partial}{\partial z} & 0 \\[2ex]
0 & 0 & \dfrac{\partial}{\partial z} & 0 & \dfrac{\partial}{\partial y} & \dfrac{\partial}{\partial x}
\end{bmatrix}
\begin{bmatrix}
-\dfrac{2f_0}{h}y \\[1.5ex] 0 \\[0.5ex] 0 \\[0.5ex] 0 \\[0.5ex] 0 \\[0.5ex] 0
\end{bmatrix}
=
\begin{bmatrix} 0 \\ 0 \\ 0 \end{bmatrix}
\tag{7.7.3}
$$

We note that the internal stresses match the surface forces at the boundaries. The strain components are

$$
\varepsilon_x = \frac{\sigma_x}{E} = -\frac{2f_0}{Eh}y \quad \varepsilon_y = -v\frac{\sigma_x}{E} = \frac{2vf_0}{Eh}y \quad \varepsilon_z = -v\frac{\sigma_x}{E} = \frac{2vf_0}{Eh}y \quad \gamma_{xy} = \gamma_{yz} = \gamma_{zx} = 0 \tag{7.7.4}
$$

Again we note that linear functions of strain satisfy compatibility (see Chapter 3, Section 3.3), that is,

$$
\frac{\partial^2 \varepsilon_x}{\partial y^2} + \frac{\partial^2 \varepsilon_y}{\partial x^2} = \frac{\partial^2 \gamma_{xy}}{\partial x \partial y} \quad \rightarrow \quad \frac{\partial^2}{\partial y^2}\left(-\frac{2f_0}{Eh}y\right) + \frac{\partial^2}{\partial x^2}\left(\frac{2vf_0}{Eh}y\right) = \frac{\partial^2}{\partial x \partial y}(0) = 0 \tag{7.7.5}
$$

Now let us find the displacements associated with this loading. If we ignore the Poisson's ratio effect and examine the displacement in the x direction we have

$$
\varepsilon_x = \frac{\partial u}{\partial x} = -\frac{2f_0}{Eh}y \quad \rightarrow \quad u(x, y) = -\frac{2f_0}{Eh}yx + f(y, z) \tag{7.7.6}
$$

Since

$$
u(0, y, z) = 0 \quad \rightarrow \quad f(y, z) = 0 \quad \rightarrow \quad u(x, y) = -\frac{2f_0}{Eh}yx \tag{7.7.7}
$$

it is confirmed that $u(x, y, z)$ is linear in y, or that plane sections remain plane.

There will be displacements in the y and z directions from the Poisson's ratio effect, but for a long slender beam these effects are small.

Numerical analysis methods allow us to solve the more exact equations of material behavior which do not make assumptions about what will happen to planar sections after loads are applied. Computer graphics allow us to present visual confirmation of this behavior. The results of the numerical analysis for a beam with the loading given in Equation 7.7.1 are shown in Figure 7.7.2. The displacement is greatly exaggerated to make it visible.

In this particular software program the origin of the axes is at the centroid of the deformed shape. Note that the shape compares favorably to Figure (b) in Example 7.3.1 if you move the axes. Note also that the end planes appear still to be plane. The contour lines between changes in color are lines of constant stress. The uniform parallel lines of color confirm that the stress distribution in the interior is the same as that applied to the ends. This infers that the plane sections do remain plane.

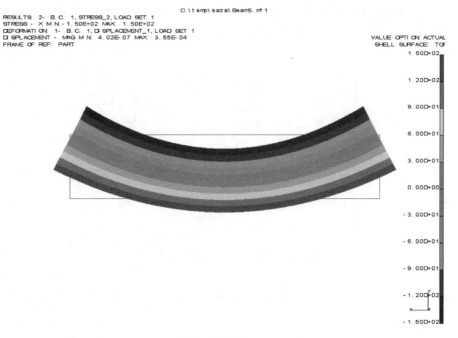

Figure 7.7.2

Once again a change in geometry, loading, or restraint will produce a problem for which it usually is impossible to find an exact analytical solution. The next demonstration applies the end stresses as shown in Figure 7.7.3. There is a parabolic distribution of $f_x(x, y, z)$ in the y direction that is compressive for positive y and tensile for negative y.

For positive y: $\quad f_x\left(\pm\dfrac{L}{2}, y, z\right) = \mp f_0 y\left(1 - \dfrac{2y}{h}\right)$

$$(7.7.8)$$

For negative y: $\quad f_x\left(\pm\dfrac{L}{2}, y, z\right) = \mp f_0 y\left(1 + \dfrac{2y}{h}\right)$

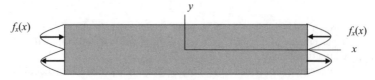

Figure 7.7.3

We cannot expect that the stress in the interior will be the same as the stress on the ends, but let us check just in case. Assume

For positive y: $\quad \sigma_x\left(x, y, z\right) = -f_0 y\left(1 - \dfrac{2y}{h}\right)$

$$(7.7.9)$$

For negative y: $\quad \sigma_x\left(x, y, z\right) = -f_0 y\left(1 + \dfrac{2y}{h}\right)$

Equilibrium is satisfied in the absence of body forces for positive y as noted in the next equation.

$$\begin{bmatrix} \dfrac{\partial}{\partial x} & 0 & 0 & \dfrac{\partial}{\partial y} & 0 & \dfrac{\partial}{\partial z} \\[2mm] 0 & \dfrac{\partial}{\partial y} & 0 & \dfrac{\partial}{\partial x} & \dfrac{\partial}{\partial z} & 0 \\[2mm] 0 & 0 & \dfrac{\partial}{\partial z} & 0 & \dfrac{\partial}{\partial y} & \dfrac{\partial}{\partial x} \end{bmatrix} \begin{bmatrix} -f_0 y \left(1 - \dfrac{2y}{h}\right) \\[2mm] 0 \\ 0 \\ 0 \\ 0 \\ 0 \end{bmatrix} = \begin{bmatrix} 0 \\ 0 \\ 0 \end{bmatrix} \tag{7.7.10}$$

It is obvious that it is also satisfied for negative y. The interior forces match the surface forces at the boundaries. So equilibrium and boundary conditions are satisfied. Continuing, the strain components would be

$$\varepsilon_x = \frac{\sigma_x}{E} = -\frac{f_0 y}{E}\left(1 - \frac{2y}{h}\right) \quad \varepsilon_y = -v\frac{\sigma_x}{E} = \frac{v f_0 y}{E}\left(1 - \frac{2y}{h}\right)$$

$$\varepsilon_z = -v\frac{\sigma_x}{E} = \frac{v f_0 y}{E}\left(1 - \frac{2y}{h}\right) \quad \gamma_{xy} = \gamma_{yz} = \gamma_{zx} = 0 \tag{7.7.11}$$

Next we check compatibility.

$$\frac{\partial^2 \varepsilon_x}{\partial y^2} + \frac{\partial^2 \varepsilon_y}{\partial x^2} = \frac{\partial^2 \gamma_{xy}}{\partial x \partial y}$$

$$\rightarrow \quad \frac{\partial^2}{\partial y^2}\left(-\frac{f_0 y}{E}\left(1 - \frac{2y}{h}\right)\right) + \frac{\partial^2}{\partial x^2}\left(\frac{v f_0 y}{E}\left(1 - \frac{2y}{h}\right)\right) - \frac{\partial^2}{\partial x \partial y}(0) \neq 0 \tag{7.7.12}$$

It fails compatibility – this cannot be a solution. So we turn to a numerical analysis.

The results of a numerical analysis are shown in Figure 7.7.4. As you can see the parabolic stresses on the ends rapidly transform to the linear distribution of beam theory away from the ends. The end loading indeed has a localized effect. What is important to the interior region is the magnitude of the moment, not its distribution on the ends.

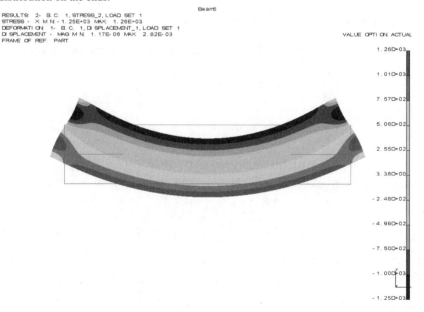

Beam5
RESULTS: 2- B.C. 1, STRESS_2, LOAD SET 1
STRESS - X M N - 1. 25E+03 MAX: 1. 26E+03
DEFORMATION: 1- B.C. 1, DISPLACEMENT_1, LOAD SET 1
DISPLACEMENT - MAG M N: 1. 17E- 06 MAX: 2. 82E- 03
FRAME OF REF: PART

VALUE OPTION: ACTUAL

1. 26D+03
1. 01D+03
7. 57D+02
5. 06D+02
2. 55D+02
3. 38D+00
- 2. 48D+02
- 4. 99D+02
- 7. 50D+02
- 1. 00D+03
- 1. 25D+03

Figure 7.7.4

You can notice a small warping at the former plane end section. You can see this better in the enlarged image in Figure 7.7.5.

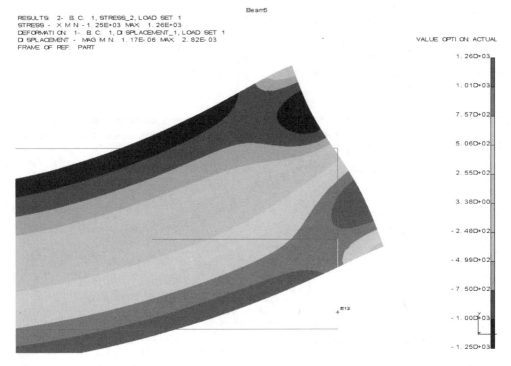

Figure 7.7.5

We may infer that in the interior away from the ends this warping effect is extremely small.

In the above examples there are no shear forces. Our next demonstration is for the beam in Example 7.3.2. The shear stress in Equation (g) of Example 7.3.2 that was obtained by beam theory is

$$\tau_{xy}(x, y, z) = \frac{F}{2I_{zz}} \left(\frac{h^2}{4} - y^2 \right) \qquad (7.7.13)$$

The bending stress is

$$\sigma_x(x, y, z) = -\frac{F(L - x)}{I_{zz}} y \qquad (7.7.14)$$

The forces at the boundaries are as shown in Figure 7.7.6. At the right end we have a distributed shear force.

$$f_y(L, y, z) = \frac{F}{2I_{zz}} \left(\frac{h^2}{4} - y^2 \right) \qquad (7.7.15)$$

At the left end there is a distributed shear force

$$f_y(0, y, z) = \frac{F}{2I_{zz}} \left(\frac{h^2}{4} - y^2 \right) \qquad (7.7.16)$$

and a distributed normal force

$$f_x(0, y, z) = -\frac{FL}{I_{zz}}y \qquad (7.7.17)$$

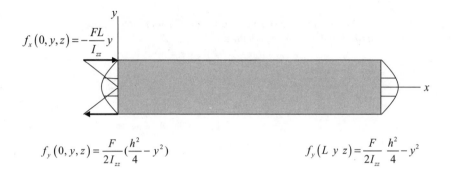

$$f_x(0, y, z) = -\frac{FL}{I_{zz}}y$$

$$f_y(0, y, z) = \frac{F}{2I_{zz}}\left(\frac{h^2}{4} - y^2\right) \qquad\qquad f_y(L \ y \ z) = \frac{F}{2I_{zz}}\frac{h^2}{4} - y^2$$

Figure 7.7.6

Assume that the stresses in the interior are exactly the same as in Equations 7.7.13 and 7.7.14 as found from beam theory. Do these stresses satisfy the three dimensional equations of equilibrium?

$$\begin{bmatrix} \dfrac{\partial}{\partial x} & 0 & 0 & \dfrac{\partial}{\partial y} & 0 & \dfrac{\partial}{\partial z} \\[2mm] 0 & \dfrac{\partial}{\partial y} & 0 & \dfrac{\partial}{\partial x} & \dfrac{\partial}{\partial z} & 0 \\[2mm] 0 & 0 & \dfrac{\partial}{\partial z} & 0 & \dfrac{\partial}{\partial y} & \dfrac{\partial}{\partial x} \end{bmatrix} \begin{bmatrix} -\dfrac{F(L-x)}{I_{zz}}y \\[2mm] 0 \\[2mm] 0 \\[2mm] \dfrac{F}{2I_{zz}}\left(\dfrac{h^2}{4}-y^2\right) \\[2mm] 0 \\[2mm] 0 \end{bmatrix} = \begin{bmatrix} ? \\ ? \\ ? \end{bmatrix} \qquad (7.7.18)$$

Expanding the equations we find

$$\frac{\partial}{\partial x}\left(-\frac{F(L-x)}{I_{zz}}y\right) + \frac{\partial}{\partial y}\left(\frac{F}{2I_{zz}}\left(\frac{h^2}{4}-y^2\right)\right) = \frac{F}{I_{zz}}y - \frac{F}{I_{zz}}y = 0$$

$$\frac{\partial}{\partial x}\left(\frac{F}{2I_{zz}}\left(\frac{h^2}{4}-y^2\right)\right) = 0 \qquad (7.7.19)$$

and so the equilibrium equations are satisfied exactly in the absence of body forces. These stresses match the surface forces at the boundaries.

Given these stresses the strains will be

$$\varepsilon_x = \frac{\sigma_x}{E} = -\frac{F(L-x)}{EI_{zz}}y \qquad \varepsilon_y = -v\frac{\sigma_x}{E} = \frac{vF(L-x)}{EI_{zz}}y \qquad \varepsilon_z = -v\frac{\sigma_x}{E} = \frac{vF(L-x)}{EI_{zz}}y$$

$$\gamma_{xy} = \frac{\tau_{xy}}{G} = \frac{F}{2GI_{zz}}\left(\frac{h^2}{4}-y^2\right) \qquad \gamma_{yz} = \gamma_{zx} = 0 \qquad\qquad (7.7.20)$$

Check for compatibility.

$$\frac{\partial^2 \varepsilon_x}{\partial y^2} + \frac{\partial^2 \varepsilon_y}{\partial x^2} = \frac{\partial^2 \gamma_{xy}}{\partial x \partial y}$$

$$\rightarrow \quad \frac{\partial^2}{\partial y^2}\left(-\frac{F(L-x)}{EI_{zz}}y\right) + \frac{\partial^2}{\partial x^2}\left(-\frac{\nu F(L-x)}{EI_{zz}}y\right) - \frac{\partial^2}{\partial x \partial y}\left(\frac{F}{2GI_{zz}}\left(\frac{h^2}{4}-y^2\right)\right) \qquad (7.7.21)$$

$$= 0 + 0 - 0 = 0$$

The other compatibility equations are satisfied – trust me. Contour plots follow.
The contour plot for σ_x bending stresses is shown in Figure 7.7.7.

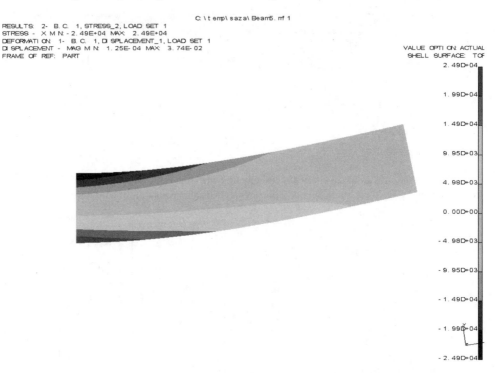

C:\temp\saza\Beam5.mf 1
RESULTS 2- B.C. 1, STRESS_2, LOAD SET 1
STRESS - X M N - 2.49E+04 MAX 2.49E+04
DEFORMATION 1- B.C. 1, DISPLACEMENT_1, LOAD SET 1
DISPLACEMENT - MAG M N 1.25E-04 MAX 3.74E-02
FRAME OF REF: PART

VALUE OPTION: ACTUAL
SHELL SURFACE: TOP
2.49D+04
1.99D+04
1.49D+04
9.95D+03
4.98D+03
0.00D+00
-4.98D+03
-9.95D+03
-1.49D+04
-1.99D+04
-2.49D+04

Figure 7.7.7

The contour plot for τ_{xy} shearing stresses is shown in Figure 7.7.8.
This confirms the accuracy of the elementary bending theory.
Now let us find the displacements associated with this loading. If we ignore the Poisson's ratio effect and examine the displacement in the x direction we have

$$\varepsilon_x = \frac{\partial u}{\partial x} = -\frac{F(L-x)}{EI_{zz}}y \quad \rightarrow \quad u(x, y) = -\frac{F\left(Lx - \dfrac{x^2}{2}\right)}{EI_{zz}}y + f(y, z) \qquad (7.7.22)$$

Apply the boundary condition

$$u(0, y, z) = 0 \quad \rightarrow \quad f(y, z) = 0 \quad \rightarrow \quad u(x, y) = -\frac{F\left(Lx - \dfrac{x^2}{2}\right)}{EI_{zz}}y \qquad (7.7.23)$$

C:\temp\saza\Beam5.nf 1

RESULTS 2- B. C. 1, STRESS_2, LOAD SET 1
STRESS - XY M N: 1. 16E+02 MAX: 1. 25E+03
DEFORMATION 1- B. C. 1, DISPLACEMENT_1, LOAD SET 1
DISPLACEMENT - MAG M N: 1. 25E-04 MAX 3. 74E-02
FRAME OF REF: PART

VALUE OPTION: ACTUAL
SHELL SURFACE: TOP

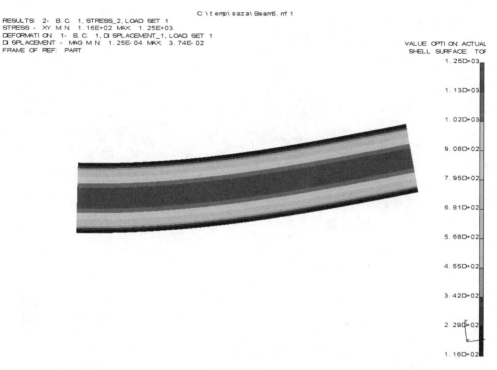

1. 25D+03
1. 13D+03
1. 02D+03
9. 08D+02
7. 95D+02
6. 81D+02
5. 68D+02
4. 55D+02
3. 42D+02
2. 29D+02
1. 16D+02

Figure 7.7.8

This confirms that $u(x, y, z)$ is linear in y, or that plane sections remain plane.

For one more example to establish St. Venant's principle consider a cantilever beam, that is, one with the left end fixed and with a constant shear force on the right end, as shown in Figure 7.7.9. Both the fixed condition and the applied load violate the beam theory stress distributions at the boundary.

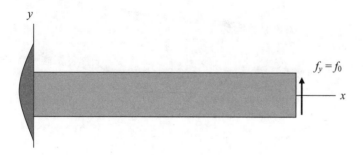

y

$f_y = f_0$

x

Figure 7.7.9

The contour plot for σ_x bending stresses is shown in Figure 7.7.10. The contour plot for τ_{xy} shearing stresses is shown in Figure 7.7.11. You will note a small distortion in the bending and shear stresses at the fixed end and at the free end for the shear stress. Over the rest of the beam the approximations are quite valid.

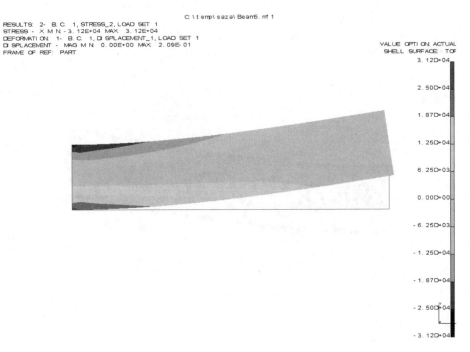

Figure 7.7.10

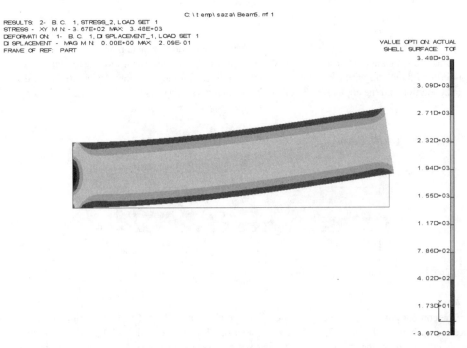

Figure 7.7.11

Finally we explore one more approximation in beam theory – the application of a distributed lateral load as shown in Figure 7.7.12 to the upper surface.

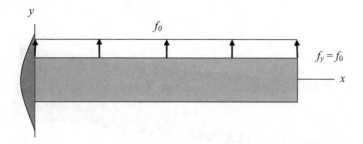

Figure 7.7.12

The bending stress contours are shown in Figure 7.7.13.

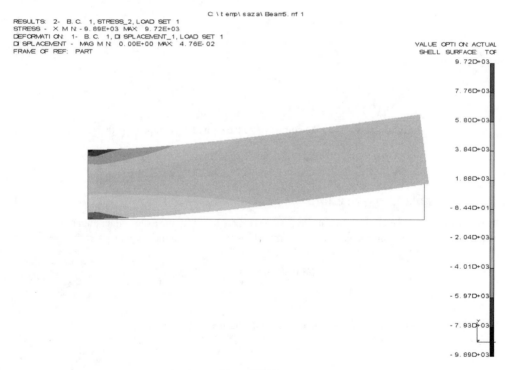

Figure 7.7.13

The shearing stress contours are shown in Figure 7.7.14. Again St. Venant's principle dashes to the rescue.

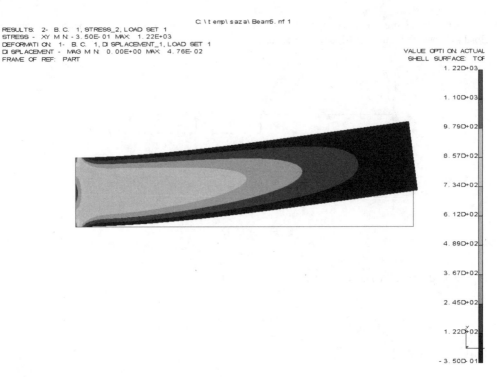

Figure 7.7.14

The σ_y component of stress is so small compared to the σ_x and τ_{xy} components that it is ignored for a slender beam.

7.8 Variable Cross Sections

If I_{zz} varies with x we can adapt the equations as we demonstrated for the axial and torsional cases. If the beam is statically determinate we find the internal shear forces and moments and from it the stresses.

$$\sigma_x(x, y) = -\frac{M(x)}{I_{zz}(x)} y \qquad \tau_{xy}(x, y) = \frac{V(x)}{I_{zz}(x)b} \int_{A_y} y \, dA_y \qquad (7.8.1)$$

Finding the displacement requires integrating with EI_{zz} inside the integral.

$$\frac{d^2v}{dx^2} = \frac{M(x)}{EI_{zz}(x)}$$

$$\rightarrow \quad \frac{dv}{dx} = \int \frac{M(x)}{EI_{zz}(x)} dx + e \qquad (7.8.2)$$

$$\rightarrow \quad v(x) = \int\int \frac{M(x)}{EI_{zz}(x)} dx + ex + f$$

If the beam is statically indeterminate we must use the fourth order equations with EI_{zz} inside the integral.

$$\frac{d^2}{dx^2} EI_{zz}(x) \frac{d^2v}{dx^2} = f_y(x)$$

$$\rightarrow \quad \frac{d}{dx} EI_{zz}(x) \frac{d^2v}{dx^2} = \int f_y(x)\,dx + a$$

$$\rightarrow \quad EI_{zz}(x) \frac{d^2v}{dx^2} = \iint f_y(x)\,dx + ax + b \qquad (7.8.3)$$

$$\rightarrow \quad \frac{dv}{dx} = \int \left(\frac{1}{EI_{zz}(x)} \iint f_y(x)\,dx \right) dx + a\frac{x^2}{2} + bx + c$$

$$\rightarrow \quad v = \iint \left(\frac{1}{EI_{zz}(x)} \iint f_y(x)\,dx \right) dx + a\frac{x^3}{6} + b\frac{x^2}{2} + cx + d$$

Once we find the displacement $v(x)$ by integrating four times and applying boundary conditions to find the constants of integration we can find the internal shear forces and moments.

$$V(x) = -\frac{d}{dx} EI_{zz}(x) \frac{d^2v}{dx^2} \qquad M(x) = EI_{zz}(x) \frac{d^2v}{dx^2} \qquad (7.8.4)$$

and the stresses are given by Equation 7.8.1.

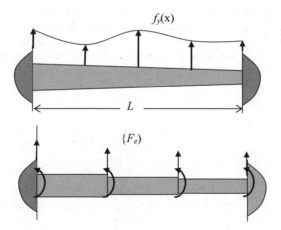

$f_y(x)$

$\{F_e\}$

Figure 7.8.1

As we learned in the axial case, except for the simplest functions for the variable cross section properties direct integration is an unwieldy process. It would be even more inconvenient when the multiple integrations for bending would be necessary. It is more common to approximate the variation by a series of constant cross sections, such as is represented in Figure 7.8.1. In practice several sections might be used, not just the three shown. The distributed load would be replaced by equivalent nodal loads and the finite element method used.

In Section 5.5 for the axial case we learned how careful interpretation of the results can achieve a high degree of accuracy. We would use the restraint moments at the ends to find the stresses there and use average values at the nodes. This will be demonstrated in Chapter 8 where the finite element method for beams is presented.

7.9   Shear Stress in Non Rectangular Cross Sections—Thin Walled Cross Sections

The formula for finding the shear stress in beams that is presented in Section 7.3 assumes that the shear stress is uniformly distributed over the width of the cross section. This is quite accurate for rectangular cross sections. For other sections that do not have large or abrupt changes in width, such as those shown in Figure 7.9.1, it may serve as a reasonable first approximation but not as a final analysis.

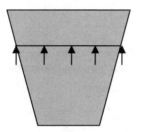
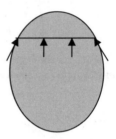

Figure 7.9.1

The assumption of a uniform shear stress across the width would produce vector components as shown in the figure on the left. Since there are no shear stresses on the side faces of the beam, the stress cannot have components normal to those surfaces at each side. A more reasonable assumption is for the shear stress at the side surfaces to be parallel to the surfaces as shown in the figure on the right.

For sections such as the thick walled I beam shown in Figure 7.8.2 the formula is inadequate except as noted below.

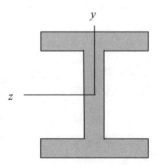

Figure 7.9.2

Using the formula from Equation 7.3.14, repeated here as Equation 7.9.1

$$\tau_{xy}(x, y) = \frac{V}{I_{zz}h} \int_{A_y} y dA_y \tag{7.9.1}$$

we would obtain the shear stress profile shown in Figure 7.9.3.

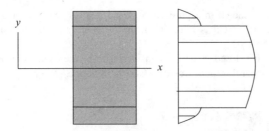

Figure 7.9.3

This shows a significant shear stress on the horizontal inside faces of the flanges. Clearly the shear stress must be zero on those faces since there is no load on those faces.

The values in the vertical web of the I beam, in particular at the midpoint where the shear stress is largest, are accurate. They are inaccurate at the junction of the vertical web and the horizontal flanges and in the horizontal flanges. The more advanced theory of elasticity is necessary for an accurate analysis.

There is one other class of cross sections, however, for which a simple theory of shear stress distribution is possible. These are the so called thin walled cross sections such as the examples shown in Figure 7.9.4.

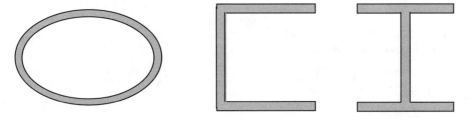

Figure 7.9.4

If the surfaces of the beam are free of shear stress then any significant component of the shear stress must be tangent to the surface. For practical purposes any shear in the thin walled cross section can be approximated as uniform across the width of the section. Once again we can define shear flow

$$q = \tau b \tag{7.9.2}$$

where b is the wall thickness. We can use the method in Section 7.3 to find the shear stress by equating the shear force necessary to place a segment of length dx in equilibrium with the normal forces acting on that segment as we have done in Figure 7.3.6 and Equations 7.3.11-13.

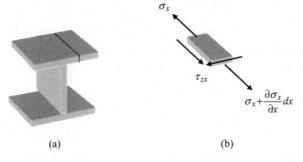

(a) (b)

Figure 7.9.5

Consider an I beam as shown in Figure 7.9.5(a) and a small segment of the upper flange in Figure 7.9.5(b). The shear at the cut is found from the differences in the normal stresses at sections x and $x + dx$.

This can be extended to a closed section by considering an element with dimensions ds by dx as shown in Figure 7.9.6(a) with shear flow and stresses as shown in Figure 7.9.6(b).

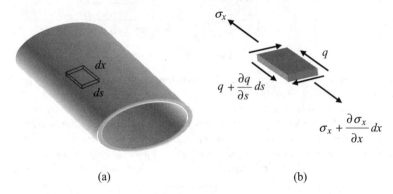

(a)   (b)

Figure 7.9.6

From equilibrium of the element

$$\sum F_x = \left(\sigma_x + \frac{\partial \sigma_x}{\partial x} dx\right) bds - \sigma_x bds - qdx + \left(q + \frac{\partial q}{\partial s} ds\right) dx = 0 \qquad (7.9.3)$$

From this we obtain

$$\frac{\partial q}{\partial s} = -b \frac{\partial \sigma_x}{\partial x} \qquad (7.9.4)$$

The bending stresses are given by

$$\sigma_x(x, y) = -\frac{My}{I_{zz}} \qquad (7.9.5)$$

When Equation 7.9.5 is substituted into Equation 7.9.4 we get

$$\frac{\partial q}{\partial s} = \frac{b}{I} \frac{\partial M}{\partial x} y = -\frac{bV}{I_{zz}} y \qquad (7.9.6)$$

This may be integrated along the path s to obtain the shear flow.

$$q = q_0 - \int_0^s \frac{V}{I_{zz}} byds \qquad (7.9.7)$$

where q_0 is the value of the shear flow at the origin of the coordinate s. A simple example will confirm the validity of this equation.

###########

Example 7.9.1

Problem: Solve for the shear stress in the simplest of cases: a thin rectangular cross section of height h shown in Figure (a). The width is b.

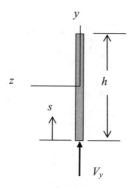

Figure (a)

Solution: Apply Equation 7.9.7.
The shear flow is

$$\tau_{xy}b = q = q_0 - \int_{-\frac{h}{2}}^{y} \frac{V}{I_{zz}} b y dy = 0 - \frac{V}{I_{zz}} b \, y^2 \big|_{-\frac{h}{2}}^{y} = \frac{Vb}{2I_{zz}} \left(\frac{h^2}{4} - y^2 \right) \tag{a}$$

where we note that $q_0 = 0$ since there are no shear forces on the upper or lower surfaces.

This is, of course, exactly what we obtained for a rectangular section in Section 7.3, Example 7.3.2, Equation (g).

###########

When the section is not symmetrical about the xy axes plane we encounter a problem with the load application. We have assumed so far that the lateral loads are along the centroidal axis of the beam. If the cross section is symmetrical about the y axis and the load is applied in the xy plane only bending occurs. If the load is off the centroidal axis both bending and torsion may occur. When the cross section is not symmetrical about the y axis this generalization is no longer true. A load in the xy plane may cause both bending and torsion. This situation is presented in detail in Chapter 10. For now we restrict all cross sections, thin walled or not, to ones with symmetry about the y axis and lateral loading only in the xy plane.

###########

Example 7.9.2

Problem: Find the shear flow in a circular thin walled cross section shown in Figure (a). The shear force is applied through the centroid of the cross section.

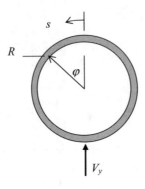

Figure (a)

Solution: Use Equation 7.9.7.
The shear flow is

$$q = q_0 - \int_0^s \frac{V}{I_{zz}} b y \, ds = q_0 - \frac{VbR^2}{I_{zz}} \int_0^\varphi \cos\varphi \, d\varphi = q_0 - \frac{VbR^2}{I_{zz}} \sin\varphi \tag{a}$$

We can argue from symmetry that the shear flow will be zero at $s = 0$ and π. Thus q_0 is zero.

Example 7.9.3

Problem: Consider the thin walled I beam shown in Figure (a). A shear force V is acting at the centroid. The vertical web and the flanges are the same thickness b. Find the shear flows in the cross section.

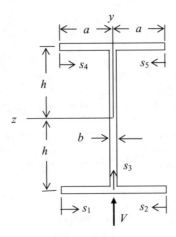

Figure (a)

Solution: Use Equation 7.9.7. The origins and directions of the local coordinates, s_1 to s_5, are shown on Figure (a).

Starting with the flange on the lower left Equation 7.9.7 becomes

$$q = q_0 - \int_0^s \frac{V}{I_{zz}} b y \, ds \quad \rightarrow \quad q_1(s_1) = 0 - \frac{Vb}{I_{zz}} \int_0^{s_1} (-h) \, ds_1 = \frac{Vbh}{I_{zz}} s_1 \quad \rightarrow \quad q_1(a) = \frac{Vbha}{I_{zz}}$$
(a)

This is plotted in Figure (b). The direction of the shear flow is shown by the arrows.
For the flange on the lower right we have

$$q = q_0 - \int_0^s \frac{V}{I_{zz}} b y \, ds \quad \rightarrow \quad q_2(s_2) = 0 - \frac{Vb}{I_{zz}} \int_0^{s_1} (-h) \, ds_2 = \frac{Vbh}{I_{zz}} s_2 \quad \rightarrow \quad q_2(a) = \frac{Vbha}{I_{zz}}$$
(b)

It also is plotted in Figure (b). The positive signs indicate that the flows are in the same direction as the local coordinates.

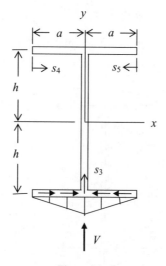

Figure (b)

At the juncture where s_1, s_2, and s_3 meet we must establish equilibrium, that is, from Figure (c) we note that

$$q_3(0) = q_1(a) + q_2(a) = \frac{2Vbha}{I_{zz}}$$
(c)

Note that this represents the state of stress at the point where the two flanges meet the vertical web.

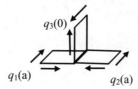

Figure (c)

For the vertical web it is convenient to stick with the y coordinate rather than s_3. The shear flow in the vertical web is then

$$q = q_0 - \int_0^s \frac{V}{I_{zz}} byds \quad \rightarrow \quad q_3(y) = \frac{2Vbha}{I_{zz}} - \frac{Vb}{I_{zz}} \int_{-h}^y ydy$$

$$= \frac{2Vbha}{I_{zz}} - \frac{Vb}{2I_{zz}} y^2\Big|_{-h}^y = \frac{2Vbha}{I_{zz}} - \frac{Vb}{2I_{zz}}(y^2 - h^2) \tag{d}$$

$$= \frac{Vb}{2I_{zz}}(4ha - y^2 + h^2)$$

Note that when $y = h$

$$q_3(h) = \frac{2Vbha}{I_{zz}} \tag{e}$$

and when $y = 0$

$$q_3(0) = \frac{Vb}{2I_{zz}}(4ha + h^2) \tag{f}$$

A plot of this shear flow is added in Figure (d). It is placed off to the side to avoid confusion.

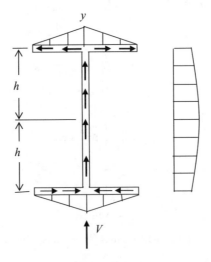

Figure (d)

Finally, the shear flows in the two upper flanges are

$$q = q_0 - \int_0^s \frac{V}{I_{zz}} byds \quad \rightarrow \quad q_4(s_4) = 0 - \frac{V}{I_{zz}} \int_0^{s_1} bhds_4 = -\frac{Vbh}{I_{zz}} s_4 \quad \rightarrow \quad q_4(a) = -\frac{Vbha}{I_{zz}}$$

$$q = q_0 - \int_0^s \frac{V}{I_{zz}} byds \quad \rightarrow \quad q_5(s_5) = 0 - \frac{V}{I_{zz}} \int_0^{s_1} bhds_1 = -\frac{Vbh}{I_{zz}} s_5 \quad \rightarrow \quad q_5(a) = -\frac{Vbha}{I_{zz}}$$

$$\tag{g}$$

These plots have also been added to Figure (d). The minus signs indicate that the positive flow is in the opposite direction of the local coordinates.

##########

7.10   Design of Beams

As noted earlier in analysis we specify the geometry, materials, loads, and restraints and ask for the displacements, strains, and stresses. In design we are more likely to be given the loads and some of the restraints but are asked what geometry and materials should be used to achieve certain acceptable values of stress and displacement. In addition to the geometric restraints we may have restraints on weight, cost, and ease of manufacture among other considerations.

The design of a beam in bending is more involved than the axial case because the shape of the cross section, in particular, the area moment of inertia, in addition to the area, has a significant influence on the design.

Once again consider one of the simplest of problems, the cantilever beam with a concentrated force on the end that was analyzed in Example 7.3.2. It is shown here in Figure 7.10.1.

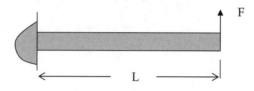

Figure 7.10.1

From static equilibrium we found the bending moment and shear force to be

$$\sum M_z = F(L - x) - M = 0 \quad \rightarrow \quad M = F(L - x)$$
$$\sum F_y = F - V = 0 \quad \rightarrow \quad V = F \tag{7.10.1}$$

The normal and shear stresses for a rectangular cross section of height h are

$$\sigma_x = -\frac{F}{I_{zz}}(L - x)y \qquad \tau_{xy}(x, y) = \frac{F}{2I_{zz}}\left(\frac{h^2}{4} - y^2\right) \tag{7.10.2}$$

The displacement is

$$v(x) = \frac{F}{EI_{zz}}\left(\frac{Lx^3}{2} - \frac{x^2}{6}\right) \tag{7.10.3}$$

Given a particular load and length we note that increasing the moment of inertia I_{zz} decreases both stress components and the displacement. Choosing a material with greater E decreases the displacement but has no effect on the stresses. For a rectangular cross section increasing the height h would appear to increase the stresses; however, this also affects the moment of inertia and so the effect is not entirely clear. Let us examine the effect of different shaped cross sections on a specific beam with a constant cross section and a load as shown in Figure 7.10.1.

###########

Example 7.10.1

Problem: The beam in Figure 7.10.1 has a length of 1000 *mm* and a load of 1600 *N*. It is made of steel with $E = 206800$ *MPa* and a yield stress $\sigma_{yield} = 240$ *MPa*. A maximum normal stress of 160 *MPa* is chosen to allow a margin of safety. A rectangular cross section is chosen initially of dimensions as shown in Figure (a).

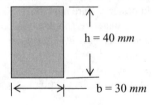

Figure (a)

Initial calculations show that the normal stress exceeds the maximum allowable value of 160 *MPa*. The moment of inertia is

$$I_{zz} = \frac{bh^3}{12} = \frac{30 \cdot 40^3}{12} = 160000 \, mm^4. \tag{a}$$

The maximum stresses are

$$\sigma_{max} = \sigma_x \left(0, \frac{h}{2}\right) = \pm \frac{FL}{I_{zz}} \frac{h}{2} = \pm \frac{1600 \cdot 1000 \cdot 20}{160000} = \pm 200 \, MPa$$

$$\tau_{max} = \tau_{xy}(x, 0) = \frac{F}{2I_{zz}} \frac{h^2}{4} = \frac{1600 \, (40)^2}{8 \cdot 160000} = 2.0 \, MPa \tag{b}$$

and the maximum displacement is

$$v_{max} = \frac{FL^3}{3EI_{zz}} = \frac{1600 \cdot (1000)^3}{3 \cdot 206800 \cdot 160000} = 16.12 \, mm \tag{c}$$

Refine the cross section size and shape to obtain an acceptable level of stress.

Solution: Given the material and the stress find a new cross section shape and size.

First let us reduce the normal stress without increasing the amount of material used by changing the dimensions of the rectangular cross section. Let us try a section with the same area where $b = 20 \, mm$ and $h = 60 \, mm$ as shown in Figure (b).

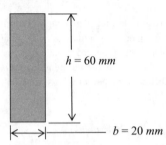

Figure (b)

The moment of inertia is now

$$I_{zz} = \frac{bh^3}{12} = \frac{20 \cdot 60^3}{12} = 360000 \, mm^4 \tag{d}$$

The maximum normal stress is now

$$\sigma_{max} = \sigma_x\left(0, \frac{h}{2}\right) = \pm\frac{FL}{I_{zz}}\frac{h}{2} = \frac{1600 \cdot 1000 \cdot 30}{360000} = 133.33\,MPa \tag{e}$$

This is below the limit we imposed. The displacement is also reduced.

$$v_{max} = \frac{FL^3}{3EI_{zz}} = \frac{1600 \cdot (1000)^3}{3 \cdot 206800 \cdot 360000} = 7.16\,mm$$

Let us say that the shear stress is so small that it does not concern us and the displacement is acceptable. We can continue to modify the dimensions of the rectangular section until the right stress levels are achieved.

Now let us see if we can actually reduce the amount of material used by changing the shape of the cross section. Let us try an I beam. Let us remove some material to produce the cross section shown in Figure (c).

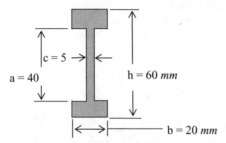

Figure (c)

The moment of inertia now is

$$I_{zz} = \frac{bh^3}{12} - \frac{(b-c)a^3}{12} = \frac{20 \cdot 60^3}{12} - \frac{15 \cdot 40^3}{12} = 280000\,mm^4 \tag{f}$$

The material removed amounts to a reduction in the moment of inertia of 80000 mm^4. The stress increases to

$$\sigma_{max} = \sigma_x\left(0, \frac{h}{2}\right) = \pm\frac{FL}{I_{zz}}\frac{h}{2} = \frac{1600 \cdot 1000 \cdot 30}{280000} = 171.4\,MPa \tag{g}$$

This is close to our target value. In the process we removed half of the material of the beam and saved weight and material cost.

Just to make sure let us check the shear stress to see if it is still small.

$$\tau_{max} = \tau_{xy}(x, 0) = \frac{V}{I_{zz}c}\int_{A_y} y\,dA_y = \frac{V}{I_{zz}5}\int_0^{20} y5\,dy + \frac{V}{I_{zz}5}\int_{20}^{30} y20\,dy$$

$$= \frac{1600}{280000}\left(\frac{y^2}{2}\bigg|_0^{20} + 4\frac{y^2}{2}\bigg|_{20}^{30}\right) = \frac{1600}{280000}(200 + 1000) = 6.86\,MPa \tag{h}$$

It is still small. The tip displacement increases to 9.21 mm.

###########

Noting the bending moment distribution for a cantilever beam one might suggest a tapered beam since the normal stress varies linearly from a maximum at the fixed end to zero at the free end. The shear stress, however, is constant over the whole length so we cannot use the bending stress as the sole criterion. Nevertheless, a tapered beam is called for in many cases. The tapered wing of an airplane is a good example. For beams supported at both ends the maximum bending moment is somewhere in the middle and so a beam with greater moment of inertia in the mid section is called for. You see this often in bridge construction. The topic of beams with varying moments of inertia will be considered in the next chapter.

############

Example 7.10.2

Problem: The uniform beam in Example 7.6.1, shown here as Figure (a), has a solid square cross section 42 mm on a side. The length is 1 meter. It is made of steel with a Young's modulus of 206800 MPa. It may carry a load that produces a normal bending stress of up to 400 MPa.

(1) Find the maximum acceptable load.
(2) Change the rectangular cross section dimensions to cut the maximum normal bending stress in half without increasing the amount of material used. Retain the uniform beam condition and the load found in (1).

$$f_y(x) = f_0$$

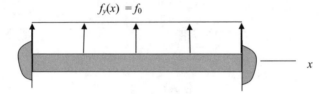

Figure (a)

Solution: Use equations from Example 7.6.1 which determine the relation between stress and load and between stress and the area moment of inertia.

(1) From Example 7.6.1, Equation (g), the maximum moment at $x = L/2$ is

$$M_{max} = \frac{f_0 L^2}{12} \tag{a}$$

The maximum stress is given by

$$\sigma_{max} = -\frac{M_{max} y_{max}}{I_{zz}} = \pm \frac{f_0 L^2}{12 I_{zz}} \frac{h}{2} \tag{b}$$

where $h/2$ is half the height of the section. The maximum y has both a plus and minus value. The area moment of inertia is

$$I_{zz} = \frac{bh^3}{12} = \frac{42 \cdot 42^3}{12} = 259308 \, mm^4 \tag{c}$$

Solve for the applied load f_0 in Equation (b).

$$f_0 = \frac{12 I_{zz} \sigma_{max}}{L^2 \frac{h}{2}} = \frac{12 \cdot 42^4 \cdot 400}{12 \cdot 1000^2 \cdot 21} = 59.27 \, N/mm \tag{d}$$

(2) We strive for a value of $\sigma_{max} = 200$ MPa. The cross section geometry figures in both the moment of inertia, I_{zz}, and in the height of the section, y_{max}. To decrease the maximum stress to 200 MPa the we

must double the value of I_{zz}/y_{max}. It must then equal

$$\frac{I_{zz}}{y_{max}} = 2 \cdot \frac{259308}{21} = 24696 \ mm^3 \tag{e}$$

Consider a rectangular section shown in Figure (b) with the same total area.

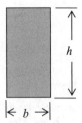

Figure (b)

To maintain the original cross section area

$$A = bh = 1764 \quad \rightarrow \quad b = \frac{1764}{h} \tag{f}$$

The moment of inertia would then be

$$I_{zz} = \frac{bh^3}{12} = \frac{1764h^2}{12} = 147h^2 \tag{g}$$

Thus we find b and h

$$\frac{I_{zz}}{y_{max}} = 24696 = \frac{147h^2}{\dfrac{h}{2}} = 294h \quad \rightarrow \quad h = \frac{24696}{294} = 84 \ mm \quad \rightarrow \quad b = \frac{1764}{84} = 21 \ mm \tag{h}$$

We can also achieve this value of I_{zz}/y_{max} with a smaller total area at the same maximum stress level and thus reduce the weight by going to, say, an I beam cross section.

##########

7.11   Large Displacements

In all the work up to now we have assumed small displacements. Displacements so small, in fact, that we have assumed that the internal forces and moments can be calculated based upon the location of the applied loads in the undeformed position. We have also assumed that stresses are small enough to stay within the linearly elastic range of the material and the strains are small enough to neglect higher order displacement terms in the strain displacement equations. It is possible to have displacements exceed the values consistent with the applied loads in the undeformed position and still not violate the conditions on stress and strain.

Consider the displacement curve of a cantilever beam with a load at the end shown in both an undeflected and a deflected position in Figure 7.11.1. Clearly the bending moment in the beam is less in the deflected position compared to what would be calculated in the undeflected position due to the leftward movement of the load by a distance Δ. There would also be an axial load induced by the component of the applied load tangent to the beam axis that would induce and axial internal force and displacement.

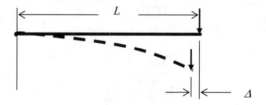

Figure 7.11.1

One solution in the literature assumes the axial load is insignificant and determines that the length of the curved beam is the same as the undeflected beam. More recently computer based analysis provides a solution based upon a series of incremental linear solutions. A small increment of the final load is applied and a linear analysis is performed. The deflected shape is used to redefine the new position of the load and the shape of the beam. The load is then increased another increment and the new deflected position is found. In this step both the lateral and axial deformations are calculated to define the location of loads and the current shape of the beam. This is repeated until the final total load is reached. With the development of efficient computer codes this is both rapid and economical.

In the past analysis methods for large displacement were inadequate or too expensive for general use. When large displacements were necessary experimental methods were used and they were often time consuming and expensive. The point being made is that it is no longer necessary to depend only upon linear analysis or testing. Beams with large displacement can be used if the situation warrants and the analysis is neither inefficient nor expensive.

7.12   Summary and Conclusions

We continue our emphasis on the displacement method. In our displacement method the classical analytical equations are summarized in Equations 7.3.16, repeated here.

$$M_c, F_c \quad \rightarrow \quad f_y(x) \quad V(x) \quad M(x) \quad \sigma_x(x, y) \quad \varepsilon_x(x, y) \quad u(x, y) \quad v(x) \leftarrow \rho$$

$$\frac{dV}{dx} = -f_y \qquad \frac{dM}{dx} = -V(x) \qquad\qquad\qquad u(x, y) = -y\frac{dv}{dx}$$

$$\varepsilon_x(x, y) = \frac{du}{dx} = -y\frac{d^2v}{dx^2}$$

$$\sigma_x(x, y) = E\varepsilon_x = -Ey\frac{d^2v}{dx^2}$$

$$M(x) = -\int_A \sigma_x y\, dA = E\frac{d^2v}{dx^2}\int_A y^2\, dA = EI_{zz}\frac{d^2v}{dx^2} \quad \rightarrow \quad \sigma_x = -\frac{M(x)y}{I_{zz}}$$

$$\frac{dM}{dx} = \frac{d}{dx}EI_{zz}\frac{d^2v}{dx^2} = -V(x) \quad \rightarrow \quad \tau_{xy}(x, y) = \frac{V(x)}{I_{zz}b}\int_{A_y} y\, dA_y$$

$$\frac{d^2M}{dx^2} = \frac{d^2}{dx^2}EI_{zz}\frac{d^2v}{dx^2} = f_y(x)$$

$$(7.12.1)$$

Several examples are provided to illustrate the classical analytical method of solution. Once again the beams can be classified as statically determinate or indeterminate; however, both can be solved using the fourth order equation.

8

A General Method (FEM) for the Bending of Beams

8.1 Introduction

As we noted in the axial and torsional cases, a new way of organizing the equations of solid mechanics evolved with the advent of the digital computer. We called this the general method because it is easily adapted to all possible materials, geometries, applied loads, and restraints. We have admitted that it is more commonly known as the *finite element method* or FEM for short. As noted in the axial and torsional cases the method depends upon dividing the structure into segments, called *elements*, and developing displacement functions for each element in terms of the *nodal* displacements at the ends of each element. In the process of satisfying Hooke's law for material properties, equilibrium, and the strain displacement equations we arrive at an *element stiffness matrix* which relates the internal forces to the nodal displacements. The elements are then assembled into a *global stiffness matrix* that relates the external applied loads and restraint forces to the nodal displacements. The result is a set of linear algebraic equations that can be solved for the nodal displacements. From these nodal displacements we can work back to the distributed displacements, strains, internal forces, and stresses. We now adapt this method to beams.

8.2 Nodes, Elements, Shape Functions, and the Element Stiffness Matrix

At the heart of the finite element method is the representation of the displacement of the element in terms of the nodal displacements and a local coordinate. These displacement functions are called *shape functions*. The shape functions for a beam were introduced in Example 7.4.2 and presented in Equation 7.4.7 for an element m based on the solution of the differential equations of beam theory applied to a segment of a beam between concentrated loads.

Consider the deformed shape of a beam that has been divided into elements as shown in Figure 8.2.1.

Let us select one element that we identify as element m between nodes n and $n+1$. Applied loads, if any, are concentrated loads located at nodes. From the solution of the differential equation for the displacement of such a beam we found that the shape of an element is a cubic polynomial. To represent this cubic polynomial we define a lateral displacement and a rotational displacement at each node. This provides four quantities needed to define the four coefficients of the cubic polynomial.

Analysis of Structures: An Introduction Including Numerical Methods, First Edition. Joe G. Eisley and Anthony M. Waas.
© 2011 John Wiley & Sons, Ltd. Published 2011 by John Wiley & Sons, Ltd.

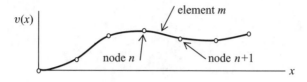

Figure 8.2.1

Let us isolate this element and define the lateral and rotational displacements at each node and the local coordinate s_m as shown in Figure 8.2.2.

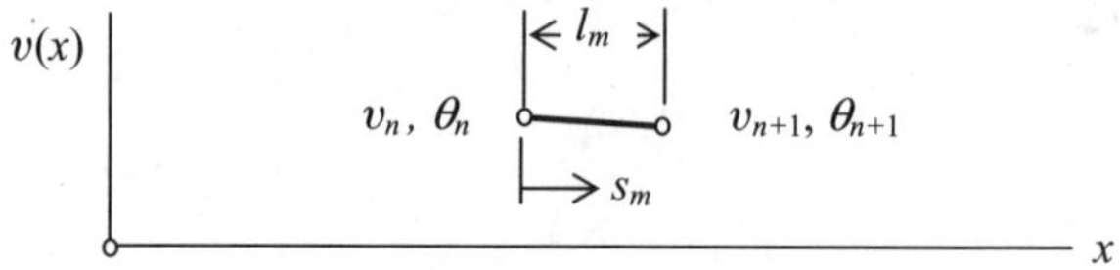

Figure 8.2.2

When $v(s_m)$ is the lateral displacement we define the rotation at each node as

$$\theta_n = \frac{dv_n}{ds_m} \quad \theta_{n+1} = \frac{dv_{n+1}}{ds_m} \tag{8.2.1}$$

In matrix form the displacements at each node are

$$\begin{bmatrix} v_n \\ \theta_n \end{bmatrix} \quad \begin{bmatrix} v_{n+1} \\ \theta_{n+1} \end{bmatrix} \tag{8.2.2}$$

Thus the nodal displacements for element m are a combination of lateral displacements and rotations at both nodes.

$$\{r_m\} = \begin{bmatrix} v_n \\ \theta_n \\ v_{n+1} \\ \theta_{n+1} \end{bmatrix} \tag{8.2.3}$$

We now write the distributed displacement in terms of the nodal displacements where $[n]$ is the matrix of shape functions.

$$v(s_m) = [n]\{r_m\} \tag{8.2.4}$$

The process of obtaining the shape functions is presented in detail in Example 7.4.2. In Equation 7.4.7 we presented

$$v(s_m) = \frac{1}{l_m^3}\left[l_m^3 - 3l_m s_m + 2s_m^3 \quad l_m^3 s_m - 2l_m^2 s_m^2 + l_m s_m^3 \quad 3l_m s_m^2 - 2s_m^3 \quad -l_m^2 s_m^2 + l_m s_m^3 \right] \begin{bmatrix} v_n \\ \theta_n \\ v_{n+1} \\ \theta_{n+1} \end{bmatrix} \tag{8.2.5}$$

Thus the shape function matrix $[n]$ is

$$[n] = \frac{1}{l_m^3}\left[l_m^3 - 3l_m s_m + 2s_m^3 \quad l_m^3 s_m - 2l_m^2 s_m^2 + l_m s_m^3 \quad 3l_m s_m^2 - 2s_m^3 \quad -l_m^2 s_m^2 + l_m s_m^3 \right] \tag{8.2.6}$$

As before, s_m is a local coordinate in the x direction and l_m is the length of the element. Each term in the shape function matrix is a cubic polynomial. This is the exact solution to the differential equation for a beam segment between concentrated loads.

There are internal forces in the form of bending moments and shear forces at each end of the element. In matrix form at each node for element m we have

$$\begin{bmatrix} V_{m,n} \\ M_{m,n} \end{bmatrix} \quad \begin{bmatrix} V_{m,n+1} \\ M_{m,n+1} \end{bmatrix} \tag{8.2.7}$$

Thus the internal force matrix for the element m is

$$\{p_m\} = \begin{bmatrix} V_{m,n} \\ M_{m,n} \\ V_{m,n+1} \\ M_{m,n+1} \end{bmatrix} \tag{8.2.8}$$

These are represented in Figure 8.2.3 for element m.

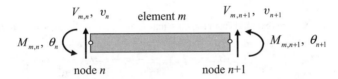

Figure 8.2.3

We must note the sign convention used here. In the traditional analysis in Chapter 7 we used the sign convention presented in Figure 8.2.4 for positive forces and moments. See Figure 7.3.5.

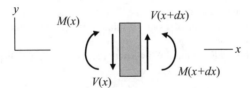

Figure 8.2.4

At node $n+1$ the two agree. At node n they are of opposite sign. In Figure 8.2.3 all internal forces are positive in the same direction as the positive coordinate axes and all internal moments are positive by the right hand rule. The purpose is to make the assembly of elements easier. Use the values at the higher node when comparing the FEM results with the traditional analytical solutions.

We now note that given Equation 8.2.4, repeated here as Equation 8.2.9,

$$v(s_m) = [n]\{r_m\} \tag{8.2.9}$$

we can find the displacement $u(s_m, y)$, assuming plane sections remain plane, in terms of the shape functions and the nodal coordinates. Thus

$$u(s_m, y) = -y\frac{dv}{ds_m} = -y\frac{d}{ds_m}[n]\{r_m\} \tag{8.2.10}$$

and from the strain displacement equation the strain is

$$\varepsilon_{xm}(s_m, y) = \frac{\partial u}{\partial s_m} = -y \frac{d^2}{ds_m^2} [n]\{r_m\} \tag{8.2.11}$$

Finally from Hooke's law the stress is

$$\sigma_{xm}(s_m, y) = E\varepsilon_x = -yE \frac{d^2}{ds_m^2} [n]\{r_m\} \tag{8.2.12}$$

From our knowledge of beam theory detailed in Chapter 7 (see Equations 7.3.4 and 7.3.8), it follows that the internal forces at the node $n+1$, where the sign conventions are the same in both the classical and FEM analysis, are

$$M_{m,n+1} = -\int_A y\sigma_x(s_m)\,dA = E_m I_m \frac{d^2 v_m}{ds_m^2} = E_m I_m \frac{d^2}{ds_m^2} [n]\{r_m\}$$

$$V_{m,n+1} = -\frac{dM_{n+1}}{ds_m} = -E_m I_m \frac{d^3 v_m}{ds_m^3} = -E_m I_m \frac{d^3}{ds_m^3} [n]\{r_m\} \tag{8.2.13}$$

At the node n the signs are reversed from the classical values and so the internal forces are

$$M_{m,n} = \int_A y\sigma_x(s_m)\,dA = -E_m I_m \frac{d^2 v_m}{ds_m^2} = -E_m I_m \frac{d^2}{ds_m^2} [n]\{r_m\}$$

$$V_{m,n} = -\frac{dM_n}{ds_m} = E_m I_m \frac{d^3 v_m}{ds_m^3} = E_m I_m \frac{d^3}{ds_m^3} [n]\{r_m\} \tag{8.2.14}$$

This relatively simple act of defining shape functions turns the problem from one of solving differential equations to that of solving linear algebraic equations. The differential equations are solved when creating the shape functions once and for all for each class of elements we examine. The genius of this approach is that all the differentiation indicated in Equations 8.2.10-14 can be performed explicitly since the shape functions $[n]$ are known. Please note that the strain displacement, Hooke's law, and equilibrium equations are satisfied without compromise.

Given that shape functions in Equation 8.2.6 we note that the distributed displacement is

$$u(s_m) = -y \frac{dv_m}{ds_m} = -y \frac{d}{ds_m} [n]\{r_m\}$$

$$= -\frac{y}{l_m^3} \left[-6l_m s_m + 6s_m^2 \quad l_m^3 - 4l_m^2 s_m + 3l_m s_m^2 \quad 6l_m s_m - 6s_m^2 \quad -2l_m^2 s_m + 3l_m s_m^2 \right]\{r_m\} \tag{8.2.15}$$

The strain and the normal stress in that element are

$$\varepsilon_{xm} = \frac{\partial u_m}{\partial s_m} = -y \frac{d^2}{ds_m^2} [n]\{r_m\}$$

$$= -\frac{y}{l_m^3} \left[-6l_m + 12s_m \quad -4l_m^2 + 6l_m s_m \quad 6l_m - 12s_m \quad -2l_m^2 + 6l_m s_m \right]\{r_m\}$$

$$\sigma_{xm} = E \frac{\partial u_m}{\partial s_m} = -yE \frac{d^2}{ds_m^2} [n]\{r_m\} \tag{8.2.16}$$

$$= -\frac{yE}{l_m^3} \left[-6l_m + 12s_m \quad -4l_m^2 + 6l_m s_m \quad 6l_m - 12s_m \quad -2l_m^2 + 6l_m s_m \right]\{r_m\}$$

As noted the internal forces are evaluated at $s_m = 0$ for node n and at $s_m = l_m$ for node $n+1$.

$$V_{m,n} = E_m I_m \frac{d^3}{ds_m^3}[n]\{r\} = \frac{E_m I_m}{l_m^3}[12 \quad 6l_m \quad -12 \quad 6l_m]\{r_m\}$$

$$M_{m,n} = -E_m I_m \frac{d^2}{ds_m^2}[n]\{r\} = \frac{E_m I_m}{l_m^3}\left[6l_m \quad 4l_m^2 \quad -6l_m \quad 2l_m^2\right]\{r_m\}$$

$$V_{m,n+1} = -E_m I_m \frac{d^3}{ds_m^3}[n]\{r\} = -\frac{E_m I_m}{l_m^3}[12 \quad 6l_m \quad -12 \quad 6l_m]\{r_m\} \tag{8.2.17}$$

$$M_{m,n+1} = E_m I_m \frac{d^2}{ds_m^2}[n]\{r\} = \frac{E_m I_m}{l_m^3}\left[6l_m \quad 2l_m^2 \quad -6l_m \quad 4l_m^2\right]\{r_m\}$$

From Equations 8.2.17 we obtain

$$\{p_m\} = [k_m]\{r_m\} = \begin{bmatrix} V_{m,n} \\ M_{m,n} \\ V_{m,n+1} \\ M_{m,n+1} \end{bmatrix} = \frac{E_m I_m}{l_m^3} \begin{bmatrix} 12 & 6l_m & -12 & 6l_m \\ 6l_m & 4l_m^2 & -6l_m & 2l_m^2 \\ -12 & -6l_m & 12 & -6l_m \\ 6l_m & 2l_m^2 & -6l_m & 4l_m^2 \end{bmatrix} \begin{bmatrix} v_n \\ \theta_n \\ v_{n+1} \\ \theta_{n+1} \end{bmatrix} \tag{8.2.18}$$

Thus the element stiffness matrix is

$$[k_m] = \frac{E_m I_m}{l_m^3} \begin{bmatrix} 12 & 6l_m & -12 & 6l_m \\ 6l_m & 4l_m^2 & -6l_m & 2l_m^2 \\ -12 & -6l_m & 12 & -6l_m \\ 6l_m & 2l_m^2 & -6l_m & 4l_m^2 \end{bmatrix} \tag{8.2.19}$$

This element stiffness matrix can be used in all subsequent beam problems that lie in the xy plane. The work we just went through need never be done again.

In summary, the path to the element stiffness matrix is

$$\{p_m\} \qquad \sigma_{xm}(s_m, y) \qquad \varepsilon_{xm}(s_m, y) \qquad u(s_m, y) \qquad v(s_m) \qquad \{r_m\}$$

$$v(s_m) = [n]\{r_m\}$$

$$u(s_m, y) = -y\frac{dv_m}{ds_m} = -y\frac{d}{ds_m}[n]\{r_m\}$$

$$\varepsilon_{xm}(s_m, y) = \frac{\partial u_m}{\partial s_m} = -y\frac{d^2}{ds_m^2}[n]\{r_m\}$$

$$\sigma_{xm}(s_m, y) = E\varepsilon_{xm} = -yE\frac{d^2}{ds_m^2}[n]\{r_m\}$$

$$M_{m,n} = \int_A y\sigma_{xm}(s_m)dA = -E_m I_m \frac{d^2}{ds_m^2}[n]\{r\} \qquad V_{m,n} = -\frac{dM_{m,n}}{ds_m} = E_m I_m \frac{d^3}{ds_m^3}[n]\{r\}$$

$$M_{m,n+1} = -\int_A y\sigma_{xm}(s_m)dA = E_m I_m \frac{d^2}{ds_m^2}[n]\{r\} \qquad V_{m,n+1} = -\frac{dM_{m,n+1}}{ds_m} = -E_m I_m \frac{d^3}{ds_m^3}[n]\{r\}$$

$$[n] = \frac{1}{l_m^3}\left[l_m^3 - 3l_m s_m + 2s_m^3 \quad l_m^3 s_m - 2l_m^2 s_m^2 + l_m s_m^3 \quad 3l_m s_m^2 - 2s_m^3 \quad -l_m^2 s_m^2 + l_m s_m^3\right]$$

$$\{p_m\} = \begin{bmatrix} V_{m,n} \\ M_{m,n} \\ V_{m,n+1} \\ M_{m,n+1} \end{bmatrix} = \frac{E_m I_m}{l_m^3} \begin{bmatrix} 12 & 6l_m & -12 & 6l_m \\ 6l_m & 4l_m^2 & -6l_m & 2l_m^2 \\ -12 & -6l_m & 12 & -6l_m \\ 6l_m & 2l_m^2 & -6l_m & 4l_m^2 \end{bmatrix} \begin{bmatrix} v_{m,n} \\ \theta_{m,n} \\ v_{m,n+1} \\ \theta_{m,n+1} \end{bmatrix} = [k_m]\{r_m\} \tag{8.2.20}$$

8.3   The Global Equations and their Solution

Refer back to the process of assembly of the element equations to form the global equations for the axial case in Chapter 5, Section 5.3. We summarize the same process for the beam equations. We seek the global equations in the form

$$\{F\} = \sum \{p_m\} = \sum [k_m]\{r_m\} = [K]\{r\} \tag{8.3.1}$$

where $[K]$ is the global stiffness matrix and $\{r\}$ is the total set of nodal displacements including known restraints and unknown values to be determined. The matrix $\{F\}$ contains both applied and restraint forces and moments.

From equilibrium of forces at each node we can define $\{F\}$ as

$$\{F\} = \begin{bmatrix} F_1 \\ M_1 \\ F_2 \\ M_2 \\ \cdots \\ F_N \\ M_N \end{bmatrix} = \begin{bmatrix} V_{1,1} \\ M_{1,1} \\ V_{1,2} + V_{2,2} \\ M_{1,2} + M_{2,2} \\ \cdots \\ V_N \\ M_N \end{bmatrix} = [K]\{r\} \tag{8.3.2}$$

As in the axial case we cannot sum the element matrices directly because they do not refer to the same nodal matrix. Thus we reformat each of the element matrices by imbedding them in a format that includes the complete set of nodal displacements. The assembly of the element matrices into a global matrix was described in detail for the axial and torsional cases and is the same process here. It is best shown by example.

##########

Example 8.3.1

Problem: Find the stresses and displacements for the beam in Example 7.4.2 by the FE method. The beam is fixed at the left end and has two transverse forces as shown in Figure (a). Let $l = L/2$. The cross section is a rectangle of height h and width b.

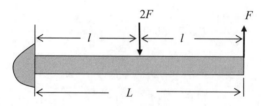

Figure (a)

Solution: Determine nodes and elements, assign values to the element stiffness matrices, and assemble the global stiffness matrix. Apply boundary conditions, partition, and solve the resulting set of linear algebraic equations.

Let us divide the beam into two elements and number the nodes and elements as shown in Figure (b).

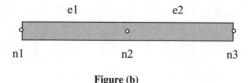

e1 e2

n1 n2 n3

Figure (b)

The process of assembling the global stiffness matrix from the element matrices is to first imbed each element matrix into the global matrix format. You may wish to review the process first outlined in Section 5.3 for the axial case.

The size of the global stiffness matrix is determined by the total number of nodal displacements. For this case

$$\{r\} = \begin{bmatrix} v_1 \\ \theta_1 \\ v_2 \\ \theta_2 \\ v_3 \\ \theta_3 \end{bmatrix} \tag{a}$$

thus the global stiffness matrix $[K]$ is of order 6×6.

For element 1

$$[k_1] = \frac{EI_{zz}}{l^3} \begin{bmatrix} 12 & 6l & -12 & 6l & 0 & 0 \\ 6l & 4l^2 & -6l & 2l^2 & 0 & 0 \\ -12 & -6l & 12 & -6l & 0 & 0 \\ 6l & 2l^2 & -6l & 4l^2 & 0 & 0 \\ 0 & 0 & 0 & 0 & 0 & 0 \\ 0 & 0 & 0 & 0 & 0 & 0 \end{bmatrix} \tag{b}$$

For element 2

$$[k_2] = \frac{EI_{zz}}{l^3} \begin{bmatrix} 0 & 0 & 0 & 0 & 0 & 0 \\ 0 & 0 & 0 & 0 & 0 & 0 \\ 0 & 0 & 12 & 6l & -12 & 6l \\ 0 & 0 & 6l & 4l^2 & -6l & 2l^2 \\ 0 & 0 & -12 & -6l & 12 & -6l \\ 0 & 0 & 6l & 2l^2 & -6l & 4l^2 \end{bmatrix} \tag{c}$$

Summing the two, that is, $[K] = [k_1] + [k_2]$, give us the global stiffness matrix.

$$[K] = \frac{EI_{zz}}{l^3} \begin{bmatrix} 12 & 6l & -12 & 6l & 0 & 0 \\ 6l & 4l^2 & -6l & 2l^2 & 0 & 0 \\ -12 & -6l & 24 & 0 & -12 & 6l \\ 6l & 2l^2 & 0 & 8l^2 & -6l & 2l^2 \\ 0 & 0 & -12 & -6l & 12 & -6l \\ 0 & 0 & 6l & 2l^2 & -6l & 4l^2 \end{bmatrix} \tag{d}$$

At node 1 there are two known support restraints, $v_1 = 0$ and $\theta_1 = 0$, and two unknown reactions, $F_1 = R_1$ and $M_1 = Q_1$. At nodes 2 and 3 there are four known loads, $F_2 = -2F$, $M_2 = 0$, $F_3 = F$, and $M_3 = 0$. The assembled equations are then

$$\{F\} = [K]\{r\} = \begin{bmatrix} R_1 \\ Q_1 \\ -2F \\ 0 \\ F \\ 0 \end{bmatrix} = \frac{EI_{zz}}{l^3} \begin{bmatrix} 12 & 6l & -12 & 6l & 0 & 0 \\ 6l & 4l^2 & -6l & 2l^2 & 0 & 0 \\ -12 & -6l & 24 & 0 & -12 & 6l \\ 6l & 2l^2 & 0 & 8l^2 & -6l & 2l^2 \\ 0 & 0 & -12 & -6l & 12 & -6l \\ 0 & 0 & 6l & 2l^2 & -6l & 4l^2 \end{bmatrix} \begin{bmatrix} 0 \\ 0 \\ v_2 \\ \theta_2 \\ v_3 \\ \theta_3 \end{bmatrix} \tag{e}$$

We can, of course, solve the assembled algebraic equations analytically; however, it is common to insert values for the various quantities and then to solve numerically using a linear equation solving software program. This is the practical approach when a large number of elements are used.

Let us assume that the material is steel with $E = 206,800 \ N/mm^2$, the length is $L=1000 \ mm$, and the cross section is rectangular with a width $b = 20 \ mm$ and a height $h = 30 \ mm$. The force is $F = 250 \ N$. Equation (e) becomes

$$\begin{bmatrix} R_1 \\ Q_1 \\ -500 \\ 0 \\ 250 \\ 0 \end{bmatrix} = 74.448 \begin{bmatrix} 12 & 3000 & -12 & 3000 & 0 & 0 \\ 3000 & 1000000 & -3000 & 500000 & 0 & 0 \\ -12 & -3000 & 24 & 0 & -12 & 3000 \\ 3000 & 500000 & 0 & 2000000 & -3000 & 500000 \\ 0 & 0 & -12 & -3000 & 12 & -3000 \\ 0 & 0 & 3000 & 500000 & -3000 & 1000000 \end{bmatrix} \begin{bmatrix} 0 \\ 0 \\ v_2 \\ \theta_2 \\ v_3 \\ \theta_3 \end{bmatrix} \tag{f}$$

We partition Equation (f) to find the displacements.

$$\begin{bmatrix} -500 \\ 0 \\ 250 \\ 0 \end{bmatrix} = 74.448 \begin{bmatrix} 24 & 0 & -12 & 3000 \\ 0 & 2000000 & -3000 & 500000 \\ -12 & -3000 & 12 & -3000 \\ 3000 & 500000 & -3000 & 1000000 \end{bmatrix} \begin{bmatrix} v_2 \\ \theta_2 \\ v_3 \\ \theta_3 \end{bmatrix} \tag{g}$$

Solving with a linear equation solver we get

$$\begin{bmatrix} v_2 \\ \theta_2 \\ v_3 \\ \theta_3 \end{bmatrix} = \begin{bmatrix} 0.5597 \\ 0.003358 \\ 3.3580 \\ 0.006716 \end{bmatrix} \begin{matrix} mm \\ rad \\ mm \\ rad \end{matrix} \tag{h}$$

Let us compare these displacements with the analytical results found in Example 7.4.2, Equations (g) and (i).

$$v(0) = 0 \qquad v\left(\frac{L}{2}\right) = \frac{FL^3}{EI_{zz}48} = 0.560 \ mm \qquad v(L) = \frac{FL^3}{EI_{zz}8} = 3.358 \ mm$$

(i)

$$\theta(0) = \frac{dv(0)}{dx} = 0 \qquad \frac{dv\left(\frac{L}{2}\right)}{dx} = \frac{FL^2}{EI_{zz}8} = 0.003358 \ rad \qquad \frac{dv(L)}{dx} = \frac{FL^2}{EI_{zz}4} = 0.006716 \ rad$$

In matrix format we have from Example 7.4.2 the same values as those from FEM.

$$\{v\} = \begin{bmatrix} v_1 \\ v_2 \\ v_3 \end{bmatrix} = \begin{bmatrix} 0 \\ 0.560 \\ 3.358 \end{bmatrix} mm \qquad \{\theta\} = \begin{bmatrix} \theta_1 \\ \theta_2 \\ \theta_3 \end{bmatrix} = \begin{bmatrix} 0 \\ 0.003358 \\ 0.006716 \end{bmatrix} rad \tag{j}$$

This is as it should be. The method used here is exactly the same since the shape functions are cubic polynomials and the analytical solution of the differential equations for the displacements are also cubic polynomials.

The other partitioned equation set for the support reactions is

$$\begin{bmatrix} R_1 \\ Q_1 \end{bmatrix} = 74.448 \begin{bmatrix} -12 & 3000 & 0 & 0 \\ -3000 & 500000 & 0 & 0 \end{bmatrix} \begin{bmatrix} v_2 \\ \theta_2 \\ v_3 \\ \theta_3 \end{bmatrix} \tag{k}$$

From which we obtain

$$\begin{bmatrix} R_1 \\ Q_1 \end{bmatrix} = \begin{bmatrix} 250.000 \\ -0.037 \end{bmatrix} \begin{matrix} N \\ N \cdot mm \end{matrix} \tag{l}$$

The exact values from static equilibrium as found in Example 7.4.2 are

$$\begin{bmatrix} R_1 \\ Q_1 \end{bmatrix} = \begin{bmatrix} 250 \\ 0 \end{bmatrix} \begin{matrix} N \\ N \cdot mm \end{matrix} \tag{m}$$

We obtained a very small round off error. The value for $Q_1 = -0.037$ is really a very small error for the moments.

The shear forces and bending moments are found from Equation 8.2.8 which is repeated here.

$$\{p_m\} = [k_m]\{r_m\} = \begin{bmatrix} V_{m,n} \\ M_{m,n} \\ V_{m,n+1} \\ M_{m,n+1} \end{bmatrix} = \frac{E_m I_m}{l_m^3} \begin{bmatrix} 12 & 6l_m & -12 & 6l_m \\ 6l_m & 4l_m^2 & -6l_m & 2l_m^2 \\ -12 & -6l_m & 12 & -6l_m \\ 6l_m & 2l_m^2 & -6l_m & 4l_m^2 \end{bmatrix} \begin{bmatrix} v_n \\ \theta_n \\ v_{n+1} \\ \theta_{n+1} \end{bmatrix} \tag{m}$$

Since all the elements are the same length the $[k_m]$ matrix is the same for all elements. For element 1:

$$\begin{bmatrix} V_{1,1} \\ M_{1,1} \\ V_{1,2} \\ M_{1,2} \end{bmatrix} = 74.448 \begin{bmatrix} 12 & 3000 & -12 & 3000 \\ 3000 & 1000000 & -3000 & 500000 \\ -12 & -3000 & 12 & -3000 \\ 3000 & 500000 & -3000 & 1000000 \end{bmatrix} \begin{bmatrix} 0 \\ 0 \\ 0.5597 \\ 0.003358 \end{bmatrix} = \begin{bmatrix} 250.000 \\ -0.037 \\ -250.000 \\ 125000 \end{bmatrix} \begin{matrix} N \\ N \cdot mm \\ N \\ N \cdot mm \end{matrix} \tag{n}$$

For element 2

$$\begin{bmatrix} V_{2,2} \\ M_{2,2} \\ V_{2,3} \\ M_{2,3} \end{bmatrix} = 74.448 \begin{bmatrix} 12 & 3000 & -12 & 3000 \\ 3000 & 1000000 & -3000 & 500000 \\ -12 & -3000 & 12 & -3000 \\ 3000 & 500000 & -3000 & 1000000 \end{bmatrix} \begin{bmatrix} 0.5597 \\ 0.003358 \\ 3.358 \\ 0.006716 \end{bmatrix} = \begin{bmatrix} -250.000 \\ -125000 \\ 250.000 \\ 0.037 \end{bmatrix} \begin{matrix} N \\ N \cdot mm \\ N \\ N \cdot mm \end{matrix} \tag{o}$$

We may also compare the shear forces and bending moments by the two methods. There may be some confusion over signs since the definition of positive shear forces and moments is different for the analytical equations and the FE method. Refer back to Figure 8.2.1 and 8.2.2. In plotting we use the convention for the analytical equations. Thus the signs at node n must be reversed for comparison.

For creating the shear diagram, then we use $V_{1,2}$ (or $-V_{1,1}$) for the shear force in element 1 and $V_{2,3}$ (or $-V_{2,2}$) for element 2. For bending moment we use the values $-M_{1,1} M_{1,2}$ from element 1 and $M_{2,3}$ from element 2. This is plotted in Figure (c).

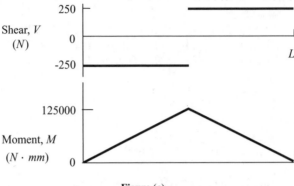

Figure (c)

These are exactly the same values as for the analytical method. See Example 7.4.2. Note that since the shape functions are cubic polynomials the moments will always be second derivatives or linear polynomials and the shear force will always be third derivatives or constants for each element. This is exact when only concentrated forces are present but will be an approximation when distributed forces are applied as we shall show in the next section.

One check on the results is that at each node the sum of the internal forces must be equal to the applied force and the sum of the internal moments must be equal the applied moments.

$$\text{At node 2} \quad -2F = V_{1,2} + V_{2,2} \quad \rightarrow \quad -500 = -250 - 250 = -500$$
$$0 = M_{1,2} + M_{2,2} \quad \rightarrow \quad 0 = 125000 - 125000 = 0 \tag{p}$$

$$\text{At node 3} \quad F = V_{2,3} \quad \rightarrow \quad 250 = 250$$
$$0 = M_{2,3} \quad \rightarrow \quad 0 = 0.037 \doteq 0$$

The maximum normal stress is at $x = L/2$ and $y = \pm h/2$.

$$\sigma_x = -\frac{M_z y}{I_{zz}} \quad \rightarrow \quad \sigma_{max} = \pm \frac{125000 \cdot 15}{45000} = \pm 41.67 N/MPa \tag{q}$$

The maximum shear stress is $y = 0$ and at all values of x.

$$\tau_{xy}(x, y) = \frac{V}{2I_{zz}} \left(\frac{h^2}{4} - y^2 \right) \quad \rightarrow \quad \tau_{max} = \pm \frac{250}{2.45000}(15)^2 = \pm 0.625 \, MPa \tag{r}$$

###########

At this point the FE method appears to be a lot more work than the analytical method. The next example is easier by FEM.

############

Example 8.3.2

Problem: Repeat Example 7.6.2 by the FEM. Plot the displacements, shear forces, and bending moments for the beam shown in Figure (a).

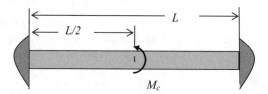

Figure (a)

Solution: Divide into two elements, assembled the global equations, partition, and solve.

The global stiffness matrix is the same as in the previous example. Only the restraints and applied forces have changed. Inserting boundary conditions we get

$$\{F\} = [K]\{r\} = \begin{bmatrix} R_1 \\ Q_1 \\ 0 \\ M_c \\ R_3 \\ Q_3 \end{bmatrix} = \frac{EI_{zz}}{l^3} \begin{bmatrix} 12 & 6l & -12 & 6l & 0 & 0 \\ 6l & 4l^2 & -6l & 2l^2 & 0 & 0 \\ -12 & -6l & 24 & 0 & -12 & 6l \\ 6l & 2l^2 & 0 & 8l^2 & -6l & 2l^2 \\ 0 & 0 & -12 & -6l & 12 & -6l \\ 0 & 0 & 6l & 2l^2 & -6l & 4l^2 \end{bmatrix} \begin{bmatrix} 0 \\ 0 \\ v_2 \\ \theta_2 \\ 0 \\ 0 \end{bmatrix} \tag{a}$$

The partitioned equations and their solution for the displacements are

$$\begin{bmatrix} 0 \\ M_c \end{bmatrix} = \frac{EI_{zz}}{l^3} \begin{bmatrix} 24 & 0 \\ 0 & 8l^2 \end{bmatrix} \begin{bmatrix} v_2 \\ \theta_2 \end{bmatrix} \rightarrow \begin{bmatrix} v_2 \\ \theta_2 \end{bmatrix} = \begin{bmatrix} 0 \\ \dfrac{M_c l}{8EI_{zz}} \end{bmatrix} \tag{b}$$

The nodal shear forces and bending moments are

$$\begin{bmatrix} V_{1,1} \\ M_{1,1} \\ V_{1,2} \\ M_{1,2} \end{bmatrix} = \frac{EI_{zz}}{l^3} \begin{bmatrix} 12 & 6l & -12 & 6l \\ 6l & 4l^2 & -6l & 2l^2 \\ -12 & -6l & 12 & -6l \\ 6l & 2l^2 & -6l & 4l^2 \end{bmatrix} \begin{bmatrix} 0 \\ 0 \\ 0 \\ \dfrac{M_c l}{8EI_{zz}} \end{bmatrix} = \frac{EI_{zz}}{l^3} \begin{bmatrix} \dfrac{3M_c l^2}{4EI_{zz}} \\ \dfrac{M_c l^3}{4EI_{zz}} \\ -\dfrac{3M_c l^2}{4EI_{zz}} \\ \dfrac{M_c l^3}{2EI_{zz}} \end{bmatrix} = \begin{bmatrix} \dfrac{3M_c}{4l} \\ \dfrac{M_c}{4} \\ -\dfrac{3M_c}{4l} \\ \dfrac{M_c}{2} \end{bmatrix} \tag{c}$$

$$
\begin{bmatrix} V_{2,2} \\ M_{2,2} \\ V_{2,3} \\ M_{2,3} \end{bmatrix} = \frac{EI_{zz}}{l^3} \begin{bmatrix} 12 & 6l & -12 & 6l \\ 6l & 4l^2 & -6l & 2l^2 \\ -12 & -6l & 12 & -6l \\ 6l & 2l^2 & -6l & 4l^2 \end{bmatrix} \begin{bmatrix} 0 \\ M_c l \\ 8EI_{zz} \\ 0 \\ 0 \end{bmatrix} = \frac{EI_{zz}}{l^3} \begin{bmatrix} \dfrac{3M_c l^2}{4EI_{zz}} \\ \dfrac{M_c l^3}{2EI_{zz}} \\ \dfrac{3M_c l^2}{4EI_{zz}} \\ \dfrac{M_c l^3}{4EI_{zz}} \end{bmatrix} = \begin{bmatrix} \dfrac{3M_c}{4l} \\ \dfrac{M_c}{2} \\ -\dfrac{3M_c}{4l} \\ \dfrac{M_c}{4} \end{bmatrix} \quad \text{(d)}
$$

Plots of shear and moment are based on our knowledge that shear forces are constant between nodes and bending moments are linear. This is exactly the same results we obtained in Example 7.6.2 by solving the differential equations.

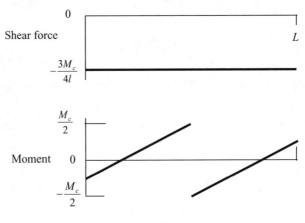

Figure (b)

The displacements are

$$
\begin{bmatrix} v_1 \\ \theta_1 \\ v_2 \\ \theta_2 \\ v_3 \\ \theta_3 \end{bmatrix} = \begin{bmatrix} 0 \\ 0 \\ 0 \\ \dfrac{M_c l}{8EI_{zz}} \\ 0 \\ 0 \end{bmatrix} \quad \text{(e)}
$$

To plot the distributed displacement we invoke the shape functions. We have dropped the subscripts from s and l since they are the same values for each element.

For element 1

$$
v_1(s) = \frac{1}{l^3} \begin{bmatrix} l^3 - 3ls^2 + 2s^3 & l^3 s - 2l^2 s^2 + ls^3 & 3ls^2 - 2s^3 & -l^2 s^2 + ls^3 \end{bmatrix} \begin{bmatrix} 0 \\ 0 \\ 0 \\ \dfrac{M_c l}{8EI_{zz}} \end{bmatrix} \quad \text{(f)}
$$

$$
= \frac{1}{l^3} \left(-l^2 s^2 + ls^3 \right) \frac{M_c l}{8EI_{zz}} = \left(-s^2 + \frac{s^3}{l} \right) \frac{M_c}{8EI_{zz}}
$$

For element 2

$$v_2(s) = \frac{1}{l^3} \left[l^3 - 3ls^2 + 2s^3 \quad l^3s - 2l^2s^2 + ls^3 \quad 3ls^2 - 2s^3 \quad -l^2s^2 + ls^3 \right] \begin{bmatrix} 0 \\ \frac{M_c l}{8EI_{zz}} \\ 0 \\ 0 \end{bmatrix} \quad (g)$$

$$= \frac{1}{l^3} \left(l^3s - 2l^2s^2 + ls^3 \right) \frac{M_c l}{8EI_{zz}} = \left(ls - 2s^2 + \frac{s^3}{l} \right) \frac{M_c}{8EI_{zz}}$$

Plot of displacements (Note the anti symmetry):

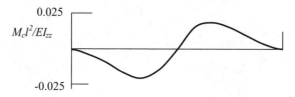

Figure (c)

############

8.4 Distributed Loads in FEM

Next we will continue with an example only slightly more complicated but where a distributed load is replaced with equivalent nodal loads. The equivalent nodal loads are obtained for each element from

$$\{F_{em}\} = \int_0^{l_m} [n]^T f_y(s_m)\, ds_m \tag{8.4.1}$$

and the total loading from equivalent nodal loads is

$$\{F_e\} = \sum \{F_{em}\} \tag{8.4.2}$$

As noted before, this expression in Equation 8.4.1 will be derived in Chapter 11.

The applied loads consist of both concentrated loads and equivalent nodal loads as expressed by

$$\{F\} = \{F_c\} + \{F_e\} \tag{8.4.3}$$

Note that applied concentrated moments are common but distributed applied moments in bending are rarely encountered and so are not included in this discussion. In any case they all would be included in equivalent nodal loads.

The shape functions consist of four cubic polynomials. It follows that there will be four equivalent nodal loads obtained from a distributed lateral load for each element. These will consist of two forces and two moments as follows.

$$\{F_{em}\} = \int_0^{l_m} [n]^T f_y(s_m)\, ds_m = \int_0^{l_m} \begin{bmatrix} l_m^3 - 3l_m s_m^2 + 2s_m^3 \\ l_m^3 s - 2l_m^2 s_m^2 + l_m s_m^3 \\ 3l_m s_m^2 - 2s_m^3 \\ -l_m^2 s_m^2 + l_m s_m^3 \end{bmatrix} \frac{f_y(s_m)}{l_m^3}\, ds_m = \begin{bmatrix} F_{em,n} \\ M_{em,n} \\ F_{em,n+1} \\ M_{em,n+1} \end{bmatrix} \tag{8.4.4}$$

It may seem strange that a lateral distributed load would produce both equivalent lateral concentrated loads and concentrated moments but this is the best possible equivalent load system.

In summary the FEM equations are

$$\{F\} \qquad \{F_c\} \qquad f_y(s_m) \to \{p_m\} \qquad \sigma_{xm}(s_m, y) \qquad \varepsilon_{xm}(s_m, y) \qquad u(s_m, y) \qquad v(s_m)\{r_m\} \leftarrow \{\rho\}$$

$$\{F_{em}\} = \int_0^l [n]^T f_y(s_m)\, ds_m \qquad\qquad \{F_e\} = \sum \{F_{em}\} \quad v(s_m) = [n]\{r_m\}$$

$$\{F\} = \{F_c\} + \{F_e\} \qquad\qquad u(s_m, y) = -y\frac{d}{ds_m}[n]\{r_m\}$$

$$\varepsilon_{xm}(s_m, y) = -y\frac{d^2}{ds_m^2}[n]\{r_m\}$$

$$\sigma_{xm}(s_m, y) = -yE\frac{d^2}{ds_m^2}[n]\{r_m\}$$

$$M_{m,n} = -E_m I_m \frac{d^2}{ds_m^2}[n]\{r\} \qquad\qquad V_{m,n} = E_m I_m \frac{d^3}{ds_m^3}[n]\{r\}$$

$$M_{m,n+1} = E_m I_m \frac{d^2}{ds_m^2}[n]\{r\} \qquad\qquad V_{m,n+1} = -E_m I_m \frac{d^3}{ds_m^3}[n]\{r\}$$

$$v(s_m) = [n]\{r_m\}$$

$$= \frac{1}{l_m^3} \begin{bmatrix} l_m^3 - 3l_m s_m + 2s_m^3 & l_m^3 s_m - 2l_m^2 s_m^2 + l_m s_m^3 & 3l_m s_m^2 - 2s_m^3 & -l_m^2 s_m^2 + l_m s_m^3 \end{bmatrix} \begin{bmatrix} v_{m,n} \\ \theta_{m,n} \\ v_{m,n+1} \\ \theta_{m,n+1} \end{bmatrix}$$

$$\{p_m\} = [k_m]\{r_m\} = \begin{bmatrix} V_{m,n} \\ M_{m,n} \\ V_{m,n+1} \\ M_{m,n+1} \end{bmatrix} = \frac{E_m I_m}{l_m^3} \begin{bmatrix} 12 & 6l_m & -12 & 6l_m \\ 6l_m & 4l_m^2 & -6l_m & 2l_m^2 \\ -12 & -6l_m & 12 & -6l_m \\ 6l_m & 2l_m^2 & -6l_m & 4l_m^2 \end{bmatrix} \begin{bmatrix} v_{m,n} \\ \theta_{m,n} \\ v_{m,n+1} \\ \theta_{m,n+1} \end{bmatrix}$$

$$\{F\} = \sum \{p_m\} = \sum [k_m]\{r_m\} = [K]\{r\} \qquad\qquad (8.4.5)$$

###########

Example 8.4.1

Problem: Repeat Example 7.6.1 as shown in Figure (a) using the FE method.

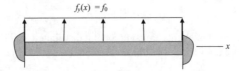

$f_y(x) = f_0$

Figure (a)

Solution: Determine nodes and elements, assign values to the element stiffness matrices, and assemble the global stiffness matrix. Find equivalent nodal loads, apply boundary conditions, and solve the resulting set of linear algebraic equations.

Let us use four elements. The nodes and elements are shown in Figure (b).

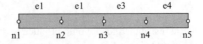

Figure (b)

Let us say that the length of each element is $l_m = l$ for a total length of $4l$. In this case the equivalent nodal loads on each element are

$$\{F_{em}\} = \frac{f_0}{l^3} \int_0^l \begin{bmatrix} l_m^3 - 3l_m s_m^2 + 2s_m^3 \\ l_m^3 s - 2l_m^2 s_m^2 + l_m s_m^3 \\ 3l_m s_m^2 - 2s_m^3 \\ -l_m^2 s_m^2 + l_m s_m^3 \end{bmatrix} ds_m = \begin{bmatrix} \dfrac{f_0 l}{2} \\ \dfrac{f_0 l^2}{12} \\ \dfrac{f_0 l}{2} \\ -\dfrac{f_0 l^2}{12} \end{bmatrix} \qquad (a)$$

Note again that while the applied load is only a transverse load the equivalent nodal loads include moments as well as forces. When we imbed just the first element stiffness matrix into the global stiffness matrix we get

$$\begin{bmatrix} \dfrac{f_0 l}{2} \\ \dfrac{f_0 l^2}{12} \\ \dfrac{f_0 l}{2} \\ -\dfrac{f_0 l^2}{12} \\ 0 \\ 0 \\ 0 \\ 0 \\ 0 \\ 0 \end{bmatrix} = \frac{EI_{zz}}{l^3} \begin{bmatrix} 12 & 6l & -12 & 6l & 0 & 0 & 0 & 0 & 0 & 0 \\ 6l & 4l^2 & -6l & 2l^2 & 0 & 0 & 0 & 0 & 0 & 0 \\ -12 & -6l & 12 & -6l & 0 & 0 & 0 & 0 & 0 & 0 \\ 6l & 2l^2 & -6l & 4l^2 & 0 & 0 & 0 & 0 & 0 & 0 \\ 0 & 0 & 0 & 0 & 0 & 0 & 0 & 0 & 0 & 0 \\ 0 & 0 & 0 & 0 & 0 & 0 & 0 & 0 & 0 & 0 \\ 0 & 0 & 0 & 0 & 0 & 0 & 0 & 0 & 0 & 0 \\ 0 & 0 & 0 & 0 & 0 & 0 & 0 & 0 & 0 & 0 \\ 0 & 0 & 0 & 0 & 0 & 0 & 0 & 0 & 0 & 0 \\ 0 & 0 & 0 & 0 & 0 & 0 & 0 & 0 & 0 & 0 \end{bmatrix} \begin{bmatrix} v_1 \\ \theta_1 \\ v_2 \\ \theta_2 \\ v_3 \\ \theta_3 \\ v_4 \\ \theta_4 \\ v_5 \\ \theta_5 \end{bmatrix} \qquad (b)$$

When we add the other three element stiffness matrices and enter the rest of the forces and the restraints we get

$$\begin{bmatrix} R_1 + \dfrac{f_0 l}{2} \\ Q_1 + \dfrac{f_0 l^2}{12} \\ f_0 l \\ 0 \\ f_0 l \\ 0 \\ f_0 l \\ 0 \\ R_5 + \dfrac{f_0 l}{2} \\ Q_5 - \dfrac{f_0 l^2}{12} \end{bmatrix} = \frac{EI_{zz}}{l^3} \begin{bmatrix} 12 & 6l & -12 & 6l & 0 & 0 & 0 & 0 & 0 & 0 \\ 6l & 4l^2 & -6l & 2l^2 & 0 & 0 & 0 & 0 & 0 & 0 \\ -12 & -6l & 24 & 0 & -12 & 6l & 0 & 0 & 0 & 0 \\ 6l & 2l^2 & 0 & 8l^2 & -6l & 2l^2 & 0 & 0 & 0 & 0 \\ 0 & 0 & -12 & -6l & 24 & 0 & -12 & 6l & 0 & 0 \\ 0 & 0 & 6l & 2l^2 & 0 & 8l^2 & -6l & 2l^2 & 0 & 0 \\ 0 & 0 & 0 & 0 & -12 & -6l & 24 & 0 & -12 & 6l \\ 0 & 0 & 0 & 0 & 6l & 2l^2 & 0 & 8l^2 & -6l & 2l^2 \\ 0 & 0 & 0 & 0 & 0 & 0 & -12 & -6l & 12 & -6l \\ 0 & 0 & 0 & 0 & 0 & 0 & 6l & 2l^2 & -6l & 4l^2 \end{bmatrix} \begin{bmatrix} 0 \\ 0 \\ v_2 \\ \theta_2 \\ v_3 \\ \theta_3 \\ v_4 \\ \theta_4 \\ 0 \\ 0 \end{bmatrix} \qquad (c)$$

Notice that the equivalent nodal moments cancel at all interior nodes in this case.

Partition the global matrix equations to find the nodal displacements.

$$
\begin{bmatrix} f_0l \\ 0 \\ f_0l \\ 0 \\ f_0l \\ 0 \end{bmatrix} = \frac{EI_{zz}}{l^3} \begin{bmatrix} 24 & 0 & -12 & 6l & 0 & 0 \\ 0 & 8l^2 & -6l & 2l^2 & 0 & 0 \\ -12 & -6l & 24 & 0 & -12 & 6l \\ 6l & 2l^2 & 0 & 8l^2 & -6l & 2l^2 \\ 0 & 0 & -12 & -6l & 24 & 0 \\ 0 & 0 & 6l & 2l^2 & 0 & 8l^2 \end{bmatrix} \begin{bmatrix} v_2 \\ \theta_2 \\ v_3 \\ \theta_3 \\ v_4 \\ \theta_4 \end{bmatrix}
$$
(d)

The equations for the support reactions are

$$
\begin{bmatrix} R_1 + \dfrac{f_0l}{2} \\ Q_1 + \dfrac{f_0l^2}{12} \\ R_5 + \dfrac{f_0l}{2} \\ Q_5 - \dfrac{f_0l^2}{12} \end{bmatrix} = \frac{EI_{zz}}{l^3} \begin{bmatrix} -12 & 6l & 0 & 0 & 0 & 0 \\ -6l & 2l^2 & 0 & 0 & 0 & 0 \\ 0 & 0 & 0 & 0 & -12 & -6l \\ 0 & 0 & 0 & 0 & 6l & 2l^2 \end{bmatrix} \begin{bmatrix} v_2 \\ \theta_2 \\ v_3 \\ \theta_3 \\ v_4 \\ \theta_4 \end{bmatrix}
$$
(e)

Let us try a numerical solution. Let the cross section be a rectangle 20 *mm* high by 10 *mm* wide. The element length is $l = 200$ *mm* and the material is aluminum with $E = 68950$ *MPa*. The loading is $f_0 = 1$ *N/mm*. Then with

$$
\frac{EI_{zz}}{l^3} = 57.458 \ \frac{N}{mm}
$$
(f)

the equations for the displacement are

$$
\begin{bmatrix} 200 \\ 0 \\ 200 \\ 0 \\ 200 \\ 0 \end{bmatrix} = 57.458 \begin{bmatrix} 24 & 0 & -12 & 1200 & 0 & 0 \\ 0 & 320000 & -1200 & 80000 & 0 & 0 \\ -12 & -1200 & 24 & 0 & -12 & 1200 \\ 1200 & 80000 & 0 & 320000 & -1200 & 80000 \\ 0 & 0 & -12 & -1200 & 24 & 0 \\ 0 & 0 & 1200 & 80000 & 0 & 320000 \end{bmatrix} \begin{bmatrix} v_2 \\ \theta_2 \\ v_3 \\ \theta_3 \\ v_4 \\ \theta_4 \end{bmatrix}
$$
(g)

From a linear solver:

$$
\begin{bmatrix} v_2 \\ \theta_2 \\ v_3 \\ \theta_3 \\ v_4 \\ \theta_4 \end{bmatrix} = \begin{bmatrix} 1.3053 \\ 0.008702 \\ 2.3205 \\ 0 \\ 1.3053 \\ -0.008702 \end{bmatrix} \begin{matrix} mm \\ rad \\ mm \\ rad \\ mm \\ rad \end{matrix}
$$
(h)

The equations for the support restraints are

$$
\begin{bmatrix} R_1 + 100 \\ Q_1 + \dfrac{40000}{12} \\ R_5 + 100 \\ Q_5 - \dfrac{40000}{12} \end{bmatrix} = 57.458 \begin{bmatrix} -12 & 1200 & 0 & 0 & 0 & 0 \\ -1200 & 80000 & 0 & 0 & 0 & 0 \\ 0 & 0 & 0 & 0 & -12 & -1200 \\ 0 & 0 & 0 & 0 & 1200 & 80000 \end{bmatrix} \begin{bmatrix} v_2 \\ \theta_2 \\ v_3 \\ \theta_3 \\ v_4 \\ \theta_4 \end{bmatrix} = \begin{bmatrix} -300 \\ -50000 \\ -300 \\ 50000 \end{bmatrix} \begin{matrix} N \\ N \cdot mm \\ N \\ N \cdot mm \end{matrix}
$$
(i)

Thus the restraint forces are

$$\begin{bmatrix} R_1 \\ Q_1 \\ R_5 \\ Q_5 \end{bmatrix} = \begin{bmatrix} -400 \\ -53333 \\ -400 \\ 53333 \end{bmatrix} \begin{matrix} N \\ N \cdot mm \\ N \\ N \cdot mm \end{matrix} \tag{j}$$

Let us check these values with the analytical solution in Example 7.5.1. From Equation (e), where $L = 800 \ mm$, $E = 68950 \ N/mm^2$ and $I = 10 \cdot (20)^3/12 \ mm^4$.

$$v(x) = \frac{f_0}{24EI_{zz}}(x^4 - 2Lx^3 + L^2 x^2) \tag{k}$$

Evaluated at the nodes

$$\begin{bmatrix} v_2 \\ \theta_2 \\ v_3 \\ \theta_3 \\ v_4 \\ \theta_4 \end{bmatrix} = \begin{bmatrix} 1.305 \\ 0.008702 \\ 2.321 \\ 0 \\ 1.305 \\ -0.008072 \end{bmatrix} \begin{matrix} mm \\ rad \\ mm \\ rad \\ mm \\ rad \end{matrix} \tag{l}$$

Compared with Equation (h) we see that the displacements at the interior nodes have excellent accuracy for such a limited number of elements. Remember that the displacement which is a fourth order polynomial in the analytical solution (Example 7.6.1, Equation (e)) is replaced by segmented cubic polynomials in each element in the approximate solution.

Now let us continue and find the moments and shear forces from which the stresses may be found. Equation 8.2.18 repeated here may be used.

$$\{p_m\} = [k_m]\{r_m\} = \begin{bmatrix} V_{m,n} \\ M_{m,n} \\ V_{m,n+1} \\ M_{m,n+1} \end{bmatrix} = \frac{E_m I_m}{l_m^3} \begin{bmatrix} 12 & 6l_m & -12 & 6l_m \\ 6l_m & 4l_m^2 & -6l_m & 2l_m^2 \\ -12 & -6l_m & 12 & -6l_m \\ 6l_m & 2l_m^2 & -6l_m & 4l_m^2 \end{bmatrix} \begin{bmatrix} v_n \\ \theta_n \\ v_{n+1} \\ \theta_{n+1} \end{bmatrix} \tag{m}$$

Since E_m, I_m, and l_m are the same for all the elements the $[k_m]$ matrix is the same for all elements. For element 1:

$$\begin{bmatrix} V_{1,1} \\ M_{1,1} \\ V_{1,2} \\ M_{1,2} \end{bmatrix} = 57.458 \begin{bmatrix} 12 & 1200 & -12 & 1200 \\ 1200 & 160000 & -1200 & 80000 \\ -12 & -1200 & 12 & -1200 \\ 1200 & 80000 & -1200 & 160000 \end{bmatrix} \begin{bmatrix} 0 \\ 0 \\ 1.3053 \\ 0.008702 \end{bmatrix} = \begin{bmatrix} -300 \\ -50000 \\ 300 \\ -10000 \end{bmatrix} \begin{matrix} N \\ N \cdot mm \\ N \\ N \cdot mm \end{matrix} \tag{n}$$

For element 2:

$$\begin{bmatrix} V_{2,2} \\ M_{2,2} \\ V_{2,3} \\ M_{2,3} \end{bmatrix} = 57.458 \begin{bmatrix} 12 & 1200 & -12 & 1200 \\ 1200 & 160000 & -1200 & 80000 \\ -12 & -1200 & 12 & -1200 \\ 1200 & 80000 & -1200 & 160000 \end{bmatrix} \begin{bmatrix} 1.305 \\ 0.008702 \\ 2.321 \\ 0 \end{bmatrix} = \begin{bmatrix} -100 \\ 10000 \\ 100 \\ -30000 \end{bmatrix} \begin{matrix} N \\ N \cdot mm \\ N \\ N \cdot mm \end{matrix} \tag{o}$$

For element 3:

$$\begin{bmatrix} V_{3,3} \\ M_{3,3} \\ V_{3,4} \\ M_{3,4} \end{bmatrix} = 57.458 \begin{bmatrix} 12 & 1200 & -12 & 1200 \\ 1200 & 160000 & -1200 & 80000 \\ -12 & -1200 & 12 & -1200 \\ 1200 & 80000 & -1200 & 160000 \end{bmatrix} \begin{bmatrix} 2.321 \\ 0 \\ 1.305 \\ -0.008702 \end{bmatrix} = \begin{bmatrix} 100 \\ 30000 \\ -100 \\ -10000 \end{bmatrix} \begin{matrix} N \\ N \cdot mm \\ N \\ N \cdot mm \end{matrix} \tag{p}$$

For element 4:

$$
\begin{bmatrix} V_{4,4} \\ M_{4,4} \\ V_{4,5} \\ M_{4,5} \end{bmatrix} = 57.458 \begin{bmatrix} 12 & 1200 & -12 & 1200 \\ 1200 & 160000 & -1200 & 80000 \\ -12 & -1200 & 12 & -1200 \\ 1200 & 80000 & -1200 & 160000 \end{bmatrix} \begin{bmatrix} 1.305 \\ -0.008702 \\ 0 \\ 0 \end{bmatrix} = \begin{bmatrix} 300 \\ 10000 \\ -300 \\ 50000 \end{bmatrix} \begin{matrix} N \\ N \cdot mm \\ N \\ N \cdot mm \end{matrix} \quad (q)
$$

Remember that, since the shape function in each element is a cubic polynomial and the shear force is the third derivative of the displacements, the shear forces are necessarily constant in each element. These are plotted as dotted horizontal lines in Figure (c) using the values on the right end of each element where the sign convention for the analytic solution agrees with the FEM solution.

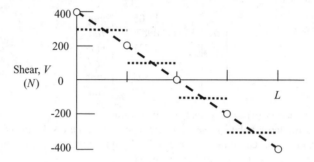

Figure (c)

But we know that the shear force at each restraint is $400\ N$ at the left end and $-400\ N$ at the right end, so let us put a point there at each end. Now we know that in the real case with a distributed load there is no jump in the shear stress at any interior node, so at each interior node average the two shear forces and put a point there. Now draw a line, shown as a dashed line, connecting the points. This is the same shear force diagram as obtained in the analytical solution as noted in Equations (r) and (s).

From the analytical equations the shear forces at the node locations are

$$
\begin{bmatrix} V_1 \\ V_2 \\ V_3 \\ V_4 \\ V_5 \end{bmatrix} = \begin{bmatrix} 400 \\ 200 \\ 0 \\ -200 \\ -400 \end{bmatrix} N \quad (r)
$$

From the FE equations adjusted by averaging at the nodes and applying support reactions the shear forces are

$$
\begin{bmatrix} V_1 = -R_1 \\ V_2 \\ V_3 \\ V_4 \\ V_5 = R_5 \end{bmatrix} = \begin{bmatrix} 400 \\ 200 \\ 0 \\ -200 \\ -400 \end{bmatrix} N \quad (s)
$$

Since the shape functions are cubic polynomials the bending moments, which are the second derivatives of the displacement are necessarily linear polynomials. These also require some interpretation when compared with the analytical solution. In Equation (t) we list the moment at each node as follows

$$\begin{bmatrix} M_1 \\ M_2 \\ M_3 \\ M_4 \\ M_5 \end{bmatrix} = \begin{bmatrix} -M_{1,1} \\ M_{1,2} \\ M_{2,3} \\ M_{3,4} \\ M_{4,5} \end{bmatrix} = \begin{bmatrix} 50000 \\ -10000 \\ -30000 \\ -10000 \\ 50000 \end{bmatrix} N \cdot mm \tag{t}$$

At the left end the sign convention for M_{11} is the opposite of that for the analytical solution and the sign must be reversed. All the others are calculated at the right end of the element and have the same sign convention as the analytical solutions. These moments are plotted as dotted lines in Figure (d).

From the analytical solution the moments at the nodes are given in Equation (u). These are plotted with a smooth curve passing through the points in Figure (d)

$$\begin{bmatrix} M_1 \\ M_2 \\ M_3 \\ M_4 \\ M_5 \end{bmatrix} = \begin{bmatrix} 53333 \\ -6667 \\ -26667 \\ -6667 \\ 53333 \end{bmatrix} N \cdot mm \tag{u}$$

What accounts for the differences? In finding the moments from Equation (n) the moments on the elements generated by Equation (a) are not taken into account. These moments are

$$\frac{f_0 l^2}{12} = \frac{40000}{12} = 3333 \ N \cdot mm \tag{v}$$

If these are added to the values for the moments in Equations (n) through (q) the nodal values from the approximate solution are identical to the analytical values. The dashed lines in Figure (d) then move up to coincide with the solid curve at the nodes. FEM codes are programmed to make these adjustments.

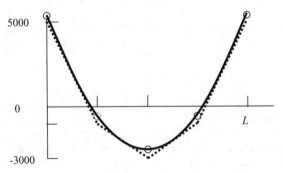

Figure (d)

We see that the values at the nodes of displacement, shear forces, and moments are remarkably accurate even when a small number of nodes are used. For values between nodes linear interpolation is usually quite satisfactory.

The maximum normal stress is at $x = 0$ and L, and at $y = \pm\frac{h}{2}$.

$$\sigma_x = -\frac{M_z y}{I_{zz}} \rightarrow \sigma_{max} = \pm\frac{53333 \cdot 10}{6667} = \pm 80 \ MPa \tag{w}$$

The maximum shear stress is at $x = 0$ and L, and at $y = 0$.

$$\tau_{xy}(x, y) = \frac{V}{2I_{zz}} \left(\frac{h^2}{4} - y^2 \right) \rightarrow \tau_{max} = \pm \frac{400}{2.6667} (10)^2 = \pm 3 \, MPa \qquad (x)$$

##########

Now let us do a problem with enough complexity to show off the advantage in using finite elements.

##########

Example 8.4.2

Problem: Find the stresses and displacements for a beam that is restrained and loaded as shown. Elements and nodes are numbered as shown in Figure (a).

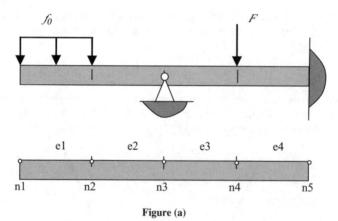

Figure (a)

Solution: Use the FE method.

The loads and restraints were conveniently located so that we can divide the beam into four elements of equal length. Let us say that the length of each element is $l_m = l$ for a total length of $4l$.
The equivalent nodal loads for the first element are

$$\{F_{e1}\} = -\frac{f_0}{l^3} \int_0^l \begin{bmatrix} l^3 - 3ls_1^2 + 2s_1^3 \\ l^3 s - 2l^2 s_1^2 + l s_1^3 \\ 3ls_1^2 - 2s_1^3 \\ -l^2 s_1^2 + l s_1^3 \end{bmatrix} ds_1 = \begin{bmatrix} -\dfrac{f_0 l}{2} \\ -\dfrac{f_0 l^2}{12} \\ -\dfrac{f_0 l}{2} \\ \dfrac{f_0 l^2}{12} \end{bmatrix} \qquad (a)$$

When we imbed just the first element stiffness matrix into the global stiffness matrix we get

$$
\begin{bmatrix}
-\dfrac{f_0 l}{2} \\[4pt]
-\dfrac{f_0 l^2}{12} \\[4pt]
-\dfrac{f_0 l}{2} \\[4pt]
\dfrac{f_0 l^2}{12} \\[4pt]
0 \\ 0 \\ 0 \\ 0 \\ 0 \\ 0
\end{bmatrix}
=
\frac{EI_{zz}}{l^3}
\begin{bmatrix}
12 & 6l & -12 & 6l & 0 & 0 & 0 & 0 & 0 & 0 \\
6l & 4l^2 & -6l & 2l^2 & 0 & 0 & 0 & 0 & 0 & 0 \\
-12 & -6l & 12 & -6l & 0 & 0 & 0 & 0 & 0 & 0 \\
6l & 2l^2 & -6l & 4l^2 & 0 & 0 & 0 & 0 & 0 & 0 \\
0 & 0 & 0 & 0 & 0 & 0 & 0 & 0 & 0 & 0 \\
0 & 0 & 0 & 0 & 0 & 0 & 0 & 0 & 0 & 0 \\
0 & 0 & 0 & 0 & 0 & 0 & 0 & 0 & 0 & 0 \\
0 & 0 & 0 & 0 & 0 & 0 & 0 & 0 & 0 & 0 \\
0 & 0 & 0 & 0 & 0 & 0 & 0 & 0 & 0 & 0 \\
0 & 0 & 0 & 0 & 0 & 0 & 0 & 0 & 0 & 0
\end{bmatrix}
\begin{bmatrix}
v_1 \\ \theta_1 \\ v_2 \\ \theta_2 \\ v_3 \\ \theta_3 \\ v_4 \\ \theta_4 \\ v_5 \\ \theta_5
\end{bmatrix}
\qquad\text{(b)}
$$

When we add the other three element stiffness matrices and enter the rest of the forces and the restraints we get

$$
\begin{bmatrix}
-\dfrac{f_0 l}{2} \\[4pt]
-\dfrac{f_0 l^2}{12} \\[4pt]
-\dfrac{f_0 l}{2} \\[4pt]
\dfrac{f_0 l^2}{12} \\[4pt]
R_3 \\[2pt]
0 \\
-F \\
0 \\
R_5 \\
Q_5
\end{bmatrix}
=
\frac{EI}{l^3}
\begin{bmatrix}
12 & 6l & -12 & 6l & 0 & 0 & 0 & 0 & 0 & 0 \\
6l & 4l^2 & -6l & 2l^2 & 0 & 0 & 0 & 0 & 0 & 0 \\
-12 & -6l & 24 & 0 & -12 & 6l & 0 & 0 & 0 & 0 \\
6l & 2l^2 & 0 & 8l^2 & -6l & 2l^2 & 0 & 0 & 0 & 0 \\
0 & 0 & -12 & -6l & 24 & 0 & -12 & 6l & 0 & 0 \\
0 & 0 & 6l & 2l^2 & 0 & 8l^2 & -6l & 2l^2 & 0 & 0 \\
0 & 0 & 0 & 0 & -12 & -6l & 24 & 0 & -12 & 6l \\
0 & 0 & 0 & 0 & 6l & 2l^2 & 0 & 8l^2 & -6l & 2l^2 \\
0 & 0 & 0 & 0 & 0 & 0 & -12 & -6l & 12 & -6l \\
0 & 0 & 0 & 0 & 0 & 0 & 6l & 2l^2 & -6l & 4l^2
\end{bmatrix}
\begin{bmatrix}
v_1 \\ \theta_1 \\ v_2 \\ \theta_2 \\ 0 \\ \theta_3 \\ v_4 \\ \theta_4 \\ 0 \\ 0
\end{bmatrix}
\qquad\text{(c)}
$$

Note that the global stiffness matrix is exactly the same as in Example 8.4.1. This is one of the strengths of the method. The stiffness matrix is independent of the loads and the restraints: thus, a variety of problems can be solved using the same global stiffness matrix as we shall demonstrate.

This can be partitioned – first, to find the nodal displacements.

$$
\begin{bmatrix}
-\dfrac{f_0 l}{2} \\[4pt]
-\dfrac{f_0 l^2}{12} \\[4pt]
-\dfrac{f_0 l}{2} \\[4pt]
\dfrac{f_0 l^2}{12} \\[4pt]
0 \\
-F \\
0
\end{bmatrix}
=
\frac{EI_{zz}}{l^3}
\begin{bmatrix}
12 & 6l & -12 & 6l & 0 & 0 & 0 \\
6l & 4l^2 & -6l & 2l^2 & 0 & 0 & 0 \\
-12 & -6l & 24 & 0 & 6l & 0 & 0 \\
6l & 2l^2 & 0 & 8l^2 & 2l^2 & 0 & 0 \\
0 & 0 & 6l & 2l^2 & 8l^2 & -6l & 2l^2 \\
0 & 0 & 0 & 0 & -6l & 24 & 0 \\
0 & 0 & 0 & 0 & 2l^2 & 0 & 8l^2
\end{bmatrix}
\begin{bmatrix}
v_1 \\ \theta_1 \\ v_2 \\ \theta_2 \\ \theta_3 \\ v_4 \\ \theta_4
\end{bmatrix}
\qquad\text{(d)}
$$

And then find the support reactions.

$$
\begin{bmatrix} R_3 \\ R_5 \\ Q_5 \end{bmatrix} = \frac{EI_{zz}}{l^3} \begin{bmatrix} 0 & 0 & -12 & -6l & 0 & -12 & 6l \\ 0 & 0 & 0 & 0 & 0 & -12 & -6l \\ 0 & 0 & 0 & 0 & 0 & 6l & 2l^2 \end{bmatrix} \begin{bmatrix} v_1 \\ \theta_1 \\ v_2 \\ \theta_2 \\ \theta_3 \\ v_4 \\ \theta_4 \end{bmatrix}
\tag{e}
$$

Let us try a numerical solution. Let the cross section be a rectangle 20 mm high by 10 mm width. The length is $l = 200$ mm and the material is aluminum with $E = 68950$ MPa. The loading is $f_0 = -1.2$ N/mm and $F = -150$ N. Then the global stiffness matrix is

$$
[K] = 57.458 \begin{bmatrix}
12 & 1200 & -12 & 1200 & 0 & 0 & 0 & 0 & 0 & 0 \\
1200 & 160000 & -1200 & 80000 & 0 & 0 & 0 & 0 & 0 & 0 \\
-12 & -1200 & 24 & 0 & -12 & 1200 & 0 & 0 & 0 & 0 \\
1200 & 80000 & 0 & 320000 & -1200 & 80000 & 0 & 0 & 0 & 0 \\
0 & 0 & -12 & -1200 & 24 & 0 & -12 & 1200 & 0 & 0 \\
0 & 0 & 1200 & 80000 & 0 & 320000 & -1200 & 80000 & 0 & 0 \\
0 & 0 & 0 & 0 & -12 & -1200 & 24 & 0 & -12 & 1200 \\
0 & 0 & 0 & 0 & 1200 & 80000 & 0 & 320000 & -1200 & 80000 \\
0 & 0 & 0 & 0 & 0 & 0 & -12 & -1200 & 12 & -1200 \\
0 & 0 & 0 & 0 & 0 & 0 & 1200 & 80000 & -1200 & 160000
\end{bmatrix}
\tag{c}
$$

The nodal forces are

$$
\{F\} = \{F_c\} + \{F_{e1}\} = \begin{bmatrix} 0 \\ 0 \\ 0 \\ 0 \\ 0 \\ 0 \\ -150 \\ 0 \\ 0 \\ 0 \end{bmatrix} + \begin{bmatrix} -120 \\ -4000 \\ -120 \\ 4000 \\ 0 \\ 0 \\ 0 \\ 0 \\ 0 \\ 0 \end{bmatrix} = \begin{bmatrix} -120 \\ -4000 \\ -120 \\ 4000 \\ 0 \\ 0 \\ -150 \\ 0 \\ 0 \\ 0 \end{bmatrix} N
\tag{d}
$$

The displacement restraints are

$$
v_3 = v_5 = \theta_5 = 0
\tag{e}
$$

The complete global equations with applied forces, restraint forces, and restraints are

$$
\begin{bmatrix} -120 \\ -4000 \\ -120 \\ 4000 \\ \hline R_3 \\ \hline 0 \\ -150 \\ 0 \\ \hline R_5 \\ Q_5 \end{bmatrix} = 57.458 \begin{bmatrix}
12 & 1200 & -12 & 1200 & 0 & 0 & 0 & 0 & 0 & 0 \\
1200 & 160000 & -1200 & 80000 & 0 & 0 & 0 & 0 & 0 & 0 \\
-12 & -1200 & 24 & 0 & -12 & 1200 & 0 & 0 & 0 & 0 \\
1200 & 80000 & 0 & 320000 & -1200 & 80000 & 0 & 0 & 0 & 0 \\
0 & 0 & -12 & -1200 & 24 & 0 & -12 & 1200 & 0 & 0 \\
0 & 0 & 1200 & 80000 & 0 & 320000 & -1200 & 80000 & 0 & 0 \\
0 & 0 & 0 & 0 & -12 & -1200 & 24 & 0 & -12 & 1200 \\
0 & 0 & 0 & 0 & 1200 & 80000 & 0 & 320000 & -1200 & 80000 \\
0 & 0 & 0 & 0 & 0 & 0 & -12 & -1200 & 12 & -1200 \\
0 & 0 & 0 & 0 & 0 & 0 & 1200 & 80000 & -1200 & 160000
\end{bmatrix} \begin{bmatrix} v_1 \\ \theta_1 \\ v_2 \\ \theta_2 \\ 0 \\ \theta_3 \\ v_4 \\ \theta_4 \\ 0 \\ 0 \end{bmatrix}
\tag{f}
$$

This can be partitioned – first, to find the nodal displacements.

$$
\begin{bmatrix} -120 \\ -4000 \\ -120 \\ 4000 \\ 0 \\ -150 \\ 0 \end{bmatrix} = 57.458 \begin{bmatrix} 12 & 1200 & -12 & 1200 & 0 & 0 & 0 \\ 1200 & 160000 & -1200 & 80000 & 0 & 0 & 0 \\ -12 & -1200 & 24 & 0 & 1200 & 0 & 0 \\ 1200 & 80000 & 0 & 320000 & 80000 & 0 & 0 \\ 0 & 0 & 1200 & 80000 & 320000 & -1200 & 80000 \\ 0 & 0 & 0 & 0 & -1200 & 24 & 0 \\ 0 & 0 & 0 & 0 & 80000 & 0 & 320000 \end{bmatrix} \begin{bmatrix} v_1 \\ \theta_1 \\ v_2 \\ \theta_2 \\ \theta_3 \\ v_4 \\ \theta_4 \end{bmatrix} \quad (g)
$$

Put this in one of the linear solvers and you get

$$
\begin{bmatrix} v_1 \\ \theta_1 \\ v_2 \\ \theta_2 \\ \theta_3 \\ v_4 \\ \theta_4 \end{bmatrix} = \begin{bmatrix} -12.748442 \\ 0.038398 \\ -5.24296 \\ 0.034917 \\ 0.014032 \\ 0.592824 \\ -0.003507997 \end{bmatrix} \begin{matrix} mm \\ rad \\ mm \\ rad \\ rad \\ mm \\ rad \end{matrix} \quad (h)
$$

The support reactions are found from

$$
\begin{bmatrix} R_3 \\ R_5 \\ Q_5 \end{bmatrix} = 57.458 \begin{bmatrix} 0 & 0 & -12 & -1200 & 0 & -12 & 1200 \\ 0 & 0 & 0 & 0 & 0 & -12 & -1200 \\ 0 & 0 & 0 & 0 & 0 & 1200 & 80000 \end{bmatrix} \begin{bmatrix} v_1 \\ \theta_1 \\ v_2 \\ \theta_2 \\ \theta_3 \\ v_4 \\ \theta_4 \end{bmatrix} = \begin{bmatrix} 556.875 \\ -166.875 \\ 24750 \end{bmatrix} \begin{matrix} N \\ N \\ N \cdot mm \end{matrix}
$$

$$(i)$$

Now that you know the support reactions and the applied loads you can find the bending moment and shear force diagrams by traditional methods, that is, by static analysis. Or you can find them by using Equations 8.2.18.

If you call within the next 5 minutes you can try another loading at virtually no increase in cost. Your equation solver is all fired up and ready to go. Just put in a new load matrix and solve again using the same stiffness matrix.

As an example let us extend the distributed load over the first two elements as shown in Figure (b).

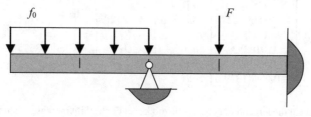

Figure (b)

The new load matrix is

$$\{F\} = \{F_c\} + \{F_{e1}\} + \{F_{e2}\} = \begin{bmatrix} 0 \\ 0 \\ 0 \\ 0 \\ 0 \\ 0 \\ -150 \\ 0 \\ 0 \\ 0 \end{bmatrix} + \begin{bmatrix} -120 \\ -4000 \\ -120 \\ 4000 \\ 0 \\ 0 \\ 0 \\ 0 \\ 0 \\ 0 \end{bmatrix} + \begin{bmatrix} 0 \\ 0 \\ -120 \\ -4000 \\ -120 \\ 4000 \\ 0 \\ 0 \\ 0 \\ 0 \end{bmatrix} = \begin{bmatrix} -120 \\ -4000 \\ -240 \\ 0 \\ -120 \\ 4000 \\ -150 \\ 0 \\ 0 \\ 0 \end{bmatrix} \tag{j}$$

The global equations are now

$$\begin{bmatrix} -120 \\ -4000 \\ -240 \\ 0 \\ R_3 - 120 \\ 4000 \\ -150 \\ 0 \\ R_5 \\ Q_5 \end{bmatrix} = 57.458 \begin{bmatrix} 12 & 1200 & -12 & 1200 & 0 & 0 & 0 & 0 & 0 & 0 \\ 1200 & 160000 & -1200 & 80000 & 0 & 0 & 0 & 0 & 0 & 0 \\ -12 & -1200 & 24 & 0 & -12 & 1200 & 0 & 0 & 0 & 0 \\ 1200 & 80000 & 0 & 320000 & -1200 & 80000 & 0 & 0 & 0 & 0 \\ 0 & 0 & -12 & -1200 & 24 & 0 & -12 & 1200 & 0 & 0 \\ 0 & 0 & 1200 & 80000 & 0 & 320000 & -1200 & 80000 & 0 & 0 \\ 0 & 0 & 0 & 0 & -12 & -1200 & 24 & 0 & -12 & 1200 \\ 0 & 0 & 0 & 0 & 1200 & 80000 & 0 & 320000 & -1200 & 80000 \\ 0 & 0 & 0 & 0 & 0 & 0 & -12 & -1200 & 12 & -1200 \\ 0 & 0 & 0 & 0 & 0 & 0 & 1200 & 80000 & -1200 & 160000 \end{bmatrix} \begin{bmatrix} v_1 \\ \theta_1 \\ v_2 \\ \theta_2 \\ 0 \\ \theta_3 \\ v_4 \\ \theta_4 \\ 0 \\ 0 \end{bmatrix} \tag{k}$$

Note that the equivalent nodal load at node 3 must be added to the restraint there. The partitioned equations for the displacement are

$$\begin{bmatrix} -120 \\ -4000 \\ -240 \\ 0 \\ 4000 \\ -150 \\ 0 \end{bmatrix} = 57.458 \begin{bmatrix} 12 & 1200 & -12 & 1200 & 0 & 0 & 0 \\ 1200 & 160000 & -1200 & 80000 & 0 & 0 & 0 \\ -12 & -1200 & 24 & 0 & 1200 & 0 & 0 \\ 1200 & 80000 & 0 & 320000 & 80000 & 0 & 0 \\ 0 & 0 & 1200 & 80000 & 320000 & -1200 & 80000 \\ 0 & 0 & 0 & 0 & -1200 & 24 & 0 \\ 0 & 0 & 0 & 0 & 80000 & 0 & 320000 \end{bmatrix} \begin{bmatrix} v_1 \\ \theta_1 \\ v_2 \\ \theta_2 \\ \theta_3 \\ v_4 \\ \theta_4 \end{bmatrix} \tag{l}$$

Run the solver again.

$$\begin{bmatrix} v_1 \\ \theta_1 \\ v_2 \\ \theta_2 \\ \theta_3 \\ v_4 \\ \theta_4 \end{bmatrix} = \begin{bmatrix} -16.055206 \\ 0.0471 \\ -6.809322 \\ 0.043619 \\ 0.019253 \\ 0.853885 \\ -0.004813298 \end{bmatrix} \begin{matrix} mm \\ rad \\ mm \\ rad \\ rad \\ mm \\ rad \end{matrix} \qquad \begin{bmatrix} R_3 - 120 \\ R_5 \\ Q_5 \end{bmatrix} = \begin{bmatrix} 766.875 \\ -256.875 \\ 0.0003675 \end{bmatrix} \begin{matrix} N \\ N \\ N \cdot mm \end{matrix} \tag{m}$$

That displacement at the left end may be a little too large. For a slight increase in shipping and handling you can put a restraint there. If we keep the distributed load on the first two elements we have the loads and restraints shown here in Figure (c).

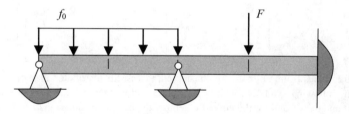

Figure (c)

The global equations are now

$$
\begin{bmatrix} R_1 - 120 \\ -4000 \\ -240 \\ 0 \\ R_3 - 120 \\ 4000 \\ -150 \\ 0 \\ R_5 \\ Q_5 \end{bmatrix} = 57.458
\begin{bmatrix}
12 & 1200 & -12 & 1200 & 0 & 0 & 0 & 0 & 0 & 0 \\
1200 & 160000 & -1200 & 80000 & 0 & 0 & 0 & 0 & 0 & 0 \\
-12 & -1200 & 24 & 0 & -12 & 1200 & 0 & 0 & 0 & 0 \\
1200 & 80000 & 0 & 320000 & -1200 & 80000 & 0 & 0 & 0 & 0 \\
0 & 0 & -12 & -1200 & 24 & 0 & -12 & 1200 & 0 & 0 \\
0 & 0 & 1200 & 80000 & 0 & 320000 & -1200 & 80000 & 0 & 0 \\
0 & 0 & 0 & 0 & -12 & -1200 & 24 & 0 & -12 & 1200 \\
0 & 0 & 0 & 0 & 1200 & 80000 & 0 & 320000 & -1200 & 80000 \\
0 & 0 & 0 & 0 & 0 & 0 & -12 & -1200 & 12 & -1200 \\
0 & 0 & 0 & 0 & 0 & 0 & 1200 & 80000 & -1200 & 160000
\end{bmatrix}
\begin{bmatrix} 0 \\ \theta_1 \\ v_2 \\ \theta_2 \\ 0 \\ \theta_3 \\ v_4 \\ \theta_4 \\ 0 \\ 0 \end{bmatrix}
$$

(n)

From the previous example you must take out the first row of all matrices and also the first column of the stiffness matrix. The partitioned equations for the displacements are

$$
\begin{bmatrix} -4000 \\ -240 \\ 0 \\ 4000 \\ -150 \\ 0 \end{bmatrix} = 57.458
\begin{bmatrix}
160000 & -1200 & 80000 & 0 & 0 & 0 \\
-1200 & 24 & 0 & 1200 & 0 & 0 \\
80000 & 0 & 320000 & 80000 & 0 & 0 \\
0 & 1200 & 80000 & 320000 & -1200 & 80000 \\
0 & 0 & 0 & -1200 & 24 & 0 \\
0 & 0 & 0 & 80000 & 0 & 320000
\end{bmatrix}
\begin{bmatrix} \theta_1 \\ v_2 \\ \theta_2 \\ \theta_3 \\ v_4 \\ \theta_4 \end{bmatrix}
$$

(o)

Solving, the displacements are

$$
\begin{bmatrix} \theta_1 \\ v_2 \\ \theta_2 \\ \theta_3 \\ v_4 \\ \theta_4 \end{bmatrix} =
\begin{bmatrix} -0.004506397 \\ -0.501919 \\ 0.0006138024 \\ 0.002051188 \\ -0.00621572 \\ -0.0005127969 \end{bmatrix}
\begin{matrix} rad \\ mm \\ rad \\ rad \\ mm \\ rad \end{matrix}
$$

(p)

The support reactions are

$$
\begin{bmatrix} R_1 - 120 \\ R_3 - 120 \\ R_5 \\ Q_5 \end{bmatrix} = 57.458
\begin{bmatrix}
1200 & -12 & 1200 & 0 & 0 & 0 \\
0 & -12 & -1200 & 0 & -12 & 1200 \\
0 & 0 & 0 & 0 & -12 & -1200 \\
0 & 0 & 0 & 0 & 1200 & 80000
\end{bmatrix}
\begin{bmatrix} -0.0045064 \\ -0.50192 \\ 0.0006138 \\ 0.0020512 \\ -0.0062157 \\ -0.0005128 \end{bmatrix}
=
\begin{bmatrix} 77.6786 \\ 272.6786 \\ 39.64286 \\ -2785, 71 \end{bmatrix}
\begin{matrix} N \\ N \\ N \cdot mm \\ N \cdot mm \end{matrix}
$$

(q)

Given the displacements we find the moments and shear forces using Equation 8.2.18, repeated here.

$$\{p_m\} = [k_m]\{r_m\} = \begin{bmatrix} V_{m,n} \\ M_{m,n} \\ V_{m,n+1} \\ M_{m,n+1} \end{bmatrix} = \frac{E_m I_m}{l_m^3} \begin{bmatrix} 12 & 6l_m & -12 & 6l_m \\ 6l_m & 4l_m^2 & -6l_m & 2l_m^2 \\ -12 & -6l_m & 12 & -6l_m \\ 6l_m & 2l_m^2 & -6l_m & 4l_m^2 \end{bmatrix} \begin{bmatrix} v_n \\ \theta_n \\ v_{n+1} \\ \theta_{n+1} \end{bmatrix} \quad (r)$$

Let us use the values from the first case considered, that is, the distributed load on the first element. Note that in the following symbols for the shear forces and moment we use two subscripts – the first for the element and the second for the node. Remember there are two values of shear force and bending moment at each interior node – one for each element at that node.

$$\begin{bmatrix} V_{1,1} \\ M_{1,1} \\ V_{1,2} \\ M_{1,2} \end{bmatrix} = 57.458 \begin{bmatrix} 12 & 1200 & -12 & 1200 \\ 1200 & 160000 & -1200 & 80000 \\ -12 & -1200 & 12 & -1200 \\ 1200 & 80000 & -1200 & 160000 \end{bmatrix} \begin{bmatrix} -12.748442 \\ 0.038398 \\ -5.24296 \\ 0.034917 \end{bmatrix} = \begin{bmatrix} -119.96 \\ -3995.537 \\ 119.96 \\ -19996.44 \end{bmatrix} \begin{matrix} N \\ N \cdot mm \\ N \\ N \cdot mm \end{matrix}$$

$$(s)$$

We are picking up a little calculation round off error here. The exact answers based on this model would be

$$\begin{bmatrix} V_1 \\ M_1 \\ V_2 \\ M_2 \end{bmatrix} = \begin{bmatrix} -120 \\ -4000 \\ 120 \\ -20000 \end{bmatrix} \begin{matrix} N \\ N \cdot mm \\ N \\ N \cdot mm \end{matrix} \quad (t)$$

Using the same matrix equation adapted to the second, third, and fourth elements we get

$$\begin{bmatrix} V_{2,2} \\ M_{2,2} \\ V_{2,3} \\ M_{2,3} \end{bmatrix} = \begin{bmatrix} -239.986 \\ 24001.82 \\ 239.986 \\ -71999.01 \end{bmatrix} \doteq \begin{bmatrix} -240 \\ 24000 \\ 240 \\ -72000 \end{bmatrix} \begin{matrix} N \\ N \cdot mm \\ N \\ N \cdot mm \end{matrix} \quad (u)$$

$$\begin{bmatrix} V_{3,3} \\ M_{3,3} \\ V_{3,4} \\ M_{3,4} \end{bmatrix} = \begin{bmatrix} 316.876 \\ 72000.13 \\ -316.876 \\ 8624.979 \end{bmatrix} \doteq \begin{bmatrix} 317 \\ 72000 \\ -317 \\ 8625 \end{bmatrix} \begin{matrix} N \\ N \cdot mm \\ N \\ N \cdot mm \end{matrix} \quad (v)$$

$$\begin{bmatrix} V_{4,4} \\ M_{4,4} \\ V_{4,5} \\ M_{4,5} \end{bmatrix} = \begin{bmatrix} 166.875 \\ 8624.979 \\ -166.875 \\ 24749.98 \end{bmatrix} \doteq \begin{bmatrix} 167 \\ 8625 \\ -167 \\ 24750 \end{bmatrix} \begin{matrix} N \\ N \cdot mm \\ N \\ N \cdot mm \end{matrix} \quad (w)$$

Now let us check these out. Compare the values at the nodes with the support reactions. You see that they agree, that is, the sum of the shear force at node 3 from elements 2 and 3 is equal to the restraint force. Also the shear and moment at node five from the element agree with the restraint values.

The maximum moment is at node 3 and so the maximum normal stress is

$$\sigma_{x,\max} = -\frac{M y_{\max}}{I_{zz}} = \pm \frac{72000 \cdot 10 \cdot 12}{80000} = \pm 108 \, MPa \quad (x)$$

and the maximum shear force is in element 3 and so the maximum shear stress is

$$\tau_{\max} = \frac{V}{I} \frac{h^2}{8} = -\frac{317 \cdot 100 \cdot 12}{80000 \cdot 2} = -2.3775 \, MPa \quad (y)$$

This may look like quite a large amount of work. Actually, in the equation solving software you have to enter only a few matrices. Many of these are used several times, so, it really is not that bad. Try solving this problem by dividing the beam into segments, solving the differential equation in each segment, and then satisfying the boundary conditions. You will soon become a convert to this method.

We should note that this method, when it requires replacing distributed loads with equivalent nodal loads, does introduce a measure of approximation into the solution. For example, we appear to have a non zero shear force at node one when, in fact, the shear force is zero there. This requires careful interpretation of the results as we have noted in Example 8.2.2. In practice this often is handled by using a large number of elements, thereby approaching the exact answer at each node with less interpretation needed.

Remember that there are FEM computer codes that do most of this work for you. We just wanted you to see what is going on inside those codes.

##########

8.5 Variable Cross Sections

It is common practice with tapered beams to break them down into a set of elements with uniform cross sections similarly to what we did in Chapter 5, Section 5.5, for the axially loaded case. An example will help to explain.

##########

Example 8.5.1

Problem: A beam has a rectangular cross section with a linear taper in width as shown in Figure (a) and a uniform height h. It has a uniform load $f_y(x) = f_0$ as shown in Figure (b). Find the displacement and stresses.

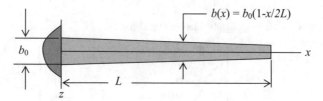

Figure (a)

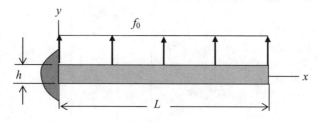

Figure (b)

Solution: Model the beam with two elements of constant cross section each. Assemble the global stiffness matrix, find the equivalent nodal loads, assign boundary restraints, partition the equations, and solve.

First we shall show the bar divided into only two elements, as in Figure (c). In reality more elements should be used to more closely approximate the exact answer but this will be sufficient to demonstrate the process. The value of I_{zz} at $x = L/4$ will be assigned to element 1 and that at $x = 3L/4$ to element 2.

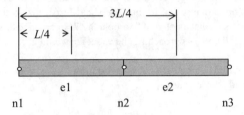

Figure (c)

The moments of inertia are based on the cross sections at $x=L/4$ and $x=3L/4$.

$$b\left(\frac{L}{4}\right) = b_0\left(1 - \frac{L}{4 \cdot 2L}\right) = \frac{7}{8}b_0 \quad b\left(\frac{3L}{4}\right) = b_0\left(1 - \frac{3L}{4 \cdot 2L}\right) = \frac{5}{8}b_0 \qquad (a)$$

The element moments of inertia are then

$$I_1 = \frac{7}{8}I_0 \quad I_2 = \frac{5}{8}I_0 \qquad (b)$$

The stiffness matrix for element 1 imbedded in global format is

$$[k_1] = \frac{EI_0}{L^3}\begin{bmatrix} 84 & 21L & -84 & 21L & 0 & 0 \\ 21L & 7L^2 & -21L & 3.5L^2 & 0 & 0 \\ -84 & -21L & 84 & -21L & 0 & 0 \\ 21L & 3.5L^2 & -21L & 7L^2 & 0 & 0 \\ 0 & 0 & 0 & 0 & 0 & 0 \\ 0 & 0 & 0 & 0 & 0 & 0 \end{bmatrix} \qquad (c)$$

The stiffness matrix for element 2 imbedded in global format is

$$[k_2] = \frac{EI_0}{L^3}\begin{bmatrix} 0 & 0 & 0 & 0 & 0 & 0 \\ 0 & 0 & 0 & 0 & 0 & 0 \\ 0 & 0 & 60 & 15L & -60 & 15L \\ 0 & 0 & 15L & 5L^2 & -15L & 2.5L^2 \\ 0 & 0 & -60 & -15L & 60 & -15L \\ 0 & 0 & 15L & 2.5L^2 & -15L & 5L^2 \end{bmatrix} \qquad (d)$$

The equivalent nodal loads for each element are

$$\{F_{em}\} = \frac{f_0}{l^3}\int_0^l \begin{bmatrix} l^3 - 3ls^2 + 2s^3 \\ l^3 s - 2l^2 s^2 + ls^3 \\ 3ls^2 - 2s^3 \\ -l^2 s^2 + ls^3 \end{bmatrix} ds = \begin{bmatrix} \dfrac{f_0 l}{2} \\ \dfrac{f_0 l^2}{12} \\ \dfrac{f_0 l}{2} \\ -\dfrac{f_0 l^2}{12} \end{bmatrix} = \begin{bmatrix} \dfrac{f_0 L}{4} \\ \dfrac{f_0 L^2}{48} \\ \dfrac{f_0 L}{4} \\ -\dfrac{f_0 L^2}{48} \end{bmatrix} \qquad (e)$$

The assembled global equations including restraints are

$$
\begin{bmatrix}
R_1 + \dfrac{f_0 L}{4} \\[2mm]
Q_1 + \dfrac{f_0 L^2}{48} \\[2mm]
\hdashline
\dfrac{f_0 L}{4} + \dfrac{f_0 L}{4} \\[2mm]
-\dfrac{f_0 L^2}{48} + \dfrac{f_0 L^2}{48} \\[2mm]
\dfrac{f_0 L}{4} \\[2mm]
-\dfrac{f_0 L^2}{48}
\end{bmatrix}
= \frac{EI_0}{L^3}
\begin{bmatrix}
84 & 21L & 84 & 21L & 0 & 0 \\
21L & 7L^2 & -21L & 3.5L^2 & 0 & 0 \\
\hdashline
-84 & -21L & 84+60 & -21L+15L & -60 & 15L \\
21L & 3.5L^2 & -21L+15L & 7L^2+5L^2 & -15L & 2.5L^2 \\
0 & 0 & -60 & -15L & 60 & -15L \\
0 & 0 & 15L & 2.5L^2 & -15L & 5L^2
\end{bmatrix}
\begin{bmatrix}
0 \\
0 \\
\hdashline
v_2 \\
\theta_2 \\
v_3 \\
\theta_3
\end{bmatrix}
$$

(f)

Partition and solve as we have been doing.

##########

As we noted before, a cantilever beam tapered in depth is suggested since the bending moment varies linearly from a maximum at the fixed end to zero at the free end. A tapered beam with half the height at the free end and constant width was considered as shown in Figure 8.5.1

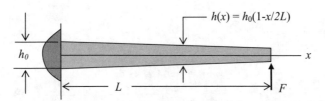

$h(x) = h_0(1 - x/2L)$

Figure 8.5.1

This, of course, removes some material compared to a uniform cross section and thus reduces the weight. The shear force is constant; however, the shear stress is highest at the free end and decreases as the cross section moment of inertia increases toward the fixed end. This has been solved with a numerical FE code and in Figure 8.5.2 we show the stress contours for the bending stress, σ_x.

The contour plots of shear stress in Figure 8.5.3 show more variation.

This shows a more efficient use of the material as far as bending goes.

It has been noted that while the shear force is constant the shear stress is much higher at the free end than at the fixed end because the moment of inertia is smaller. In fact, the peak value at $y = 0$ is double the value on the tapered beam compared to the uniform beam. Although the contour plots do not make this clear the distribution is nearly parabolic all along the length.

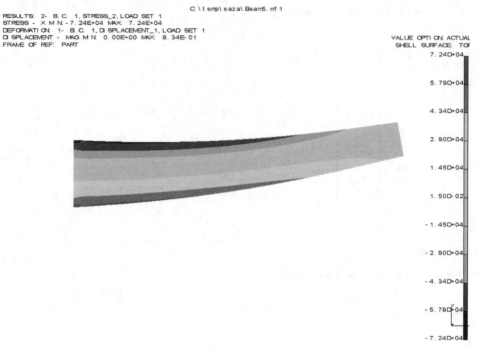

C:\ t emp\ saza\ Beam5. nf 1

RESULTS: 2- B. C. 1, STRESS_2, LOAD SET 1
STRESS - X M N: - 7. 24E+04 MAX: 7. 24E+04
DEFORMATI ON: 1- B. C. 1, DI SPLACEMENT_1, LOAD SET 1
DI SPLACEMENT - MAG M N: 0. 00E+00 MAX: 8. 34E- 01
FRAME OF REF: PART

VALUE OPTI ON: ACTUAL
SHELL SURFACE: TOP

7. 24D+04
5. 79D+04
4. 34D+04
2. 90D+04
1. 45D+04
1. 50D 02
- 1. 45D+04
- 2. 90D+04
- 4. 34D+04
- 5. 79D+04
- 7. 24D+04

Figure 8.5.2

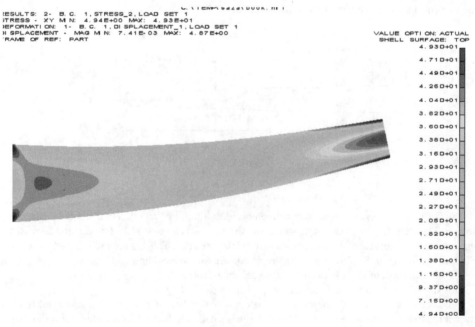

C: \ TEMP\ saza\ book. nf 1

RESULTS: 2- B. C. 1, STRESS_2, LOAD SET 1
STRESS - XY M N: 4. 94E+00 MAX: 4. 93E+01
DEFORMATI ON: 1- B. C. 1, DI SPLACEMENT_1, LOAD SET 1
DI SPLACEMENT - MAG M N: 7. 41E- 03 MAX: 4. 87E+00
FRAME OF REF: PART

VALUE OPTI ON: ACTUAL
SHELL SURFACE: TOP

4. 93D+01
4. 71D+01
4. 49D+01
4. 26D+01
4. 04D+01
3. 82D+01
3. 60D+01
3. 38D+01
3. 16D+01
2. 93D+01
2. 71D+01
2. 49D+01
2. 27D+01
2. 05D+01
1. 82D+01
1. 60D+01
1. 38D+01
1. 16D+01
9. 37D+00
7. 15D+00
4. 94D+00

Figure 8.5.3

8.6 Summary and Conclusions

As the problems become more complex the classical method becomes cumbersome. A general (finite element) method based on the definition of shape functions is then developed. The equations for this method are summarized in Equation 8.4.5 and repeated here.

$$\{F\} \quad \{F_c\} \quad f_y(s_m) \to \{p_m\} \quad \sigma_{xm}(s_m, y) \quad \varepsilon_{xm}(s_m, y) \quad u(s_m, y) \quad v(s_m) \quad \{r_m\} \leftarrow \{\rho\}$$

$$\{F_{em}\} = \int_0^l [n]^T f_y(s_m)\, ds_m \qquad \{F_e\} = \sum \{F_{em}\} \qquad v(s_m) = [n]\{r_m\}$$

$$\{F\} = \{F_c\} + \{F_e\} \qquad\qquad u(s_m, y) = -y\frac{d}{ds_m}[n]\{r_m\}$$

$$\varepsilon_{xm}(s_m, y) = -y\frac{d^2}{ds_m^2}[n]\{r_m\}$$

$$\sigma_{xm}(s_m, y) = -yE\frac{d^2}{ds_m^2}[n]\{r_m\}$$

$$M_{m,n} = -E_m I_m \frac{d^2}{ds_m^2}[n]\{r\} \qquad V_{m,n} = E_m I_m \frac{d^3}{ds_m^3}[n]\{r\}$$

$$M_{m,n+1} = E_m I_m \frac{d^2}{ds_m^2}[n]\{r\} \qquad V_{m,n+1} = -E_m I_m \frac{d^3}{ds_m^3}[n]\{r\}$$

$$v(s_m) = [n]\{r_m\}$$

$$= \frac{1}{l_m^3}\left[l_m^3 - 3l_m s_m + 2s_m^3 \quad l_m^3 s_m - 2l_m^2 s_m^2 + l_m s_m^3 \quad 3l_m s_m^2 - 2s_m^3 \quad -l_m^2 s_m^2 + l_m s_m^3 \right]\begin{bmatrix} v_{m,n} \\ \theta_{m,n} \\ v_{m,n+1} \\ \theta_{m,n+1} \end{bmatrix}$$

$$\{p_m\} = [k_m]\{r_m\} = \begin{bmatrix} V_{m,n} \\ M_{m,n} \\ V_{m,n+1} \\ M_{m,n+1} \end{bmatrix} = \frac{E_m I_m}{l_m^3}\begin{bmatrix} 12 & 6l_m & -12 & 6l_m \\ 6l_m & 4l_m^2 & -6l_m & 2l_m^2 \\ -12 & -6l_m & 12 & -6l_m \\ 6l_m & 2l_m^2 & -6l_m & 4l_m^2 \end{bmatrix}\begin{bmatrix} v_{m,n} \\ \theta_{m,n} \\ v_{m,n+1} \\ \theta_{m,n+1} \end{bmatrix}$$

$$\{F\} = \sum \{p_m\} = \sum [k_m]\{r_m\} = [K]\{r\} \tag{8.6.1}$$

9

More about Stress and Strain, and Material Properties

9.1 Introduction

For the most part we have presented the results for stress and strain with respect to a fixed rectangular Cartesian coordinate system. One exception is torsion where we have used cylindrical coordinates. For the axially loaded slender bar the state of stress has only one non zero component, σ_x, with respect to the chosen coordinate system. Briefly we did note that shear stress is present in an axially load bar if the axes of reference are rotated with respect to the original axes.

We must ask whether changes in orientation of the axes will produce different states of stress in other cases. States of stress that we have examined so far include a rectangular solid loaded by uniform hydrostatic pressure in Chapter 2, Example 2.6.2, resulting in a three dimensional state of stress. We have examined briefly rectangular plates with uniform and linear planar edge loads in Chapter 3, Example 3.5.1, and in Chapter 4, Section 4.2, and concluded that the stresses in the interior matched the edge loads. In Chapter 6 we examined the state of shear stress in torsion and in Chapter 7 the state of both normal and shear stress in bending. If we put these together in various combinations, and we can by the principle of superposition, we can solve for some quite complicated states of stress. Several cases of this type are examined briefly in Chapter 10. It is time to extend our discussion to a more general state of stress and strain and examine what happens when different orientations of the axes are used.

A restriction imposed until this chapter is to consider that all stresses and strains are below the proportional limit of a homogenous, isotropic, linearly elastic solid except for Chapter 3, Section 3.8, where fiber reinforced composite laminates were very briefly considered. Might it be that the stresses with respect to the axes used so far are not the ones that must be used to satisfy our restrictions on material properties? And are there other material properties that must be considered in the analysis and design of solid bodies? All this and more will be coming right up.

9.2 Transformation of Stress in Two Dimensions

The stress components we find with respect to the chosen rectangular Cartesian coordinates in any case tell only part of the story. Let us reexamine what happens in an axially loaded bar when the axes are rotated as shown in Figures 4.8.1 and 4.8.2, repeated here in Figure 9.2.1.

Analysis of Structures: An Introduction Including Numerical Methods, First Edition. Joe G. Eisley and Anthony M. Waas.
© 2011 John Wiley & Sons, Ltd. Published 2011 by John Wiley & Sons, Ltd.

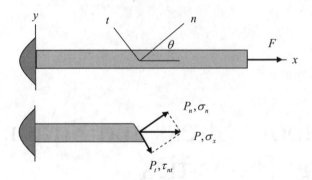

Figure 9.2.1

We saw briefly in Chapter 4, Section 4.8, that if we consider a coordinate system nt with rotation θ about the z axis the stress components normal and tangential to the face normal to the n axis are

$$\sigma_n = \frac{P}{A}\cos^2\theta = \sigma_x\cos^2\theta \qquad \tau_{nt} = -\frac{P}{A}\sin\theta\cos\theta = -\sigma_x\sin\theta\cos\theta \qquad (9.2.1)$$

In Figure 9.2.2 (repeat of Figure 4.8.3) we plot the variation in the normal stress and shearing stress as the angle changes.

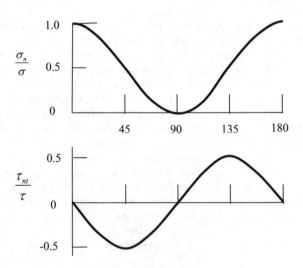

Figure 9.2.2

The upper plot is

$$\frac{\sigma_n}{\sigma_x} = \cos^2\theta \qquad (9.2.2)$$

and the lower plot is

$$\frac{\tau_{nt}}{\tau_{xy}} = \sin\theta\cos\theta \qquad (9.2.3)$$

We note that when $\theta = 0°$ then

$$\sigma_n = \sigma_x = \sigma_{max} \qquad (9.2.4)$$

and as θ increases the normal stress decreases until it reaches zero at $\theta = 90°$. The shear stress is zero at $\theta = 0°$ and $\theta = \pm 90°$ and reaches a maximum at $\theta = \pm 45°$ or

$$\tau_{max} = \pm \frac{P}{2A} \tag{9.2.5}$$

Now let us consider the general case of a two dimensional stress at a point where there are non zero components

$$\{\sigma\} = \begin{bmatrix} \sigma_x \\ \sigma_y \\ \tau_{xy} \end{bmatrix} \tag{9.2.6}$$

and zero components

$$\sigma_z = \tau_{zx} = \tau_{yz} = 0 \tag{9.2.7}$$

In Figure 9.2.3(a) we show symbolically the stresses at a point with respect to the xy axes and in Figure 9.2.3(b) we show them with respect to the sr axes which are rotated an amount θ.

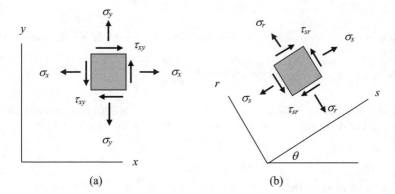

(a)   (b)

Figure 9.2.3

We wish to express the stress components with respect to the rotated sr axes in terms of the components of stress with respect to the xy axes. Let us find the equilibrium of the stresses at a point as shown in Figure 9.2.4.

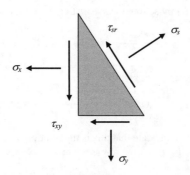

Figure 9.2.4

Let the area of the face on which σ_s acts be A. Then on the face normal to the x axis the area is $A \cos \theta$ and on the y normal face the area is $A \sin \theta$.

Equilibrium of forces in the r and s direction provides

$$\sum F_s = \sigma_s A - \tau_{xy} A \sin \theta \cos \theta - \sigma_y A \sin \theta \sin \theta - \tau_{xy} A \cos \theta \sin \theta - \sigma_x A \cos \theta \cos \theta = 0$$

$$\rightarrow \; \sigma_s = \sigma_x \cos^2 \theta + \sigma_y \sin^2 \theta + \tau_{xy} 2 \sin \theta \cos \theta \tag{9.2.8}$$

$$\sum F_r = \tau_{sr} A + \tau_{xy} A \sin \theta \sin \theta - \sigma_y A \sin \theta \cos \theta - \tau_{xy} A \cos \theta \cos \theta + \sigma_x A \cos \theta \sin \theta = 0$$

$$\rightarrow \; \tau_{sr} = (\sigma_y - \sigma_x) \sin \theta \cos \theta + \tau_{xy} (\cos^2 \theta - \sin^2 \theta) \tag{9.2.9}$$

Using the trigonometric identities

$$\sin 2\theta = 2 \sin \theta \cos \theta, \quad \sin^2 \theta = (1 - \cos 2\theta)/2 \quad \text{and} \quad \cos^2 \theta = (1 + \cos 2\theta)/2 \tag{9.2.10}$$

we can write

$$\sigma_s = \frac{\sigma_x + \sigma_y}{2} + \frac{\sigma_x - \sigma_y}{2} \cos 2\theta + \tau_{xy} \sin 2\theta \tag{9.2.11}$$

$$\tau_{rs} = -\frac{\sigma_x - \sigma_y}{2} \sin 2\theta + \tau_{xy} \cos 2\theta \tag{9.2.12}$$

We can get the normal stress in the r direction by substituting $\theta + 90°$ for θ in the expression for σ_s. Thus

$$\sigma_r = \frac{\sigma_x + \sigma_y}{2} - \frac{\sigma_x - \sigma_y}{2} \cos 2\theta + \tau_{xy} \sin 2\theta \tag{9.2.13}$$

Now that we can find the state of stress for any orientation of the axes of particular interest are rotations of axes that find the maximum values of normal and shear stress components. That will be treated in the next section.

9.3   Principal Axes and Principal Stresses in Two Dimensions

If we differentiate Equation 9.2.11 with respect to θ and set the result to zero we can find the angle for maximum and minimum normal stresses. This angle identifies axes called the *principal axes* and the stresses with respect to those axes are called the *principal stresses.*

$$\frac{d\sigma_s}{d\theta} = -\frac{\sigma_x - \sigma_y}{2} 2 \sin 2\theta + \tau_{xy} 2 \cos 2\theta = 0 \tag{9.3.1}$$

Note that we get the same result by setting the shear stress to zero, that is, from Equation 9.2.12

$$\tau_{rs} = -\frac{\sigma_x - \sigma_y}{2} \sin 2\theta + \tau_{xy} \cos 2\theta = 0 \tag{9.3.2}$$

From either Equation 9.3.1 or 9.3.2 we obtain

$$\tan 2\theta_p = \frac{\tau_{xy}}{(\sigma_x - \sigma_y)/2} \tag{9.3.3}$$

We have added the subscript p to indicate that this is the rotation necessary to define a principal axis.

To find the rotation for which the normal stresses are a maximum or a minimum the following Figure 9.3.1 is helpful.

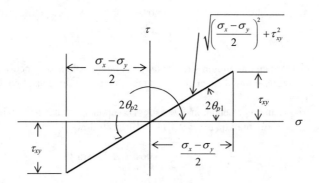

Figure 9.3.1

There are two different rotations that are 90° apart that define the axes of principal stress. Given $\tan 2\theta_p$ from the figure you can identify the sine and cosine of $2\theta_p$.

$$\sin 2\theta = \frac{\tau_{xy}}{\sqrt{\left(\dfrac{\sigma_x - \sigma_y}{2}\right)^2 + \tau_{xy}^2}} \qquad \cos 2\theta = \frac{\dfrac{\sigma_x - \sigma_y}{2}}{\sqrt{\left(\dfrac{\sigma_x - \sigma_y}{2}\right)^2 + \tau_{xy}^2}} \qquad (9.3.4)$$

If we substitue these into Equation 9.2.11 we obtain

$$\sigma_{p\,\max} = \frac{\sigma_x + \sigma_y}{2} + \sqrt{\left(\frac{\sigma_x - \sigma_y}{2}\right)^2 + \tau_{xy}^2} \qquad (9.3.5)$$

for the rotation that produces the maximum principal stress and

$$\sigma_{p\,\min} = \frac{\sigma_x + \sigma_y}{2} - \sqrt{\left(\frac{\sigma_x - \sigma_y}{2}\right)^2 + \tau_{xy}^2} \qquad (9.3.6)$$

for the rotation that produces the minimum principal stress.

The maximum and minimum principal stresses are found at the angle of rotation for which the shear stresses are zero, thus we can say

$$\tau_{p1} = \tau_{p2} = 0 \qquad (9.3.7)$$

If we differentiate Equation 9.2.12 with respect to θ and set the result to zero we can find the angle for maximum shearing stresses.

$$\frac{d\tau_{rs}}{d\theta} = \frac{\sigma_x - \sigma_y}{2} 2\cos 2\theta + \tau_{xy} 2 \sin 2\theta = 0$$

$$\tan 2\theta_{sh} = -\frac{\sigma_x - \sigma_y}{2\tau_{xy}} \qquad (9.3.8)$$

The maximum shear is then

$$\tau_{\max} = \sqrt{\left(\frac{\sigma_x - \sigma_y}{2}\right)^2 + \tau_{xy}^2} \qquad (9.3.9)$$

The normal stress at $\tau_{\max}$ is the same in both directions

$$\sigma = \frac{\sigma_x + \sigma_y}{2} \tag{9.3.10}$$

We see that the maximum and minimum stresses may not be the values obtained with respect to the original axes chosen for representing the equations and their solutions. One must find those values from the above equations.

There is a widely used visual representation of the stress components that is helpful in interpreting the results of an analysis. Let us rewrite Equations 9.2.11 and 12 as follows

$$\sigma_s - \frac{\sigma_x + \sigma_y}{2} = \frac{\sigma_x - \sigma_y}{2} \cos 2\theta + \tau_{xy} \sin 2\theta$$

$$\tau_{rs} = -\frac{\sigma_x - \sigma_y}{2} \sin 2\theta + \tau_{xy} \cos 2\theta \tag{9.3.11}$$

We can eliminate θ by squaring both Equations 9.3.11 and adding the results. Then

$$\left(\sigma_s - \frac{\sigma_x + \sigma_y}{2}\right)^2 + \tau_{rs}^2 = \left(\frac{\sigma_x - \sigma_y}{2}\right)^2 + \tau_{xy}^2 = R^2 \tag{9.3.12}$$

where we have set the right hand side equal to R^2. This equation can be recognized as the equation of a circle. Such a circle is plotted in Figure 9.3.2 and is known as *Mohr's circle*.

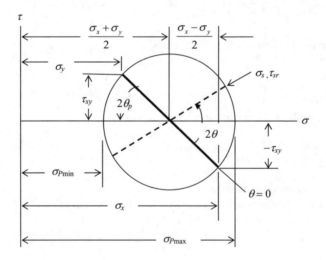

Figure 9.3.2

This allows us to see the values of stress at any rotation θ at a glance.

##########

Example 9.3.1

Problem: The state of plane stress at a point is found to be as shown in Figure (a). Find the principal stresses and the angle of the principal axes. Then find the maximum shear stress and the values of the normal stress at maximum shear.

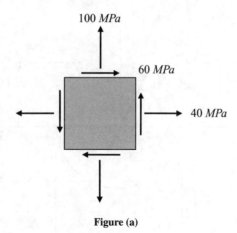

Figure (a)

Solution: Substitute the values of stress into the appropriate formulas.
For the principal stresses

$$\sigma_{p\,max,\,p\,min} = \frac{\sigma_x + \sigma_y}{2} \pm \sqrt{\left(\frac{\sigma_x - \sigma_y}{2}\right)^2 + \tau_{xy}^2} = \frac{100 + 40}{2} \pm \sqrt{\left(\frac{40 - 100}{2}\right)^2 + 60^2}$$

$$\sigma_{p\,max} = 137.082\ MPa \qquad \sigma_{p\,min} = 2.918\ MPa$$

(a)

The rotation angle for the principal axes is

$$\tan 2\theta_p = \frac{2\tau_{xy}}{\sigma_x - \sigma_y} = \frac{2 \cdot 60}{40 - 100} = -2 \quad \rightarrow \quad \theta_p = -\frac{63.435°}{2} = -31.718°$$

(b)

The maximum shear stress is

$$\tau_{max} = \sqrt{\left(\frac{\sigma_x - \sigma_y}{2}\right)^2 + \tau_{xy}^2} = \sqrt{\left(\frac{40 - 100}{2}\right)^2 + 60^2} = 67.082\ MPa$$

(c)

The rotation angle for maximum shear is

$$\tan 2\theta_{sh} = -\frac{\sigma_x - \sigma_y}{2\tau_{xy}} = -\frac{40 - 100}{2 \cdot 60} = 0.5 \quad \rightarrow \quad 2\theta_{sh} = 26.57° \quad \rightarrow \quad \theta_{sh} = 13.28°$$

(d)

The normal stress at the angle of maximum shear is the same in both directions.

$$\sigma = \frac{\sigma_x + \sigma_y}{2} = \frac{40 + 100}{2} = 70$$

(e)

Note that the rotation angle between maximum principal stress and maximum shear is 45 degrees. This is represented on a Mohr's circle in Figure (b).

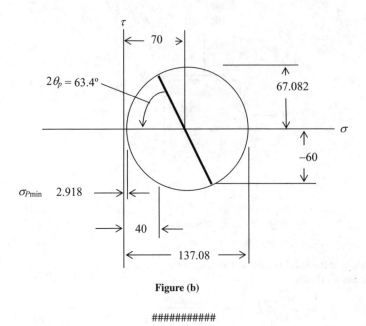

Figure (b)

###########

9.4   Transformation of Strain in Two Dimensions

Plane strain has a similar transformation. Let us relate the strain at a point with respect to the *sr* axes to the strain with respect to the *xy* axes. The *sr* axes are rotated by an angle θ from the *xy* axes as shown in Figure 9.4.1.

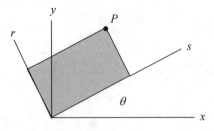

Figure 9.4.1

With respect to the *xy* axes we have the strain displacement relations

$$\varepsilon_x = \frac{\partial u}{\partial x} \qquad \varepsilon_y = \frac{\partial v}{\partial y} \qquad \gamma_{xy} = \frac{\partial u}{\partial y} + \frac{\partial v}{\partial x} \qquad (9.4.1)$$

With respect to the *sr* axes the strain displacement relations are

$$\varepsilon_s = \frac{\partial u_s}{\partial s} \qquad \varepsilon_r = \frac{\partial v_s}{\partial r} \qquad \gamma_{sr} = \frac{\partial u_s}{\partial r} + \frac{\partial v_s}{\partial s} \qquad (9.4.2)$$

From Figure 9.4.1 we note the coordinates of a point P are related in the two coordinate systems by

$$x = s \cos \theta - r \sin \theta \qquad\qquad y = s \sin \theta + r \cos \theta \qquad\qquad (9.4.3)$$

and displacements of the point P are related by

$$u_s = u \cos \theta + v \sin \theta \qquad\qquad v_r = v \cos \theta - u \sin \theta \qquad\qquad (9.4.4)$$

Using Equation 9.4.3 we can write

$$\varepsilon_s = \frac{\partial u_s}{\partial s} = \frac{\partial u_s}{\partial x}\frac{\partial x}{\partial s} + \frac{\partial u_s}{\partial y}\frac{\partial y}{\partial s} = \frac{\partial u_s}{\partial x}\cos \theta + \frac{\partial u_s}{\partial y}\sin \theta \qquad\qquad (9.4.5)$$

Then with Equation 9.4.4 we can obtain

$$\varepsilon_s = \frac{\partial u}{\partial x}\cos^2 \theta + \left(\frac{\partial v}{\partial x} + \frac{\partial u}{\partial y}\right)\cos \theta \sin \theta + \frac{\partial v}{\partial y}\sin^2 \theta \qquad\qquad (9.4.6)$$

Finally using Equation 9.4.1 we can write

$$\varepsilon_s = \varepsilon_x \cos^2 \theta + \varepsilon_y \sin^2 \theta + \gamma_{xy} \cos \theta \sin \theta \qquad\qquad (9.4.7)$$

or

$$\varepsilon_s = \frac{\varepsilon_x + \varepsilon_y}{2} + \frac{\varepsilon_x - \varepsilon_y}{2}\cos 2\theta + \frac{\gamma_{xy}}{2}\sin 2\theta \qquad\qquad (9.4.8)$$

By the same process we can also write

$$\varepsilon_r = \frac{\varepsilon_x + \varepsilon_y}{2} - \frac{\varepsilon_x - \varepsilon_y}{2}\cos 2\theta + \frac{\gamma_{xy}}{2}\sin 2\theta \qquad\qquad (9.4.9)$$

$$\frac{\gamma_{rs}}{2} = -\frac{\varepsilon_x - \varepsilon_y}{2}\sin 2\theta + \frac{\gamma_{xy}}{2}\cos 2\theta \qquad\qquad (9.4.10)$$

Just as we did in Equation 9.3.12 for stress we can do for strain.

$$\left(\varepsilon_x - \frac{\varepsilon_x + \varepsilon_y}{2}\right)^2 + \left(\frac{\gamma_{xy}}{2}\right)^2 = \left(\frac{\varepsilon_x - \varepsilon_y}{2}\right)^2 + \left(\frac{\gamma_{xy}}{2}\right)^2 = R^2 \qquad\qquad (9.4.11)$$

A Mohr's circle for strain can be drawn.

We can use these relations to establish a relationship between the shear modulus and Young's modulus by considering an element of material in a state of pure shear, that is

$$\sigma_x = \sigma_y = 0 \qquad\qquad \tau_{xy} \neq 0 \qquad\qquad (9.4.12)$$

The maximum and minimum principal stresses are then

$$\sigma_{p\,max} = \tau_{xy} \qquad\qquad \sigma_{p\,min} = -\tau_{xy} \qquad\qquad (9.4.13)$$

The corresponding maximum principal normal strain is

$$\varepsilon_{p\,max} = \frac{1}{E}(\sigma_{p\,max} - v\sigma_{p\,min}) = \frac{1+v}{E}\tau_{xy} \qquad\qquad (9.4.14)$$

From Equation 9.4.10 and Hooke's law we get

$$\varepsilon_{p\,max} = \frac{\gamma_{xy}}{2} = \frac{\tau_{xy}}{2G} \qquad\qquad (9.4.15)$$

By equating Equation 9.4.14 and 9.4.15 we can conclude that

$$G = \frac{E}{2(1+\nu)} \qquad (9.4.16)$$

We used this back in Chapter 3, Section 3.4, to simplify writing the stress-strain relations in two and three dimensions. Now we have evidence that it is true.

9.5   Strain Rosettes

In experimental mechanics stress cannot normally be measured directly; however, strain can be measured using a *strain guage* based on the electrical resistance of a wire or foil bonded to the surface of the member under strain. When the wire is stretched its length and cross sectional area change and the resitance to electrical current changes. This effect can be calibrated.

These guages are usually arranged in a pattern in a group of three. This group is called a *strain rosette*. They may be arranged in a general pattern such as shown in Figure 9.5.1.

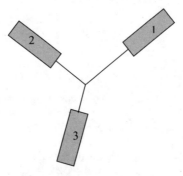

Figure 9.5.1

More commonly, they are arranged in a group at 45° intervals as shown in Figure 9.5.2(a) or at 60° intervals as shown in Figure 9.5.2(b).

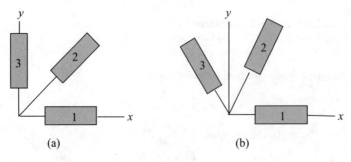

(a)   (b)

Figure 9.5.2

In the general case we can relate the strain measured by each guage to the strain in xy coordinates by Equation 9.4.7.

$$\varepsilon_1 = \varepsilon_x \cos^2 \theta_1 + \varepsilon_y \sin^2 \theta_1 + \gamma_{xy} \cos \theta_1 \sin \theta_1$$
$$\varepsilon_2 = \varepsilon_x \cos^2 \theta_2 + \varepsilon_y \sin^2 \theta_2 + \gamma_{xy} \cos \theta_2 \sin \theta_2 \qquad (9.5.1)$$
$$\varepsilon_3 = \varepsilon_x \cos^2 \theta_3 + \varepsilon_y \sin^2 \theta_3 + \gamma_{xy} \cos \theta_3 \sin \theta_3$$

or in matrix form

$$
\begin{bmatrix}
\cos^2 \theta_1 & \sin^2 \theta_1 & \sin \theta_1 \cos \theta_1 \\
\cos^2 \theta_2 & \sin^2 \theta_2 & \sin \theta_2 \cos \theta_2 \\
\cos^2 \theta_2 & \sin^2 \theta_3 & \sin \theta_3 \cos \theta_3
\end{bmatrix}
\begin{bmatrix}
\varepsilon_x \\
\varepsilon_y \\
\gamma_{xy}
\end{bmatrix}
=
\begin{bmatrix}
\varepsilon_1 \\
\varepsilon_2 \\
\varepsilon_3
\end{bmatrix}
\qquad (9.5.2)
$$

Solve these equations for the strains in the xy coordinates and from those strains find the stresses

$$\{\sigma\} = [G]\{\varepsilon\} \qquad (9.5.3)$$

In the case of a 45° rosette $\theta_1 = 0$, $\theta_2 = 45°$, and $\theta_3 = 90°$ and so relations in Equation 9.5.2 reduce to

$$
\begin{bmatrix}
\varepsilon_x \\
\varepsilon_y \\
\gamma_{xy}
\end{bmatrix}
=
\begin{bmatrix}
\varepsilon_1 \\
\varepsilon_3 \\
2\varepsilon_2 - (\varepsilon_1 + \varepsilon_3)
\end{bmatrix}
\qquad (9.5.4)
$$

In the case of a 60° rosette $\theta_1 = 0$, $\theta_2 = 60°$, and $\theta_3 = 120°$ and so the relations reduce to

$$
\begin{bmatrix}
\varepsilon_x \\
\varepsilon_y \\
\gamma_{xy}
\end{bmatrix}
=
\begin{bmatrix}
\varepsilon_1 \\
\dfrac{1}{3}(2\varepsilon_2 + 2\varepsilon_3 - \varepsilon_1) \\
\dfrac{2}{\sqrt{3}}(\varepsilon_2 - \varepsilon_3)
\end{bmatrix}
\qquad (9.5.5)
$$

###########

Example 9.5.1

Problem: A 45° strain rosette on the surface of a thin flat plate measures the following strains. The material is aluminum with $E = 68950\ MPa$ and $v = 0.3$.

$$
\begin{bmatrix}
\varepsilon_1 \\
\varepsilon_2 \\
\varepsilon_3
\end{bmatrix}
=
\begin{bmatrix}
0.0012 \\
0.0002 \\
0.0005
\end{bmatrix}
\qquad (a)
$$

The rosette is oriented with the xy axes as shown in Figure 9.5.2(a). Find the stresses in the xy coordinates, the maximum principal stress, and the rotation angle of the principal axes.

Solution: Use Equation 9.5.4 to find the strains with respect to the xy axes. Use Hooke's law to find the stresses. Then find the principal stresses and axes using the formulas from Section 9.3.
From Equation 9.5.4

$$
\begin{bmatrix}
\varepsilon_x \\
\varepsilon_y \\
\gamma_{xy}
\end{bmatrix}
=
\begin{bmatrix}
\varepsilon_1 \\
\varepsilon_3 \\
2\varepsilon_2 - (\varepsilon_1 + \varepsilon_3)
\end{bmatrix}
=
\begin{bmatrix}
0.0012 \\
0.0005 \\
-0.0013
\end{bmatrix}
\qquad (b)
$$

The stresses from Equation 3.4.22 are

$$
\begin{bmatrix} \sigma_x \\ \sigma_y \\ \tau_{xy} \end{bmatrix} = \begin{bmatrix} \dfrac{E}{1-v^2} & \dfrac{vE}{1-v^2} & 0 \\ \dfrac{vE}{1-v^2} & \dfrac{E}{1-v^2} & 0 \\ 0 & 0 & G \end{bmatrix} \begin{bmatrix} \varepsilon_x \\ \varepsilon_y \\ \gamma_{xy} \end{bmatrix} = \begin{bmatrix} 75769 & 22731 & 0 \\ 22731 & 75769 & 0 \\ 0 & 0 & 26519 \end{bmatrix} \begin{bmatrix} 0.0012 \\ 0.0005 \\ -0.0013 \end{bmatrix} = \begin{bmatrix} 102.29 \\ 65.16 \\ -34.47 \end{bmatrix}
$$

(c)

The maximum principal stress from Equation 9.3.5 is

$$
\sigma_{p\,max} = \frac{\sigma_x + \sigma_y}{2} + \sqrt{\left(\frac{\sigma_x - \sigma_y}{2}\right)^2 + \tau_{xy}^2} = 122.877 \; MPa
$$

(d)

The rotation angle for the principal axes from Equation 9.3.3 is

$$
\tan 2\theta_p = \frac{2\tau_{xy}}{\sigma_x - \sigma_y} = -1.8567 \quad \rightarrow \quad \theta_p = -30.85°
$$

(e)

###########

9.6   Stress Transformation and Principal Stresses in Three Dimensions

Transformation of stress and strain in three dimensions and the principal axes and stresses and strains will be presented here without proof. The full development may be found in books on the theory of elasticity.

We show two sets of axes – one rotated with respect to the other – in Figure 9.6.1. The components of the vector F are shown with respect to both the axes xyz and $x'y'z'$.

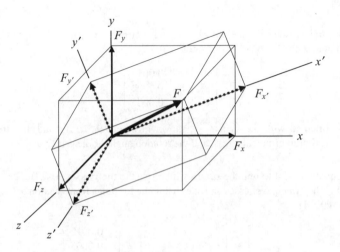

Figure 9.6.1

To simplify the notation we define a set of direction cosines in the following table.

	x	y	z
x'	l_1	m_1	n_1
y'	l_2	m_2	n_2
z'	l_3	m_3	n_3

where, for example, $l_1 = \cos xx'$, where xx' signifies the angle between the x axis and the x' axis. Similarly, $m_3 = \cos yz'$, where yz' signifies the angle between the y axis and the z' axis.

We define a transformation matrix

$$[T] = \begin{bmatrix} l_1 & m_1 & n_1 \\ l_2 & m_2 & n_2 \\ l_3 & m_3 & n_3 \end{bmatrix} \tag{9.6.1}$$

You can verify that the components of the force F are given by the matrix equation

$$\{F'\} = [T]\{F\} \tag{9.6.2}$$

Thus given the force components with respect to a particular orientation of the axes we can find the components with respect to any rotated axes. For the transformation of stress we must recall the definition of stress must specify both the direction of the force and the orientation of the surface to which it is referred. It can be shown that the transformation of stress is given by

$$[\sigma'] = [T][\sigma][T]^T \tag{9.6.3}$$

where $[\sigma]$ is written in the form

$$[\sigma] = \begin{bmatrix} \sigma_x & \tau_{xy} & \tau_{zx} \\ \tau_{xy} & \sigma_y & \tau_{yz} \\ \tau_{zx} & \tau_{yz} & \sigma_z \end{bmatrix} \tag{9.6.4}$$

It can also be shown that there are three orthogonal planes on each of which the stress at a point consists of a normal stress and no shearing stress. These are called the *principal stresses* and the rotated coordinate axes that are normal to each of these planes are called *principal axes*.

Let us look at the stresses at a point on a small tetrahedron as shown in Figure 9.6.2. The face of the tetrahedon is oriented so that the normal stress σ_p is the only stress acting on that face, thus it is a principal stress. It follows from Equation 9.6.3 that

$$[\sigma_p] = [T_p][\sigma][T_p]^T \tag{9.6.5}$$

where $[T_p]$ is the appropriate set of direction cosines for the principal axes and $[\sigma_p]$ are the principal stresses.

The orientation of the principal axes for the normal stress in Figure 9.6.2 is unknown. Let us identify them as

$$\begin{bmatrix} l \\ m \\ n \end{bmatrix} \tag{9.6.6}$$

It can be shown that

$$\begin{bmatrix} \sigma_x & \tau_{xy} & \tau_{zx} \\ \tau_{xy} & \sigma_y & \tau_{yz} \\ \tau_{zx} & \tau_{yz} & \sigma_z \end{bmatrix} \begin{bmatrix} l \\ m \\ n \end{bmatrix} = \sigma_p \begin{bmatrix} l \\ m \\ n \end{bmatrix} \tag{9.6.7}$$

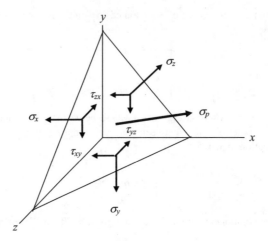

Figure 9.6.2

This can be rewritten as a classical eigenvalue problem.

$$\begin{bmatrix} \sigma_x - \sigma_p & \tau_{xy} & \tau_{zx} \\ \tau_{xy} & \sigma_y - \sigma_p & \tau_{yz} \\ \tau_{zx} & \tau_{yz} & \sigma_z - \sigma_p \end{bmatrix} \begin{bmatrix} l \\ m \\ n \end{bmatrix} = \begin{bmatrix} 0 \\ 0 \\ 0 \end{bmatrix} \tag{9.6.8}$$

One possible solutions is

$$\begin{bmatrix} l \\ m \\ n \end{bmatrix} = \begin{bmatrix} 0 \\ 0 \\ 0 \end{bmatrix} \tag{9.6.9}$$

however, equations of this form have solutions other than zero for discrete values of s determined by

$$\begin{vmatrix} \sigma_x - \sigma & \tau_{xy} & \tau_{zx} \\ \tau_{xy} & \sigma_y - \sigma & \tau_{yz} \\ \tau_{zx} & \tau_{yz} & \sigma_z - \sigma \end{vmatrix} = 0 \tag{9.6.10}$$

When expanded it yields the following cubic equation

$$\sigma^3 - S_1 \sigma^2 + S_2 \sigma - S_3 = 0 \tag{9.6.11}$$

where

$$\begin{aligned} S_1 &= \sigma_x + \sigma_y + \sigma_z \\ S_2 &= \sigma_x \sigma_y + \sigma_x \sigma_y + \sigma_x \sigma_y - \tau_{xy}^2 - \tau_{yz}^2 - \tau_{zx}^2 \\ S_3 &= \sigma_x \sigma_y \sigma_z + 2\tau_{xy}\tau_{yz}\tau_{zx} - \sigma_x \tau_{yz}^2 - \sigma_y \tau_{zx}^2 - \sigma_z \tau_{xy}^2 \end{aligned} \tag{9.6.12}$$

The three roots of Equation 9.6.11 are the three principal stresses. The direction of the principal stresses are found by substituting one of the principle stresses into Equation 9.6.13.

$$\begin{bmatrix} \sigma_x - \sigma_p & \tau_{xy} & \tau_{zx} \\ \tau_{xy} & \sigma_y - \sigma_p & \tau_{yz} \\ \tau_{zx} & \tau_{yz} & \sigma_z - \sigma_p \end{bmatrix} \begin{bmatrix} l \\ m \\ n \end{bmatrix} = \begin{bmatrix} 0 \\ 0 \\ 0 \end{bmatrix} \tag{9.6.13}$$

Since this is a set of homogenous equations we can only solve for two of the direction cosines in terms of the third. The third one is found from the property of direction cosines that

$$l^2 + m^2 + n^2 = 1 \tag{9.6.14}$$

The software products, such as Mathematica, Maple, Matlab, and Mathcad, among others have eigenvalue solvers for exactly this form of the equations.

###########

Example 9.6.1

Problem: Find the principal normal stresses and the maximum shear stress at a point in the three dimensional solid where the stresses are

$$
\begin{array}{lll}
\sigma_x = 20 \ MPa & \sigma_y = -40 \ MPa & \sigma_z = 100 \ MPa \\
\tau_{xy} = -40 \ MPa & \tau_{yz} = 60 \ MPa & \tau_{zx} = 0 \ MPa
\end{array} \tag{a}
$$

Solution: Use Equation 9.6.13
With the stress values inserted the eigenvalue equations become

$$
\begin{bmatrix}
20 - \sigma_p & -40 & 0 \\
-40 & -40 - \sigma_p & 60 \\
0 & 60 & 100 - \sigma_p
\end{bmatrix}
\begin{bmatrix} l \\ m \\ n \end{bmatrix}
=
\begin{bmatrix} 0 \\ 0 \\ 0 \end{bmatrix} \tag{b}
$$

The three eigenvalues, or principal stresses, are

$$
\begin{bmatrix} \sigma_{p1} \\ \sigma_{p2} \\ \sigma_{p3} \end{bmatrix}
=
\begin{bmatrix} 124.2 \\ -76.9 \\ 32.7 \end{bmatrix} MPa \tag{c}
$$

The eigenvector, or orientation of the principal stresses, is found by inserting one of the principal stresses into Equation (a) and solving for

$$
\begin{bmatrix}
l_1 & l_2 & l_3 \\
m_1 & m_2 & m_3 \\
n_1 & n_2 & n_3
\end{bmatrix}
=
\begin{bmatrix}
-0.1421 & 0.3701 & 0.9181 \\
-0.3642 & -0.8820 & 0.2992 \\
0.9204 & -0.2918 & 0.2601
\end{bmatrix} \tag{d}
$$

Each column presents the eigenvectors or direction cosines for the corresponding principal stress.

###########

9.7 Allowable and Ultimate Stress, and Factors of Safety

Up to now we have limited the discussion to levels of stress and strain that do not exceed the linearly elastic range of the material. Most load bearing structures are designed so that the levels of stress and strain always remain in this linear range. The point at which higher values of stress causes the stress-strain curve to deviate from a straight line is called the *proportional limit*. We need to examine what happens when stress values exceed this limit. A generic stress-strain curve is shown as obtained from a uniaxial test in Figure 9.7.1.

The point at which the curve is no longer elastic is called the *yield stress*. In many typical structural metals the yield stress is very near the proportional limit and so we do not have to take into account any small region of non linear elasticity.

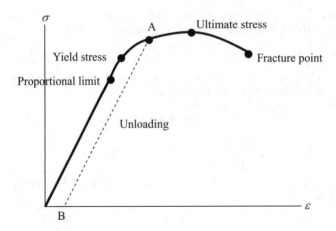

Figure 9.7.1

In metals, when stresses are below the yield value and when loads are reduced the stresses and strains follow the curve down on the same path that was followed when increasing the load. If, in fact the stress does exceed the yield value, say at point A, then unloading follows the path AB as shown and a permanent strain results. The details of the stress-strain curve for such loading paths for other materials like fiber reinforced composites can be different, influenced by the polymer stress-strain curve. These details are beyond the scope of this text that provides an introduction.

Going back to the typical response for a metal as shown in Figure 9.7.1, at some point the stress reaches a maximum value called the *ultimate stress* and there is continued strain without an increase in stress, in fact there is a relaxation in which strain increases even with a decrease in stress. Finally, the material fractures.

In design, a certain value of stress is determined as permissible under normal conditions of loading. This is called the *allowable stress*. In most cases in the design of reusable structures the allowable stress is well below the yield point. After all, we do not want the structure to have permanent deformation under normal use. If we define failure as the value of stress where permanent deformation occurs the ratio of that value to the allowable value is called the *factor of safety*.

$$F.S. = \frac{\sigma_{fail}}{\sigma_{allow}} \qquad (9.7.1)$$

A suitable value for *F.S.* depends on many factors, such as, degree of accuracy of loads and restraints, approximations used in analysis, deteriorizations over lifetime of use, repetitive and dynamic loading, accuracy of material properties, and many more consideration.

Just what value to use in design requires careful consideration. Structures such as bridges and building must have fairly high factors of safety because they have extended lives and must survive often under severe environmental conditions and human safety is a paramount concern. Aircraft must have much lower values in order to keep the weight down to be economically viable; however, they have shorter lives, frequent maintenance, and carefully trained operators. But always human safety is a paramount concern. Unmanned space craft can have very low values of *F.S.* because they are most carefully operated and human life is not at risk.

In the case of uniaxial stress it is quite clear that when the stress excedes the yield point the material will achieve permanent deformation. In the case of plane stress and a state of three dimensional stress it might seem logical that when the maximum principal stress exceeds that yield point the material will deform permanently. Actually it is not as simple as that. Apparently the mechanism for yield is more complicated than that. There are in fact several criteria offered for when a material will yield under multiaxial stress. We shall present only one called the *von Mises yield stress*, which is widely accepted for designing with metals.

Without further explanation, the von Mises criteria is based on the maximum strain energy of distortion. In the calculation of strain energy, part of that energy is a result of change in volume; another part is a result of distortion or change in shape. The distortion energy per unit volume of an isotropic material in plane stress is

$$U_d = \frac{1}{6G} \left(\sigma_{p_1}^2 - \sigma_{p_1}\sigma_{p_2} + \sigma_{p_2}^2 \right) \tag{9.7.2}$$

In the case of a uniaxial test specimen, such as used to find the stress strain curve in Chapter 3, Section 3.4, we note that at yield

$$\sigma_{p_1} = \sigma_Y \qquad\qquad \sigma_{p_2} = 0 \tag{9.7.3}$$

Thus

$$(U_d)_Y = \frac{1}{6G} \sigma_Y^2 \tag{9.7.4}$$

For plane stress the von Mises criterion stated that the general 2D state of stress is safe as long as, U_d represented by (9.7.2) is less than U_d that one finds in a uniaxial test, i.e. given by (9.7.4). Thus, it follows that,

$$\sigma_{p_1}^2 - \sigma_{p_1}\sigma_{p_2} + \sigma_{p_2}^2 \leq \sigma_Y^2 \tag{9.7.5}$$

Based on (9.7.5), we define the von Mises stress in two dimensions as

$$\sigma_{VM} = \sqrt{\sigma_{p_1}^2 - \sigma_{p_1}\sigma_{p_2} + \sigma_{p_2}^2} \tag{9.7.6}$$

and compare it to the yield stress based on a one dimensional test of the axially loaded slender bar, as stated earlier.

The distortion energy for a state of three dimensional stress is

$$U_d = \frac{1}{12G} [(\sigma_{p_1} - \sigma_{p_2})^2 + (\sigma_{p_2} - \sigma_{p_3})^2 + (\sigma_{p_3} - \sigma_{p_1})^2] \tag{9.7.7}$$

For three dimensional stress, using a similar reasoning to that followed for the 2D case, the von Mises criterion states that the state of stress is safe as long as

$$(\sigma_{p_1} - \sigma_{p_2})^2 + (\sigma_{p_2} - \sigma_{p_3})^2 + (\sigma_{p_3} - \sigma_{p_1})^2 \leq 2\sigma_Y^2, \tag{9.7.8}$$

which is obtained from (9.7.8) and (9.7.4). We define the von Mises stress for three dimensions as

$$\sigma_{VM} = \sqrt{\frac{(\sigma_{p_1} - \sigma_{p_2})^2 + (\sigma_{p_2} - \sigma_{p_3})^2 + (\sigma_{p_3} - \sigma_{p_1})^2}{2}} \tag{9.7.9}$$

and compare it to the yield stress based on a one dimensional test of the axially loaded slender bar.

9.8 Fatigue

Before we move onto examples, we must note a few things regarding designing against yield using a factor of safety for those structures like wings and bridges which experience repeated loading or cyclic loads. In those instances, a structure can experience failure at a level of load that is below the load which causes permanent set. This is due to fatigue. In fatigue, a material can fail at a load level that is below that which is needed to cause yield. Failure due to repeated loading can be seen when one bends a paper clip back and forth until failure. Consider a structural component that is subjected to sinusoidal cylic loading. We can describe such loading with an amplitude and a period (or frequency). Even though the stress amplitude may not be as large to cause failure in one cycle, with repeated loading, it is found that the structure undergoes fatigue failure by the growth of small flaws (cracks) that can accumulate damage over each cycle and reach a critical state of damage that can lead to large and abrupt crack growth. In general, the number of cylces until failure increases with a diminishing stress amplitude and there exists a stress amplitude below which the structure has infinite fatigue life, the latter being what is ideally desirable. For each material, the resistance to fatigue can be measured by doing tests at cyclic loading of constant amplitude and frequency. In those cases two quantities are needed to characterize the loading, the average stress, defined as $\sigma_{ave.} = \dfrac{\sigma_{max} - \sigma_{min}}{2}$, and the mean stress, defined as, $\sigma_{ave.} = \dfrac{\sigma_{max} + \sigma_{min}}{2}$. To find the fatigue stress level for loading with zero mean value, the number of cycles to failure, N, is recorded as a function of stress amplitude. A plot of stress amplitude vs. N produces a S-N diagram which is used as a material property, the fatigue resistance curve. In general, the S-N curve is a monotonically decreasing curve and below a certain stress amplitude the material shows no limit in number of cycles to failure – referred to as the fatgue limit. The notion of "safe life design" requires that the part sees service stresses below this fatigue limit, ensuring an infinite life!

Fatiuge life is also affected by mean stress and usually, it is found that the larger the mean stress, the more diminished is the S-N diagram. For many aerospace structural materials S-N diagrams are available in handbooks. Designing against fatigue is an involved subject that goes beyond the scope of this introductory text and the reader is referred to Bathias and Pineau (2010) for more details.

###########

Example 9.8.1

Problem: Find the von Mises stress for the state of stress in Example 9.3.1.

Solution: Use Equation 9.7.5.

The principal stresses are from Example 9.3.1.

$$\sigma_{p_1} = 2.918 \, MPa \qquad\qquad \sigma_{p_2} = 137.082 \, MPa \qquad\qquad (a)$$

Substituting the values

$$\sigma_{VM} = \sqrt{\sigma_{p_1}^2 - \sigma_{p_1}\sigma_{p_2} + \sigma_{p_2}^2} = \sqrt{(2.918)^2 - 2.918 \cdot 137.082 + (137.082)^2} = 135.65 \, MPa \qquad (b)$$

Note that in this case it is slightly smaller than the maximum principle stress given in Equation (a).

###########

These relations are programmed and automatically calculated in the FEM computer codes. Factors of safety are often based on the von Mises stresses.

9.9 Creep

In some material and environmental conditions the displacement may increase slowly with time although the load is constant and the stress is below the normal yield stress. In particular some materials behave this way when the temperature is increased beyond some appropriate level. This phenomenon is known as *creep*. For most aerospace metals, creep is significant only at high temperatures and in those situations, such as in engine structures, special metal alloys have been developed that exhibit high creep resistance. Creep is a specialized subject that is beyond the scope of this introductory text.

9.10 Orthotropic Materials—Composites

There is growing use of materials for structures that are neither homogenous nor istropic. Plywood, for example has been around for a long time. Built up layers of cloth glued together is another. For some time these were used primarily in low load situations. Today there are a number of laminated and filament reinforced materials used in very high tech structures, for example, in spacecraft. A new generation of aircraft make substantial use of composites. In many cases these can be modeled as orthotropic materials, that is, the materials have specific and different properties in each of the three coordinate directions.

Without proof the strain stress relations for an orthotropic material are

$$
\begin{Bmatrix} \varepsilon_x \\ \varepsilon_y \\ \varepsilon_z \\ \gamma_{xy} \\ \gamma_{yz} \\ \gamma_{zx} \end{Bmatrix} =
\begin{bmatrix}
\dfrac{1}{E_x} & \dfrac{-v_{yx}}{E_y} & \dfrac{-v_{zx}}{E_z} & 0 & 0 & 0 \\[2mm]
\dfrac{-v_{xy}}{E_x} & \dfrac{1}{E_y} & \dfrac{-v_{zy}}{E_z} & 0 & 0 & 0 \\[2mm]
\dfrac{-v_{xz}}{E_x} & \dfrac{-v_{yz}}{E_y} & \dfrac{1}{E_z} & 0 & 0 & 0 \\[2mm]
0 & 0 & 0 & \dfrac{1}{G_{xy}} & 0 & 0 \\[2mm]
0 & 0 & 0 & 0 & \dfrac{1}{G_{yz}} & 0 \\[2mm]
0 & 0 & 0 & 0 & 0 & \dfrac{1}{G_{zx}}
\end{bmatrix}
\begin{Bmatrix} \sigma_x \\ \sigma_y \\ \sigma_z \\ \tau_{xy} \\ \tau_{yz} \\ \tau_{zx} \end{Bmatrix}
\tag{9.10.1}
$$

It can be shown that

$$
\frac{v_{xy}}{E_x} = \frac{v_{yx}}{E_y} \qquad \frac{v_{xz}}{E_x} = \frac{v_{zx}}{E_z} \qquad \frac{v_{zy}}{E_z} = \frac{v_{yz}}{E_y}
\tag{9.10.2}
$$

This set of equations can be inverted to provide

$$
\{\sigma\} = [G_{ortho}]\{\varepsilon\}
\tag{9.10.3}
$$

or

$$
\begin{Bmatrix} \sigma_x \\ \sigma_y \\ \sigma_z \\ \tau_{xy} \\ \tau_{yz} \\ \tau_{zx} \end{Bmatrix} = \frac{1}{\Lambda}
\begin{bmatrix}
E_{x0}(1 - v_{zy}v_{yz}) & E_y(v_{xy} + v_{xz}v_{zy}) & E_z(v_{xz} + v_{xy}v_{yz}) & 0 & 0 & 0 \\
E_y(v_{xy} + v_{xz}v_{zy}) & E_y(1 - v_{zx}v_{xz}) & E_z(v_{yz} + v_{yx}v_{xz}) & 0 & 0 & 0 \\
E_z(v_{xz} + v_{xy}v_{yz}) & E_z(v_{yz} + v_{yx}v_{xz}) & E_z(1 - v_{xy}v_{yx}) & 0 & 0 & 0 \\
0 & 0 & 0 & G_{xy} & 0 & 0 \\
0 & 0 & 0 & 0 & G_{yz} & 0 \\
0 & 0 & 0 & 0 & 0 & G_{zx}
\end{bmatrix}
\begin{Bmatrix} \varepsilon_x \\ \varepsilon_y \\ \varepsilon_z \\ \gamma_{xy} \\ \gamma_{yz} \\ \gamma_{zx} \end{Bmatrix}
\tag{9.10.4}
$$

where

$$
\Lambda = 1 - v_{xy}v_{yx} - v_{xz}v_{zx} - v_{yz}v_{zy} - 2v_{xz}v_{zy}v_{yx}
\tag{9.10.5}
$$

This expression for $[G_{ortho}]$ can be used in place of $[G]$ wherever it is found in any of the previous equations provided the xyz axes are alligned with the material axes in an orthotropic material.

9.11   Summary and Conclusions

In Chapter 1 a brief introduction was provided for the concepts of displacement, stress, strain, equilibrium, and material properties. Then in Chapters 2 through 6 methods for finding the values of displacement, stress, and strain consistent with the laws of equilibrium and material properties. Now, in this chapter we have examined these quantities in more detail.

10

Combined Loadings on Slender Bars—Thin Walled Cross Sections

10.1 Introduction

The geometry of the slender bar, its material properties, the applied loads, and the restraints determine the displacements and stresses in the bar. In previous chapters we have limited the geometry, loading, and restraints to allow uncoupled axial, torsional, and bending displacements and stresses. Each case is considered independently. In the case of bending we have further restricted the analysis to displacements in the xy plane. We now remove some of these restrictions to consider simultaneous multiple application of loading. We also remove both the restriction of bending in one plane and the requirement that $I_{yz} = 0$.

Of particular interest is the extension of the theory of thin walled cross sections to include simultaneous bending and torsion and the addition of concentrated stiffeners to the sections.

We start with a brief review of slender bar theory as presented in previous chapters and then embark on the extensions of the analysis just noted.

10.2 Review and Summary of Slender Bar Equations

It is convenient to summarize the equations used in the preceding chapters to analyze slender bars. In the following sections of this chapter we shall apply these equations to combined loadings and in some cases extend them to more complicated geometry.

10.2.1 Axial Loading

We found in Chapter 4 that the differential equation for the displacement of an axially loaded slender bar is

$$\frac{d}{dx}EA\frac{du}{dx} = -f_x(x) \tag{10.2.1}$$

and when EA is constant

$$EA\frac{d^2u}{dx^2} = -f_x(x) \quad \rightarrow \quad u(x) = -\frac{1}{EA}\iint f_x(x)dx + ex + f \tag{10.2.2}$$

Analysis of Structures: An Introduction Including Numerical Methods, First Edition. Joe G. Eisley and Anthony M. Waas.
© 2011 John Wiley & Sons, Ltd. Published 2011 by John Wiley & Sons, Ltd.

The equation is integrated to find $u(x)$ and boundary conditions are used to find the two constants of integration. Once the displacements are known the internal force is found from

$$P = EA\frac{du}{dx} \tag{10.2.3}$$

and the stress is

$$\sigma_x = \frac{P}{A} \tag{10.2.4}$$

We note that when the bar is statically determinate we can find P directly from equilibrium and immediately solve for the stress. To find the displacement we then solve the first order equation, that is, Equation 10.2.3.

An alternative is to use the finite element method as presented in Chapter 5. The first step is to divide the bar into elements and number the elements and the nodes. Shape functions $[n]$ that represent the distributed displacement $u(s_m)$ in terms of the nodal displacements $\{r_m\}$ of each element are shown in Equation 10.2.5.

$$u(s_m) = [n]\{r_m\} = \begin{bmatrix} 1 - \dfrac{s_m}{l_m} & \dfrac{s_m}{l_m} \end{bmatrix} \begin{bmatrix} u_n \\ u_{n+1} \end{bmatrix} \tag{10.2.5}$$

From these shape functions we find the element stiffness matrix that represents the internal forces $\{p_m\}$ in terms of the nodal displacements.

$$\{p_m\} = \begin{bmatrix} p_{m,n} \\ p_{m,n+1} \end{bmatrix} = \frac{E_m A_m}{l_m} \begin{bmatrix} 1 & -1 \\ -1 & 1 \end{bmatrix} \begin{bmatrix} u_n \\ u_{n+1} \end{bmatrix} = [k_m]\{r_m\} \tag{10.2.6}$$

The element stiffness matrices $[k_m]$ are assembled to form the global stiffness matrix $[K]$.

$$[K] = \sum [k_m] \tag{10.2.7}$$

For each element the distributed applied loads are converted to equivalent nodal loads by

$$\{F_{em}\} = \int_0^{l_m} [n]^T f_x(s)ds_m = \int_0^{l_m} \begin{bmatrix} 1 - \dfrac{s_m}{l_m} \\ \dfrac{s_m}{l_m} \end{bmatrix} f_x(s)ds_m = \begin{bmatrix} F_{em,n} \\ F_{em,n+1} \end{bmatrix} \tag{10.2.8}$$

These are assembled to form the total equivalent nodal loads $\{F_e\}$ and are added to the actual applied concentrated loads $\{F_c\}$ to form global load matrix $\{F\}$.

$$\{F\} = \{F_c\} + \sum \{F_{em}\} = \{F_c\} + \{F_e\} \tag{10.2.9}$$

The global equations for the nodal displacements are

$$\{F\} = \{F_c\} + \{F_e\} = [K]\{r\} \tag{10.2.10}$$

The known nodal restraints and the unknown restraint forces are inserted in the global equations and then the equations are partitioned into one set for solving the unknown nodal displacements and another set for solving the unknown restraint forces. Once the nodal displacements are known the distributed displacements are found for each element by Equation 10.2.5 repeated here as Equation 10.2.11.

$$u(s_m) = [n]\{r_m\} = \begin{bmatrix} 1 - \dfrac{s_m}{l_m} & \dfrac{s_m}{l_m} \end{bmatrix} \begin{bmatrix} u_n \\ u_{n+1} \end{bmatrix} \tag{10.2.11}$$

The internal forces are found by Equations 10.2.6 and the stresses are found for each element by

$$\sigma_{xm} = E_m \frac{d}{ds_m}[n]\{r_m\} = \frac{E_m}{l_m}\begin{bmatrix} -1 & 1 \end{bmatrix}\begin{bmatrix} u_n \\ u_{n+1} \end{bmatrix} \tag{10.2.12}$$

As explained in Chapter 5 the axial forces and stresses are exact values for bars with concentrated loads only, but must be interpreted when equivalent nodal loads are used (such as when you have distributed axial loads). In particular forces are averaged at the nodes and curves are fitted through the resulting values. Chapter 5, Section 5.4, has an example demonstrating this.

10.2.2 Torsional Loading

We found in Chapter 6 that the differential equation for the displacement of a torsionally loaded slender bar is

$$\frac{d}{dx}GJ\frac{d\beta}{dx} = -t_x(x) \tag{10.2.13}$$

and when GJ is constant

$$GJ\frac{d^2\beta}{dx^2} = -t_x(x) \quad \rightarrow \quad \beta(x) = -\frac{1}{GJ}\iint t(x)dx + ex + f \tag{10.2.14}$$

The equation is integrated to find $\beta(x)$ and boundary conditions are used to find the two constants of integration. Once the rotational displacement is known the internal torque is found from

$$T = GJ\frac{d\beta}{dx} \tag{10.2.15}$$

and the stress is

$$\tau = \frac{T}{J}r \tag{10.2.16}$$

where τ is expressed in polar coordinates.

These equations are restricted to circular cross sections and J is the area polar moment of inertia of the cross section. In fact notice that these equations for torsion are identical in form to the equations that govern the axial deformation of slender bars, Equations 10.2.1 – 10.2.4.

We note that when the bar is statically determinate we can find T directly from equilibrium and immediately solve for the stress. To find the displacement we then solve the first order equation, that is, Equation 10.2.15.

An alternative is to use the finite element method as presented in Chapter 6. The first step is to divide the bar into elements and number the elements and the nodes. Shape functions $[n]$ that represent the distributed rotational displacement $\beta_m(s)$ in terms of the nodal displacements $\{\phi_m\}$ of each element are shown in Equation 10.2.17 (see similarity between Equation 10.2.17 and 10.2.5).

$$\beta(s_m) = [n]\{\phi_m\} = \begin{bmatrix} 1 - \dfrac{s_m}{l_m} & \dfrac{s_m}{l_m} \end{bmatrix}\begin{bmatrix} \beta_n \\ \beta_{n+1} \end{bmatrix} \tag{10.2.17}$$

From these shape functions we find the element stiffness matrix that represents the internal torques $\{\psi_m\}$ in terms of the nodal displacements.

$$\{\psi_m\} = \begin{bmatrix} \psi_{m,n} \\ \psi_{m,n+1} \end{bmatrix} = \frac{G_m J_m}{l_m}\begin{bmatrix} 1 & -1 \\ -1 & 1 \end{bmatrix}\begin{bmatrix} \beta_n \\ \beta_{n+1} \end{bmatrix} = [k_m]\{\phi_m\} \tag{10.2.18}$$

Again Equation 10.2.18 is similar to Equation 10.2.6.

The element stiffness matrices $[k_m]$ are assembled to form the global stiffness matrix $[K]$.

$$[K] = \sum [k_m] \tag{10.2.19}$$

For each element the distributed applied moments, $t_x(s_m)$, are converted to equivalent nodal moments by

$$\{M_{em}\} = \int [n]^T t_x ds_m = \int_0^{l_m} \begin{bmatrix} 1 - \dfrac{s_m}{l_m} \\ \dfrac{s_m}{l_m} \end{bmatrix} t_x ds_m = \begin{bmatrix} M_{m,n} \\ M_{m,n+1} \end{bmatrix} \tag{10.2.20}$$

These are assembled to form the total equivalent nodal moments $\{M_e\}$ and are added to the actual applied concentrated moments $\{M_c\}$ for form global moment matrix $\{M\}$.

$$\{M\} = \{M_c\} + \sum \{M_{em}\} = \{M_c\} + \{M_e\} \tag{10.2.21}$$

The global equations for the nodal displacements are

$$\{M\} = \{M_c\} + \{M_e\} = [K]\{\phi\} \tag{10.2.22}$$

The known nodal restraints and the unknown restraint forces are inserted in the global equations and then the equations are partitioned into one set for solving the unknown nodal displacements and another set for solving the unknown restraint forces. Once the nodal displacements are known the distributed displacements are found for each element by

$$\beta(s_m) = [n]\{\phi_m\} = \begin{bmatrix} 1 - \dfrac{s_m}{l_m} & \dfrac{s_m}{l_m} \end{bmatrix} \begin{bmatrix} \beta_n \\ \beta_{n+1} \end{bmatrix} \tag{10.2.23}$$

For each element the internal moments are found by Equation 10.2.18 and the only unknown shear stresses by

$$\tau_m = \frac{G_m}{l_m} \begin{bmatrix} -1 & 1 \end{bmatrix} \begin{bmatrix} \beta_n \\ \beta_{n+1} \end{bmatrix} r \tag{10.2.24}$$

As explained in Chapter 6 the internal torques and stresses are exact values for bars loaded by concentrated torques but must be interpreted when the applied loads are distributed and equivalent nodal loads are used. In particular torques are averaged at the nodes and curves are fitted through the resulting values. Chapter 6, Section 6.9, has examples and details of this stress recovery process.

10.2.3   Bending in One Plane

We found in Chapter 7 that the differential equation for the displacement of a laterally loaded slender bar when $I_{yz} = 0$, that is, when at least one of the cross sectional axes, y or z, is a principal axis (symmetrical cross sections are an example) and the load entirely in the xy plane is

$$\frac{d^2}{dx^2} EI_{zz} \frac{d^2 v}{dx^2} = f_y(x) \tag{10.2.25}$$

and when EI_{zz} is constant the solution is

$$EI_{zz} \frac{d^4 v}{dx^4} = f_y(x) \quad \rightarrow \quad v(x) = \frac{1}{EI_{zz}} \iiiint f_y(x) dx + \frac{a}{6}x^3 + \frac{b}{2}x^2 + cx + d \tag{10.2.26}$$

The boundary conditions are used to find the four constants of integration. Once the displacement is known the internal bending moment and shear force are found from

$$M = EI_{zz} \frac{d^2v}{dx^2} \qquad V = -EI_{zz} \frac{d^3v}{dx^3} \tag{10.2.27}$$

The normal stresses are

$$\sigma_x = -\frac{My}{I_{zz}} \tag{10.2.28}$$

and for a rectangle cross section the shear stresses are

$$\tau_{xy}(x, y) = \frac{V}{I_{zz}b} \int_{A_y} y \, dA \tag{10.2.29}$$

where b is the width of the cross section.

We note that when the beam is statically determinate we can find V and M directly from equilibrium and immediately solve for the stresses. To find the displacement we then solve the second order equation, that is, Equation 10.2.27.

An alternative is to use the finite element method as presented in Chapter 8. The first step is to divide the bar into elements and number the elements and the nodes.

Shape functions $[n]$ that represent the distributed displacement $v(s_m)$ in terms of the nodal displacements $\{r_m\}$ of each element are shown in Equation 10.2.30.

$$v(s_m) = [n]\{r_m\}$$

$$= \frac{1}{l_m^3} \left[l_m^3 - 3l_m s_m + 2s_m^3 \quad l_m^3 s_m - 2l_m^2 s_m^2 + l_m s_m^3 \quad 3l_m s_m^2 - 2s_m^3 \quad -l_m^2 s_m^2 + l_m s_m^3 \right] \{r_m\} \tag{10.2.30}$$

From these shape functions we find the element stiffness matrix that represents the internal shear forces and moments, $\{p_m\}$, in terms of the nodal displacements, $\{r_m\}$.

$$\{p_m\} = [k_m]\{r_m\} = \begin{bmatrix} V_{m,n} \\ M_{m,n} \\ V_{m,n+1} \\ M_{m,n+1} \end{bmatrix} = \frac{E_m I_m}{l_m^3} \begin{bmatrix} 12 & 6l_m & -12 & 6l_m \\ 6l_m & 4l_m^2 & -6l_m & 2l_m^2 \\ -12 & -6l_m & 12 & -6l_m \\ 6l_m & 2l_m^2 & -6l_m & 4l_m^2 \end{bmatrix} \begin{bmatrix} v_{m,n} \\ \theta_{m,n} \\ v_{m,n+1} \\ \theta_{m,n+1} \end{bmatrix} \tag{10.2.31}$$

The element stiffness matrices $[k_m]$ are assembled to form the global stiffness matrix $[K]$.

$$[K] = \sum [k_m] \tag{10.2.32}$$

For each element the distributed applied loads are converted to equivalent nodal loads by

$$\{F_{em}\} = \int_0^{l_m} [n]^T f_y(s_m) \, ds_m = \int_0^{l_m} \begin{bmatrix} l_m^3 - 3l_m s_m^2 + 2s_m^3 \\ l_m^3 s - 2l_m^2 s_m^2 + l_m s_m^3 \\ 3l_m s_m^2 - 2s_m^3 \\ -l_m^2 s_m^2 + l_m s_m^3 \end{bmatrix} \frac{f_y(s_m)}{l_m^3} \, ds_m = \begin{bmatrix} F_{em,n} \\ M_{em,n} \\ F_{em,n+1} \\ M_{em,n+1} \end{bmatrix} \tag{10.2.33}$$

These are assembled to form the total equivalent nodal loads $\{F_e\}$ and are added to the actual applied concentrated loads $\{F_c\}$ for form the global load matrix $\{F\}$.

$$\{F\} = \{F_c\} + \sum \{F_{em}\} = \{F_c\} + \{F_e\} \tag{10.2.34}$$

The global equations for the nodal displacements are

$$\{F\} = \{F_c\} + \{F_e\} = [K]\{r\} \tag{10.2.35}$$

The known nodal restraints and the unknown restraint forces are inserted in the global equations and then the equations are partitioned into one set for solving the unknown nodal displacements and another set for solving the unknown restraint forces. Once the nodal displacements are known the distributed displacements are found for each element by Equation 10.2.30 repeated here as Equations 10.2.36.

$$v(s_m) = [n]\{r_m\}$$
$$= \frac{1}{l_m^3} \left[l_m^3 - 3l_m s_m + 2s_m^3 \quad l_m^3 s_m - 2l_m^2 s_m^2 + l_m s_m^3 \quad 3l_m s_m^2 - 2s_m^3 \quad -l_m^2 s_m^2 + l_m s_m^3 \right] \{r_m\} \tag{10.2.36}$$

The internal moments and shear forces are found by Equation 10.2.31. The normal and shear stresses are found at the nodes for each element using Equations 10.2.28 and 10.2.29 repeated here as Equation 10.2.37

$$\sigma_x = -\frac{My}{I_{zz}} \qquad \tau_{xy}(x, y) = \frac{V}{I_{zz}b} \int_{A_y} y \, dA \tag{10.2.37}$$

The normal stresses are linear functions between nodes and the distributed values are found for each element by

$$\sigma_{xm} = -\frac{yE}{l_m^3} \left[-6l_m + 12s_m \quad -4l_m^2 + 6l_m s_m \quad 6l_m - 12s_m \quad -2l_m^2 + 6l_m s_m \right] \begin{bmatrix} v_{m,n} \\ \theta_{m,n} \\ v_{m,n+1} \\ \theta_{m,n+1} \end{bmatrix} \tag{10.2.38}$$

The shear force in *each element* is a constant and consequently the shear stress is the same at all values of s. They are equal to the values calculated at either node.

As explained in Chapter 8 the shear forces and stresses and the bending moments and stresses are exact values for beams with only concentrated loads but must be interpreted when equivalent nodal loads are used. In particular, forces and moments are averaged at the nodes and curves are fitted through the resulting values. Chapter 8, Section 8.4, has examples and details for this process, but we will give examples to clarify this.

10.3   Axial and Torsional Loads

For bars with certain restrictions on the cross sections, combined axial and torsional loads can produce both axial and torsional displacements that are uncoupled In these instances each case can be solved separately and superposition can be used to find answers for the case of combined loading. In the FEM method the equations can be combined into one set. A propeller shaft (circular) is a good example. The drag on the propeller blades produces a torque while the thrust of the propeller produces an axial force. Here is a simple example.

###########

Example 10.3.1

Problem: A shaft with a circular cross section is loaded as shown in Figure (a). This simulates the loads on a propeller shaft. Find the displacement and stress in the shaft.

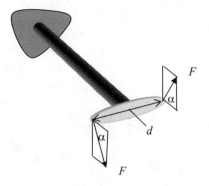

Figure (a)

Solution: The axial and torsional loads can be applied separately and the results for displacement and stress simply added.

The net effect is an axial load (thrust) which we shall designate F_T and a moment (from drag on the propeller blades) which we shall designate M_D.

$$F_T = 2F \cos \alpha \qquad M_D = 2dF \sin \alpha \tag{a}$$

From Example 4.3.1 the displacement and stress for the axial case are

$$u_a(x) = \frac{F_T}{EA} x \qquad \sigma_{xa} = \frac{F_T}{A} \tag{b}$$

From Example 6.3.1 the displacement and stress for the torsional case are

$$\beta(x) = \frac{M_D}{GJ} x \qquad \tau = \frac{M_D}{J} r \tag{c}$$

###########

The finite element equations can be solved separately for axial and torsional effects also. Often they are combined into one set. In the case where an element lies along the x axis we simply place the axial and torsion nodal displacements in the same matrix. Let the nodal displacements and internal forces be

$$\{r_m\} = \begin{bmatrix} u_n \\ \beta_n \\ u_{n+1} \\ \beta_{n+1} \end{bmatrix} \qquad \{p_m\} = \begin{bmatrix} p_{m,n} \\ \psi_{m,n} \\ p_{m,n+1} \\ \psi_{m,n+1} \end{bmatrix} \tag{10.3.1}$$

Combining Equations 10.2.6 and 10.2.18 the element matrix equations become

$$\{p_m\} = [k_m]\{r_m\} = \begin{bmatrix} p_{m,n} \\ \psi_{m,n} \\ p_{m,n+1} \\ \psi_{m,n+1} \end{bmatrix} = \begin{bmatrix} \dfrac{E_m A_m}{l_m} & 0 & -\dfrac{E_m A_m}{l_m} & 0 \\ 0 & \dfrac{G_m J_m}{l_m} & 0 & -\dfrac{G_m J_m}{l_m} \\ -\dfrac{E_m A_m}{l_m} & 0 & \dfrac{E_m A_m}{l_m} & 0 \\ 0 & -\dfrac{G_m J_m}{l_m} & 0 & \dfrac{G_m J_m}{l_m} \end{bmatrix} \begin{bmatrix} u_{m,n} \\ \beta_{m,n} \\ u_{m,n+1} \\ \beta_{m,n+1} \end{bmatrix} \tag{10.3.2}$$

##########

Example 10.3.2

Problem: Solve Example 10.3.1 by finite elements.

Solution: Use Equation 10.3.2.
We can use just one element. Insert restraints, restraint forces, and applied loads.

$$
\begin{bmatrix} R_{x1} \\ Q_{x1} \\ F_T \\ M_D \end{bmatrix} =
\begin{bmatrix}
\dfrac{EA}{L} & 0 & -\dfrac{EA}{L} & 0 \\
0 & \dfrac{GJ}{L} & 0 & -\dfrac{GJ}{L} \\
-\dfrac{EA}{L} & 0 & \dfrac{EA}{L} & 0 \\
0 & -\dfrac{GJ}{L} & 0 & \dfrac{GJ}{L}
\end{bmatrix}
\begin{bmatrix} 0 \\ 0 \\ u_2 \\ \beta_2 \end{bmatrix}
\tag{a}
$$

The partitioned equations for the displacement are

$$
\begin{bmatrix} F_T \\ M_D \end{bmatrix} =
\begin{bmatrix}
\dfrac{EA}{L} & 0 \\
0 & \dfrac{GJ}{L}
\end{bmatrix}
\begin{bmatrix} u_2 \\ \beta_2 \end{bmatrix}
\tag{b}
$$

This is easily solved.

$$
\begin{bmatrix} u_2 \\ \beta_2 \end{bmatrix} =
\begin{bmatrix} \dfrac{F_T L}{EA} \\ \dfrac{M_D L}{GJ} \end{bmatrix}
\tag{c}
$$

The internal forces are

$$
\begin{bmatrix} p_{1,2} \\ \psi_{1,2} \end{bmatrix} =
\begin{bmatrix}
\dfrac{EA}{L} & 0 \\
0 & \dfrac{GJ}{L}
\end{bmatrix}
\begin{bmatrix} u_2 \\ \beta_2 \end{bmatrix} =
\begin{bmatrix} F_T \\ M_D \end{bmatrix}
\tag{d}
$$

and the stresses are

$$
\sigma_x = \frac{F_T}{A} \qquad \tau_a = \frac{M_D}{J} r
\tag{e}
$$

The distributed displacements are

$$
u(x) = \begin{bmatrix} 1 - \dfrac{x}{L} & \dfrac{x}{L} \end{bmatrix} \begin{bmatrix} 0 \\ u_2 \end{bmatrix} = \frac{x}{L} \frac{F_T L}{EA} = \frac{F_T}{EA} x
$$
$$
\beta(x) = \begin{bmatrix} 1 - \dfrac{x}{L} & \dfrac{x}{L} \end{bmatrix} \begin{bmatrix} 0 \\ \beta_2 \end{bmatrix} = \frac{x}{L} \frac{M_D L}{GJ} = \frac{M_D}{GJ} x
\tag{f}
$$

You can compare these results with those in Example 10.3.1.

##########

10.4 Axial and Bending Loads—2D Frames

When loads produce both bending and axial component forces in a bar, the problems may be treated as uncoupled and each case solved separately when the axial force is *sufficiently small*. The results can then be found using the principle of superposition. There are circumstances, however, particularly when the axial forces are large and compressive, when the bending and axial effects are coupled. This is considered in detail, later, in Chapter 14. In this section we limit consideration to cross sections which are symmetrical about the xy plane, for which $I_{yz} = 0$, where the loads are in the xy plane, and when the axial loads are sufficiently small. An example will explain what must be done.

##########

| Example 10.4.1 |

Problem: Consider the case when the load on a slender bar is at an angle to the x axis other than 0 or 90 degrees as shown in Figure (a). The cross section is rectangular with a height h and a width b. Find the displacements and stresses.

Solution: The applied force can be broken into components as shown and the axial and bending problems solved separately. Fortunately the two equations are uncoupled. The displacements and stresses can be found by adding the two results. Superposition holds.

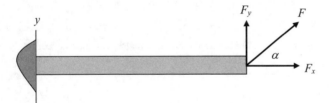

Figure (a)

From Example 4.3.1 the displacement and stress for the axial case are

$$u_a(x) = \frac{F_x}{EA}x = \frac{F\cos\alpha}{EA}x \qquad \sigma_{xa} = \frac{F_x}{A} = \frac{F\cos\alpha}{A} \qquad (a)$$

where $A = hb$.

From Example 7.3.2 the displacement and stress for the bending case are

$$v_b(x) = \frac{F_y}{EI_{zz}}\left(\frac{Lx^2}{2} - \frac{x^3}{6}\right) = \frac{F\sin\alpha}{EI_{zz}}\left(\frac{Lx^2}{2} - \frac{x^3}{6}\right)$$

$$\sigma_{xb}(x, y) = -\frac{F_y(L - x)}{I_{zz}}y = -\frac{F\sin\alpha(L - x)}{I_{zz}}y \qquad (b)$$

$$\tau_{xyb}(x, y) = \frac{F_y}{I_{zz}b}\int_{A_y} ydA_y = \frac{F\sin\alpha}{I_{zz}b}\int_{A_y} ydA_y$$

where $I_{zz} = bh^3/12$. Since in bending the axial displacement is

$$u_b(x, y) = -y\frac{dv}{dx} \qquad (c)$$

The total axial displacement at any point in the bar is given by

$$u(x, y) = u_a + u_b = \frac{F\cos\alpha}{EA}x - \frac{F\sin\alpha}{EI_{zz}}\left(Lx - \frac{x^2}{2}\right)y \qquad (d)$$

The combined stresses are

$$\sigma_x(x, y) = \sigma_a + \sigma_{xb} = \frac{F\cos\alpha}{A} - \frac{F\sin\alpha(L-x)}{I_{zz}}y$$

$$\tau(x, y) = \tau_{xy}(x, y) = \frac{F\sin\alpha}{I_{zz}b}\int_{A_y} ydA_y \qquad (e)$$

###########

We can solve finite element equations independently and add the results or we can adapt them to work with combined loads. In the case where an element lays along the x axis we simply place the axial and bending nodal displacements in the same matrix. The element nodal displacement matrix and the internal force matrix are

$$\{r_m\} = \begin{bmatrix} u_n \\ v_n \\ \theta_n \\ u_{n+1} \\ v_{n+1} \\ \theta_{n+1} \end{bmatrix} \qquad \{p_m\} = \begin{bmatrix} p_{m,n} \\ V_{m,n} \\ M_{m,n} \\ p_{m,n+1} \\ V_{m,n+1} \\ M_{m,n+1} \end{bmatrix} \qquad (10.4.1)$$

Combining Equations 10.2.6 and 10.2.31 the element matrix equations become

$$\begin{bmatrix} p_{m,n} \\ V_{m,n} \\ M_{m,n} \\ p_{m,n+1} \\ V_{m,n+1} \\ M_{m,n+1} \end{bmatrix} = \begin{bmatrix} \dfrac{E_m A_m}{l_m} & 0 & 0 & -\dfrac{E_m A_m}{l_m} & 0 & 0 \\ 0 & \dfrac{12E_m I_m}{l_m^3} & \dfrac{6E_m I_m}{l_m^2} & 0 & -\dfrac{12E_m I_m}{l_m^3} & \dfrac{6E_m I_m}{l_m^2} \\ 0 & \dfrac{6E_m I_m}{l_m^2} & \dfrac{4E_m I_m}{l_m} & 0 & -\dfrac{6E_m I_m}{l_m^2} & \dfrac{2E_m I_m}{l_m} \\ -\dfrac{E_m A_m}{l_m} & 0 & 0 & \dfrac{E_m A_m}{l_m} & 0 & 0 \\ 0 & -\dfrac{12E_m I_m}{l_m^3} & -\dfrac{6E_m I_m}{l_m^2} & 0 & \dfrac{12E_m I_m}{l_m^3} & -\dfrac{6E_m I_m}{l_m^2} \\ 0 & \dfrac{6E_m I_m}{l_m^2} & \dfrac{2E_m I_m}{l_m} & 0 & -\dfrac{6E_m I_m}{l_m^2} & \dfrac{4E_m I_m}{l_m} \end{bmatrix} \begin{bmatrix} u_n \\ v_n \\ \theta_n \\ u_{n+1} \\ v_{n+1} \\ \theta_{n+1} \end{bmatrix}$$

$$(10.4.2)$$

###########

Example 10.4.2

Problem: Solve Example 10.4.1 by finite elements.

Solution: Use Equation 10.3.2.
We can use just one element as numbered in Figure (a).

node 1   node 2

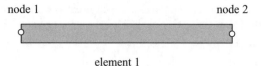

element 1

Figure (a)

The restraints, restraint forces, and loads have been inserted. We have added nodal coordinate subscripts to better keep track of the forces.

$$
\begin{bmatrix} R_{1x} \\ R_{1y} \\ Q_1 \\ \hline F_{2x} \\ F_{2y} \\ 0 \end{bmatrix}
=
\left[\begin{array}{ccc:ccc}
\dfrac{EA}{L} & 0 & 0 & -\dfrac{EA}{L} & 0 & 0 \\
0 & \dfrac{12EI}{L^3} & \dfrac{6EI}{L^2} & 0 & -\dfrac{12EI}{L^3} & \dfrac{6EI}{L^2} \\
0 & \dfrac{6EI}{L^2} & \dfrac{4EI}{L} & 0 & -\dfrac{6EI}{L^2} & \dfrac{2EI}{L} \\
\hdashline
-\dfrac{EA}{L} & 0 & 0 & \dfrac{EA}{L} & 0 & 0 \\
0 & -\dfrac{12EI}{L^3} & -\dfrac{6EI}{L^2} & 0 & \dfrac{12EI}{L^3} & -\dfrac{6EI}{L^2} \\
0 & \dfrac{6EI}{L^2} & \dfrac{2EI}{L} & 0 & -\dfrac{6EI}{L^2} & \dfrac{4EI}{L}
\end{array}\right]
\begin{bmatrix} 0 \\ 0 \\ 0 \\ \hline u_2 \\ v_2 \\ \theta_2 \end{bmatrix}
\tag{a}
$$

Partition and solve for the displacements.

$$
\begin{bmatrix} F_{2x} \\ F_{2y} \\ 0 \end{bmatrix}
=
\begin{bmatrix}
\dfrac{EA}{L} & 0 & 0 \\
0 & \dfrac{12EI}{L^3} & -\dfrac{6EI}{L^2} \\
0 & -\dfrac{6EI}{L^2} & \dfrac{4EI}{L}
\end{bmatrix}
\begin{bmatrix} u_2 \\ v_2 \\ \theta_2 \end{bmatrix}
\tag{b}
$$

The displacements are

$$
\begin{bmatrix} u_2 \\ v_2 \\ \theta_2 \end{bmatrix}
=
\begin{bmatrix} \dfrac{F_{2x}L}{EA} \\ \dfrac{F_{2y}L^3}{3EI} \\ \dfrac{F_{2y}L^2}{2EI} \end{bmatrix}
=
\begin{bmatrix} \dfrac{F\cos\alpha L}{EA} \\ \dfrac{F\sin\alpha L^3}{3EI} \\ \dfrac{F\sin\alpha L^2}{2EI} \end{bmatrix}
\tag{c}
$$

Once the nodal displacements are known the distributed displacements can be found using the shape functions. The restraint forces are found from the other partitioned equation

$$
\begin{bmatrix} R_{1x} \\ R_{1y} \\ Q_1 \end{bmatrix}
=
\begin{bmatrix}
-\dfrac{EA}{L} & 0 & 0 \\
0 & -\dfrac{12EI}{L^3} & \dfrac{6EI}{L^2} \\
0 & -\dfrac{6EI}{L^2} & \dfrac{2EI}{L}
\end{bmatrix}
\begin{bmatrix} u_2 \\ v_2 \\ \theta_2 \end{bmatrix}
=
\begin{bmatrix}
-\dfrac{EA}{L} & 0 & 0 \\
0 & -\dfrac{12EI}{L^3} & \dfrac{6EI}{L^2} \\
0 & -\dfrac{6EI}{L^2} & \dfrac{2EI}{L}
\end{bmatrix}
\begin{bmatrix} \dfrac{F\cos\alpha L}{EA} \\ \dfrac{F\sin\alpha L^3}{3EI} \\ \dfrac{F\sin\alpha L^2}{2EI} \end{bmatrix}
\tag{d}
$$

The stresses are obtained from the nodal forces using the usual formulas. In this case of using just one element the element nodal forces are just the applied loads and the restraint forces, that is,

$$
\begin{bmatrix}
p_{1,1} \\
V_{1,1} \\
M_{1,1} \\
p_{1,2} \\
V_{1,2} \\
M_{1,2}
\end{bmatrix}
=
\begin{bmatrix}
R_{1x} \\
R_{1y} \\
Q_1 \\
F_{2x} \\
F_{2y} \\
0
\end{bmatrix}
\tag{e}
$$

###########

For a simple frame structure such as the one shown in Figure 10.4.1 we are concerned about both axial and bending displacements and stresses in the same member.

The differential equation solution can be found by dividing the frame into two members, the vertical part and the horizontal part, and matching boundary conditions where the two parts meet. It becomes a bit of a chore, especially when the frames become more complicated. We shall go directly to the FEM formulation and find that it is straight forward to obtain a solution.

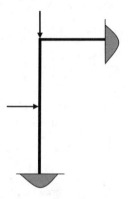

Figure 10.4.1

For a simple frame, like the one in Figure 10.4.1, we must set up a transformation matrix to find the element stiffness matrices for different orientations of the elements. In Section 5.6 we developed the transformation matrix for axial displacements only for use in pin jointed truss problems. It becomes more complicated when both axial and bending displacements are combined.

We show an element oriented at an angle α to the x axis in Figure 10.4.2.

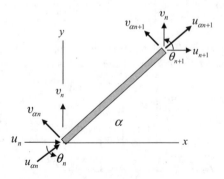

Figure 10.4.2

Here we have both u and v components at each node and have added the rotational nodal displacements. From the notation in Figure 10.4.2 we note that

$$\{p_{\alpha m}\} = [k_{\alpha m}]\{r_{\alpha m}\} = \frac{E_m I_m}{l_m^3} \begin{bmatrix} \dfrac{A_m l_m^2}{I_m} & 0 & 0 & -\dfrac{A_m l_m^2}{I_m} & 0 & 0 \\ 0 & 12 & 6l_m & 0 & -12 & 6l_m \\ 0 & 6l_m & 4l_m^2 & 0 & -6l_m & 2l_m^2 \\ -\dfrac{A_m l_m^2}{I_m} & 0 & 0 & \dfrac{A_m l_m^2}{I_m} & 0 & 0 \\ 0 & -12 & -6l_m & 0 & 12 & -6l_m \\ 0 & 6l_m & 2l_m^2 & 0 & -6l_m & 4l_m^2 \end{bmatrix} \begin{bmatrix} u_{\alpha n} \\ v_{\alpha n} \\ \theta_{\alpha n} \\ u_{\alpha n+1} \\ v_{\alpha n+1} \\ \theta_{\alpha n+1} \end{bmatrix}$$

(10.4.3)

Following the steps in the axial case (see Chapter 5, Section 5.6, Equations 5.6.1-5) we can see that the transformation relations at node n are

$$\begin{bmatrix} u_{\alpha n} \\ v_{\alpha n} \end{bmatrix} = \begin{bmatrix} \cos\alpha & \sin a \\ -\sin\alpha & \cos\alpha \end{bmatrix} \begin{bmatrix} u_n \\ v_n \end{bmatrix} \qquad \theta_{\alpha n} = \theta_n$$

(10.4.4)

The full set of transformations for both nodes is presented in Equation 10.4.5.

$$[r_{\alpha m}] = \begin{bmatrix} u_{\alpha n} \\ v_{\alpha n} \\ \theta_{\alpha n} \\ u_{\alpha n+1} \\ v_{\alpha n+1} \\ \theta_{\alpha n+1} \end{bmatrix} = \begin{bmatrix} \cos\alpha & \sin\alpha & 0 & 0 & 0 & 0 \\ -\sin\alpha & \cos\alpha & 0 & 0 & 0 & 0 \\ 0 & 0 & 1 & 0 & 0 & 0 \\ 0 & 0 & 0 & \cos\alpha & \sin\alpha & 0 \\ 0 & 0 & 0 & -\sin\alpha & \cos\alpha & 0 \\ 0 & 0 & 0 & 0 & 0 & 1 \end{bmatrix} \begin{bmatrix} u_n \\ v_n \\ \theta_n \\ u_{n+1} \\ v_{n+1} \\ \theta_{n+1} \end{bmatrix} = [T]\{r_m\}$$

(10.4.5)

The same transformation applies to the internal force matrix, thus

$$\{p_{\alpha m}\} = [T]\{p_m\}$$

(10.4.6)

As explained in Chapter 5 a transformation matrix has the interesting property that its inverse is equal to its transpose, that is, $[T]^{-1} = [T]^T$. You can verify that $[T]^T[T] = [1]$. Following the same steps used in Section 5.6, Equations 5.6.7 -10, we can state that

$$\{p_m\} = [T]^T\{p_{\alpha m}\}$$

(10.4.7)

and since

$$\{p_{\alpha m}\} = [k_{\alpha m}]\{r_{\alpha m}\}$$
$$\rightarrow [T]\{p_m\} = [k_{\alpha m}][T]\{r_m\}$$
$$\rightarrow [T]^T[T]\{p_m\} = \{p_m\} = [T]^T[k_{\alpha m}][T]\{r_m\} = [k_m]\{r_m\}$$

(10.4.8)

Thus

$$[k_m] = [T]^T[k_{\alpha m}][T]$$

(10.4.9)

################

Example 10.4.3

Problem: The frame in Figure 10.4.1 has the dimensions and loads shown in Figure (a). The cross section is 24 *mm* by 24 *mm*. It is made of aluminum, $E = 68950$ *MPa*. Solve for the displacements and stresses.

Solution: Choose three elements as shown in Figure (a). Number the nodes and elements and insert values in the appropriate equations. Note that since the loads are concentrated this will obtain exactly the same answer as the differential equation solution.

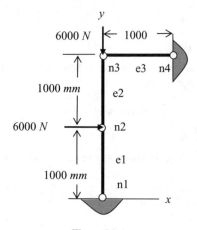

Figure (a)

For elements 1 and 2 the angle α is $90°$, thus, $\cos 90° = 0$ and $\sin 90° = 1$. The stiffness matrix for these two elements is

$$[k_1] = [k_2] = \begin{bmatrix} 0 & -1 & 0 & 0 & 0 & 0 \\ 1 & 0 & 0 & 0 & 0 & 0 \\ 0 & 0 & 1 & 0 & 0 & 0 \\ 0 & 0 & 0 & 0 & -1 & 0 \\ 0 & 0 & 0 & 1 & 0 & 0 \\ 0 & 0 & 0 & 0 & 0 & 1 \end{bmatrix} [k_{\alpha m}] \begin{bmatrix} 0 & 1 & 0 & 0 & 0 & 0 \\ -1 & 0 & 0 & 0 & 0 & 0 \\ 0 & 0 & 1 & 0 & 0 & 0 \\ 0 & 0 & 0 & 0 & 1 & 0 \\ 0 & 0 & 0 & -1 & 0 & 0 \\ 0 & 0 & 0 & 0 & 0 & 1 \end{bmatrix} \quad \text{(a)}$$

where

$$[k_{\alpha m}] = \frac{E_m I_m}{l_m^3} \begin{bmatrix} \dfrac{A_m l_m^2}{I_m} & 0 & 0 & -\dfrac{A_m l_m^2}{I_m} & 0 & 0 \\ 0 & 12 & 6l_m & 0 & -12 & 6l_m \\ 0 & 6l_m & 4l_m^2 & 0 & -6l_m & 2l_m^2 \\ -\dfrac{A_m l_m^2}{I_m} & 0 & 0 & \dfrac{A_m l_m^2}{I_m} & 0 & 0 \\ 0 & -12 & -6l_m & 0 & 12 & -6l_m \\ 0 & 6l_m & 2l_m^2 & 0 & -6l_m & 4l_m^2 \end{bmatrix} \quad \text{(b)}$$

Since all elements are the same length and have the same cross sections we will drop the subscripts *m*.

When Equation (b) is inserted in Equation (a) and the multiplications are carried out the first two element stiffness matrices are

$$[k_1] = [k_2] = \frac{EI}{l^3} \begin{bmatrix} 12 & 0 & -6l & -12 & 0 & -6l \\ 0 & \dfrac{Al^2}{I} & 0 & 0 & -\dfrac{Al^2}{I} & 0 \\ -6l & 0 & 4l^2 & 6l & 0 & 2l^2 \\ -12 & 0 & 6l & 12 & 0 & 6l \\ 0 & -\dfrac{Al^2}{I} & 0 & 0 & \dfrac{Al^2}{I} & 0 \\ -6l & 0 & 2l^2 & 6l & 0 & 4l^2 \end{bmatrix} \tag{c}$$

Now we must be careful in the interpretation of the internal nodal forces in the equation for each element.

For element 1

$$\{p_1\} = [k_1]\{r_1\} \tag{d}$$

For element 1, the x direction displacements and forces are lateral forces associated with bending and the y direction displacements and forces are axial. We need a suitable notation to identify the direction of the internal and external forces at each node. Note in Equation (e) we have associated $V_{1,1}$ with u_1 and $P_{1,1}$ with v_1, that is, shear forces with the x direction displacement and axial forces with the y direction displacement. Note also that the matrix transformation exhibited in Equation (a) produces the axial stiffness components is the proper row and column.

$$\{p_1\} = \begin{Bmatrix} V_{1,1} \\ P_{1,1} \\ M_{1,1} \\ V_{1,2} \\ P_{1,2} \\ M_{1,2} \end{Bmatrix} = \frac{EI}{l^3} \begin{bmatrix} 12 & 0 & -6l & -12 & 0 & -6l \\ 0 & \dfrac{Al^2}{I} & 0 & 0 & -\dfrac{Al^2}{I} & 0 \\ -6l & 0 & 4l^2 & 6l & 0 & 2l^2 \\ -12 & 0 & 6l & 12 & 0 & 6l \\ 0 & -\dfrac{Al^2}{I} & 0 & 0 & \dfrac{Al^2}{I} & 0 \\ -6l & 0 & 2l^2 & 6l & 0 & 4l^2 \end{bmatrix} \begin{Bmatrix} u_1 \\ v_1 \\ \theta_1 \\ u_2 \\ v_2 \\ \theta_2 \end{Bmatrix} \tag{e}$$

The same transformation occurs for element 2; however, element 3 needs no transformation and the usual displacements and forces are associated. We have for element 3 with $\alpha = 0$

$$[k_3] = \frac{EI}{l^3} \begin{bmatrix} \dfrac{Al^2}{I} & 0 & 0 & -\dfrac{Al^2}{I} & 0 & 0 \\ 0 & 12 & 6l & 0 & -12 & 6l \\ 0 & 6l & 4l^2 & 0 & -6l & 2l^2 \\ -\dfrac{Al^2}{I} & 0 & 0 & \dfrac{Al^2}{I} & 0 & 0 \\ 0 & -12 & -6l & 0 & 12 & -6l \\ 0 & 6l & 2l^2 & 0 & -6l & 4l^2 \end{bmatrix} \tag{f}$$

We must extend this notation to the global matrices. The applied force matrix with restraint forces and applied load added and the nodal displacement matrix with the restraints added become

$$
\{F\} =
\begin{bmatrix}
F_{1x} \\
F_{1y} \\
M_1 \\
F_{2x} \\
F_{2y} \\
M_2 \\
F_{3x} \\
F_{3y} \\
M_3 \\
F_{4x} \\
F_{4y} \\
M_4
\end{bmatrix}
=
\begin{bmatrix}
R_{1x} \\
R_{1y} \\
Q_1 \\
6000 \\
0 \\
0 \\
0 \\
-6000 \\
0 \\
R_{4x} \\
R_{4y} \\
Q_4
\end{bmatrix}
\qquad
\{r\} =
\begin{bmatrix}
u_1 \\
v_1 \\
\theta_1 \\
u_2 \\
v_2 \\
\theta_2 \\
u_3 \\
v_3 \\
\theta_3 \\
u_4 \\
v_4 \\
\theta_4
\end{bmatrix}
=
\begin{bmatrix}
0 \\
0 \\
0 \\
u_2 \\
v_2 \\
\theta_2 \\
u_3 \\
v_3 \\
\theta_3 \\
0 \\
0 \\
0
\end{bmatrix}
\tag{g}
$$

The complete global matrix equation is shown in Equation (h).

$$
[F] =
\begin{bmatrix}
R_{1x} \\
R_{1y} \\
Q_1 \\
\hline
6000 \\
0 \\
0 \\
0 \\
-6000 \\
0 \\
\hline
R_{4x} \\
R_{4y} \\
Q_4
\end{bmatrix}
= [K]\{r\}
$$

$$
= \frac{EI}{l^3}
\begin{bmatrix}
12 & 0 & -6l & -12 & 0 & -6l & 0 & 0 & 0 & 0 & 0 & 0 \\
0 & \dfrac{Al^2}{I} & 0 & 0 & -\dfrac{Al^2}{I} & 0 & 0 & 0 & 0 & 0 & 0 & 0 \\
-6l & 0 & 4l^2 & 6l & 0 & 2l^2 & 0 & 0 & 0 & 0 & 0 & 0 \\
-12 & 0 & 6l & 24 & 0 & 0 & -12 & 0 & -6l & 0 & 0 & 0 \\
0 & -\dfrac{Al^2}{I} & 0 & 0 & \dfrac{2Al^2}{I} & 0 & 0 & -\dfrac{Al^2}{I} & 0 & 0 & 0 & 0 \\
-6l & 0 & 2l^2 & 0 & 0 & 8l^2 & 6l & 0 & 2l^2 & 0 & 0 & 0 \\
0 & 0 & 0 & -12 & 0 & 6l & 12+\dfrac{Al^2}{I} & 0 & 6l & -\dfrac{Al^2}{I} & 0 & 0 \\
0 & 0 & 0 & 0 & -\dfrac{Al^2}{I} & 0 & 0 & \dfrac{Al^2}{I}+12 & 6l & 0 & -12 & 6l \\
0 & 0 & 0 & -6l & 0 & 2l^2 & 6l & 6l & 8l^2 & 0 & -6l & 2l^2 \\
0 & 0 & 0 & 0 & 0 & 0 & -\dfrac{Al^2}{I} & 0 & 0 & \dfrac{Al^2}{I} & 0 & 0 \\
0 & 0 & 0 & 0 & 0 & 0 & 0 & -12 & -6l & 0 & 12 & -6l \\
0 & 0 & 0 & 0 & 0 & 0 & 0 & 6l & 2l^2 & 0 & -6l & 4l^2
\end{bmatrix}
\begin{bmatrix}
0 \\
0 \\
0 \\
u_2 \\
v_2 \\
\theta_2 \\
u_3 \\
v_3 \\
\theta_3 \\
0 \\
0 \\
0
\end{bmatrix}
\tag{h}
$$

Insert $l = 1000\ mm$, $A = 576\ mm^2$, $E = 68950\ Mpa$ and the partitioned equations for the deflections are

$$
\begin{bmatrix} 6000 \\ 0 \\ 0 \\ 0 \\ -6000 \\ 0 \end{bmatrix} = 1.9063
\begin{bmatrix}
24 & 0 & 0 & -12 & 0 & -6000 \\
0 & 41667 & 0 & 0 & -20833 & 0 \\
0 & 0 & 8000000 & 6000 & 0 & 2000000 \\
-12 & 0 & 6000 & 20845 & 0 & 6000 \\
0 & -20833 & 0 & 0 & 20845 & 6000 \\
-6000 & 0 & 2000000 & 6000 & 6000 & 8000000
\end{bmatrix}
\begin{bmatrix} u_2 \\ v_2 \\ \theta_2 \\ u_3 \\ v_3 \\ \theta_3 \end{bmatrix}
\tag{i}
$$

Using one of the linear equation solvers (see, for example, Appendix C) we get

$$
\begin{bmatrix} u_2 \\ v_2 \\ \theta_2 \\ u_3 \\ v_3 \\ \theta_3 \end{bmatrix}
=
\begin{bmatrix} 164.05 \\ -0.1887 \\ -0.0329 \\ 0.0661 \\ -0.3775 \\ 0.1315 \end{bmatrix}
\begin{matrix} mm \\ mm \\ rad \\ mm \\ mm \\ rad \end{matrix}
\tag{j}
$$

The internal forces are

$$
\begin{bmatrix} V_{1,1} \\ P_{1,1} \\ M_{1,1} \\ V_{1,2} \\ P_{1,2} \\ M_{1,2} \end{bmatrix}
= 1.9063
\begin{bmatrix}
12 & 0 & -6000 & -12 & 0 & -6000 \\
0 & 20833 & 0 & 0 & -20833 & 0 \\
-6000 & 0 & 4000000 & 6000 & 0 & 2000000 \\
-12 & 0 & 6000 & 12 & 0 & 6000 \\
0 & -20833 & 0 & 0 & 20833 & 0 \\
-6000 & 0 & 2000000 & 6000 & 0 & 4000000
\end{bmatrix}
\begin{bmatrix} 0 \\ 0 \\ 0 \\ 164.05 \\ -0.1887 \\ -0.0329 \end{bmatrix}
=
\begin{bmatrix} -3376.2 \\ 7495.5 \\ 1750877 \\ 3376.2 \\ -7495.5 \\ 1625344 \end{bmatrix}
\begin{matrix} N \\ N \\ N \cdot mm \\ N \\ N \\ N \cdot mm \end{matrix}
\tag{k}
$$

$$
\begin{bmatrix} V_{2,2} \\ P_{2,2} \\ M_{2,2} \\ V_{2,3} \\ P_{2,3} \\ M_{2,3} \end{bmatrix}
= 1.9063
\begin{bmatrix}
12 & 0 & -6000 & -12 & 0 & -6000 \\
0 & 20833 & 0 & 0 & -20833 & 0 \\
-6000 & 0 & 4000000 & 6000 & 0 & 2000000 \\
-12 & 0 & 6000 & 12 & 0 & 6000 \\
0 & -20833 & 0 & 0 & 20833 & 0 \\
-6000 & 0 & 2000000 & 6000 & 0 & 4000000
\end{bmatrix}
\begin{bmatrix} 164.05 \\ -0.1887 \\ -0.0329 \\ 0.0661 \\ -0.3775 \\ 0.1315 \end{bmatrix}
=
\begin{bmatrix} 2623.8 \\ 7495.5 \\ -1625344 \\ -2623.8 \\ -7495.5 \\ -998435 \end{bmatrix}
\begin{matrix} N \\ N \\ N \cdot mm \\ N \\ N \\ N \cdot mm \end{matrix}
\tag{l}
$$

$$
\begin{bmatrix} P_{3,3} \\ V_{3,3} \\ M_{3,3} \\ P_{3,4} \\ V_{3,4} \\ M_{3,4} \end{bmatrix}
= 1.9063
\begin{bmatrix}
20833 & 0 & 0 & -20833 & 0 & 0 \\
0 & 12 & 6000 & 0 & -12 & 6000 \\
0 & 6000 & 4000000 & 0 & -6000 & 2000000 \\
-20833 & 0 & 0 & 20833 & 0 & 0 \\
0 & -12 & -6000 & 0 & 12 & -6000 \\
0 & 6000 & 2000000 & 0 & -6000 & 4000000
\end{bmatrix}
\begin{bmatrix} 0.0661 \\ -0.3775 \\ 0.1315 \\ 0 \\ 0 \\ 0 \end{bmatrix}
=
\begin{bmatrix} 2623.8 \\ 1495.5 \\ 998435 \\ -2623.8 \\ -1495.5 \\ 497059 \end{bmatrix}
\begin{matrix} N \\ N \\ N \cdot mm \\ N \\ N \\ N \cdot mm \end{matrix}
\tag{m}
$$

To find the stresses you must combine the bending and axial terms, that is

$$
\sigma = -\frac{My}{I} + \frac{P}{A} \qquad \tau = \frac{V}{2I}\left(\frac{12^2}{4} - y^2\right)
\tag{n}
$$

To find the distributed displacements use the shape functions for each element. The final shape (not to scale) would be approximately as shown in Figure (b) with the bending displacements dominating the axial ones. For example, the vertical displacement of node 2 ($-0.1887\ mm$) due to the axial displacement of element 1 is so small that it can reasonably be neglected; however, that lateral displacement due to bending is quite large. At node 3 both the vertical and horizontal displacement resulting from the axial displacements of the elements also is very small; however, the rotation at node 3 ($0.1315\ rad$) is large enough to produce a significant bending displacement in the member.

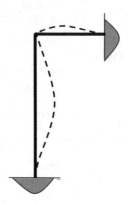

Figure (b)

##########

The solution of even such a simple frame problem like this by either the classical differential equation method or the modern finite element method is quite a chore. Fortunately, the finite element method has been programmed for graphical input of the geometry, automatic creation of nodes and elements, assembly of the matrices, and direct application of loads and restraints to the assembled equations. Partitioning of the equation and their solution is also automated. An additional bonus is the graphical as well as numerical presentation of the results. A tutorial on using a finite element code for solving this and more complicated frame problems is provided separately.

10.5 Bending in Two Planes

10.5.1 When I_{yz} is Equal to Zero

So far we have only considered beams that have the yz axes oriented as principal axes of inertia ($I_{yz} = 0$). This is true for all beams with cross sections symmetrical about at least one of the axes, y or z. We also have limited the discussion to loads only in the xy plane. The simple act of placing a load at an angle to the y axis will induce displacements in both the y and z directions. Consider a beam with a distributed loading represented by both $f_y(x)$ and $f_z(x)$. If $I_{yz} = 0$ the problem in each component direction is uncoupled from the other. In addition to the equations we have been using (xy plane) we must solve those in the xz plane. The displacement in the z direction is denoted by $w(x)$. The width of the cross section in the y direction is h.

To summarize, these equations for bending in the xz plane when $I_{yz} = 0$ are

$$M_{cy}, F_{cz} \quad \rightarrow \quad f_z(x) \quad V_z(x) \quad M_y(x) \quad \sigma_x(x,z) \quad \varepsilon_x(x,z) \quad u(x,z) \quad w(x) \leftarrow \rho$$

$$\frac{dV_z}{dx} = -f_z(x) \qquad \frac{dM_y}{dx} = -V_z(x) \qquad\qquad u(x,z) = -z\frac{dw}{dx}$$

$$\varepsilon_x(x,z) = \frac{du}{dx} = -y\frac{d^2w}{dx^2}$$

$$\sigma_x(x,z) = E\varepsilon_x = -Ey\frac{d^2w}{dx^2}$$

$$M_y(x) = -\int_A \sigma_x z\, dA = E\frac{d^2w}{dx^2}\int_A z^2\, dA = EI_{yy}\frac{d^2w}{dx^2} \quad \rightarrow \quad \sigma_x = -\frac{M_y z}{I_{yy}} \qquad (10.5.1)$$

$$\frac{dM_y}{dx} = \frac{d}{dx}EI_{yy}\frac{d^2w}{dx^2} = -V_z(x) \qquad \tau_{zx}(x,z) = \frac{V_z}{I_{yy}h}\int_{A_z} z\, dA_z$$

$$\frac{d^2M_y}{dx^2} = \frac{d^2}{dx^2}EI_{yy}\frac{d^2w}{dx^2} = f_z(x)$$

Since we know how to deal with similar equations in the xy plane it should present no particular difficulty to solve these equations for specific cases. The results for the two sets are simply added together, that is, the total magnitude of the displacement is the vector sum of $v(w)$ and $w(x)$ and the normal stresses are

$$\sigma_x(x, y, z) = -\frac{M_z}{I_{zz}}y - \frac{M_y}{I_{yy}}z \tag{10.5.2}$$

In a similar way we write the finite element equations for the two planes. There is a small problem with sign conventions. In the classical differential equations the sign convention in the xz plane is as shown in Figure 10.5.1. Subscripts have been added to be sure which components are in effect.

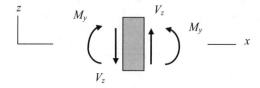

Figure 10.5.1

Note that the shear forces are positive by the same rule as shear stresses and the moments are positive if they produce compressive normal stresses in the positive z space (this is the convention we have chosen). This is the same rule as we used in the xy plane and so the equations come out with the same signs.

In finite elements the sign convention is to have all internal forces positive according to the right hand rule as shown in Figure 10.5.2. This is desirable for the addition, that is, the assembly, of the element matrices to form the global matrix equations.

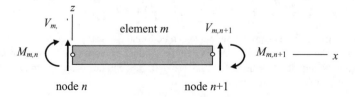

Figure 10.5.2

Note that the moments are in the opposite direction from those in Chapter 8, Figure 8.2.3. This is because in Figure 8.2.3 the z axis is positive in the direction toward the observer and in Figure 10.5.2 the y axis is positive away from the observer. The result is that the element stiffness matrix in the xz plane has similar components but some different signs from those in the xy plane.

$$\{p_m\} = [k_m]\{r_m\} = \begin{bmatrix} V_{m,n} \\ M_{m,n} \\ V_{m,n+1} \\ M_{m,n+1} \end{bmatrix} = \frac{E_m I_m}{l_m^3} \begin{bmatrix} 12 & -6l_m & -12 & -6l_m \\ -6l_m & 4l_m^2 & 6l_m & 2l_m^2 \\ -12 & -6l_m & 12 & 6l_m \\ -6l_m & 2l_m^2 & 6l_m & -4l_m^2 \end{bmatrix} \begin{bmatrix} v_{m,n} \\ \theta_{m,n} \\ v_{m,n+1} \\ \theta_{m,n+1} \end{bmatrix} \tag{10.5.3}$$

In any case the problem can be solved in each plane and the results added for the total effect. The two stiffness matrices can be combined into one as we did for combined axial and bending. In fact the axial, torsional, and two bending matrices can be combined into one. This gets to be a lot of writing. The FEM computer codes do this nicely for us and there is no need to explicitly write the lengthy equations.

10.5.2   When I_{yz} is Not Equal to Zero

When $I_{yz} \neq 0$ we must modify the governing equations. We can still assume that plane sections remain plane but now consider the rotation of the cross section about both the y and z axes at the same time. By neglecting the Poisson's ratio effect and shear deformation we can assume $v(x)$ and $w(x)$ are the lateral displacements. The displacement in the x direction becomes

$$u(x, y, z) = -y\frac{dv}{dx} - z\frac{dw}{dx} \tag{10.5.4}$$

The strain component of interest is (remember we neglect shear deformation)

$$\varepsilon_x(x, y, z) = \frac{\partial u}{\partial x} = -y\frac{d^2v}{dx^2} - z\frac{d^2w}{dx^2} \tag{10.5.5}$$

The axial stress in the interior of the beam is

$$\sigma_x(x, y, z) = E\varepsilon_x = -yE\frac{d^2v}{dx^2} - zE\frac{d^2w}{dx^2} \tag{10.5.6}$$

The stress resultant moments are defined by

$$M_y(x) = -\int_A \sigma_x(x, y, z)z dA = E\int_A \left(y\frac{d^2v}{dx^2} + z\frac{d^2w}{dx^2}\right)z dA$$
$$= E\left(\frac{d^2v}{dx^2}\int_A yz dA + \frac{d^2w}{dx^2}\int_A z^2 dA\right) = E\left(\frac{d^2v}{dx^2}I_{yz} + \frac{d^2w}{dx^2}I_{yy}\right) \tag{10.5.7}$$

$$M_z(x) = -\int_A \sigma_x(x, y, z)y dA = E\int_A \left(y\frac{d^2v}{dx^2} + z\frac{d^2w}{dx^2}\right)y dA$$
$$= E\left(\frac{d^2v}{dx^2}\int_A y^2 dA + \frac{d^2w}{dx^2}\int_A yz dA\right) = E\left(\frac{d^2v}{dx^2}I_{zz} + \frac{d^2w}{dx^2}I_{yz}\right) \tag{10.5.8}$$

For a statically determinate beam the moments can be found from static equilibrium and therefore are known quantities. The area moments of inertia and the product of inertia are known for any given cross section. Thus the two equations, Equations 10.5.7 and 10.5.8, can be solved for the second derivatives of the displacements in terms of the moments.

$$\frac{d^2v}{dx^2} = \frac{1}{E}\frac{I_{yy}M_z - I_{yz}M_y}{I_{yy}I_{zz} - I_{yz}^2} \qquad \frac{d^2w}{dx^2} = -\frac{1}{E}\frac{I_{yz}M_z - I_{zz}M_y}{I_{yy}I_{zz} - I_{yz}^2} \tag{10.5.9}$$

The stress is found by substituting Equations 10.5.9 into Equation 10.5.6.

$$\sigma_x(x, y, z) = E\varepsilon_x = -yE\frac{d^2v}{dx^2} - zE\frac{d^2w}{dx^2} = -\left(\frac{I_{yy}M_z - I_{yz}M_y}{I_{yy}I_{zz} - I_{yz}^2}\right)y + \left(\frac{I_{yz}M_z - I_{zz}M_y}{I_{yy}I_{zz} - I_{yz}^2}\right)z \tag{10.5.10}$$

These equations can be made to look less intimidating by the following notation, let

$$I = I_{yy}I_{zz} - I_{yz}^2 \tag{10.5.11}$$

and

$$I_{zz} = \frac{I_{yy}I_{zz} - I_{yz}^2}{I_{yy}} = \frac{I}{I_{yy}} \qquad I_{yy} = \frac{I_{yy}I_{zz} - I_{yz}^2}{I_{zz}} = \frac{I}{I_{zz}} \qquad I_{yz} = \frac{I_{yy}I_{zz} - I_{yz}^2}{I_{yz}} = \frac{I}{I_{yz}} \tag{10.5.12}$$

Equations 10.5.9 and 10.5.10 may then be written

$$\frac{d^2v}{dx^2} = \frac{1}{E}\frac{I_{yy}M_z - I_{yz}M_y}{I_{yy}I_{zz} - I_{yz}^2}$$

$$\rightarrow EI\frac{d^2v}{dx^2} = I_{yy}M_z - I_{yz}M_y \quad \rightarrow \quad \frac{d^2v}{dx^2} = \frac{M_z}{EI_{zz}} - \frac{M_y}{EI_{yz}} \qquad (10.5.13)$$

$$\frac{d^2w}{dx^2} = -\frac{1}{E}\frac{I_{yz}M_z - I_{zz}M_y}{I_{yy}I_{zz} - I_{yz}^2}$$

$$\rightarrow EI\frac{d^2w}{dx^2} = I_{yz}M_z - I_{zz}M_y \quad \rightarrow \quad \frac{d^2w}{dx^2} = -\frac{M_z}{EI_{yz}} + \frac{M_y}{EI_{yy}}$$

$$\sigma_x(x, y, z) = -\left(\frac{I_{yy}M_z - I_{yz}M_y}{I_{yy}I_{zz} - I_{yz}^2}\right)y + \left(\frac{I_{yz}M_z - I_{zz}M_y}{I_{yy}I_{zz} - I_{yz}^2}\right)z$$

$$\rightarrow \sigma_x(x, y, z) = -\left(\frac{M_z}{I_{zz}} - \frac{M_y}{I_{yz}}\right)y + \left(\frac{M_z}{I_{yz}} - \frac{M_y}{I_{yy}}\right)z \qquad (10.5.14)$$

Just as a check, if $I_{yz} = 0$, these equations reduce to

$$\frac{d^2v}{dx^2} = \frac{M_z}{EI_{zz}} \qquad \frac{d^2w}{dx^2} = \frac{M_y}{EI_{yy}}$$

$$\sigma_x(x, y, z) = -\frac{M_z}{I_{zz}}y - \frac{M_y}{I_{yy}}z \qquad (10.5.15)$$

These are the same as the ones we derived before for $I_{yz} = 0$ in Chapter 7 and presented in the first part of this section.

###########

Example 10.5.1

Problem: A cantilever beam has a cross section as shown in Figure (a). It is oriented with respect to rectangular Cartesian coordinates as shown. The x axis is the centroidal axis.

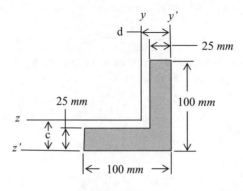

Figure (a)

A concentrated moment about the z axis is applied at the free end as shown in Figure (b). Find the normal stresses in and the displacement of the beam.

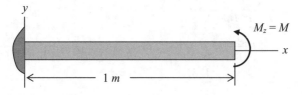

Figure (b)

Solution: First find the centroid of the cross section. Then we have two choices. One is to find the principal axes of inertia of the cross section with respect to centroidal axes and use Equations 10.5.15. The other is to find the moments and product of inertia with respect to the centroidal axes as shown in Figure (a) and use Equations 10.5.9 and 10.5.10. We will do the latter.

First we must find the centroid of the cross section and the centroidal moments and product of inertia. The centroid is found by dividing the cross section into two rectangular portions as shown in Figure (c) and use the transfer formula from Appendix B, Equation B.2.3.

The centroid is measured from the $y'z'$ axes and is located at

$$d = \frac{\sum z_s A_s}{\sum A_s} = \frac{12.5 \cdot 75 \cdot 25 + 50 \cdot 100 \cdot 25}{75 \cdot 25 + 100 \cdot 25} = 33.93 \ mm \tag{a}$$

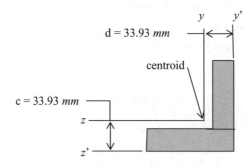

Figure (c)

From symmetry

$$c = 33.93 \ mm \tag{b}$$

Now we must find the moments of inertia with respect to the yz centroidal axes. See Appendix B, Equations B.3.5-7. The dimensions from the centroid of the angle to the centroids of the two segments are shown in Figure (d). The centroidal moments of inertia are

$$I_{yy} = \sum \left(I_{y_c y_c} + A z_c^2 \right) = \frac{75 \cdot 25^3}{12} + 75 \cdot 25 \cdot (21.43)^2 + \frac{25 \cdot 100^3}{12} + 100 \cdot 25 \cdot (16.07)^2$$
$$= 3.688 \cdot 10^6 \ mm^4 \tag{c}$$

From symmetry

$$I_{zz} = 3.688 \cdot 10^6 \ mm^4 \tag{d}$$

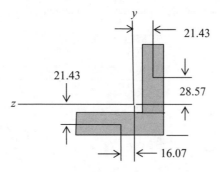

Figure (d)

To find the product of inertia we note that, for each rectangular area, one of the offset distances in negative and one is positive. The product of inertia is

$$I_{yz} = \sum (I_{y_c z_c} + A y_c z_c)$$
$$= -75 \cdot 25 \cdot (28.57)(21.43) - 100 \cdot 25 \cdot (21.43)(16.07) \tag{e}$$
$$= -2.100 \cdot 10^6 \ mm^4$$

Next we find the following quantities.

$$I = I_{yy} I_{zz} - I_{yz}^2 = (3.688 \cdot 10^6)^2 - (-2.100 \cdot 10^6)^2 = 9.565 \cdot 10^{12} \tag{f}$$

and noting that the new equivalent quantities are non-italics, I_{zz}, I_{yy} and I_{yz}

$$I_{zz} = \frac{I_{yy} I_{zz} - I_{yz}^2}{I_{yy}} = \frac{I}{I_{yy}} = \frac{9.565 \cdot 10^{12}}{3.688 \cdot 10^6} = 2.594 \cdot 10^6$$

$$I_{yy} = \frac{I_{yy} I_{zz} - I_{yz}^2}{I_{zz}} = \frac{I}{I_{zz}} = \frac{9.565 \cdot 10^{12}}{3.688 \cdot 10^6} = 2.594 \cdot 10^6 \tag{g}$$

$$I_{yz} = \frac{I_{yy} I_{zz} - I_{yz}^2}{I_{yz}} = \frac{I}{I_{yz}} = \frac{9.565 \cdot 10^{12}}{-2.100 \cdot 10^6} = -4.761 \cdot 10^6$$

The normal stress is

$$\sigma_x(x, y, z) = -\left(\frac{M_z}{I_{zz}} - \frac{M_y}{I_{yz}} \right) y + \left(\frac{M_z}{I_{yz}} - \frac{M_y}{I_{yy}} \right) z$$

$$= -\left(\frac{M}{2.594 \cdot 10^6} \right) y - \left(\frac{M}{4.761 \cdot 10^6} \right) z \tag{h}$$

The displacements are found by integrating the following two equations.

$$\frac{d^2 v}{dx^2} = \frac{M_z}{EI_{zz}} - \frac{M_y}{EI_{yz}} = \frac{M}{E \cdot 2.594 \cdot 10^6} = 0.3855 \cdot 10^{-6} \frac{M}{E}$$

$$\frac{d^2 w}{dx^2} = -\frac{M_z}{EI_{yz}} + \frac{M_y}{EI_{yy}} = \frac{M}{E \cdot 4.761 \cdot 10^6} = 0.2100 \cdot 10^{-6} \frac{M}{E} \tag{i}$$

Integrate each equation twice and apply the boundary conditions.

$$v(x) = \frac{0.3855 \cdot 10^{-6} M}{2E} x^2 + a_1 x + b_1$$

$$w(x) = \frac{0.2100 \cdot 10^{-6} M}{2E} x^2 + a_2 x + b_2$$

(j)

The boundary conditions are

$$v(0) = 0 \qquad \frac{dv(0)}{dx} = 0 \qquad w(0) = 0 \qquad \frac{dw(0)}{dx} = 0 \qquad \text{(k)}$$

You can quickly see that

$$a_1 = b_1 = a_2 = b_2 = 0 \qquad \text{(l)}$$

The displacements are

$$v(x) = \frac{0.3855 \cdot 10^{-6} M}{2E} x^2 \qquad w(x) = \frac{0.2100 \cdot 10^{-6} M}{2E} x^2 \qquad \text{(m)}$$

The consequence of a non zero product or inertia is that a moment about the z axis produces displacements in both the y and z directions. In fact the displacement in the z direction is quite large.

Example 10.5.2

Problem: Repeat Example 10.5.1 by finding the principal axes of inertia and using the uncoupled equations for bending in two planes.

Solution: Use Equations 10.2.26-28 and Equations 10.5.15.
From Appendix B, the principal axes are rotated by an amount

$$\tan 2\theta = \frac{I_{yz}}{I_{zz} - I_{yy}} = \frac{I_{yz}}{0} \quad \rightarrow \quad 2\theta = 90° \quad \rightarrow \quad \theta = 45° \qquad \text{(a)}$$

This might have been anticipated from the symmetry as shown in Figure (a).

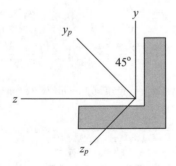

Figure (a)

The principal moments of inertia are

$$I_{\substack{max \\ min}} = \frac{I_{yy} + I_{zz}}{2} \pm \sqrt{\left(\frac{I_{yy} - I_{zz}}{2}\right)^2 + I_{yz}^2}$$

$$= 3.688 \cdot 10^6 \pm \sqrt{\left(\frac{0}{2}\right)^2 + (2.100 \cdot 10^6)^2} = 3.688 \cdot 10^6 \pm 2.100 \cdot 10^6$$

(b)

Thus

$$I_{yy_p} = 3.688 \cdot 10^6 + 2.100 \cdot 10^6 = 5.788 \cdot 10^6 mm^4$$

$$I_{zz_p} = 3.688 \cdot 10^6 - 2.100 \cdot 10^6 = 1.588 \cdot 10^6 mm^4$$

(c)

The moments with respect to the principal axes are

$$M_{y_p} = -M \cos 45° = -0.7071 \cdot M \qquad M_{z_p} = M \sin 45° = 0.7071 \cdot M$$

(d)

The normal stress is

$$\sigma_x(x, y_p, z_p) = -\frac{M_{y_p}}{I_{yy_p}} z' - \frac{M_{z'}}{I_{zz_p}} y' = \frac{0.7071 \cdot M}{5.788 \cdot 10^6} z_p - \frac{0.7071 \cdot M}{1.588 \cdot 10^6} y_p$$

$$= (0.1222 \cdot z_p - 0.4453 \cdot y_p) M \cdot 10^{-6}$$

(e)

The displacements are

$$EI_{z_p} \frac{d^2 v_p}{dx^2} = M_{z_p} = 0.7071 M$$

$$\to \quad \frac{d^2 v_p}{dx^2} = \frac{M_{z_p}}{EI_{z_p}} = \frac{0.7071 M}{E \cdot 1.588 \cdot 10^6} = 0.4453 \cdot 10^{-6} \frac{M}{E}$$

$$\to \quad \frac{dv_p}{dx} = 0.420 \cdot 10^{-6} \frac{M}{E} x + a_1$$

$$\to \quad v_p = 0.420 \cdot 10^{-6} \frac{M}{E} \frac{x^2}{2} + a_1 x + b_1$$

(f)

$$EI_{y_p} \frac{d^2 w_p}{dx^2} = M_{y_p} = -0.7071 M$$

$$\to \quad \frac{d^2 w_p}{dx^2} = \frac{M_{y_p}}{EI_{y_p}} = -\frac{0.7071 M}{E \cdot 5.788 \cdot 10^6} = -0.1222 \cdot 10^{-6} \frac{M}{E}$$

$$\to \quad \frac{dw_p}{dx} = -0.124 \cdot 10^{-6} \frac{M}{E} \frac{x^2}{2} + a_2$$

$$\to \quad w_p = -0.124 \cdot 10^{-6} \frac{M}{E} \frac{x^2}{2} + a_2 x + b_2$$

From the boundary conditions we can see immediately that

$$a_1 = b_1 = a_2 = b_2 = 0$$

(g)

And so the displacements are

$$v_p = 0.4453 \cdot 10^{-6} \frac{M}{E} \frac{x^2}{2} \qquad w_p = -0.1222 \cdot 10^{-6} \frac{M}{E} \frac{x^2}{2}$$

(h)

For a quick check let us calculate the stress and displacement at the lower right corner of the angle. From Example 10.5.1 the yz coordinates at the corner are

$$y = -33.9 \qquad z = -33.9 \tag{i}$$

From Example 10.5.1, Equation (f), the stress is

$$\sigma_x(x, y, z) = (-0.3846y - 0.2092z)M \cdot 10^6$$
$$\rightarrow \sigma_x(x, -33.9, -33.9) = (0.3846 \cdot 33.9 + 0.2092 \cdot 33.9)M \cdot 10^{-6} \tag{j}$$
$$= 20.13 \cdot M \cdot 10^{-6} \ N/mm^2$$

The $y_p z_p$ coordinates at the corner are

$$y_p = -47.98 \qquad z_p = 0 \tag{k}$$

The stress from Equation (e) in this example is

$$\sigma_x(x, y_p, z_p) = \sigma_x(x, -47.98, 0) = (0.4211 \cdot 47.98)M \cdot 10^{-6}$$
$$= 20.20 \cdot M \cdot 10^{-6} \ N/mm^2 \tag{l}$$

A little round off error has crept in. You can check and find that the displacements also agree.

##########

The above equations may be adapted to solve statically indeterminate cases for $I_{yz} \neq 0$. We can use equilibrium to obtain the fourth order equations necessary to solve indeterminate problems that may also be used with determinate cases. We have already noted in Equations 7.3.17 and 10.5.1 that

$$\frac{dV_y}{dx} = -f_y \qquad \frac{dM_z}{dx} = -V_y \qquad \frac{dV_z}{dx} = -f_z \qquad \frac{dM_y}{dx} = -V_z \tag{10.5.16}$$

By differentiating Equations 10.5.13 and inserting Equations 10.5.16 we obtain

$$\frac{d^3v}{dx^3} = -\frac{V_y}{EI_{zz}} + \frac{V_z}{EI_{yz}} \qquad \frac{d^3w}{dx^3} = -\frac{V_z}{EI_{yy}} + \frac{V_y}{EI_{yz}} \tag{10.5.17}$$

and

$$\frac{d^4v}{dx^4} = \frac{f_y}{EI_{zz}} - \frac{f_z}{EI_{yz}} \qquad \frac{d^4w}{dx^4} = \frac{f_z}{EI_{yy}} - \frac{f_y}{EI_{yz}} \tag{10.5.18}$$

Once Equations 10.5.18 are solved for the displacements the bending moments are obtained from Equations 10.5.7 and 10.5.8.

To find the shear forces

$$V_y = -\frac{dM_z}{dx} = -\frac{d}{dx}E\left(\frac{d^2v}{dx^2}I_{zz} + \frac{d^2w}{dx^2}I_{yz}\right) = -E\left(\frac{d^3v}{dx^3}I_{zz} + \frac{d^3w}{dx^3}I_{yz}\right)$$
$$V_z = -\frac{dM_y}{dx} = -\frac{d}{dx}E\left(\frac{d^2v}{dx^2}I_{yz} + \frac{d^2w}{dx^2}I_{yy}\right) = -E\left(\frac{d^3v}{dx^3}I_{yz} + \frac{d^3w}{dx^3}I_{yy}\right) \tag{10.5.19}$$

To find the shear stresses is not so easy. The method we used in Chapter 7, Section 7.3, is accurate, strictly speaking, for rectangular cross sections and can obtain reasonable results for some other sections as discussed in Section 7.9. A true three dimensional elasticity solution may be needed in many cases. We will offer some methods in Chapter 12. Because of the ability to solve fully 3D problems has been made easy with the finite element methods, we will see in due course that we can get accurate solutions for fairly complicated cross section shapes.

When lateral loads are applied to beams with $I_{yz} \neq 0$ it is common to get displacements in both the y and z directions. An additional problem often arises. A lateral load may cause twist as well as bending in two planes. The shear stress is then a combination of that found in bending and that found in torsion. For sections or arbitrary cross section this may become a problem for 3D elasticity theory and not amenable to solving with the equations we have developed so far. There is hope, however, to obtain solutions in the analysis of thin walled cross sections using slender bar theory. That will be considered in Sections 10.6 and 10.7.

The finite element formulation for $I_{yz} \neq 0$ is available and has been captured in FEM software. The same shape functions are used together with the equations we develop in this section. We shall not write it down here but trust that it has been appropriately captured in the finite element computer codes

10.6   Bending and Torsion in Thin Walled Open Sections—Shear Center

In Chapter 7, Section 7.9, we considered the shear flow due to transverse shear forces in thin walled closed and open sections. We determined that the shear flow can be calculated by Equation 7.9.7 repeated here as Equation 10.6.1.

$$q = q_0 - \int_0^s \frac{V}{I_{zz}} byds \qquad (10.6.1)$$

The analysis was restricted to sections that were symmetrical about the xy axes plane. Let us extend the theory of thin walled sections to those with $I_{yz} = 0$ and with loading in the y direction but without symmetry about the y axis. Two examples of such cross sections are shown in Figure 10.6.1.

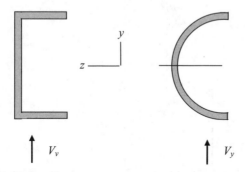

Figure 10.6.1

Depending upon where the lateral load is applied, twist as well as bending may occur. There is a position along the z axes on each cross section through which the load must be applied to produce only bending but no torsion. This is defined as the point about which the moment of the applied load is equal to the moment of the shear flow. This point is called the *shear center*. The locus of such points along the axis is called the *elastic axis*. The shear center location on the cross section is defined by Equation 10.6.2.

$$V_y e_z = \int rqds \qquad (10.6.2)$$

where e_z is the moment arm of the V_y force about some point of reference. This is best understood by an example.

##########

Example 10.6.1

Problem: A thin walled cantilever beam has a C shaped cross section. A rigid plate is fastened at the free end so that a concentrated load can be applied at different locations along the z axis as shown in Figure (a). The origin of the yz axes is at the centroid of the cross section. The force is shown applied at the shear center, that is, at a point where the beam will bend without twisting. The problem here is to find the shear flow and the shear center, that is, the location of the force along the z axis.

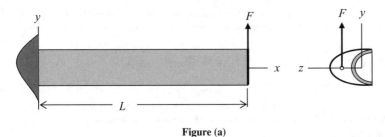

Figure (a)

Solution: Find the shear flow using Equation 10.6.1 and then find the shear center using Equation 10.6.2. The details of the cross section are shown in Figure (b). The uniform wall thickness is b.

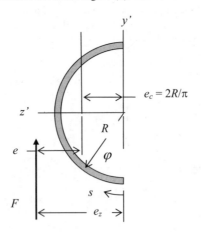

Figure (b)

We define a coordinate s that starts at the bottom edge and follows the centerline of the curved section. It is convenient to use the $y'z'$ coordinates to find the centroid and the shear flows. In defining beam bending behavior we switch to centroidal axes. In terms of the radius R and the angle φ as shown in Figure (b) we have

$$y' = -R\cos\varphi \qquad z' = R\sin\varphi \qquad ds = Rd\varphi \tag{a}$$

The centroid is found to be

$$e_c = \frac{\int z'\, dA}{\int dA} = \frac{\int_0^\pi (R\sin\varphi)\, bR d\varphi}{\pi Rb} = \frac{-R\cos\varphi|_0^\pi}{\pi} = \frac{2R}{\pi} \tag{b}$$

The moment of inertia of the cross section is

$$I_{zz} = \int_A y'^2 dA = \int_0^\pi \left(R^2 \cos^2 \varphi\right) Rbd\varphi = R^3 b \left(\frac{\varphi}{2} + \frac{\sin\varphi\cos\varphi}{2}\right)\Big|_0^\pi = \frac{\pi R^3 b}{2} \tag{c}$$

The shear flow is defined as positive in the positive direction of the coordinate s. The value of the shear flow at $s = 0$ is zero, that is, $q_0 = 0$. This can be stated because there is no shear on the edge at 90° to the cross section. The shear flow in the curved section is

$$q = q_0 - \int_0^s \frac{V_y}{I_{zz}} y'bds = 0 + \frac{F}{I_{zz}} \int_0^\varphi (R\cos\varphi) Rbd\varphi = \frac{FbR^2}{I_{zz}} \sin\varphi = \frac{2F}{\pi R} \sin\varphi \tag{d}$$

To find the shear center equate the moment of the applied force to the moment of the shear flow (principle of equivalence). Using Equation 10.6.2 we get

$$V_y e_z = \int rqds \quad \rightarrow$$

$$Fe_z = \int_0^\pi qRRd\varphi = \frac{2F}{\pi R} \int_0^\pi (\sin\varphi) R^2 d\varphi = -\frac{2FR}{\pi} \cos\varphi\Big|_0^\pi = \frac{4FR}{\pi} \tag{e}$$

Therefore the shear center is located at

$$e_z = \frac{4R}{\pi} \tag{f}$$

or from the centroid of the section.

$$e = e_z - e_c = \frac{4R}{\pi} - \frac{2R}{\pi} = \frac{2R}{\pi} \tag{g}$$

The shear center is to the left of the section, approximately as shown in Figure (b).

With the moment of inertia for this cross section added the displacement in bending only, as was found in Example 7.3.2, Equation (d), is repeated here.

$$v(x) = \frac{F}{EI_{zz}} \left(\frac{Lx^2}{2} - \frac{x^3}{6}\right) = \frac{F}{\pi R^3 bE} \left(Lx^2 - \frac{x^3}{3}\right) \quad \rightarrow \quad v(L) = \frac{2FL^3}{3\pi R^3 bE} \tag{h}$$

The formula for the normal bending stresses is the same as that found in Example 7.3.2, Equation (f). The normal stresses are

$$\sigma_x(x, y) = -\frac{M}{I_{zz}} y = -\frac{F(L-x)}{I_{zz}} y = \frac{2F(L-x)}{\pi R^3 b} y \quad \rightarrow \quad \sigma_x(0, R) = \frac{2FL}{\pi R^2 b} \tag{i}$$

Let us add some values to the quantities to get a better feel of the actual behavior of a cantilever beam. Let

$$L = 1000 \ mm \qquad R = 50 \ mm \qquad b = 2 \ mm \qquad E = 206800 \ N/mm^2 \qquad F = 1000 \ N \tag{j}$$

The centroid and the shear center are located at

$$e_c = \frac{2R}{\pi} = 0.6366R = 31.831 \ mm \qquad e_z = \frac{4R}{\pi} = 1.273R = 63.66 \ mm \tag{k}$$

The shear flow and the maximum values of the shear flow and shear stress at $\varphi = \pi/2$ are

$$q(\varphi) = \frac{2F}{\pi R} \sin \varphi = \frac{2000}{\pi 50} \sin \varphi = 12.732 \sin \varphi \; N/mm$$

$$\rightarrow q\left(\frac{\pi}{2}\right) = 12.732 \; N/mm \tag{l}$$

$$\rightarrow \tau\left(\frac{\pi}{2}\right) = \frac{q}{b} = \frac{12.732}{2} = 6.366 \; N/mm^2$$

The bending stress and its maximum value at $x = 0$, $y = 50$ are

$$\sigma_x(x, y) = \frac{2F(L-x)}{\pi R^3 b} y = \frac{2 \cdot 1000}{\pi \cdot 50^3 \cdot 2}(1000 - x)y = 0.0025465(1000 - x)y$$

$$\rightarrow \sigma(0, 50) = 0.0025465(1000)50 = 127.324 \; N/mm^2 \tag{m}$$

The tip displacement is

$$v(L) = \frac{2FL^3}{3\pi R^3 bE} = \frac{2 \cdot 1000 \cdot 1000^3}{3 \cdot \pi \cdot 50^3 \cdot 2 \cdot 206800} = 4.105 \; mm \tag{n}$$

For the load given these appear to be quite modest values. We shall find some interesting comparisons with the results of a similar problem in Example 10.6.2 coming right up.

###########

Having found the shear center and the shear flow associated with a force through the shear center let us ask what happens when the shear force is acting through some point other than the shear center. Clearly some twist will occur. To solve this problem we replace the actual applied force with a statically equivalent applied force of equal value through the shear center and a moment about the shear center. The effect of the moment is a shear stress and a twist as determined from torsional theory for open thin walled sections.

In Chapter 6, Section 6.8, we extended the torsional analysis to an open thin walled cross section. We found the stress to be

$$\tau = \frac{T}{J_{eff}} \delta \tag{10.6.3}$$

where δ can be interpreted to be a coordinate normal to the line midway between the two outer surfaces. If the wall thickness is b then the maximum shear stress is on the two outer edges, that is,

$$\tau_{max} = \pm \frac{T}{J_{eff}} \frac{b}{2} \tag{10.6.4}$$

The effective torsional stiffness is given as

$$J_{eff} = \frac{hb^3}{3} \tag{10.6.5}$$

where h is the length of the midline and b is the wall thickness. This effective stiffness can be used in the differential equation for the displacement.

$$GJ_{eff} \frac{d\beta}{dx} = T \tag{10.6.6}$$

This equation can be integrated to find the displacement.
Once again an example will make this all very clear.

############

Example 10.6.2

Problem: The beam in Example 10.6.1 now has a force F applied at the centroid of the cross section at the free end as shown in Figure (a). Find the stresses and the displacements of the beam.

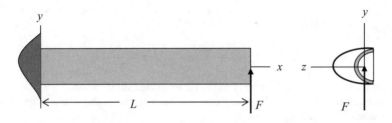

Figure (a)

Solution: Resolve the applied load at the centroid into a lateral force applied through the shear center and a torque about the shear center. Add the torsional stress using Equation 10.6.3 to the shear flow found in Example 10.6.1.

The equivalent loading for purposes of analysis is to have the force F at the shear center for bending shear analysis and the moment M about the shear center for torsional analysis.

When the load is applied through centroid a moment about the shear center is

$$M = F\left(\frac{4R}{\pi} - \frac{2R}{\pi}\right) = F \cdot \frac{2R}{\pi} \tag{a}$$

These loads are shown in Figure (b).

The results for lateral displacement and normal stress for this beam are the same as presented in Example 10.6.1.

$$v(x) = \frac{F}{EI_{zz}}\left(\frac{Lx^2}{2} - \frac{x^3}{6}\right) = \frac{F}{\pi R^3 bE}\left(Lx^2 - \frac{x^3}{3}\right) \tag{b}$$

$$\sigma_x(x, y) = -\frac{M}{I_{zz}}y = -\frac{F(L-x)}{I_{zz}}y = \frac{2F(L-x)}{\pi R^3 b}y \tag{c}$$

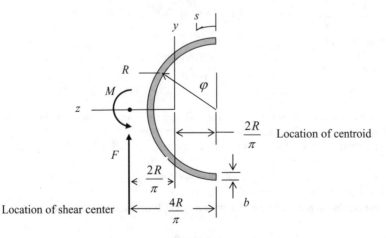

Figure (b)

The shear stress in bending was found in Example 10.6.1 to be

$$q = q_0 - \int_0^s \frac{V_y}{I_{zz}} y'b \, ds = \frac{FbR^2}{I_{zz}} \sin\varphi = \frac{2F}{\pi R} \sin\varphi \qquad (d)$$

To this must be added the shear stress of torsion. From Equation 10.6.5 the effective torsional constant is

$$J_{eff} = \frac{hb^3}{3} = \frac{\pi R b^3}{3} \qquad (e)$$

From Equation 10.6.3 the torsional stress is

$$\tau = \frac{T}{J_{eff}}\delta = M\frac{3}{hb^3}\delta = \frac{2RF}{\pi} \cdot \frac{3}{\pi R b^3}\delta = \frac{6F}{\pi^2 b^3}\delta \qquad (f)$$

where δ is measured normal to the mid radius and has the maximum values at $\pm b/2$.

The torsional displacement for a cantilever beam with a moment at the free end was found in Example 6.3.1. to be

$$\beta(x) = \frac{M}{GJ_{eff}}x \qquad (g)$$

For this case we have

$$\beta(x) = \frac{M}{GJ_{eff}}x = \frac{2RF}{\pi} \cdot \frac{3}{G\pi R b^3}x = \frac{6F}{\pi^2 G b^3}x \qquad (h)$$

Let us put in the same numbers, add the shear modulus, and compare. Let

$$L = 1000 \ mm \qquad R = 50 \ mm \qquad b = 2 \ mm \qquad G = 79538 \ N/mm^2 \qquad F = 1000 \ N \qquad (i)$$

The bending stress and the lateral tip displacement will be the same as in Example 10.6.1. To the shear stress we must add the effect of the torsional moment and now we will have torsional displacement as well. The distributed torsional shear stress and its maximum value at $\delta = \pm 1 \ mm$ will be

$$\tau(\delta) = \frac{6F}{\pi^2 b^3}\delta = \frac{6 \cdot 1000}{\pi^2 \cdot 2^3}\delta = 75.991\delta \quad \rightarrow \quad \tau(\pm 1) = \pm 75.991 \ N/mm^2 \qquad (j)$$

This must be added to the shear stress of bending (see Example 10.6.1, Equation(1)). The maximum value will be at $\varphi = \pi/2$, $\delta = \pm 1$.

$$\tau\left(\varphi, \pm\frac{b}{2}\right) = \tau\left(\frac{\pi}{2}, \pm 1\right) = (6.366 \pm 75.991) \ N/mm^2 \qquad (k)$$

Now let us look at the twist.

$$\beta(x) = \frac{6F}{\pi^2 G b^3}x = \frac{6 \cdot 1000}{\pi^2 \cdot 79538 \cdot 2^3}x = 0.0009554x \ rad$$
$$\rightarrow \beta(L) = 0.0009554 \cdot L = 0.9554 \ rad = 54.744° \qquad (l)$$

This is quite a large rotational displacement and may very well be unacceptable. In fact, using the undeformed geometry to define the moment is no longer valid.

Open sections are very weak in torsion and are generally avoided as load carrying members.

###########

We can extend this to thin walled beams with $I_{yz} \neq 0$. From Equation 7.9.4 we have

$$\frac{\partial q}{\partial s} = -b\frac{\partial \sigma_x}{\partial x} \qquad (10.6.8)$$

When we insert the value of the normal stress in Equation 10.5.14 into Equation 10.6.8 we get

$$\frac{\partial q}{\partial s} = b\frac{d}{dx}\left(\left(\frac{M_z}{I_{zz}} - \frac{M_y}{I_{yz}}\right)y - \left(\frac{M_z}{I_{yz}} - \frac{M_y}{I_{yy}}\right)z\right) = \left(-\frac{V_z}{I_{zz}} + \frac{V_y}{I_{yz}}\right)by - \left(-\frac{V_z}{I_{yz}} + \frac{V_y}{I_{yy}}\right)bz \quad (10.6.9)$$

The shear forces to be used in Equation 10.6.9 are found using Equations 10.5.19.

############

10.7 Bending and Torsion in Thin Walled Closed Sections—Shear Center

Let us extend the theory of thin walled closed sections to those with $I_{yz} = 0$ and with loading in the y direction but without symmetry about the y axis. An example of such a cross section is shown in Figure 10.7.1.

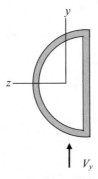

Figure 10.7.1

Depending upon where the lateral load is applied, twist as well as bending may occur. As in the case of the open section we shall first find the shear for pure bending and the location of the shear center. When the transverse force is not through the shear center we will resolve it into a force at the shear center and a moment about the shear center.

The same equations for the shear flow and shear center used in Section 10.6 apply here.

$$q = q_0 - \int_0^s \frac{V_y}{I_{zz}}byds \qquad V_ye_z = \int rqds \qquad (10.7.1)$$

In this case of bending only we shall use the condition of no twist to find the value of q_0 in Equation 10.7.1. Consider the element under strain as shown in Figure 10.7.2.

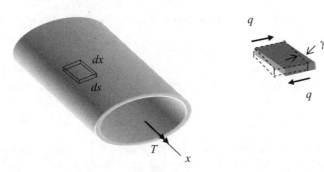

Figure 10.7.2

Since we are using the shear flow q we assume the stress is constant through the thickness and we can identify the geometry of the element by its mid plane surface as shown in Figure 10.7.3.

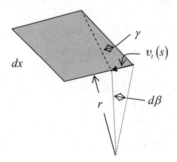

Figure 10.7.3

The displacement $v_t(s)$ is in the direction of the coordinate s. From Hooke's law

$$\gamma = \frac{\tau}{G} = \frac{q}{Gb} \tag{10.7.2}$$

From the Figure 10.7.2

$$v_t = \gamma dx = r d\beta \qquad \rightarrow \qquad \gamma = r\frac{d\beta}{dx} = \frac{q}{Gb} \tag{10.7.3}$$

If we integrate around the section

$$\int \frac{q}{Gb}ds = \int r\frac{d\beta}{dx}ds \tag{10.7.4}$$

and assume that $d\beta/dx$ is a constant around the section then from Equations 6.5.6-8

$$\int rds = 2A_T \qquad \rightarrow \qquad \frac{d\beta}{dx}\int rds = \frac{d\beta}{dx}\cdot 2A_T \tag{10.7.5}$$

where A_T is the enclosed area of the section. Thus we find the rate of twist is

$$\frac{d\beta}{dx} = \frac{1}{2A_T}\int \frac{q}{Gb}ds \tag{10.7.6}$$

This applies to any value of q. When q is constant, as in the case of pure torque, from Equation 6.5.9

$$q = \frac{T}{2A_T} \qquad \rightarrow \qquad \frac{d\beta}{dx} = \frac{1}{2A_T}\int \frac{q}{Gb}ds = \frac{T}{4A_T^2}\int \frac{1}{Gb}ds \tag{10.7.7}$$

Equation 10.7.7 is the result we got in Chapter 6, Section 6.6. From Equation 10.7.7 we defined

$$GJ_{eff}\frac{d\beta}{dx} = T \tag{10.7.8}$$

where

$$J_{eff} = \frac{4A_T^2}{\int \dfrac{ds}{b}} \tag{10.7.9}$$

Equation 10.7.6 will be a valuable tool in finding the shear flow and shear center for a beam with a closed thin walled cross section. The next example illustrates this.

$$\#\#\#\#\#\#\#\#\#\#$$

Example 10.7.1

Problem: Find the shear flow and shear center in the closed single cell thin walled cross section shown in Figure (a). The shear force is applied at the shear center of the cross section in the y direction.

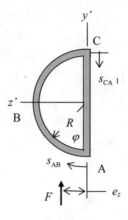

Figure (a)

Solution: Use Equations 10.7.1 and 10.7.6.
For the curved portion use

$$y' = -R\cos\varphi \qquad z' = R\sin\varphi \qquad ds_{AB} = Rd\varphi \tag{a}$$

just as we did in Example 10.6.1. The shear flow in the curved part is

$$q_{ABC} = q_0 - \int_0^s \frac{F}{I_{zz}} by' ds_{AB} = q_A + \frac{FbR^2}{I_{zz}} \int_0^{\varphi} \cos\varphi d\varphi = q_A + \frac{FbR^2}{I_{zz}} \sin\varphi \tag{b}$$

Note that q_{ABC} at $\varphi = \pi$, that is at point C, equals q_A. Then, along the vertical web $q_C = q_A$ and $ds_{CA} = -dy'$.

$$q_{CA} = q_C + \int_{-R}^{y'} \frac{F}{I_{zz}} by' dy' = q_A + \frac{Fb}{2I_{zz}}(y'^2 - R^2) \tag{c}$$

To find q_A assume the section does not twist. Use Equation 10.7.6.

$$\frac{d\beta}{dx} = \frac{1}{2A_T} \int \frac{q}{Gb} ds = 0 \quad \rightarrow \quad \int q ds = 0$$

$$\int q ds = \int_0^{\pi} \left(q_A + \frac{FbR^2}{I_{zz}} \sin\varphi\right) R d\varphi + \int_R^{-R} \left(q_A + \frac{Fb}{2I_{zz}}(y'^2 - R^2)\right)(-dy')$$

$$= Rq_A\varphi|_0^{\pi} - \frac{FbR^3}{I_{zz}} \cos\varphi|_0^{\pi} - q_A y'|_R^{-R} - \frac{Fb}{2I_{zz}}\left(\frac{y'^3}{3} - R^2 y'\right)\Big|_R^{-R}$$

$$= Rq_A\pi - \frac{FbR^3}{I_{zz}}(-2) - q_A(-2R) - \frac{Fb}{I_{zz}}\left(-\frac{2R^3}{3} + 2R^3\right) \tag{d}$$

$$= (R\pi + 2R)q_A + \frac{FbR^3}{I_{zz}}\left(\frac{4}{3}\right) = 0 \quad \rightarrow \quad q_A = -\frac{4}{3(\pi + 2)}\frac{FbR^2}{I_{zz}}$$

The total shear flow is

$$q_{ABC} = q_A + \frac{FbR^2}{I_{zz}} \sin\varphi = -\frac{4FbR^2}{3(\pi+2)I_{zz}} + \frac{FbR^2}{I_{zz}} \sin\varphi = \frac{FbR^2}{I_{zz}} \left(-\frac{4}{3(\pi+2)} + \sin\varphi \right) \qquad (e)$$

$$q_{CA} = q_A + \frac{Fb}{2I_{zz}}(y^2 - R^2) = -\frac{4FbR^2}{3(\pi+2)I_{zz}} + \frac{Fb}{2I_{zz}}(y^2 - R^2)$$

$$= \frac{Fb}{I_{zz}} \left(-\frac{4R^2}{3(\pi+2)} + \frac{1}{2}(y^2 - R^2) \right) \qquad (f)$$

To find the shear center, let

$$Fe_z = \int_s qrds = \int_0^\pi \left(q_A + \frac{FbR^2}{I_{zz}} \sin\varphi \right) R^2 d\varphi$$

$$= \int_0^\pi \frac{FbR^2}{I_{zz}} \left(-\frac{4}{3(\pi+2)} + \sin\varphi \right) R^2 d\varphi \qquad (g)$$

$$= \frac{FbR^4}{I_{zz}} \left(-\frac{4\varphi}{3(\pi+2)} - \cos\varphi \right) \Big|_0^\pi = \frac{FbR^4}{I_{zz}} \frac{2(\pi+6)}{3(\pi+2)}$$

or

$$e_z = \frac{bR^4}{I_{zz}} \frac{2(\pi+6)}{3(\pi+2)} \qquad (h)$$

Let us assign some numerical values.

$$L = 1000 \ mm \qquad R = 50 \ mm \qquad b = 2 \ mm \qquad E = 206800 \ N/mm^2 \qquad F = 1000 \ N \qquad (i)$$

The moment of inertia is

$$I_{zz} = \frac{\pi R^3 b}{2} + \frac{b(2R)^3}{12} = \frac{\pi \cdot 50^3 \cdot 2}{2} + \frac{2 \cdot 100^3}{12}$$

$$= 392699.08 + 166666.67 = 559365.75 \ mm^4 \qquad (j)$$

The shear center is located at

$$e_z = \frac{bR^4}{I_{zz}} \frac{2(\pi+6)}{3(\pi+2)} = \frac{2 \cdot 50^4 \cdot 2(\pi+6)}{559365.75 \cdot 3(\pi+2)} = 26.488 \ mm \qquad (k)$$

The shear flows are

$$q_{ABC} = \frac{FbR^2}{I_{zz}} \left(-\frac{4}{3(\pi+2)} + \sin\phi \right) = 8.939(-0.2593 + \sin\varphi)$$

$$= -2.3180 + 8.9387 \sin\varphi \qquad (l)$$

$$q_{CA} = \frac{Fb}{I_{zz}} \left(-\frac{4R^2}{3(\pi+2)} + \frac{1}{2}(y^2 - R^2) \right) = \frac{FbR^2}{I_{zz}} \left(-\frac{4}{3(\pi+2)} + \frac{1}{2}\left(\frac{y^2}{R^2} - 1\right) \right)$$

$$= 8.939 \left(-0.2593 + 0.5\left(\frac{y^2}{2500} - 1\right) \right) = -6.787 + 0.0017878y^2$$

In the above analysis the positive direction of the shear flow in the vertical web was in the negative y direction, that is, from point C to A. Let us redefine it as positive in the positive y direction and label it q_{AC}, then

$$q_{AC} = 6.787 - 0.0017878y^2 \qquad (m)$$

The shear flows q_{ABC} and q_{AC} are plotted in Figure (b). Note that the shear flow q_{ABC} follows the curved web and the vertical component of the shear flow in the curved web and that in the vertical web add up to a total force equal to the applied load.

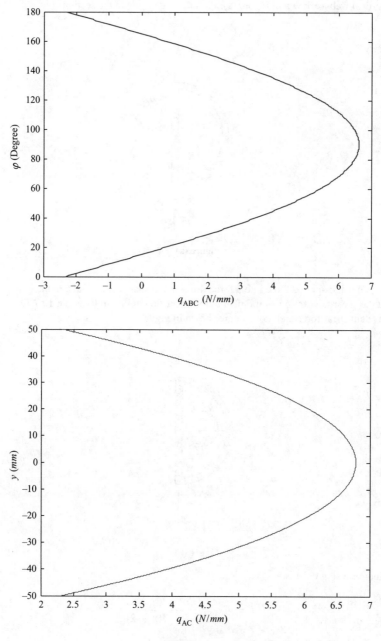

Figure (b)

Example 10.7.2

Problem: The shear force for the beam with the D shaped cross section in Example 10.7.1 is applied along the vertical web as shown in Figure (a). Find the shear flow in the section.

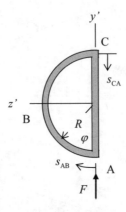

Figure (a)

Solution: Resolve the applied force into a force through the shear center and a moment about the shear center. Add the shear flow from that moment to the shear flow found in Example 10.7.1.

The equivalent shear force and moment are shown in Figure (b).

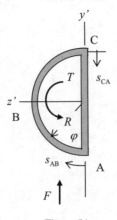

Figure (b)

The moment causing twist is

$$T = F e_z = \frac{FbR^4}{I_{zz}} \frac{2(\pi + 6)}{3(\pi + 2)} \tag{a}$$

The shear flow from this moment is

$$q_T = -\frac{T}{2A_T} = -\frac{FbR^4}{2\pi R^2 I_{zz}} \frac{2(\pi + 6)}{3(\pi + 2)} = -\frac{FbR^2}{\pi I_{zz}} \frac{(\pi + 6)}{3(\pi + 2)} \tag{b}$$

The minus sign in needed because the shear flow in Example 10.7.1 was positive in the counter clockwise direction.

Add this to the values found in Example 10.7.1, Equations (e) and (f).

############

10.8   Stiffened Thin Walled Beams

Thin walled sections are often combined with stiffeners such as shown in Figure 10.8.1a. Consider that this is a section of a beam with a lateral load acting in the y direction through the centroid of the section. Assume the bending moment M_z and the shear force V_y have been found.

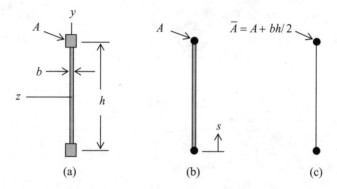

Figure 10.8.1

The discussion of the solution for shear flow in the web of an I beam in Section 7.9 applies here. There are two addition assumptions to simplify the analysis in common use. In Figure 10.8.1 (b) we depict the stiffener area as acting at a point but maintain the web thickness. The moment of inertia contributed by the two stiffeners is then approximated as

$$(I_{zz})_{st} = 2\left(\frac{h}{2}\right)^2 A = \frac{h^2}{2}A \tag{10.8.1}$$

where A is the area of the stiffener. This neglects the moment of inertia of the stiffeners about their own centroids. This is a common approximation when h is sufficiently large. The total moment of inertia in case (b) is then

$$I_{zz} = \frac{h^2}{2}A + \frac{bh^3}{12} \tag{10.8.2}$$

In Figure 10.8.1 (c) if the moment of inertia contributed by the web is small compared to the contribution of the stiffener we can neglect it entirely or we can lump the web area in with the stiffener area by adding half of the web area to each stiffener.

This new area that we label $\overline{A}$ is then used in the various equations.

$$\overline{A} = A + \frac{bh}{2} \tag{10.8.3}$$

The corresponding moment of inertia is

$$(\overline{I}_{zz})_{st} = 2\left(\frac{h}{2}\right)^2 \overline{A} = \frac{h^2}{2}\overline{A} \tag{10.8.4}$$

The normal stress on the stiffener is

$$\sigma_x = -\frac{M}{I_{zz}}y \qquad \rightarrow \qquad (\sigma_x)_{st} = -\frac{M}{I_{zz}}y = -\frac{M}{I_{zz}}\frac{h}{2} = \frac{Mh}{\overline{A}} \tag{10.8.5}$$

The shear flow in then a constant

$$q = \frac{V_y}{h} \tag{10.8.6}$$

Another way to approach the solution is to consider the equilibrium of a free body of the segment of the top stiffener as shown in Figure 10.8.2.

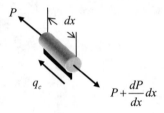

Figure 10.8.2

Summing forces we get

$$\sum F_x = P + \frac{dP}{dx}dx - P - qdx = 0 \tag{10.8.7}$$

$$\rightarrow q_c = \frac{dP}{dx} = \frac{d}{dx}(\sigma_x A) = -\frac{d}{dx}\left(\frac{M}{I_{zz}}yA\right) = -\frac{\overset{dM}{\overbrace{\frac{dx}{I_{zz}}}}\frac{h}{2}A = \frac{V_y}{I_{zz}}\frac{h}{2}A$$

We can use either A or $\overline{A}$ in Equation 10.8.6. When we insert the moment of inertia in Equation 10.8.6 we get

$$q_c = \frac{V_y}{I_{zz}}\frac{h}{2}A = \frac{2V_y}{h^2 A}\frac{h}{2}A = \frac{V_y}{h} \tag{10.8.8}$$

We can replace A with $\overline{A}$ and get the same result for the shear flow. Using $\overline{A}$ the normal stress will be lower owing to the larger value of the moment of inertia. Actually since the shear flow is a constant in the web we can go directly to

$$q_c = \frac{V_y}{h} \tag{10.8.9}$$

This is depicted in Figure 10.8.3 (c). If instead we use the approximation in Figure (b) the shear flow in the web is

$$q_b = \frac{V_y}{I_{zz}}\frac{h}{2}A - \frac{V_y}{I_{zz}}\int_0^s ybds = \frac{V_y}{I_{zz}}\frac{h}{2}A - \frac{V_y}{I_{zz}}\int_{-\frac{h}{2}}^y ybdy$$

$$= \frac{V_y}{I_{zz}}\frac{h}{2}A - \frac{V_y b}{2I_{zz}}y^2\Big|_{-\frac{h}{2}}^y = \frac{V_y}{2I_{zz}}\left[hA + b\left(\frac{h^2}{4} - y^2\right)\right] \tag{10.8.10}$$

where the moment of inertia is given by Equation 10.8.2. This distribution of shear flow is depicted in Figure 10.8.3 (b).

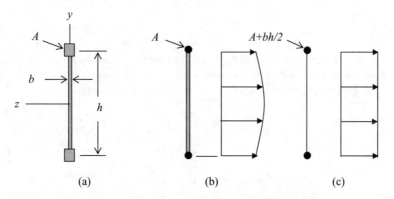

y

A b z h

A

$A+bh/2$

(a)   (b)   (c)

Figure 10.8.3

A numerical example will provide a measure of the validity of these approximations.

##########

Example 10.8.1

Problem: The cross sections in Figure 10.8.1 are given the following numerical parameters:
The stiffeners are 10 *mm* by 10 *mm*. The height is 100 *mm* and the web thickness is 2 *mm*.
Compare the two approximations in Figures 10.8.3 (b) and (c).

Solution: Calculate the normal and shear stresses by the two approximations.
First let us check the accuracy of neglecting the moment of inertia of the stiffeners about their own
centroids. Including the terms neglected

$$(I_{zz})_{st} = 2\frac{a^4}{12} + 2\left(\frac{h}{2}\right)^2 A = 2\frac{10^4}{12} + \frac{100^2}{2}100$$

$$= 1,666.67 + 500,000 = 501,666.67 \; mm^4 \tag{a}$$

Including the moments of inertia about the stiffeners' centroids increases the total by only a factor of
0.0033. It is reasonable to neglect them. In this case if we add the shear web to the moment of inertia
we have

$$I_{zz} = \frac{h^2}{2}A + \frac{bh^3}{12} = \frac{100^2}{2}100 + \frac{2 \cdot 100^3}{12} = 500,000 + 166,666.67 = 666666.67 \; mm^4 \tag{b}$$

The shear stress in Figure 10.8.2 (b) is

$$q_b = \frac{V_y}{2I_{zz}}\left[hA + b\left(\frac{h^2}{4} - y^2\right)\right] = \frac{V_y}{1333333}[15000 - 2y^2] \tag{c}$$

At the stiffeners ($y = \pm 50 \; mm$) the value is

$$q_b = \frac{V_y}{1333333}[15000 - 2 \cdot 50^2] = 0.0075 \cdot V_y \tag{d}$$

At the centroid of the whole section ($y = 0$) the value is

$$q_b = \frac{V_y}{1333333}[15000 - 2 \cdot 0^2] = 0.01125 \cdot V_y \tag{e}$$

Now let us try the approximation of Figure 10.8.1 (c). We know that the shear flow is constant and
therefore

$$q_c = \frac{V_y}{h} = 0.01 \cdot V_y \tag{f}$$

Example 10.8.2

Problem: Compare the shear flow magnitudes and distributions in a thin walled rectangular cross section without and with stringers. Compare the cross section in Example 7.9.1 with two solutions in Example 10.8.1. From Example 7.9.2 the cross section is shown in Figure (a).

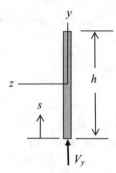

Figure (a)

In Example 7.9.1 we found the shear flow to be

$$q = \frac{Vb}{2I_{zz}}\left(\frac{h^2}{4} - y^2\right) \tag{a}$$

Given the same dimension and properties as in Example 10.8.1 and $V_y = 1000\,N$ we have

$$q = 0.006(2500 - y^2) \tag{b}$$

In Example 10.8.1 we found the shear flow to be

$$q_b = \frac{V_y}{1333333}[15000 - 2y^2] = 0.0015(7500 - y^2) \tag{c}$$

These two values along with a uniform shear flow are plotted in Figure (b).

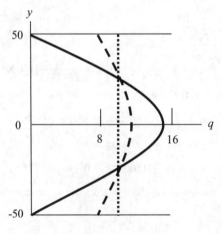

Figure (b)

The solid line refers to the section with no stiffeners, Equation (b); the dashed line refers to the section with stiffeners, Equation (c); and the dotted line refers to the case where a uniform shear flow is assumed. A section with a larger stiffener area would approach the constant case.

Note that adding stiffeners decreases the normal stress and decreases the maximum value of the shear flow; however, the integral of the shear flow will always add up to the applied load. Thus, the area under each of those curves in Figure (b) will be the same.

##########

This approximation may be extended to more complex shapes and is especially helpful for closed sections including multi cell sections. The next example considers a single cell thin walled section with stiffeners.

##########

Example 10.8.3

Problem: Find the stresses in the single cell thin walled section shown in Figure (a). The dimensions are the same as in Example 10.8.1. The load is in the y direction and is acting through the shear center. Assume the bending moment M_z and the shear force V_y have been found. Compare approximations in Figures (b) and (c).

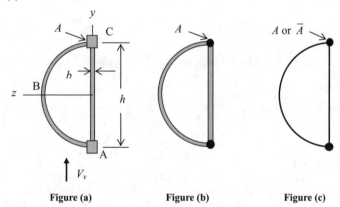

Figure (a)   Figure (b)   Figure (c)

Solution: Find the shear flows using the same equations as in Example 10.7.1. Use equilibrium of the shear flows at the stiffener and the condition of no twist to find the unknown shear flows at the start of each web.

First consider the case in Figure (b) as shown in Figure (d).

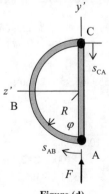

Figure (d)

The shear flows are

$$q_{ABC} = q_A + \frac{FbR^2}{I_{zz}}\sin\varphi \qquad q_{CA} = q_C + \frac{Fb}{2I_{zz}}(y^2 - R^2) \tag{a}$$

To find q_A and q_C, place the upper stiffener in equilibrium as shown in Figure (e).

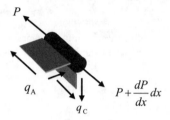

Figure (e)

Summing forces we get

$$\sum F_x = P + \frac{dP}{dx}dx - P + (q_C - q_A)dx = 0 \tag{b}$$

$$\rightarrow q_A - q_C = \frac{dP}{dx} = \frac{d}{dx}(\sigma_x A) = -\frac{d}{dx}\left(\frac{M}{I_{zz}}yA\right) = -\frac{\frac{dM}{dx}}{I_{zz}}RA = \frac{F}{I_{zz}}RA$$

This is one equation in two unknowns, q_A and q_C. To find another equation use the condition of no twist.

$$\int qds = \int_0^\pi \left(q_A + \frac{FbR^2}{I_{zz}}\sin\varphi\right)Rd\varphi + \int_R^{-R}\left(q_C + \frac{Fb}{2I_{zz}}(y^2 - R^2)\right)(-dy')$$

$$= Rq_A\,\varphi|_0^\pi - \frac{FbR^3}{I_{zz}}\cos\varphi|_0^\pi - q_C\,y|_R^{-R} - \frac{Fb}{2I_{zz}}\left(\frac{y^3}{3} - R^2y\right)\Big|_R^{-R}$$

$$= Rq_A\pi - \frac{FbR^3}{I_{zz}}(-2) - q_C(-2R) - \frac{Fb}{2I_{zz}}\left(-\frac{2R^3}{3} + 2R^3\right) \tag{c}$$

$$= R\pi q_A + 2Rq_C + \frac{FbR^3}{I_{zz}}\left(\frac{4}{3}\right) = 0 \quad \rightarrow \quad \pi q_A + 2q_C = -\frac{4FbR^2}{3I_{zz}}$$

Solving the two equations provides

$$q_A = \frac{F}{(\pi + 2)I_{zz}}\left(hA - \frac{4bR^2}{3}\right) \qquad q_C = q_A - \frac{F}{I_{zz}}\frac{h}{2}A \tag{d}$$

Let us put some numbers in since all these symbolic quantities are difficult to interpret. Using the same values from Example 10.7.1 and with stiffener areas of 100 mm^2 we add the stiffeners to the moment of inertia.

$$I_{zz} = 559365.75 + 2R^2A = 559365.75 + 2\cdot 50^2\cdot 100 = 1059365.75\ mm^4 \tag{e}$$

Given these values

$$q_A - q_C = \frac{F}{I_{zz}}RA = \frac{1000\cdot 50\cdot 100}{1059365.77} = 4.7198\ N/mm$$

$$\pi q_A + 2q_C = -\frac{4FbR^2}{3I_{zz}} = -\frac{4\cdot 1000\cdot 2\cdot 2500}{3\cdot 1059365.75} = -6.2930\ N/mm \tag{f}$$

Solving for q_A and q_C we get

$$q_A = 0.6120 \ N/mm \qquad q_C = -4.1078 \ N/mm \tag{g}$$

The shear flows in the webs are

$$q_{ABC} = q_A + \frac{FbR^2}{I_{zz}} \sin\varphi = 0.6120 + \frac{1000 \cdot 2 \cdot 2500}{1059365.75} \sin\varphi = 0.6120 + 4.7198 \sin\varphi$$

$$q_{CA} = q_C + \frac{Fb}{2I_{zz}}(y^2 - R^2) = -4.1078 + \frac{1000 \cdot 2}{2 \cdot 1059365.75}(y^2 - 2500) \tag{h}$$

$$= -4.1078 + 0.000944(y^2 - 2500) \quad \rightarrow \quad q_{AC} = 4.1078 - 0.000944(y^2 - 2500)$$

Now find the shear center.

$$Fe_z = \int_0^\pi q_{ABC}R^2 d\varphi = \int_0^\pi (0.6120 + 4.7198 \sin\varphi)50^2 d\varphi$$

$$= (0.6120\varphi - 4.7198\cos\varphi)|_0^\pi \ 50^2 = (0.6120 \cdot \pi - 4.7198(-2))50^2 = 1000e_z \tag{i}$$

$$\rightarrow e_z = 28.4057 \ mm$$

The shear flows are plotted as dashed lines in Figure (f). The solid lines are the values obtained for the section with no stiffeners from Example 10.7.2.

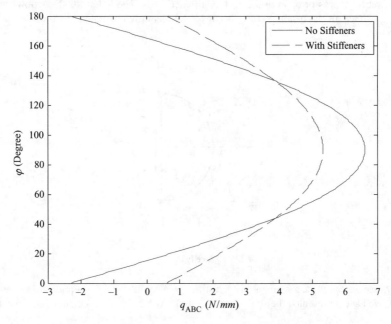

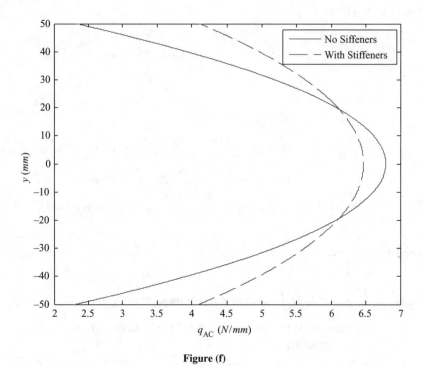

Figure (f)

The idealization in Figure (c) assumes that the bending stresses are carried entirely by the stiffeners. Depending on the relative size of the stiffeners and the thickness of the webs we can either ignore the web thickness or apportion it to the stiffener areas. The unknown shear flows are labeled in Figure (g). Since all the bending stress is carried by the stiffeners the shear flows in the webs are constant values.

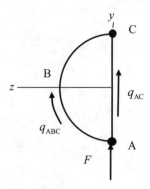

Figure (g)

In either case we can place a segment of each stiffener in equilibrium. The upper stiffener is shown in Figure (h). We have represented the shear flow q_{AC} in the vertical web in the direction we anticipate is positive.

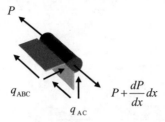

Figure (h)

Summing forces

$$P + \frac{dP}{dx}dx - P - q_{ABC}dx - q_{AC}dx = 0 \tag{j}$$

$$\rightarrow \quad q_{ABC} + q_{AC} = \frac{dP}{dx} = \frac{d}{dx}(\sigma_x A) = -\frac{\frac{dM}{dx}}{I_{zz}}yA = \frac{F}{I_{zz}}RA$$

Note that

$$I_{zz} = 2R^2 A \tag{k}$$

and so

$$q_{ABC} + q_{AC} = \frac{F}{I_{zz}}RA = \frac{FRA}{2R^2 A} = \frac{F}{2R} \quad \rightarrow \quad (q_{ABC} + q_{AC})2R = F \tag{l}$$

This is one equation containing the two unknown shear flow components. We need another equation. Setting twist to zero and finding the shear center is one way.

$$\int q\,ds = q_{ABC}\pi R - q_{AC}2R = 0 \quad \rightarrow \quad \pi q_{ABC} - 2q_{AC} = 0 \tag{m}$$

Solving the two simultaneous Equations (d) and (e) we get

$$q_{ABC} = \frac{F}{(\pi + 2)R} \qquad q_{AC} = \frac{\pi F}{2(\pi + 2)R} \tag{n}$$

Equate the applied moment to the moment of the shear flow to find the shear center.

$$Fe_z = q_{ABC}\pi R \cdot R = \frac{F\pi R}{(\pi + 2)} \quad \rightarrow \quad e_z = \frac{\pi R}{(\pi + 2)} \tag{o}$$

Now let us put some numbers in.

$$q_{ABC} = \frac{F}{(\pi + 2)R} = \frac{1000}{(\pi + 2)50} = 3.890\ N/mm \qquad q_{AC} = \frac{\pi F}{2(\pi + 2)R} = \frac{\pi 1000}{2(\pi + 2)50} = 6.110\ N/mm$$

$$e_z = \frac{\pi R}{(\pi + 2)} = \frac{\pi 50}{(\pi + 2)} = 30.55\ mm \tag{p}$$

These shear flow values are added to the previous figures as shown in Figure (i).

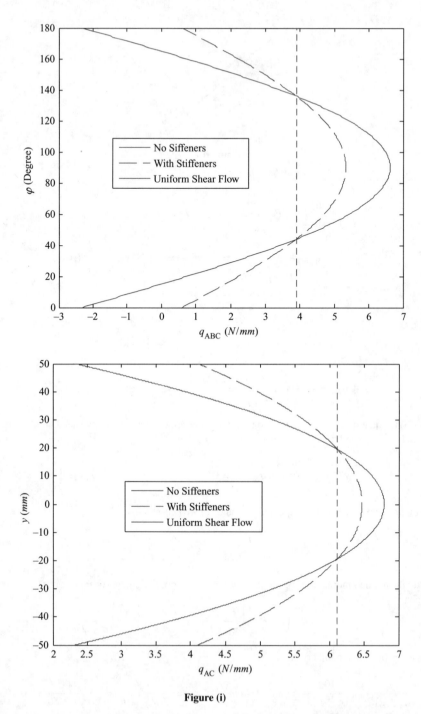

Figure (i)

Note that when the assumption of constant shear flows in the web is invoked the size of the stringers has no effect on the shear. Normal stress will be quite different, of course.

###########

The analysis of multi cell sections proceeds in the same manner. It gets to be quite complicated and so we will just outline the procedure in the next example for the case where the shear flows are constant in each web. In Chapter 12 we will provide a computer based method of analysis for solving for shear flow in thin walled sections that takes some of the pain out of the effort.

###########

Example 10.8.4

Problem: Consider the multi cell section shown in Figure (a). Given the shear force find the shear flow in the section and the shear center.

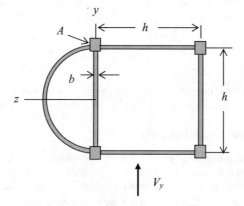

Figure (a)

Solution: Use the approximation that all bending stresses are carried in the stiffeners and the shear flow is constant in each web between stiffeners.

The section is idealized as in Figure (b).

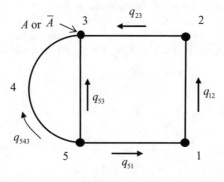

Figure (b)

Set up free body diagrams of each stiffener as shown in Figure (c).

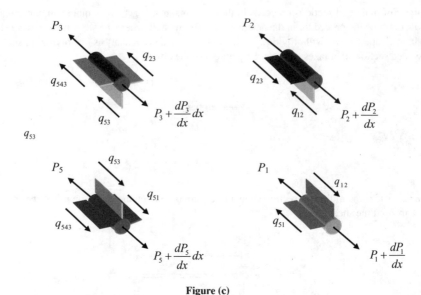

Figure (c)

Summing forces on each of the stiffeners will provide four equations in five unknowns. To get the fifth equation equate the moment of the applied load to the moment of the shear flows.

10.9   Summary and Conclusions

In this chapter we have been exploring ever increasing complexity of structural problems. In Section 10.3 we examined combined torsional and axial loading which, fortunately, are uncoupled and can be combined by simple superposition. The combined axial and bending problems in Section 10.4 are a bit more complicated because of the variety of geometrical arrangement of the frame members. And bending in two planes in Section 10.5, especially when $I_{xy} \neq 0$, presents some challenges. In all three cases the FEM provides advantages in solving problems.

The analysis of thin walled cross sections of Sections 10.6-8 extends our under standing of beam bending and torsional theory. It soon becomes evident that this analysis becomes tedious and cumbersome beyond the simple examples given there.

Clearly we need to develop ever more powerful methods. We have formulated problems that use the capabilities of linear equation solvers in Chapters 5, 8, and this chapter. Now we must go on to still more powerful methods. Actually we have been using them without formulation to illustrate 3D solutions that justify our 1D approximate methods. These will be introduced in more detail in the next two chapters.

11

Work and Energy
Methods—Virtual Work

11.1 Introduction

There is an alternative way of arriving at Newton's laws of static equilibrium different from the method of summation of forces and moments using a free body diagram. These are methods based on work and energy. Equating work done directly to strain energy stored has limited usefulness as described in Chapter 4, Section 4.11. The displacement of an axially loaded bar or pin jointed truss can be found at the point of application of a single force in the direction of that force if the body is statically determinate. Castigliano's second theorem, introduced in the same section (Chapter 4), extends the finding of the displacement at the point of application of a concentrated force to include multiple forces but the body must still be statically determinate.

In Chapter 6, Section 6.6, we used the argument that work done equals strain energy stored to define the effective torsional stiffness of a thin wall section. In Chapter 7, Section 7.5 we extended these arguments and Castigliano's second theorem to beams in bending.

Now we extend consideration of work and strain energy principles to provide general solutions to indeterminate as well as determinate bodies. There are several different formulations of the work and energy methods that may be found in the various references that extend these methods. We shall present here only the *principle of virtual work*. This is sufficient for our immediate needs.

11.2   Introduction to the Principle of Virtual Work

Setting the summation of forces and moments to zero on a free body diagram as a statement of equilibrium is intuitively believable and simple to understand. The *principle of virtual work* is an equivalent statement for bodies that are in equilibrium. To cast it is its simplest possible form (in the hope that this will gain your confidence in believing it in more complicated cases) we consider a single particle in equilibrium. Quite obviously the summation of forces acting on the particle must be zero if it is in equilibrium. The principle of virtual work (PVW) states the very same thing in the following way:

A particle is in equilibrium if and only if the virtual work of all the forces acting on the particle is zero during an arbitrary virtual displacement.

Analysis of Structures: An Introduction Including Numerical Methods, First Edition. Joe G. Eisley and Anthony M. Waas.
© 2011 John Wiley & Sons, Ltd. Published 2011 by John Wiley & Sons, Ltd.

Let us first examine what a *virtual displacement* is. We are not speaking of an actual displacement caused by the action of forces but of an imagined (virtual) displacement with no change in the forces that are acting. Instead we ask what work (virtual) would be done if the particle was displaced (virtually) with no changes in the values of the forces. If the virtual work is zero then that body is in equilibrium.

Consider a single particle acted upon by several forces as shown in Figure 11.2.1.

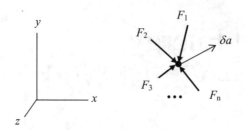

Figure 11.2.1

If the particle is in equilibrium the imagined (virtual) work δW done by the imagined (virtual) displacement δa (with components δu, δv, δw in the three coordinate directions) would be

$$\delta W = \sum F_{ix}\delta u + \sum F_{iy}\delta v + \sum F_{iz}\delta w = 0 \qquad (11.2.1)$$

and the PVW states that $\delta W = 0$. Thus, for arbitrary values of δu, δv, δw it follows that for $\delta W = 0$ it must be that

$$\sum F_{ix} = 0 \qquad \sum F_{iy} = 0 \qquad \sum F_{iz} = 0 \qquad (11.2.2)$$

So, in this case at least, the principle of virtual work and Newton's laws are equivalent. Will it be harder to believe in more complicated cases? Will you just accept it on faith for a while until a number of illustrations of the use of it have succeeded? After all you are taking Newton's laws on faith in much the same way. You believe them because you have had so many illustrations of their truth; that is, they are verified in practice.

Now let us consider a slightly more complicated case. Let there be a linear spring with spring constant k attached to the particle and fixed at the other end as shown in Figure 11.2.2. For convenience of illustration let us restrict all forces to lie in the xy plane and have the x-direction coincide with the direction of the spring force. Now imagine the particle to have achieved a state of equilibrium under a spring extension u, that is, under the action of the actual forces the particle is displaced until the spring force exactly equals the sum of the applied forces, and at that state the spring has stretched and amount u.

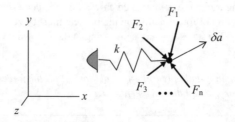

Figure 11.2.2

Now, the total virtual work is defined as the sum of the external virtual work δW^e and the internal virtual work δW^i where the internal virtual work is equal to the negative of the virtual strain energy δU, that is,

$$\delta W = \delta W^e + \delta W^i = \delta W^e - \delta U \tag{11.2.3}$$

Now, give the particle a virtual displacement with components δu and δv. We know that, at equilibrium, the linear spring has a force

$$F = ku \tag{11.2.4}$$

The spring will store strain energy by an amount shown in the crosshatched region in Figure 11.2.3 as a result of the initial displacement.

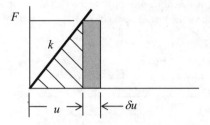

Figure 11.2.3

That strain energy stored is

$$U = \frac{1}{2}Fu = \frac{1}{2}ku^2 \tag{11.2.5}$$

When a virtual displacement is imposed, the virtual strain energy stored is shown by the shaded rectangular area in Figure 11.2.3, that is, by neglecting higher order terms

$$\begin{aligned}
\delta W^i &= -\delta U = -\left[U\left(u + \delta u\right) - U\left(u\right)\right] \\
&= -\left[\frac{1}{2}k\left(u + \delta u\right)^2 - \frac{1}{2}ku^2\right] \\
&= -\left[\frac{1}{2}k\left(u^2 + 2u\delta u + \delta u^2\right) - \frac{1}{2}ku^2\right] \\
&= -(ku)\delta u
\end{aligned} \tag{11.2.6}$$

Remember that the forces including the spring force do not change under a virtual displacement therefore the external virtual work is

$$\delta W^e = \sum\left(F_{ix}\delta u + F_{iy}\delta v\right) \tag{11.2.7}$$

Now according to the principle of virtual work δW is zero if the body is in equilibrium or

$$\delta W = \delta W^e + \delta W^i = \delta W^e - \delta U = \sum\left(F_{ix}\delta u + F_{iy}\delta v\right) - (ku)\delta u = 0 \tag{11.2.8}$$

By regrouping terms we have

$$\delta W = \left(\sum F_{ix} - ku\right)\delta u + \left(\sum F_{iy}\right)\delta v = 0 \tag{11.2.9}$$

To obtain equilibrium for arbitrary virtual displacements δu and δv it must be that

$$\sum F_{ix} - ku = 0 \qquad \sum F_{iy} = 0 \tag{11.2.10}$$

These are the familiar equations of equilibrium obtained from Newton's laws; however, this time obtained from virtual work.

Let us now extend the PVW to an elastic body such as a slender bar. The principle of virtual work, in this case, is stated as

> *A deformable solid body is in equilibrium if and only if the total virtual work is zero for any virtual displacement that does not violate the geometric constraints. Such a virtual displacement is called kinematically admissible.*

Now consider the principle of virtual work applied to the slender bar in Figure 11.2.4. We wish to find the displacement at the end where the load is applied.

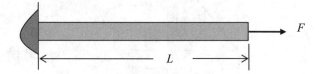

Figure 11.2.4

Since the stress and strain in the bar are constant we can define the strain in the bar to be

$$\varepsilon_x = \frac{u_L}{L} \tag{11.2.11}$$

and from Hooke's law the stress is

$$\sigma_x = E\varepsilon_x \tag{11.2.12}$$

Thus the strain energy in the bar in terms of displacement u_L is

$$U_A = \frac{1}{2} \int_V \varepsilon_x \sigma_x dV = \frac{1}{2} \int_0^L \varepsilon_x^2 E A dx = \frac{1}{2} \int_0^L \frac{u_L^2}{L^2} E A dx = \frac{E A}{2L} u_L^2 \tag{11.2.13}$$

Impose a virtual displacement δu_L, that is, the virtual displacement of the bar at $u(L)$. The external virtual work would be

$$\delta W^e = F \delta u_L \tag{11.2.14}$$

The virtual strain energy would be (neglecting higher order terms)

$$\delta U_A = U_A(u_L + \delta u_L) - U_A(u_L) = \frac{E A}{2L}(u_L + \delta u_L)^2 - \frac{E A}{2L}(u_L)^2 = \frac{E A}{L} u_L \delta u_L \tag{11.2.15}$$

The body is in equilibrium if

$$\delta W = \delta W^e - \delta U_A = F \delta u_L - \frac{E A}{L} u_L \delta u_L = \delta u_L \left(F - \frac{E A}{L} u_L \right) = 0 \tag{11.2.16}$$

Since δu_L is an arbitrary virtual displacement the above equation is true only if

$$F - \frac{E A}{L} u_L = 0 \quad \rightarrow \quad u_L = \frac{F L}{E A} \tag{11.2.17}$$

This is exactly what we get using Newton's laws for equilibrium. It is also exactly what we got equating work to strain energy in Chapter 4, Section 4.11. It is also what we get using Castigliano's second theorem.

By now you should be a little more convinced that the PVW gets the same results as Newton's laws for equilibrium for our purposes.

This was done for a statically determinate bar with a special case of axial loading. In the next section we shall extend the results of the principle of virtual work to the general case of a slender bar in axial, torsional, and bending displacement with distributed and concentrated loadings.

11.3 Static Analysis of Slender Bars by Virtual Work

We shall repeat the derivation of the general (FEM) method for slender bars using the principle of virtual work. This is an excellent use of virtual work and will lead us to an extension of this approach to more complicated problems.

11.3.1 Axially Loading

Consider a slender bar with axial concentrated loads, a constant cross section A, and Young's modulus E. The bar is divided into nodes and elements as shown in Figure 11.3.1. There must be a node at the location of each concentrated load and at other points where you wish to define a nodal displacement or have the stress reported.

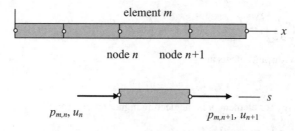

Figure 11.3.1

We define the nodal displacements and internal nodal forces for element m as shown in Figure 11.3.1.

$$\{r_m\} = \begin{bmatrix} u_n \\ u_{n+1} \end{bmatrix} \quad \{p_m\} = \begin{bmatrix} p_{m,n} \\ p_{m,n+1} \end{bmatrix} \tag{11.3.1}$$

The virtual displacements are

$$\{\delta r_m\} = \begin{bmatrix} \delta u_n \\ \delta u_{n+1} \end{bmatrix} \tag{11.3.2}$$

Now, we treat the element shown in Figure 11.3.1 as the "system" with the external loads being the force resultants $\{p_m\}$. Then, the external virtual work on element m is

$$\delta W_m^e = \{\delta r_m\}^T \{p_m\} = \begin{bmatrix} \delta u_n & \delta u_{n+1} \end{bmatrix} \begin{bmatrix} p_{m,n} \\ p_{m,n+1} \end{bmatrix} \tag{11.3.3}$$

Let us assume that the distributed displacement for element m can be written in terms of the nodal displacements $\{r_m\}$, that is, we can define the shape functions $[n]$ so that

$$u_m(s) = [n]\{r_m\} \tag{11.3.4}$$

Given the shape functions the strain displacement and Hooke's law equations can be written in terms of the nodal displacements.

$$\varepsilon_{sm} = \frac{du_m}{ds} = \frac{d}{ds}[n]\{r_m\} \qquad \sigma_{sm} = E\varepsilon_{sm} = E\frac{du_m}{ds} = E\frac{d}{ds}[n]\{r_m\} \tag{11.3.5}$$

The strain energy is

$$
\begin{aligned}
U_m &= \frac{1}{2}\int_V \varepsilon_{sm}\sigma_{sm}dV = \frac{1}{2}\int_V E\varepsilon_{sm}^2 dV \\
&= \frac{1}{2}\int_0^{l_m}\left(E\left(\frac{du}{ds}\right)^2\int_A dA\right)ds \\
&= \frac{1}{2}\int_0^{l_m} EA\left(\frac{du}{ds}\right)^2 ds = \frac{1}{2}\int_0^{l_m} EA\left(\frac{d}{ds}[n]\{r_m\}\right)^2 ds \\
&= \frac{EA}{2}\int_0^{l_m}\left(\{r_m\}^T\frac{d}{ds}[n]^T\right)\left(\frac{d}{ds}[n]\{r_m\}\right)ds
\end{aligned} \tag{11.3.6}
$$

Note that to square a matrix you must take the transpose of a matrix times the matrix (See Appendix A), that is,

$$\left(\frac{du}{ds}\right)^2 = \left(\frac{d}{ds}[n]\{r_m\}\right)^2 = \left(\{r_m\}^T\frac{d}{ds}[n]^T\right)\left(\frac{d}{ds}[n]\{r_m\}\right) \tag{11.3.7}$$

The distributed virtual displacements are

$$\delta u = [n]\{\delta r_m\} \tag{11.3.8}$$

The virtual strain energy for element m is (neglecting higher order terms)

$$
\begin{aligned}
\delta U_m &= U_m(u + \delta u) - U_m(u) \\
&= \frac{E}{2}\int_0^{l_m}\left(\frac{d(u+\delta u)}{ds}\right)^2 Ads - \frac{E}{2}\int_0^{l_m}\left(\frac{du}{ds}\right)^2 Ads \\
&= EA\int_0^{l_m}\frac{d\delta u}{ds}\frac{du}{ds}ds = \{\delta r_m\}^T EA\int_0^{l_m}\frac{d}{ds}[n]^T\frac{d}{ds}[n]ds\{r_m\}
\end{aligned} \tag{11.3.9}
$$

Now set the virtual work to zero to establish equilibrium.

$$
\begin{aligned}
\delta W_m &= \delta W_m^e - \delta U_m \\
&= \{\delta r_m\}^T\{p_m\} - \{\delta r_m\}^T EA\int_0^{l_m}\frac{d}{ds}[n]^T\frac{d}{ds}[n]ds\{r_m\} \\
&= \{\delta r_m\}^T\left(\{p_m\} - EA\int_0^{l_m}\frac{d}{ds}[n]^T\frac{d}{ds}[n]ds\{r_m\}\right) = 0
\end{aligned} \tag{11.3.10}
$$

With $\{\delta r_m\}^T$ an arbitrary virtual displacement the only way this can be true is if

$$\{p_m\} - EA\int_0^{l_m}\frac{d}{ds}[n]^T\frac{d}{ds}[n]ds\{r_m\} = 0 \tag{11.3.11}$$

or

$$\{p_m\} = [k_m]\{r_m\} \tag{11.3.12}$$

where

$$[k_m] = EA \int_0^{l_m} \frac{d}{ds}[n]^T \frac{d}{ds}[n]\,ds \tag{11.3.13}$$

We remind you that the arbitrary virtual displacement for an elastic body must be kinematically admissible. This means that the shape functions must be chosen to satisfy any geometric constraints whether internal or at the boundaries. In this case it means that the axial displacement at a node where two elements are joined must be the same for both elements and any realizable displacement constraints at the boundaries must be satisfied.

Fortunately the shape functions used in our previous work, namely, Equation 5.2.2, meet these requirements.

$$u_m(s) = [n]\{r_m\} = \begin{bmatrix} 1 - \dfrac{s}{l_m} & \dfrac{s}{l_m} \end{bmatrix} \begin{bmatrix} u_n \\ u_{n+1} \end{bmatrix} \tag{11.3.14}$$

Substituting $[n]$ into Equation 11.3.13 we get

$$[k_m] = EA \int_0^{l_m} \frac{d}{ds}[n]^T \frac{d}{ds}[n]\,ds = E_m A_m \int_0^{l_m} \frac{d}{ds}\begin{bmatrix} 1 - \dfrac{s}{l_m} \\ \dfrac{s}{l_m} \end{bmatrix} \frac{d}{ds}\begin{bmatrix} 1 - \dfrac{s}{l_m} & \dfrac{s}{l_m} \end{bmatrix} ds \tag{11.3.15}$$

which reduces to

$$[k_m] = E_m A_m \int_0^{l_m} \begin{bmatrix} -\dfrac{1}{l_m} \\ \dfrac{1}{l_m} \end{bmatrix} \begin{bmatrix} -\dfrac{1}{l_m} & \dfrac{1}{l_m} \end{bmatrix} ds = \frac{E_m A_m}{l_m} \begin{bmatrix} 1 & -1 \\ -1 & 1 \end{bmatrix} \tag{11.3.16}$$

Thus

$$\{p_m\} = [k_m]\{r_m\} = \begin{bmatrix} p_{m,n} \\ p_{m,n+1} \end{bmatrix} = \frac{E_m A_m}{l_m} \begin{bmatrix} 1 & -1 \\ -1 & 1 \end{bmatrix} \begin{bmatrix} u_n \\ u_{n+1} \end{bmatrix} \tag{11.3.17}$$

This is exactly the same result that we obtained in Chapter 5, Equation 5.2.9, and also summarized in Chapter 10. There are differences, however. We used the PVW and not Newton's laws for equilibrium and the polynomials used for the shape functions were chosen without knowledge of the exact answer (or could be chosen so). For all we know (if we had done it this way first) this is an approximate relation for the distributed displacement and so for the stiffness. Later on you will learn that this method will work to obtain approximate equations in cases where the exact shape function has never been found. Fortunately, it can be shown that you can approach the exact answer as closely as you wish with this method by using smaller and smaller elements even if the shape function itself is not the exact shape of the element displacement.

Having found the element stiffness matrix we can proceed to assemble the global stiffness matrix as before.

$$\{F\} = \sum \{p_m\} = \sum [k_m]\{r_m\} = [K]\{r\} \tag{11.3.18}$$

Of course the element matrices must be imbedded in the global format before summing.

As in Chapter 5 concentrated loads at nodes are added directly. To these we add distributed loads converted to equivalent nodal loads. In Chapter 5 (and in Chapter 10) we presented the formula for converting distributed loads into equivalent concentrated loads without proof. Now let us derive the formula we introduced there (Equation 5.4.4).

The distributed external load $f_s(s)$ contributes to the external virtual work term for element m as follows

$$\delta W_m^e = \int_0^{l_m} (\delta u_m)^T f_s(s)ds = \{\delta r_m\}^T \int_0^{l_n} [n]^T f_s(s)ds$$

$$= \{\delta r_m\}^T \int_0^{l_m} \begin{bmatrix} 1 - \dfrac{s}{l_m} \\[2mm] \dfrac{s}{l_m} \end{bmatrix} f_s(s)ds = \{\delta r_m\}^T \{F_{em}\} \tag{11.3.19}$$

Thus the equivalent nodal loads contributed by a distributed load on element m are found from

$$\{F_{em}\} = \int_0^{l_m} [n]^T f_s(s)ds = \int_0^{l_m} \begin{bmatrix} 1 - \dfrac{s}{l_m} \\[2mm] \dfrac{s}{l_m} \end{bmatrix} f_s(s)ds \tag{11.3.20}$$

This is our justification for finding the equivalent nodal loads using Equation 5.4.4. For example, a uniform load f_0 would contribute

$$\{F_{em}\} = \begin{bmatrix} F_{em,n} \\ F_{em,n+1} \end{bmatrix} = \int_0^{l_m} \begin{bmatrix} 1 - \dfrac{s}{l_m} \\[2mm] \dfrac{s}{l_m} \end{bmatrix} f_0 ds = \begin{bmatrix} \dfrac{f_0 l_m}{2} \\[2mm] \dfrac{f_0 l_m}{2} \end{bmatrix} \tag{11.3.21}$$

while a triangular load $f_s(s) = f_0 \dfrac{s}{l_m}$ would contribute

$$\{F_{em}\} = \begin{bmatrix} F_{em,n} \\ F_{em,n+1} \end{bmatrix} = \int_0^{l_m} \begin{bmatrix} 1 - \dfrac{s}{l_m} \\[2mm] \dfrac{s}{l_m} \end{bmatrix} f_0 \dfrac{s}{l_m} ds = \begin{bmatrix} \dfrac{f_0 l_m}{6} \\[2mm] \dfrac{f_0 l_m}{3} \end{bmatrix} \tag{11.3.22}$$

We now have derived a formal way for defining equivalent nodal loads that ensures the best possible distribution for satisfying equilibrium. For the global equivalent nodal loads we have

$$\{F_e\} = \sum \{F_{em}\} \tag{11.3.23}$$

Putting this all together, the global equations would be

$$\{F\} = \{F_c\} + \{F_e\} = [K]\{r\} \tag{11.3.24}$$

where the global stiffness $[K]$ is assembled from the element stiffness matrices as described in Chapter 5.

To complete the solution the restraints are added to the nodal displacement matrix and the restraint forces are added the nodal force matrix, the equations are partitioned and solved.

In summary

$$\{F\} \qquad \{F_c\} \qquad f_s(s) \;\rightarrow\; \{p_m\} \qquad U_m \qquad \sigma_{sm} \qquad \varepsilon_{sm} \qquad u_m(s) \qquad \{r_m\} \leftarrow \rho$$

$$\{F_{em}\} = \int_0^{l_m} [n]^T f_s(s)\,ds \qquad \{F_e\} = \sum \{F_{em}\} \qquad\qquad u_m(s) = [n(s)]\{r_m\}$$

$$\{F\} = \{F_c\} + \{F_e\} \qquad\qquad\qquad\qquad \varepsilon_{sm} = \frac{du}{ds} = \frac{d}{ds}[n(s)]\{r_m\}$$

$$\sigma_{sm} = E\varepsilon_{sm} = E\frac{d}{ds}[n(s)]\{r_m\} \qquad (11.3.25)$$

$$U_m = \frac{1}{2}\int_V \varepsilon_{sm}\sigma_{sm}\,dV = \frac{1}{2}\int_V E\varepsilon_{sm}^2\,dV = \frac{EA}{2}\int_0^{l_m}\left(\frac{du}{ds}\right)^2 ds$$

$$= \frac{EA}{2}\int_0^{l_m}\left(\{r_m\}^T\frac{d}{ds}[n]^T\right)\left(\frac{d}{ds}[n]\{r_m\}\right)ds$$

$$\delta W_m = \delta W_m^e - \delta U_m = \{\delta r_m\}^T\left(\{p_m\} - EA\int_0^{l_m}\frac{d}{ds}[n]^T\frac{d}{ds}[n]\,ds\,\{r_m\}\right) = 0$$

$$\{p_m\} = EA\int_0^{l_m}\frac{d}{ds}[n]^T\frac{d}{ds}[n]\,ds\,\{r_m\} = [k_m]\{r_m\}$$

$$[n] = \left[1 - \frac{s}{l_m} \quad \frac{s}{l_m}\right]$$

$$\{p_m\} = \left[\begin{array}{c} p_{m,n} \\ p_{m,n+1} \end{array}\right] = \frac{E_m A_m}{L_m}\left[\begin{array}{cc} 1 & -1 \\ -1 & 1 \end{array}\right]\left[\begin{array}{c} u_n \\ u_{n+1} \end{array}\right] = [k_m]\{r_m\}$$

$$[K] = \sum [k_m]$$

$$\{F\} = \{F_c\} + \{F_e\} = [K]\{r\}$$

In the first row of Equation 11.3.25 we have the typical unknowns in the formulation of a problem as applied to a single element m.

$\{r_m\}$ - nodal displacements

$u_m(s)$ - distributed displacement

ε_{sm} - normal strain

σ_{sm} - normal stress

U_m - strain Energy

$\{p_m\}$ - internal forces

and the known quantities

$$f_s(s)\text{ - distributed force}$$

$$\{F_c\}\text{ - concentrated forces}$$

$$\{F\}\text{ - total applied forces}$$

$$\rho\text{ - restraints}$$

In the final global set of equations the nodal restraints ρ are added to the nodal displacement matrix$\{r\}$, the corresponding nodal restraint forces are added to the global applied load matrix $\{F\}$, the resulting equations are partitioned and solved.

So far we have found what we already knew; however, in the very next chapter we shall put the principle of virtual work to good use in deriving FEM equations for 2D and 3D structures for which the methods we used in Chapters 5, 6, and 8 are inadequate.

11.3.2 Torsional Loading

Just as in Chapter 6, torsional equations can be derived by direct analogy to the axial case. As a reminder the nodal rotational displacements and internal nodal torques are shown in Figure 11.3.2 for element m.

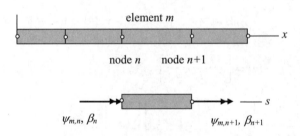

Figure 11.3.2

In matrix form the nodal displacements and torques are

$$\{\phi_m\} = \begin{bmatrix} \beta_n \\ \beta_{n+1} \end{bmatrix} \qquad \{\psi_m\} = \begin{bmatrix} \psi_n \\ \psi_{n+1} \end{bmatrix} \tag{11.3.26}$$

The rotational displacement $\beta_m(s)$ is given in terms of the nodal displacements $\{\phi_m\}$ and shape functions $[n]$.

$$\beta_m(s) = [n(s)]\{\phi_m\} \tag{11.3.27}$$

In this case the shape functions must satisfy compatible rotations where two elements are joined and any realizable displacement rotational constraints at the boundaries to be kinematically admissible.

We shall just summarize the equations here.

$$\{M\} \quad \{M_c\} \quad t_s(s) \quad \rightarrow \quad \{\psi_m\} \quad U_m \quad \tau_m \quad \gamma_m \quad \beta_m(s) \quad \{\phi_m\} \quad \leftarrow \quad \rho$$

$$\{M_{em}\} = \int_0^{l_m} [n]^T t_s(s) ds \qquad \{M_e\} = \sum \{M_{em}\} \qquad\qquad \beta_m(s) = [n(s)]\{\phi_m\}$$

$$\{M\} = \{M_c\} + \{M_e\} \qquad\qquad\qquad \gamma_m = \frac{d\beta}{ds} r = \frac{d}{ds}[n(s)]\{\phi_m\} r$$

$$\tau_m = G\gamma_m = G\frac{d}{ds}[n(s)]\{\phi_m\} r$$

$$U_m = \frac{1}{2}\int_V \gamma\tau dV = \frac{1}{2}\int_V G\gamma^2 dV = \frac{1}{2}\int_0^{l_m}\left(G\left(\frac{d\beta}{ds}\right)^2\int_A r^2 dA\right)ds$$

$$= \frac{GJ}{2}\int_0^{l_m}\left(\frac{d\beta}{ds}\right)^2 ds = \frac{GJ}{2}\int_0^{l_m}\left(\{\phi_m\}^T\frac{d}{ds}[n]^T\right)\left(\frac{d}{ds}[n]\{\phi_m\}\right)ds$$

$$\delta W_m = \delta W_m^e - \delta U_m = \{\delta\phi_m\}\left(\{\psi_m\} - GJ\int_0^{l_m}\frac{d}{ds}[n]^T\frac{d}{ds}[n]ds\{\phi_m\}\right) = 0$$

$$\{\psi_m\} = GJ\int_0^{l_m}\frac{d}{ds}[n]^T\frac{d}{ds}[n]ds\{\phi_m\} = [k_m]\{\phi_m\}$$

$$[n] = \left[1 - \frac{s}{l_m} \quad \frac{s}{l_m}\right]$$

$$\{\psi_m\} = \begin{bmatrix}\psi_{m,n}\\\psi_{m,n+1}\end{bmatrix} = \frac{G_m J_m}{L_m}\begin{bmatrix}1 & -1\\-1 & 1\end{bmatrix}\begin{bmatrix}\beta_n\\\beta_{n+1}\end{bmatrix} = [k_m]\{\phi_m\}$$

$$[K] = \sum [k_m]$$

$$\{M\} = \{M_c\} + \{M_e\} = [K]\{\phi\}$$

(11.3.28)

The point of all this is that the PVW provides the same set of FEM equations that we obtained in Chapter 6, Section 6.9, by enforcing equilibrium in an alternative manner. The solution of the equations proceeds in exactly the same way as before.

11.3.3   Beams in Bending

The displacements and internal forces on a beam element m are shown in Figure 11.3.3.

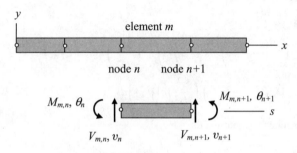

Figure 11.3.3

In matrix form the nodal displacements and internal forces are

$$\{r_m\} = \begin{bmatrix} v_n \\ \theta_n \\ v_{n+1} \\ \theta_{n+1} \end{bmatrix} \qquad \{p_m\} = \begin{bmatrix} V_{m,n} \\ M_{m,n} \\ V_{m,n+1} \\ M_{m,n+1} \end{bmatrix} \tag{11.3.29}$$

The strain energy for beams in bending consists of a part based on normal stresses and a part based on shear stresses.

$$U_B = \frac{1}{2}\int \sigma_x \varepsilon_x dV + \frac{1}{2}\int \gamma_{xy}\tau_{xy}dV \tag{11.3.30}$$

It was noted, however, that the strain energy contribution of the shear stress for a slender bar is quite small and so can be neglected. In Example 7.5.2 the rationale for neglecting it is demonstrated. So we shall proceed using only the normal stress part.

Remember that we did not include shear deformation in the previous development of the FEM equations in Chapter 8.

The distributed displacements for an element m are now presented in terms of the nodal displacements and a shape function matrix.

$$v_m(s) = [n]\{r_m\} \qquad u_m(s, y) = -y\frac{dv}{ds} = -y\frac{d}{ds}[n]\{r_m\} \tag{11.3.31}$$

The strain displacement and Hooke's law equations are

$$\varepsilon_{sm}(s, y) = \frac{du}{ds} = -y\frac{d^2}{ds^2}[n]\{r_m\} \qquad \sigma_{xm}(s, y) = E\varepsilon_{sm} = -yE\frac{d^2}{ds^2}[n]\{r_m\} \tag{11.3.32}$$

The strain energy of bending for a beam element m in terms of displacement is found to be

$$U_m = \frac{1}{2}\int_V (\varepsilon_{sm}\sigma_{sm})dV = \frac{1}{2}\int_V E(\varepsilon_{sm})^2 dV = \frac{1}{2}\int_V E\left(-y\frac{d^2v}{ds^2}\right)^2 dV$$

$$= \frac{1}{2}\int_0^{l_m}\left(E\left(\frac{d^2v}{ds^2}\right)^2\int_A y^2 dA\right)ds = \frac{EI_m}{2}\int_0^{l_m}\left(\frac{d^2v}{ds^2}\right)^2 ds \tag{11.3.33}$$

$$= \frac{EI_m}{2}\int_0^{l_m}\left(\{r_m\}^T\frac{d^2}{ds^2}[n]^T\right)\left(\frac{d^2}{ds^2}[n]\{r_m\}\right)ds$$

The virtual strain energy of bending for a beam element with a uniform cross section is then (again neglecting higher order terms)

$$\delta U_m = U_m(v + \delta v) - U_m(v)$$

$$= \frac{EI_m}{2}\int_0^{l_m}\left(\frac{d^2(v + \delta v)}{ds^2}\right)^2 ds - \frac{EI_m}{2}\int_0^{l_m}\left(\frac{d^2v}{ds^2}\right)^2 ds = EI_m\int_0^{l_m}\frac{d^2\delta v}{ds^2}\frac{d^2v}{ds^2}ds \tag{11.3.34}$$

$$= EI_m\int_0^{l_m}\left(\{\delta r_m\}^T\frac{d^2}{ds^2}[n]^T\right)\left(\frac{d^2}{ds^2}[n]\{r_m\}\right)ds$$

The nodal matrices can be taken outside the integral.

$$\delta U_m = \{\delta r_m\}^T EI_m\int_0^{l_m}\frac{d^2}{ds^2}[n]^T\frac{d^2}{ds^2}[n]ds\{r_m\} \tag{11.3.35}$$

For element m we apply the principle of virtual work.

$$\delta W_m = \delta W_m^e - \delta U_m = \{\delta r_m\}^T \{p_m\} - \{\delta r_m\}^T EI_m \int_0^{l_m} \frac{d^2}{ds^2}[n]^T \frac{d^2}{ds^2}[n]\, ds\, \{r_m\} = 0 \qquad (11.3.36)$$

Since $\{\delta r_m\}^T$ is an arbitrary virtual displacement it follows that

$$\{p_m\} = EI_m \int_0^{l_m} \frac{d^2}{ds^2}[n]^T \frac{d^2}{ds^2}[n]\, ds\, \{r_m\} = [k_m]\{r_m\} \qquad (11.3.37)$$

In this case to be kinematically admissible the shape functions must satisfy both displacement and slope compatibility at all the nodes.

For an element m, by the same arguments used in introducing Equation 8.2.6, we select the following shape functions.

$$[n] = \frac{1}{l_m^3}\left[\, l_m^3 - 3l_m s^2 + 2s^3 \quad l_m^3 s - 2l_m^2 s^2 + l_m s^3 \quad 3l_m s^2 - 2s^3 \quad -l_m^2 s^2 + l_m s^3 \,\right] \qquad (11.3.38)$$

These are cubic polynomials that represent the true shape of a uniform beam segment between concentrated loads. The second derivative is

$$\frac{d^2}{ds^2}[n] = \frac{1}{l_m^3}\left[\, -6l_m + 12s \quad -4l_m^2 + 6l_m s \quad 6l_m - 12s \quad -2l_m^2 + 6l_m s \,\right] \qquad (11.3.39)$$

Upon substituting Equation 11.3.39 into Equation 11.3.37 and simplifying the result we get

$$\{p_m\} = [k_m]\{r_m\} = \begin{bmatrix} V_{m,n} \\ M_{m,n} \\ V_{m,n+1} \\ M_{m,n+1} \end{bmatrix} = \frac{E_m I_m}{l_m^3} \begin{bmatrix} 12 & 6l_m & -12 & 6l_m \\ 6l_m & 4l_m^2 & -6l_m & 2l_m^2 \\ -12 & -6l_m & 12 & -6l_m \\ 6l_m & 2l_m^2 & -6l & 4l_m^2 \end{bmatrix}\begin{bmatrix} v_n \\ \theta_n \\ v_{n+1} \\ \theta_{n+1} \end{bmatrix} \qquad (11.3.40)$$

Thus the element stiffness matrix is

$$[k_m] = \frac{E_m I_m}{l_m^3}\begin{bmatrix} 12 & 6l_m & -12 & 6l_m \\ 6l_m & 4l_m^2 & -6l_m & 2l_m^2 \\ -12 & -6l_m & 12 & -6l_m \\ 6l_m & 2l_m^2 & -6l_m & 4l_m^2 \end{bmatrix} \qquad (11.3.41)$$

This is exactly what we got in Chapter 8, Equations 8.2.19. Furthermore, from the external virtual work, we obtain

$$\{F_e\} = \int_0^{l_m} [n]^T f_y(s)\, ds \qquad (11.3.42)$$

In summary

$$\{F\} \quad \{F_c\} \quad f_y(s) \quad \rightarrow \quad \{p_m\} \qquad U_m \qquad \sigma_{sm}(s, y) \qquad \varepsilon_{sm}(s, y) \qquad u_m(s, y) \qquad v_m(s) \qquad \{r_m\} \leftarrow \rho$$

$$\{F_{em}\} = \int_0^{l_m} [n]^T f_y(s)ds \qquad \{F_e\} = \sum \{F_{em}\} \qquad\qquad\qquad v_m(s) = [n(s)]\{r_m\}$$

$$\{F\} = \{F_c\} + \{F_{enl}\} \qquad\qquad\qquad u_m(s, y) = -y\frac{dv}{ds} = -y\frac{d}{ds}[n(s)]\{r_m\}$$

$$\varepsilon_{sm}(s, y) = \frac{du}{ds} = -y\frac{d^2}{ds^2}[n(s)]\{r_m\}$$

$$\sigma_{sm}(s, y) = E\varepsilon_{sm} = -yE\frac{d^2}{ds^2}[n(s)]\{r_m\} \qquad\qquad (11.3.43)$$

$$U_m = \frac{1}{2}\int_V (\varepsilon_{sm}\sigma_{sm})dV = \frac{1}{2}\int_V E(\varepsilon_{sm})^2 dV = \frac{E}{2}\int_V \left(-y\frac{d^2v}{ds^2}\right)^2 dV$$

$$= \frac{EI_m}{2}\int_0^{l_m} \left(\frac{d^2v}{ds^2}\right)^2 ds = \frac{EI_m}{2}\int_0^{l_m} \left(\{r_m\}^T \frac{d^2}{ds^2}[n]^T\right)\left(\frac{d^2}{ds^2}[n]\{r_m\}\right) ds$$

$$\delta W_m = \delta W_m^e - \delta U_m = \{\delta r_m\}\left(\{p_m\} = EI_m \int \frac{d^2}{ds^2}[n]^T \frac{d^2}{ds^2}[n]ds \{r_m\}\right) = 0$$

$$\{p_m\} = EI_m \int \frac{d^2}{ds^2}[n]^T \frac{d^2}{ds^2}[n]ds \{r_m\} = [k_m]\{r_m\}$$

$$[n] = \frac{1}{l_m^3}\left[l_m^3 - 3l_m s^2 + 2s^3 \quad l_m^3 s - 2l_m^2 s^2 + l_m s^3 \quad 3l_m s^2 - 2s^3 \quad -l_m^2 s^2 + l_m s^3\right]$$

$$\{p_m\} = \begin{bmatrix} V_{m,n} \\ M_{m,n} \\ V_{m,n+1} \\ M_{m,n+1} \end{bmatrix} = \frac{E_m I_m}{l_m^3} \begin{bmatrix} 12 & 6l_m & -12 & 6l_m \\ 6l_m & 4l_m^2 & -6l_m & 2l_m^2 \\ -12 & -6l_m & 12 & -6l_m \\ 6l_m & 2l_m^2 & -6l_m & 4l_m^2 \end{bmatrix} \begin{bmatrix} v_n \\ \theta_n \\ v_{n+1} \\ \theta_{n+1} \end{bmatrix} = [k_m]\{r_m\}$$

$$[K] = \sum [k_m]\{r_m\}$$

$$\{F\} = \{F_c\} + \{F_{enl}\} = [K]\{r\}$$

The solution proceeds in exactly the same way as in Chapter 8.

11.3.4 Combined Axial, Torsional, and Bending Behavior

The principal of virtual work can be applied to bending in the xz plane, to beams with $I_{yz} \neq 0$, and to combined axial, torsional, and bending behavior. The results are exactly the same as described in Chapter 10.

11.4 Static Analysis of 3D and 2D Solids by Virtual Work

Let us extend virtual work to the case for a general state of stress, that is, for 3D solid bodies without restrictions on geometry and to 2D thin plates within plane loads. We need the strain displacement equations and the Hooke's law equations for the 3D and 2D cases. These are taken from Chapter 3 and repeated here for convenience.

The 3D strain displacement equations are

$$\{\varepsilon\} = \begin{bmatrix} \varepsilon_x \\ \varepsilon_y \\ \varepsilon_z \\ \gamma_{xy} \\ \gamma_{yz} \\ \gamma_{zx} \end{bmatrix} = \begin{bmatrix} \dfrac{\partial}{\partial x} & 0 & 0 \\ 0 & \dfrac{\partial}{\partial y} & 0 \\ 0 & 0 & \dfrac{\partial}{\partial z} \\ \dfrac{\partial}{\partial y} & \dfrac{\partial}{\partial x} & 0 \\ 0 & \dfrac{\partial}{\partial z} & \dfrac{\partial}{\partial y} \\ \dfrac{\partial}{\partial z} & 0 & \dfrac{\partial}{\partial x} \end{bmatrix} \begin{bmatrix} u \\ v \\ w \end{bmatrix} = [D]\{u\} \tag{11.4.1}$$

and in 2D plane stress are

$$\{\varepsilon\} = \begin{bmatrix} \varepsilon_x \\ \varepsilon_y \\ \gamma_{xy} \end{bmatrix} = \begin{bmatrix} \dfrac{\partial}{\partial x} & 0 \\ 0 & \dfrac{\partial}{\partial y} \\ \dfrac{\partial}{\partial y} & \dfrac{\partial}{\partial x} \end{bmatrix} \begin{bmatrix} u \\ v \end{bmatrix} = [D]\{u\} \tag{11.4.2}$$

Hooke's law in 3D is

$$\{\sigma\} = \begin{bmatrix} \sigma_x \\ \sigma_y \\ \sigma_z \\ \tau_{xy} \\ \tau_{yz} \\ \tau_{zx} \end{bmatrix} = \begin{bmatrix} \lambda+2G & \lambda & \lambda & 0 & 0 & 0 \\ \lambda & \lambda+2G & \lambda & 0 & 0 & 0 \\ \lambda & \lambda & \lambda+2G & 0 & 0 & 0 \\ 0 & 0 & 0 & G & 0 & 0 \\ 0 & 0 & 0 & 0 & G & 0 \\ 0 & 0 & 0 & 0 & 0 & G \end{bmatrix} \begin{bmatrix} \varepsilon_x \\ \varepsilon_y \\ \varepsilon_z \\ \gamma_{xy} \\ \gamma_{yz} \\ \gamma_{zx} \end{bmatrix} = [G]\{\varepsilon\} \tag{11.4.3}$$

where

$$\lambda = \frac{vE}{(1+v)(1-2v)} \tag{11.4.4}$$

And, in 2D it is

$$\{\sigma\} = \begin{bmatrix} \sigma_x \\ \sigma_y \\ \tau_{xy} \end{bmatrix} = \begin{bmatrix} \dfrac{E}{1-v^2} & \dfrac{vE}{1-v^2} & 0 \\ \dfrac{vE}{1-v^2} & \dfrac{E}{1-v^2} & 0 \\ 0 & 0 & G \end{bmatrix} \begin{bmatrix} \varepsilon_x \\ \varepsilon_y \\ \gamma_{xy} \end{bmatrix} = [G]\{\varepsilon\} \tag{11.4.5}$$

For an element m we define nodal displacements $\{r_m\}$ and if we can find some shape functions $[n]$ such that

$$\{u_m\} = [n]\{r_m\} \tag{11.4.6}$$

where $\{u\}$ is a matrix of the distributed displacement components then

$$\{\varepsilon_m\} = [D]\{u_m\} = [D][n]\{r_m\} \quad \rightarrow \quad \{\sigma_m\} = [G]\{\varepsilon_m\} = [G][D]\{u_m\} = [G][D][n]\{r_m\} \quad (11.4.7)$$

The strain energy is

$$U_m = \int_V \{\varepsilon\}^T \{\sigma\} dV = \int_V ([D][n]\{r_m\})^T ([G][D][n]\{r_m\}) dV \tag{11.4.8}$$

Symbolically the virtual work is the sum of the external and internal virtual work or

$$\delta W_m = \delta W_m^e + \delta W_m^i = \delta W_m^e - \delta U_m = 0 \tag{11.4.9}$$

The external work is done by forces acting on the surfaces of the body whatever its shape. These surfaces forces are represented symbolically by $\{F_s\}$ and will be given specific values in example problems in later chapters.

$$\delta W_m^e = \int_A \{\delta u\}^T \{F_s\} dA \tag{11.4.10}$$

The virtual internal strain energy is

$$\delta U_m = \int_V \{\delta \varepsilon\}^T \{\sigma\} dV \tag{11.4.11}$$

Setting the virtual work equal to zero for equilibrium obtains

$$\delta W_m = \delta W_m^e - \delta U_m = \int_A \{\delta u\}^T \{F_s\} dA - \int_V \{\delta \varepsilon\}^T \{\sigma\} dV = 0 \tag{11.4.12}$$

Following our lead with the slender bar, perhaps it is possible to divide the solid body or the thin plate into elements, define shape functions, and turn this integral equation into a set of linear algebraic equations.

We designate the applied forces on the surfaces of a 3D solid or on the edges of a thin plate element m as f_{sm}. They may be forces per unit length in plane stress cases or forces per unit area in 3D solids.

The virtual work becomes

$$\delta W_m = \int [\delta r_m]^T [n]^T \{f_{sm}\} dA - \int \{\delta r\}^T ([D][n])^T [G][D][n]\{r_m\} dV = 0 \tag{11.4.13}$$

Since the virtual nodal displacements $\{\delta r_m\}$ and the real nodal displacements $\{r_m\}$ are constants they can be brought outside the integrals to obtain

$$\delta W_m = [\delta r_m]^T \left(\int [n]^T \{f_{sm}\} dA - \int ([D][n])^T [G][D][n] dV \{r_m\} \right) = 0 \tag{11.4.14}$$

For arbitrary values of the virtual nodal displacements $\{\delta r\}$ the quantity in the parenthesis must be zero.

$$\int [n]^T \{f_{sm}\} dA - \int ([D][n])^T [G][D][n] dV \{r_m\} = 0 \tag{11.4.15}$$

We define for each element

$$\{p_m\} = \int [n]^T \{f_{sm}\} dA \tag{11.4.16}$$

which we recognize as the equivalent nodal forces. And we define the element stiffness matrix for the element m.

$$[k_m] = \int ([D][n])^T [G][D][n] dV \tag{11.4.17}$$

The internal element forces are

$$\{p_m\} = [k_m]\{r_m\} \tag{11.4.18}$$

These are assembled into our final set of algebraic equations, where $[K]$ is the global stiffness matrix.

$$\{F\} = \sum \{p_m\} = [K]\{r\} \tag{11.4.19}$$

In summary

$$\{F\} \quad \{F_c\} \quad \{F_s\} \quad \rightarrow \quad \{p_m\} \qquad \{\sigma_m\} \qquad \{\varepsilon_m\} \qquad \{u_m\} \qquad \{r_m\} \quad \leftarrow \quad \rho$$

$$\{F_{em}\} = \int_{A_m} [n]^T \{F_s\}\, dA_m \qquad \{F_e\} = \sum \{F_{em}\} \qquad\qquad \{u_m\} = [n]\{r_m\}$$

$$\{F\} = \{F_c\} + \{F_e\} \qquad\qquad\qquad\qquad \{\varepsilon_m\} = [D]\{u_m\} = [D][n]\{r_m\}$$

$$\{\sigma_m\} = [G]\{\varepsilon_m\} = [G][D][n]\{r_m\}$$

$$U_m = \int_V \{\varepsilon_m\}^T \{\sigma_m\}dV = \int_V ([D][n]\{r_m\})^T ([G][D][n]\{r_m\})\, dV$$

$$\delta W_m = \delta W_m^e - \delta U_m = \{\delta r_m\} \left(\{p_m\} - \int ([D][n])^T [G][D][n]\, dV \{r_m\} \right) = 0$$

$$\{p_m\} = \int ([D][n])^T [G][D][n]\, dV \{r_m\} = [k_m]\{r_m\}$$

$$[K] = \sum \{k_m\}$$

$$\{F\} = [K]\{r\} \tag{11.4.20}$$

11.5   The Element Stiffness Matrix for Plane Stress

There is no simple polynomial expression for the plane stress shape functions for any but the simplest geometry and loading. We cannot conveniently turn to analytical solutions for suggestions to what the shape functions should be as we did for axial, torsional, and bending cases. The displacement functions are continuous, however, and so approximate polynomial shape functions can be derived that, when used in a large number of small elements, closely approximate the exact solution.

The earliest attempt to implement the finite element method for plane stress was to divide the plate into triangular segments, or elements, resulting in three nodes for each element. A triangular element is shown in Figure 11.5.1 with the notations for element displacements and forces. For plane stress there are two displacement components and two force components at each node.

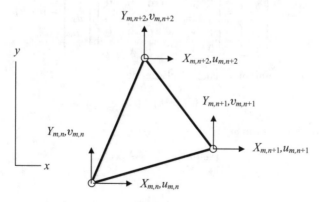

Figure 11.5.1

The nodal displacement and internal force matrices are

$$\{r_m\} = \begin{bmatrix} u_n \\ v_n \\ u_{n+1} \\ v_{n+1} \\ u_{n+2} \\ v_{n+2} \end{bmatrix} \qquad \{p_m\} = \begin{bmatrix} X_{m,n} \\ Y_{m,n} \\ X_{m,n+1} \\ Y_{m,n+1} \\ X_{m,n+2} \\ Y_{m,n+2} \end{bmatrix} \qquad (11.5.1)$$

In an early development simple linear polynomial displacements were assumed of the form

$$\begin{aligned} u(x, y) &= a_1 + a_2 x + a_3 y \\ v(x, y) &= a_4 + a_5 x + a_6 y \end{aligned} \qquad (11.5.2)$$

Since the strains are first derivatives of the displacements the resulting strains are constants, thus, this element is called the *constant strain triangle*. Remember that the axial bar element is a constant strain element. A difference is that the shape functions are exact for the axial elements with truly concentrated loads but are only approximate for the plane stress element under nearly all conditions. However, the PVW assures us the best possible result for equilibrium within the limits of the approximate shape function.

The development of shape functions from the polynomial form of the displacements is straight forward but a bit of a mess. In fact, the process is described in detail in many books but the actual shape functions usually are not even written down explicitly. Briefly, the process is given in the next few equations.

Equation 11.5.2 can be rewritten in this form.

$$\{u\} = \begin{bmatrix} u \\ v \end{bmatrix} = \begin{bmatrix} 1 & x & y & 0 & 0 & 0 \\ 0 & 0 & 0 & 1 & x & y \end{bmatrix} \begin{bmatrix} a_1 \\ a_2 \\ a_3 \\ a_4 \\ a_5 \\ a_6 \end{bmatrix} = [\Phi]\{a\} \qquad (11.5.3)$$

The nodal displacements $\{r_m\}$ can be written in terms of $\{a\}$ as shown in Equation 11.5.4

$$\{r_m\} = \begin{bmatrix} u_1 \\ v_1 \\ u_2 \\ v_2 \\ u_3 \\ v_3 \end{bmatrix} = \begin{bmatrix} 1 & x_1 & y_1 & 0 & 0 & 0 \\ 0 & 0 & 0 & 1 & x_1 & y_1 \\ 1 & x_2 & y_2 & 0 & 0 & 0 \\ 0 & 0 & 0 & 1 & x_2 & y_2 \\ 1 & x_3 & y_3 & 0 & 0 & 0 \\ 0 & 0 & 0 & 1 & x_3 & y_3 \end{bmatrix} \begin{bmatrix} a_1 \\ a_2 \\ a_3 \\ a_4 \\ a_5 \\ a_6 \end{bmatrix} = [h]\{a\} \qquad (11.5.4)$$

Invert $[h]$ to obtain $\{a\}$ in terms of the nodal displacements.

$$\{a\} = [h]^{-1}\{r_m\} \qquad (11.5.5)$$

Insert Equation 11.5.5 into Equation 11.5.3.

$$\{u\} = [\Phi][h]^{-1}\{r\} = [n]\{r_m\} \qquad (11.5.6)$$

Thus

$$[n] = [\Phi][h]^{-1} \qquad (11.5.7)$$

The shape functions are of the form

$$\{u\} = \begin{bmatrix} u(x, y) \\ v(x, y) \end{bmatrix} = \begin{bmatrix} n_{11} & n_{12} & n_{13} & n_{14} & n_{15} & n_{16} \\ n_{21} & n_{22} & n_{23} & n_{24} & n_{25} & n_{26} \end{bmatrix} \begin{bmatrix} u_n \\ v_n \\ u_{n+1} \\ v_{n+1} \\ u_{n+2} \\ v_{n+2} \end{bmatrix} = [n]\{r_m\} \qquad (11.5.8)$$

We can find $[h]^{-1}$ and the shape functions $[n]$ explicitly and write them down; however, they are rarely used in explicit format. What is actually done is the above process is programmed and only the coordinates of the nodes are actually entered and the program then generates the shape functions for each element for use in the next steps.

Once you have defined the shape functions you can insert them in the expression for virtual work. The result is the element stiffness matrix

$$[k_m] = \int ([D][n])^T [G][D][n]\, dV \qquad (11.5.9)$$

and the formula for equivalent nodal loads, namely.

$$\{F_{em}\} = \int [n]^T \{f_s\}\, dA \qquad (11.5.10)$$

Note that strain displacement, Hooke's law, and equilibrium are all satisfied in the process. Since we are using kinematically admissible shape functions compatibility is always satisfied.

The element stiffness matrix is of the form (where we entered just the order of the $[k_m]$ matrix)

$$\{p_m\} = \begin{bmatrix} X_{m,n} \\ Y_{m,n} \\ X_{m,n+1} \\ Y_{m,n+1} \\ X_{m,n+2} \\ Y_{m,n+2} \end{bmatrix} = [6x6] \begin{bmatrix} u_n \\ v_n \\ u_{n+1} \\ v_{n+1} \\ u_{n+2} \\ v_{n+2} \end{bmatrix} = [k_m]\{r_m\} \qquad (11.5.11)$$

The element matrices are then assembled to find the global equations. The process of assembly requires imbedding the element matrices in the global matrix format and summing. This has quite naturally led to commercial computer based programs that generate the shape functions from input of the nodal coordinates, add material properties and continue on to create the element stiffness matrices, and assemble them for solution. These programs also create the equivalent nodal loads from specific distributed loads on the edges and, of course, create nodal restraints based on distributed restraints.

The constant strain triangle was never very successful because the restriction to constant strain was too severe and so it required too many elements to converge to an accurate answer. Eventually additional elements including quadrilateral configurations were derived using higher order polynomials. These will be discussed in the next chapter. The above explanation does give you some idea of how the problem of creating the element stiffness may be approached.

We should mention that among the conditions imposed on the element is that two adjacent elements using two nodes in common must be compatible in that they share a common edge, that is, no gaps or overlaps.

11.6   The Element Stiffness Matrix for 3D Solids

Deriving the element stiffness matrix for 3D solids encounters problems similar to those for plane stress. One of the earliest attempts was for a four noded tetrahedron as shown in Figure 11.6.1. At node n

$$\{r_m\}_n = \begin{bmatrix} u_n \\ v_n \\ w_n \end{bmatrix} \tag{11.6.1}$$

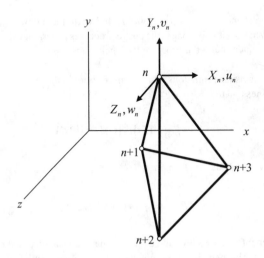

Figure 11.6.1

There are four nodes for element m with three nodal displacements at each node. In an early development simple linear polynomial displacements were assumed of the form

$$u(x, y) = a_1 + a_2 x + a_3 y + a_4 z$$
$$v(x, y) = a_5 + a_6 x + a_7 y + a_8 z$$
$$w(x, y) = a_9 + a_{10} x + a_{11} y + a_{12} z$$

Steps similar to those described for plane stress eventually lead to a shape function matrix of the form

$$\begin{bmatrix} u(x, y, z) \\ v(x, y, z) \\ w(x, y, z) \end{bmatrix} = [3x12][12x1] = [n]\{r_m\} \tag{11.6.2}$$

The element stiffness matrix is formed from the same virtual work equations as summarized in Equation 11.4.20.

We enter the order of the matrices rather than type them in full.

$$[p_m] = [12x1] = [12x12][12x1] = [k_m]\{r_m\} \tag{11.6.3}$$

The assembly process to produce the global matrices proceeds are described before.

This element was never very successful. Eventually additional elements including hexahedral configurations were derived using higher order polynomials. Currently a ten noded tetrahedral element with mid edge nodes is the favorite. These will be discussed in the next chapter.

11.7   Summary and Conclusions

The PVW supplies a straightforward way to extend the general method to other structural forms. It also provides the optimum method for converting distributed applied loads into equivalent nodal loads. This is summarized in Equation 11.3.25 for axially loaded slender bars, in Equation 11.3.28 for torsionally loaded shafts, in Equation 11.3.43 for beams in bending, and in Equation 11.4.20 for 3D and 2D solids.

In the case of slender bars the shape functions for axial, torsional, and bending can be found from the exact analytical equations when the bar is restricted to concentrated loads; however, this is not possible for many other structural forms of interest as we shall see in the remaining chapters. The extension to 2D and 3D solids is presented in the next chapter.

In the case of slender bars with truly concentrated loads the FEM is an exact analytical method, that is, it satisfies the differential equation for slender bars exactly. In succeeding chapters the PVW and FEM will be used to find approximate solutions to problems for which exact analytical solutions are more difficult or impossible to find.

12

Structural Analysis in Two and Three Dimensions

12.1 Introduction

We have studied in great detail the stress, strain, and displacement of slender bars (1D) with axial, torsional, and bending loads.

1. In the axial case the problem is formulated in rectangular Cartesian coordinates with just one component of displacement, u, normal stress, σ_x, and normal strain, ε_x. All other displacement, stress, and strain components are zero or are ignored. This is possible because of the assumption, supported by St. Venant's principle, that the stress is uniform across the cross section and the plane cross section remains plane when displaced. The x axis is the loci of centroids of the cross section area. A simple relation defines strain as a first derivative of displacement and another simple relation exists between the stress and strain represented by Young's modulus E. Applying equilibrium provides a second order differential equation in terms of the displacement that can be solved by direct integration. From this solution for the displacement we can find the stresses and strains.

2. In the torsional case with respect to cylindrical coordinates there is just one component of displacement, β, shearing stress, τ, and shearing strain, γ. All other displacement, stress, and strain components are zero or are ignored. This is possible because of the assumption, supported by St. Venant's principle, that for solid circular cross sections the stress varies linearly with the radius of the cross section and the plane cross section rotates as a plane when displaced. The x axis is the loci of centroids of the cross section area. The shearing strain is defined in terms of the first derivative of the displacement. A simple relation exists between the stress and strain as represented by the shear modulus G. Applying equilibrium provides a second order differential equation in terms of the displacement that can be solved by direct integration. From this solution for the displacement we can find the stresses and strains. A similar formulation is provided for thin walled open and closed sections.

3. In bending in the xy plane, with $I_{yz} = 0$ and with respect to rectangular Cartesian coordinates, we have two displacement components, u and v, two stress components, σ_x and τ_{xy}, and two strain components, ε_x and γ_{xy}. The x axis is the loci of centroids of the cross section area and the yz axes are principal axes of the cross sectional area. A simplified relation exists between the u and v displacements and the normal and shear components based on the assumption, supported by St. Venant's principle, that

---

Analysis of Structures: An Introduction Including Numerical Methods, First Edition. Joe G. Eisley and Anthony M. Waas.
© 2011 John Wiley & Sons, Ltd. Published 2011 by John Wiley & Sons, Ltd.

plane sections remain plane. This formulation considers bending displacement but neglects shear displacement. Applying equilibrium provides a fourth order differential equation in terms of the displacement that can be solved by direct integration. From this solution for the displacement we can find the stresses and strains. Similar equations can be found for bending in xz plane.

Since the governing displacement is a function of a single independent variable in each case, that is, $u(x)$ for axial, $\beta(x)$ for torsional, and $v(x)$ for bending in the xy plane, and $w(x)$ for bending in the xz plane, these are often called *one dimensional* structures. These are very effective structures as far as they go, but load carrying bodies are not such simple shapes for the most part. At one time many structures were designed as slender bars or as assemblies of slender bars, such as trusses and frames, because they could be analyzed easily while more complicated shapes were avoided because they were difficult to analyze. As more powerful methods of analysis have been developed the geometry of structures has become more complex. This has resulted in lower costs for manufacture and assembly, reduced material costs, lighter weight and greater efficiency, and other benefits.

We shall now consider more general cases of stress and strain in two and three dimensions for which simplifying assumptions are much harder to find. Actually we have considered some cases for simple geometry and loading. We introduced a simple three dimensional problem in Chapter 2, Example 2.6.2, a cube of material under hydrostatic pressure. In Chapter 3, Example 3.5.2, we examined a thin plate with edge loads in the plane of the plate, and in Chapter 4, Section 4.2, we examined a slender bar loaded with uniform and varying end loads. In all these cases solutions were based on the three dimensional equations of elasticity. Also in Chapter 3, Example 3.5.2, we examined a particular two dimensional state of stress that satisfied equilibrium but concluded that it could not be a solution. In Chapter 6, Section 6.4, we found that the linearly radial distribution of torsional stress in a circular cross section did, indeed, satisfy the three dimensional equations of elasticity under certain circumstances. Finally in Chapter 7, Section 7.7, we examined the circumstances under which the stresses in beams did satisfy the three dimensional equations.

The results of these previous examples suggest that some exact solutions are available for certain simple geometrical shapes if the surface loading is simple enough, such as, uniform or linearly varying, and the boundary restraints are idealized. The attempt at solutions becomes more difficult with minor changes in geometry, loading, and restraint. We need to extend our methods to handle these more difficult cases.

A more complete study of such problems is the subject of the theory of elasticity in more advanced texts and is much beyond the scope of this effort; however, we shall introduce the finite element method as applied to two and three dimensional solid bodies. This is the principal way such problems are solved today.

12.2 The Governing Equations in Two Dimensions—Plane Stress

Consider a thin flat plate of arbitrary shape and *uniform* thickness h restrained to prohibit rigid body motion and loaded on unrestrained edges in the plane of the plate, for example, as shown in Figure 12.2.1. The plate is made of homogeneous and isotropic material.

A small element in the interior of the body of dimensions dx by dy is shown in Figure 12.2.2 under a state of stress. The equilibrium of this element in the x-y plane was considered previously in Chapter 2, Section 2.6.

We shall assume that the loads and restraints are such that

$$\sigma_z = \tau_{zx} = \tau_{yz} = 0, \tag{12.2.1}$$

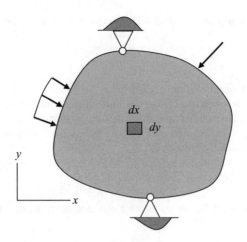

Figure 12.2.1

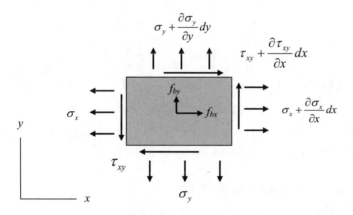

Figure 12.2.2

which corresponds to state of plane stress. Equilibrium was presented in matrix form in Chapter 2, Equation 2.6.16 and is repeated here as Equation 12.2.2.

$$[E]\{\sigma\} = \begin{bmatrix} \dfrac{\partial}{\partial x} & 0 & \dfrac{\partial}{\partial y} \\[2ex] 0 & \dfrac{\partial}{\partial y} & \dfrac{\partial}{\partial x} \end{bmatrix} \begin{bmatrix} \sigma_x \\ \sigma_y \\ \tau_{xy} \end{bmatrix} = - \begin{bmatrix} f_{bx} \\ f_{by} \end{bmatrix} = -\{f_b\} \tag{12.2.2}$$

The strain displacement equations were presented in Chapter 3, Equation 3.2.11, and are repeated here as Equation 12.2.3.

$$\{\varepsilon\} = \begin{bmatrix} \varepsilon_x \\ \varepsilon_y \\ \gamma_{xy} \end{bmatrix} = \begin{bmatrix} \dfrac{\partial}{\partial x} & 0 \\[2ex] 0 & \dfrac{\partial}{\partial y} \\[2ex] \dfrac{\partial}{\partial y} & \dfrac{\partial}{\partial x} \end{bmatrix} \begin{bmatrix} u \\ v \end{bmatrix} = [D]\{u\} \tag{12.2.3}$$

As shown in Chapter 3, Equations 3.4.20–3.4.23, it is true that there is a non zero component of strain in the z direction given by

$$\varepsilon_z = -\frac{v}{E}\left(\sigma_x + \sigma_y\right) \tag{12.2.4}$$

The result of the ε_z component of strain is a change in the thickness of the plate depending upon the local value of the stresses; however, the change has little effect on the behavior of the plate if the plate is sufficiently thin compared to its other two dimensions and so it is generally neglected.

In Chapter 3, Section 3.3, it was noted that the strains, ε_x, ε_y, and γ_{xy} are not independent. They satisfy the compatibility equation (Equation 3.3.3),

$$\frac{\partial^2 \varepsilon_x}{\partial y^2} + \frac{\partial^2 \varepsilon_y}{\partial x^2} = \frac{\partial^2 \gamma_{xy}}{\partial x \partial y} \tag{12.2.5}$$

From Chapter 3, Equation 3.4.20, the plane stress, strain-stress relations for a linear, homogeneous, and isotropic material are,

$$\begin{bmatrix} \varepsilon_x \\ \varepsilon_y \\ \gamma_{xy} \end{bmatrix} = \begin{bmatrix} \dfrac{1}{E} & \dfrac{-v}{E} & 0 \\ \dfrac{-v}{E} & \dfrac{1}{E} & 0 \\ 0 & 0 & \dfrac{1}{G} \end{bmatrix} \begin{bmatrix} \sigma_x \\ \sigma_y \\ \tau_{xy} \end{bmatrix} \tag{12.2.6}$$

It is convenient to invert Equation 12.2.6 to obtain

$$\begin{bmatrix} \sigma_x \\ \sigma_y \\ \tau_{xy} \end{bmatrix} = \begin{bmatrix} \dfrac{E}{1-v^2} & \dfrac{vE}{1-v^2} & 0 \\ \dfrac{vE}{1-v^2} & \dfrac{E}{1-v^2} & 0 \\ 0 & 0 & G \end{bmatrix} \begin{bmatrix} \varepsilon_x \\ \varepsilon_y \\ \gamma_{xy} \end{bmatrix} = [G]\{\varepsilon\} \tag{12.2.7}$$

In summary the set of equations to be satisfied are

$$\{F\} \quad \rightarrow \quad \{f_b\} \qquad\qquad \{\sigma\} \qquad\qquad \{\varepsilon\} \qquad\qquad \{u\} \quad \leftarrow \{\rho\}$$
$$\{\varepsilon\} = [D]\{u\}$$
$$\{\sigma\} = [G]\{\varepsilon\} \tag{12.2.8}$$
$$[E]\{\sigma\} = -\{f_b\}$$
$$[E][G][D]\{u\} = -\{f_b\}$$

In the first row we have the typical internal unknowns in the formulation of a problem.

$$\{u\} - \text{distributed displacements}$$
$$\{\varepsilon\} - \text{strains}$$
$$\{\sigma\} - \text{stresses}$$

and the known external quantities

$$\{f_b\} - \text{body forces}$$
$$\{F\} - \text{surface forces}$$
$$\{\rho\} - \text{restraints}$$

The arrows are there to indicate that both $\{F\}$ and $\{\rho\}$ are external *boundary conditions* to be satisfied in the process of solving the equations.

The equations connecting these quantities are in the second, third, and fourth rows.

$$\{\varepsilon\} = [D]\{u\} - \text{strain displacement}$$
$$\{\sigma\} = [G]\{\varepsilon\} - \text{Hooke's law}$$
$$[E]\{\sigma\} = -\{f_b\} - \text{static equilibrium}$$

In the fifth row we have combined these equations to form a set of equation in terms of the displacements.

$$[E][G][D]\{u\} = -\{f_b\} \tag{12.2.9}$$

There are many books written on the subject of finding analytical solutions to Equations 12.2.8 for various combinations of geometry, applied loads, and restraints. We can find some solutions to the equations for very simple geometry, loading, and restraint. The number of exact analytical solutions is very limited. Most often stresses are found that satisfy equilibrium, compatibility, and boundary conditions then the displacements are calculated. The displacement based set of equations, Equations 12.2.9, is seldom used directly; however, one such solution is found in the next example.

###########

Example 12.2.1

Problem: A thin flat rectangular plate has uniform loads on edges as shown in Figure (a). The plate has a thickness h, has a length in the x direction of a and a width in the y direction of b. The origin of the coordinates is in the center of the plate. The edge loads are constant and have the units of N/mm^2. Find the stresses, strains, and displacements of the plate.

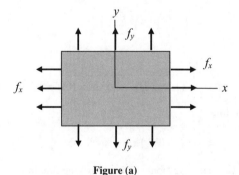

Figure (a)

Solution: Uniform loads are likely to provide a uniform interior stress which would suggest a linear displacement. So let us assume

$$u = cx \qquad v = dy \tag{a}$$

where c and d are constants to be determined if, in fact, this is a valid assumption.

Substituting into the strain displacement relations we get

$$
\begin{bmatrix} \varepsilon_x \\ \varepsilon_y \\ \gamma_{xy} \end{bmatrix} = \begin{bmatrix} \dfrac{\partial}{\partial x} & 0 \\ 0 & \dfrac{\partial}{\partial y} \\ \dfrac{\partial}{\partial y} & \dfrac{\partial}{\partial x} \end{bmatrix} \begin{bmatrix} u \\ v \end{bmatrix} = \begin{bmatrix} \dfrac{\partial}{\partial x} & 0 \\ 0 & \dfrac{\partial}{\partial y} \\ \dfrac{\partial}{\partial y} & \dfrac{\partial}{\partial x} \end{bmatrix} \begin{bmatrix} cx \\ dy \end{bmatrix} = \begin{bmatrix} c \\ d \\ 0 \end{bmatrix}
\tag{b}
$$

These values of constant strain can then be entered in the Hooke's law equations, Equations 12.2.7.

$$
\begin{bmatrix} \sigma_x \\ \sigma_y \\ \tau_{xy} \end{bmatrix} = \begin{bmatrix} \dfrac{E}{1-v^2} & \dfrac{vE}{1-v^2} & 0 \\ \dfrac{vE}{1-v^2} & \dfrac{E}{1-v^2} & 0 \\ 0 & 0 & G \end{bmatrix} \begin{bmatrix} \varepsilon_x \\ \varepsilon_y \\ \gamma_{xy} \end{bmatrix} = \begin{bmatrix} \dfrac{E}{1-v^2} & \dfrac{vE}{1-v^2} & 0 \\ \dfrac{vE}{1-v^2} & \dfrac{E}{1-v^2} & 0 \\ 0 & 0 & G \end{bmatrix} \begin{bmatrix} c \\ d \\ 0 \end{bmatrix}
$$

$$
= \frac{E}{1-v^2} \begin{bmatrix} c + vd \\ vc + d \\ 0 \end{bmatrix}
\tag{c}
$$

One can quickly note that constant values of stress satisfy the equilibrium equations, Equation 12.2.2, identically if the body forces are zero.

$$
[E]\{\sigma\} = \begin{bmatrix} \dfrac{\partial}{\partial x} & 0 & \dfrac{\partial}{\partial y} \\ 0 & \dfrac{\partial}{\partial y} & \dfrac{\partial}{\partial x} \end{bmatrix} \begin{bmatrix} \sigma_x \\ \sigma_y \\ \tau_{xy} \end{bmatrix} = \begin{bmatrix} \dfrac{\partial}{\partial x} & 0 & \dfrac{\partial}{\partial y} \\ 0 & \dfrac{\partial}{\partial y} & \dfrac{\partial}{\partial x} \end{bmatrix} \frac{E}{1-v^2} \begin{bmatrix} c + vd \\ vc + d \\ 0 \end{bmatrix} = \begin{bmatrix} 0 \\ 0 \\ 0 \end{bmatrix}
\tag{d}
$$

What remains is to satisfy the boundary conditions.

$$
\sigma_x \left(\pm \frac{a}{2} \right) = \frac{E}{1-v^2}(c + vd) = f_x
$$

$$
\sigma_y \left(\pm \frac{b}{2} \right) = \frac{E}{1-v^2}(vc + d) = f_y
\tag{e}
$$

Solving for c and d

$$
c = \frac{1}{E}(f_x - vf_y) \qquad d = \frac{1}{E}(f_y - vf_x)
\tag{f}
$$

The displacement is therefore

$$
u(x, y) = \frac{1}{E}(f_x - vf_y)x \qquad v(x, y) = \frac{1}{E}(f_y - vf_x)y
\tag{g}
$$

Now we learned earlier (Chapter 3, Section 3.3) that satisfying equilibrium is not enough. Let us check to see if compatibility (Equation 3.3.3) is satisfied.

$$
\frac{\partial^2 \varepsilon_x}{\partial y^2} + \frac{\partial^2 \varepsilon_y}{\partial x^2} = \frac{\partial^2 \gamma_{xy}}{\partial x \partial y} \quad \rightarrow \quad \frac{\partial^2 c}{\partial y^2} + \frac{\partial^2 d}{\partial x^2} = \frac{\partial^2 0}{\partial x \partial y} \quad \rightarrow \quad 0 + 0 = 0
\tag{h}
$$

So the assumed form is a valid solution. It satisfies strain displacement, Hooke's law, equilibrium, and all boundary conditions. Note that by starting with displacements compatibility is satisfied.

##########

Solutions are also possible for linearly varying edge loads; however, as soon as we restrain an edge, have more complicated edge loads, or change the geometry of the plate the search for an analytic solution becomes very difficult. Known useful results are limited. What we really need is a method for all combinations of geometry, loading, and restraint. That method is based on virtual work as was first introduced in Chapter 11. It is called the finite element method. It will be extended to plane stress problems in the next section.

12.3   Finite Elements and the Stiffness Matrix for Plane Stress

The equations for the finite element method are developped in Chapter 11, Section 4, and are summarized here in Equations 12.3.1. These are repeated here for convenience.

$$\{F\} \qquad \{F_s\} \qquad \{p_m\} \qquad \{\sigma_m\} \qquad \{\varepsilon_m\} \qquad \{u_m\} \qquad \{r_m\} \quad \leftarrow \quad \rho$$

$$\{F_{em}\} = \int_{A_m} [n]^T \{F_s\}\, dA_m \qquad \{F\} = \{F_e\} = \sum \{F_{em}\} \qquad \{u_m\} = [n]\{r_m\}$$

$$\{\varepsilon_m\} = [D]\{u_m\} = [D][n]\{r_m\}$$

$$\{\sigma_m\} = [G]\{\varepsilon_m\} = [G][D][n]\{r_m\}$$

$$U_m = \int_V \{\varepsilon_m\}^T \{\sigma_m\} dV = \int_V ([D][n]\{r_m\})^T ([G][D][n]\{r_m\})\, dV \qquad\qquad (12.3.1)$$

$$\delta W_m = \delta W_m^e - \delta U_m = \{\delta r_m\}\left(\{p_m\} - \int ([D][n])^T [G][D][n]\, dV \{r_m\} \right) = 0$$

$$\{p_m\} = \int ([D][n])^T [G][D][n]\, dV \{r_m\} = [k_m]\{r_m\}$$

$$[K] = \sum \{k_m\}$$

$$\{F\} = [K]\{r\}$$

There are no exact polynomial expressions for the plane stress shape functions for any but the simplest geometry and loading. We cannot conveniently turn to analytical solutions for suggestions to what the shape functions should be as we did for axial, torsional, and bending cases. The displacement functions are continuous, however, and so approximate polynomial shape functions can be derived that, when used in a large number of elements, closely approximate the exact solution.

We have briefly described in Chapter 11, Section 11.5, the quest for shape functions for the constant strain triangle element. Even for this simplest of plane stress elements it is impossible to actually write down the shape functions and the resulting stiffness matrices for each element and assemble them by hand as we have illustrated for slender bars. We must depend on computer programs that do this quickly and neatly upon given our instructions.

As noted in Chapter 11 this has quite naturally led to commercial programs that generate the shape functions from input of the nodal coordinates, add material properties, and continue on to create the element stiffness matrices and assemble them for solution. These programs also create the equivalent nodal loads from specific distributed loads on the edges and, of course, create nodal restraints based on distributed restraints.

We also noted that the constant strain triangle was never very successful, so great effort was expended to develop more advanced element stiffness matrices using higher order polynomials for the shape functions.

The actual process was to take a square element and map it into a general quadrilateral as shown in Figure 12.3.1.

Still later two other popular elements were developped – the six noded triangle and the eight noded quadrilateral each with curved edges as illustrated in Figure 12.3.2.

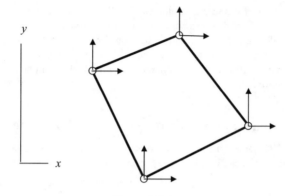

Figure 12.3.1

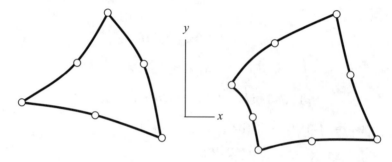

Figure 12.3.2

The derivation of these elements is well beyond the scope of this work, but you get some idea of what is involved by the example given in Chapter 11, Section 11.5.

Plane stress elements were used in the solutions presented in Chapter 4, Section 4.2, and in Chapter 7, Section 7.7. They were also adapted to use on thin walled shell structures such as the hollow tubes in Chapter 6, Section 6.6.

Because of the size and complexity of the element stiffness matrices we do not attempt to write them out for each element. Because a large number of elements are needed we do not assemble the elements, find equivalent nodal loads, partition the matrices, and insert the resulting matrices into a linear equation a linear equation solver. Instead all this has been programmed and offered in a number of commercial FE codes. In addition there are a number of pre and post processors that work directly with surface and solid models to create automatically the nodes and elements and to present the results in the form of color contour plots. We shall use one such set of commercial codes to create solid models, attach boundary restraints and loads to the geometry of the models, to create nodes and elements (called a mesh), to solve for the displacements and stresses, and finally to present the results in color contour plots. One such set of codes was already used in Chapters 4, 6, and 7 to demonstrate the validity of the simplifying assumptions of slender bar theory. Additional examples will be offered of ever increasing complexity. Steps in the process of solution will be presented.

We shall now offer a plane stress solution based on FEM for one of the classic problems of structural analysis – the plate with a hole. The analysis starts with a surface created by solid modeling or computer aided design (CAD) software. The applied loads and restraints are applied directly to the CAD model. The nodes and elements are created by a modern automatic mesh generator. The elements used are sophisticated quadrilaterals. The software converts the loads and restraints to equivalent nodal loads and restraints. A powerful linear equation solver completes the solution. Finally, the results are presented in the form of contour plots.

############

Example 12.3.1

Problem: A square plate with a hole is loaded with uniform edge loads as shown in Figure (a). Find the stresses and displacements.

The plate has a dimension of 100 *mm* on a side and is 2 *mm* thick. The hole is centered and has a diameter of 30 *mm*. The material is aluminum with $E = 68950$ N/mm^2. A uniform load of 2 N/mm is applied to the two opposite edges as shown. This provides an edge stress of 1 N/mm^2.

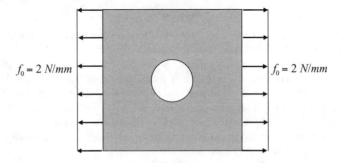

Figure (a)

Solution: Create a CAD model; add loads to the edges; use the automatic mesh generator to create nodes and elements; invoke the solver; and present contour plots of stresses and displacements.

A solid model of the plate was created in the I-DEAS software as shown in Figure (b).

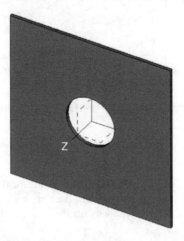

Figure (b)

The plane stress elements are based on surfaces and the thickness is added as a scalar quantity in a property table. A surface was selected and loads were added to the edges. These are shown in Figure (c).

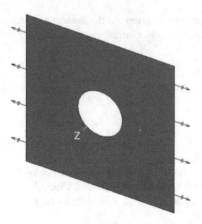

Figure (c)

Nodes and elements were then created by the automatic mesh generator. We have the choice of 3 or 6 noded triangular, or 4 or 8 noded quadrilateral elements. We have chosen the 4 noded quadrilateral for this example. We have the choice of specifying the density of the mesh, that is, the number of nodes and elements, in several ways. One way is to specify the nominal edge length of each element. An edge length of 10 mm was chosen. In anticipation of high stress in the vicinity of the hole we have used a feature of the mesh generator that allows us to increase the number of nodes on the edge of the hole. This provides smaller element edge lengths there. Sixteen equally spaced nodes were placed on the edge of the hole. For visual reasons the element size was kept fairly large for the first solution attempt. In creating the elements a thickness of 2 mm was specified.

The nodes and elements are shown in Figure (d). Nodes are at the corners of each element.

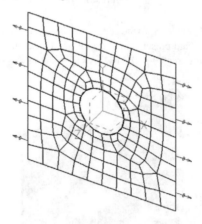

Figure (d)

You will note that while the geometry and loading are symmetrical about both the x and y axes the elements are not, that is, the automatic mesh generator does not maintain perfect symmetry. There are mesh generator functions that will enforce symmetry but they require a little more work. We will address this problem shortly.

The statement of the problem is now complete. The solver was invoked. Tables of the values of the displacements and stresses at the nodes are available; however, more useful are contour plots of these quantities throughout the plate. We have the option of displaying contour plots of the three Cartesian components of stress, the maximum and minimum principle stresses, the maximum shear stresses, and the von Mises stresses. In Figure (e) we show a contour plot of the σ_x stress on an undeformed plate.

We note the large increase in stress at the top and bottom of the hole. The stress at the boundary is 1 N/mm^2 while the maximum stress is 3.54 N/mm^2. This phenomenon is known as *stress concentration*.

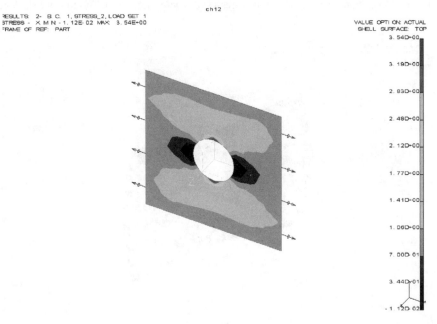

Figure (e)

In Figure (f) we show a contour plot of the displacements placed on a deformed model. What is plotted is the maximum displacement, that is, the vector sum of the x and y displacements.

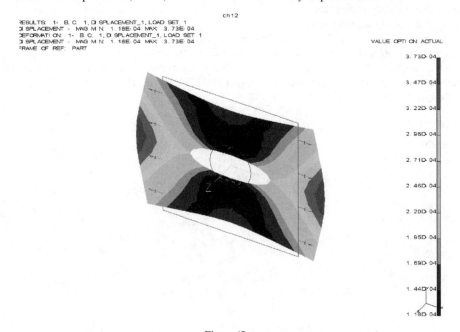

Figure (f)

The displacements are greatly exaggerated for visual reasons. The actual displacements are quite small, 0.000373 *mm* is the maximum.

In Figure (g) we show a contour plot of the stresses on a deformed model. This time we show the von Mises stresses.

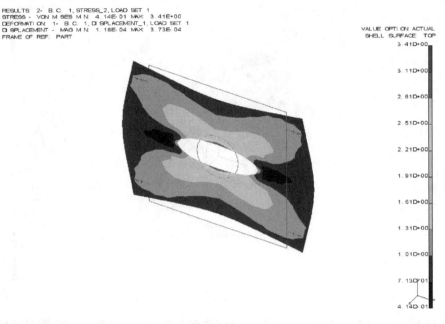

Figure (g)

Increased accuracy is obtained by using a finer and finer mesh until convergence is observed. Not long ago the cost of computation was so high that various means were used to keep the number of nodes and elements to a minimum. Computation has become so fast and cheap that for simple problems like this it is not a problem to increase the density substantially. We now repeat the whole process using a much finer mesh density to examine the accuracy of the method. The nodes and elements are shown in Figure (h). There are 4100 nodes and 3960 elements in this FE model. The mesh in the vicinity of the hole was made much finer deliberately.

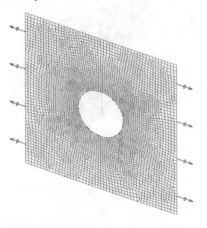

Figure (h)

A contour plot of the von Mises stresses are shown in Figure (i) on a deformed model to compare with those in Figure (g). Note the symmetry of the solution – it must certainly have converged. The maximum von Mises stress has changed from 3.41 to 3.84. Nowhere near this number of elements are needed for an accurate answer.

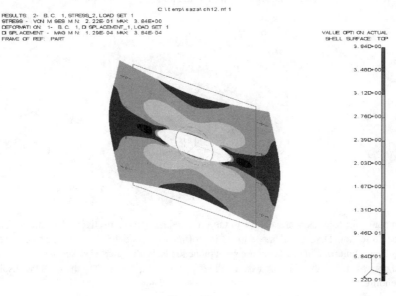

Figure (i)

The displacement contour plot is shown in Figure (j). The maximum displacement has changed from 0.000373 *mm* to 0.000383 *mm*.

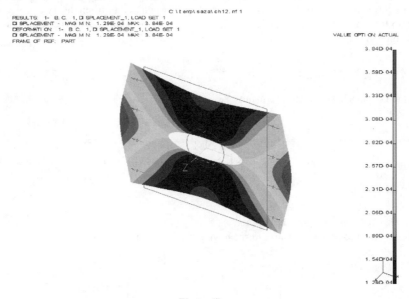

Figure (j)

###########

12.4 Thin Flat Plates—Classical Analysis

When a thin plate is subject to lateral loads and is restrained at the edges to prevent rigid body motion, bending of the plate occurs. We shall orient the plate in the xy plane as shown in Figure 12.4.1. A rectangular plate of *uniform* thickness is shown but the plate can be any shape.

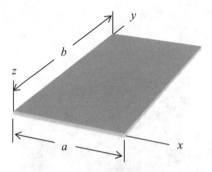

Figure 12.4.1

The in plane displacements are u and v in the x and y directions respectively. The lateral displacement w is in the z direction. Distributed lateral loads are normal to the undeformed plate surface; are designated $f_z(x,y)$; and have units of N/mm^2. There are no in plane loads. With the x and y axes in the plane of the plate and the z axis normal to the plate we assume the following simplification to the in-plane displacements, due to bending,

$$u(x, y, z) = -z\frac{\partial w\,(x, y)}{\partial x} \qquad v(x, y, z) = -z\frac{\partial w\,(x, y)}{\partial y} \qquad (12.4.1)$$

Note that $w\,(x, y)$ is a function of only x and y while $u(x, y, z)$ and $v(x, y, z)$ are functions of $x, y,$ and z. This is equivalent to the "plane sections remain plane" assumption of beam theory and is verified in practice. In fact, this can be thought of as the 2D extension of this assumption. The strain components corresponding to (12.4.1) become,

$$\varepsilon_x(x, y, z) = -z\frac{\partial^2 w}{\partial x^2} \qquad \varepsilon_y(x, y, z) = -z\frac{\partial^2 w}{\partial y^2} \qquad \gamma_{xy}(x, y, z) = -2z\frac{\partial^2 w}{\partial x\partial y} \qquad (12.4.2)$$

The strain components in matrix form are

$$\{\varepsilon\} = \begin{bmatrix} \varepsilon_x \\ \varepsilon_y \\ \gamma_{xy} \end{bmatrix} = \begin{bmatrix} -z\dfrac{\partial^2}{\partial x^2} \\[2mm] -z\dfrac{\partial^2}{\partial y^2} \\[2mm] -2z\dfrac{\partial^2}{\partial x\partial y} \end{bmatrix} \{w\} = [D]\{w\} \qquad (12.4.3)$$

where $\{w\}$ has only one row and one column.

The stress-strain equations in matrix form are given in Equation 12.2.7, as

$$\begin{bmatrix} \sigma_x \\ \sigma_y \\ \tau_{xy} \end{bmatrix} = \begin{bmatrix} \dfrac{E}{1-v^2} & \dfrac{vE}{1-v^2} & 0 \\[2mm] \dfrac{vE}{1-v^2} & \dfrac{E}{1-v^2} & 0 \\[2mm] 0 & 0 & G \end{bmatrix} \begin{bmatrix} \varepsilon_x \\ \varepsilon_y \\ \gamma_{xy} \end{bmatrix} = [G]\{\varepsilon\} \qquad (12.4.4)$$

The stresses in terms of the displacements are,

$$\sigma_x(x, y, z) = \frac{E}{1 - v^2}(\varepsilon_x + v\varepsilon_y) = -z\frac{E}{1 - v^2}\left(\frac{\partial^2 w}{\partial x^2} + v\frac{\partial^2 w}{\partial y^2}\right)$$

$$\sigma_y(x, y, z) = \frac{E}{1 - v^2}(\varepsilon_y + v\varepsilon_x) = -z\frac{E}{1 - v^2}\left(\frac{\partial^2 w}{\partial y^2} + v\frac{\partial^2 w}{\partial x^2}\right) \qquad (12.4.5)$$

$$\tau_{xy}(x, y, z) = G\gamma_{xy} = -z\frac{E}{1 + v}\frac{\partial^2 w}{\partial x\partial y}$$

When dealing with equilibrium it is convenient to define stress resultants in terms of moments per unit length. With reference to Figure 12.4.2, and in the sense as indicated there (using the double arrow convention and the right hand cork-screw rule),

$$M_x(x, y) = -\int_{-\frac{h}{2}}^{\frac{h}{2}} \sigma_x z \, dz = \frac{E}{1 - v^2}\int_{-\frac{h}{2}}^{\frac{h}{2}}\left(\frac{\partial^2 w}{\partial x^2} + v\frac{\partial^2 w}{\partial y^2}\right)z^2 dz = \frac{Eh^3}{1 - v^2}\left(\frac{\partial^2 w}{\partial x^2} + v\frac{\partial^2 w}{\partial y^2}\right)$$

$$M_y(x, y) = -\int_{-\frac{h}{2}}^{\frac{h}{2}} \sigma_y z \, dz = \frac{E}{1 - v^2}\int_{-\frac{h}{2}}^{\frac{h}{2}}\left(\frac{\partial^2 w}{\partial y^2} + v\frac{\partial^2 w}{\partial x^2}\right)z^2 dz = \frac{Eh^3}{1 - v^2}\left(\frac{\partial^2 w}{\partial y^2} + v\frac{\partial^2 w}{\partial x^2}\right)$$

$$M_{xy}(x, y) = -\int_{-\frac{h}{2}}^{\frac{h}{2}} \tau_{xy} z \, dz = \frac{E}{1 + v}\int_{-\frac{h}{2}}^{\frac{h}{2}}\left(\frac{\partial^2 w}{\partial x\partial y}\right)z^2 dz = \frac{Eh^3(1 - v)}{1 - v^2}\left(\frac{\partial^2 w}{\partial x\partial y}\right)$$

$$(12.4.6)$$

and the transverse shear forces per unit length are,

$$Q_x(x, y) = \int_{-\frac{h}{2}}^{\frac{h}{2}} \tau_{zx} dz \qquad Q_y(x, y) = \int_{-\frac{h}{2}}^{\frac{h}{2}} \tau_{yz} dz \qquad (12.4.7)$$

The forces and moments on a small element of the plate are shown in Figure 12.4.2.

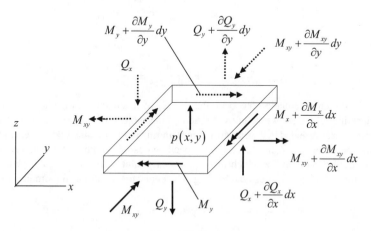

Figure 12.4.2

From summation of forces in the z direction we get

$$F_z = \frac{\partial Q_x}{\partial x} + \frac{\partial Q_y}{\partial y} + p(x, y) = 0 \qquad (12.4.8)$$

From the summation of moments about the x and y axes we get

$$\sum M_x = \frac{\partial M_{xy}}{\partial x} + \frac{\partial M_y}{\partial y} + Q_y = 0 \qquad \sum M_y = \frac{\partial M_{xy}}{\partial y} + \frac{\partial M_x}{\partial x} + Q_x = 0 \tag{12.4.9}$$

To simplify the notation let

$$D_h = \frac{Eh^3}{12(1 - v^2)} \tag{12.4.10}$$

where, h is the uniform plate thickness. Now find the forces in terms of the displacements by combining Equations 12.4.6 and 12.4.9.

$$Q_x = -\frac{\partial M_x}{\partial x} - \frac{\partial M_{xy}}{\partial y} = -D_h \frac{\partial}{\partial x} \left(\frac{\partial^2 w}{\partial x^2} + v \frac{\partial^2 w}{\partial y^2} \right) - D_h (1 - v) \frac{\partial}{\partial y} \left(\frac{\partial^2 w}{\partial x \partial y} \right)$$

$$= -D_h \frac{\partial}{\partial x} \left(\frac{\partial^2 w}{\partial x^2} + \frac{\partial^2 w}{\partial y^2} \right)$$

$$\tag{12.4.11}$$

$$Q_y = -\frac{\partial M_y}{\partial y} - \frac{\partial M_{xy}}{\partial x} = -D_h \frac{\partial}{\partial y} \left(\frac{\partial^2 w}{\partial y^2} + v \frac{\partial^2 w}{\partial x^2} \right) - D_h (1 - v) \frac{\partial}{\partial x} \left(\frac{\partial^2 w}{\partial x \partial y} \right)$$

$$= -D_h \frac{\partial}{\partial y} \left(\frac{\partial^2 w}{\partial y^2} + \frac{\partial^2 w}{\partial x^2} \right)$$

Finally Equation 12.4.8 in terms of displacement becomes

$$\frac{\partial Q_x}{\partial x} + \frac{\partial Q_y}{\partial y} + p(x, y) = -D_h \left(\frac{\partial^4 w}{\partial x^4} + v \frac{\partial^4 w}{\partial x^2 \partial y^2} \right) - D_h \left(\frac{\partial^4 w}{\partial y^4} + v \frac{\partial^4 w}{\partial x^2 \partial y^2} \right) + p(x, y) = 0$$

$$\tag{12.4.12}$$

This may be written in the standard form

$$\nabla^4 w = \frac{\partial^4 w}{\partial x^4} + 2 \frac{\partial^4 w}{\partial x^2 \partial y^2} + \frac{\partial^4 w}{\partial y^4} = \frac{p}{D_h} \tag{12.4.13}$$

where the operator ∇^4 is

$$\nabla^4 = \frac{\partial^4}{\partial x^4} + 2 \frac{\partial^4}{\partial x^2 \partial y^2} + \frac{\partial^4}{\partial y^4} \tag{12.4.14}$$

This partial differential equation has been the subject of much study in mathematics as well as in applied engineering. There are whole books written on the subject of exact and approximate analytical solutions. Several are presented in the classic text by Timoshenko and Woinowsky-Krieger (1987). Exact solutions have been found for only a very few special cases of geometry, boundary conditions, and loading. One set of conditions for which there is an exact solution is a rectangular plate with simple support on all four edges and a sinusoidal loading. The boundary conditions for simple support are

$$w(0, y) = 0 \qquad \frac{\partial^2 w(0, y)}{\partial^2 x} = 0 \qquad w(a, y) = 0 \qquad \frac{\partial^2 w(a, y)}{\partial^2 x} = 0$$

$$\tag{12.4.15}$$

$$w(x, 0) = 0 \qquad \frac{\partial^2 w(x, 0)}{\partial^2 y} = 0 \qquad w(x, b) = 0 \qquad \frac{\partial^2 w(x, b)}{\partial^2 y} = 0$$

The sinusoidal loading is

$$p(x, y) = c_{ij} \sin \frac{i \pi x}{a} \sin \frac{i \pi y}{b} \tag{12.4.16}$$

The displacement equation becomes

$$\frac{\partial^4 w}{\partial x^4} + 2\frac{\partial^4 w}{\partial x^2 \partial y^2} + \frac{\partial^4 w}{\partial y^4} = \frac{c_{ij}}{D_h} \sin\frac{i\pi x}{a} \sin\frac{i\pi y}{b} \qquad (12.4.17)$$

A solution is suggested of the form

$$w(x, y) = C_{ij} \sin\frac{i\pi x}{a} \sin\frac{i\pi y}{b} \qquad (12.4.18)$$

One can quickly verify by substitution that this satisfies the boundary conditions. Upon substitution the trial solution into Equation (12.4.16) we obtain

$$C\left(\frac{i^4\pi^4}{a^4} + 2\frac{i^2 j^2\pi^4}{a^2 b^2} + \frac{j^4\pi^4}{b^4}\right) = \frac{c_{ij}}{D_h} \qquad (12.4.19)$$

The trouble is that plates usually are not rectangular, seldom, if ever, are simply supported, and do not have such a convenient form of loading. There are other cases for which approximate analytical solutions have been found; however, the idealizations in geometry, loading, and restraint necessary to obtain analytical solutions have largely given way to the methods used in the next section which do not require these idealizations. The main usefulness of the few exact solutions that have been obtained is to verify the validity of the approximate methods that have been devised. Before we dive into attacking problems using the finite element method, one point regarding boundary conditions for plate analysis is warranted. Suppose, in the previous problem, imagine that three edges are simply supported and the edge $x = a$ is a free edge. Then the boundary conditions on the free edge, in terms of the forces and moments denoted in Figure 12.4.2 would become, $M_x(a, y) = M_{xy}(a, y) = Q_x(a, y) = 0$. But, when these are expressed in terms of displacements, there is one extra condition on this edge, since for the fourth order symmetric governing equation, one only needs two boundary conditions on each edge. This apparent paradox was solved by Kirchhoff who combined two of the conditions in defining the effective Kirchhoff transverse shear resultant on an edge. Thus, introduce the Kirchhoff shear resultants, V_x, V_y, which are operational on edges with shear resultants, Q_x, Q_y, respectively, and are defined as,

$$V_x = Q_x + \frac{\partial M_{xy}}{\partial y}; \qquad V_y = Q_y + \frac{\partial M_{xy}}{\partial x} \qquad (12.4.20)$$

Now, the boundary conditions on the edge $x = a$ for a free edge are,

$$V_x(a, y) = Q_x(a, y) + \frac{\partial M_{xy}}{\partial y}(a, y) = 0 \quad \text{and} \quad M_x(a, y) = 0 \qquad (12.4.21)$$

and, if in a problem, the edge, $y = b$ is a free edge, then,

$$V_y(x, b) = Q_y(x, b) + \frac{\partial M_{xy}}{\partial x}(x, b) = 0 \quad \text{and} \quad M_y(x, b) = 0 \qquad (12.4.22)$$

Equations 12.4.20 and 12.4.21 are the Kirchoff plate boundary conditions at a free edge.

12.5   Thin Flat Plates—FEM Analysis

The plate shape functions have been defined for the same configurations as for plane stress, namely, three and six noded triangles, and four and eight noded quadrilaterals. We shall illustrate just the four noded quadrilateral. We show the internal forces and displacements at just one of the nodes in Figure 12.5.1. The nodal force is designated Z and the nodal moments M_x and M_y.

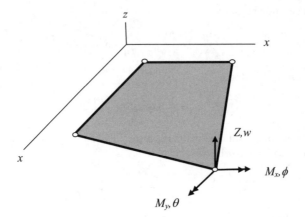

Figure 12.5.1

The displacements at one node consist of the lateral displacement w and two rotations

$$\theta = \frac{\partial w}{\partial x} \qquad \phi = \frac{\partial w}{\partial y} \tag{12.5.1}$$

At node n

$$\{r_m\}_n = \begin{bmatrix} w_n \\ \theta_n \\ \phi_n \end{bmatrix} \tag{12.5.2}$$

There are four nodes for element m with three nodal displacements at each node. The shape function matrix is of the form

$$w_m(x, y) = [n]\{r_m\} = [1x12][12x1] \tag{12.5.3}$$

The element stiffness matrix is formed from the same virtual work equation as before.

$$\{p_m\} = [k_m]\{r_m\} = \int ([D][n])^T [G][D][n]\, dV \{r_m\} \tag{12.5.4}$$

The matrix $[D]$ is given in Equation 12.4.3 and $[G]$ is given in 12.4.4.
The formula for equivalent nodal loads is

$$\{F\}_{enl} = \int [n]^T \{f_s\}\, dA \tag{12.5.5}$$

The order of the matrices are shown in Equation 12.5.6

$$[p_m] = [12x1] = [12x12][12x1] = [k_m]\{r_m\} \tag{12.5.6}$$

The assembly process to produce the global matrices proceeds are described above.

The steps for a solution are the same as those described for Example 12.3.1. To emphasize the fact that ease of solution does not depend upon simplified geometry, loading, and restraints we shall give an example that would be very difficult to solve by any of the exact or approximate analytical methods that have preceded the FEM.

###########

Example 12.5.1

Problem: A trapezoidal shaped plate has a fixed restraint on three edges and is free on the fourth edge as shown in Figure (a). The coordinates of the four corners are given in millimeters in the parentheses. The plate is 10 *mm* thick. The material is steel with $E = 206800 \ N/mm^2$. The load consists of a uniform distributed load of $0.2 \ N/mm^2$ in the positive z direction. Find the stresses and displacements.

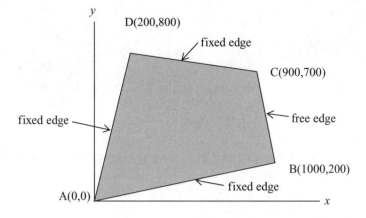

Figure (a)

Solution: Create a CAD model; add restraints to the edges; add loads to the surface; use the automatic mesh generator to create nodes and elements; invoke the solver; and present contour plots of stresses and displacements.

A solid model of the plate was created in the I-DEAS software. Restraints were added to the edges and loads to the surface. These are shown in Figure (b).

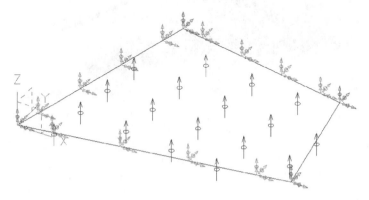

Figure (b)

Nodes and elements were then created by the automatic mesh generator. The plate elements are based on surfaces and the thickness is added as a scalar quantity in a property table. Once again we have the choice of specifying the density of the mesh, that is, the number of nodes and elements, in several ways. We have specified nominal edge length of 25 *mm*. For visual reasons the element size was kept fairly large. The nodes and elements are shown in Figure (c).

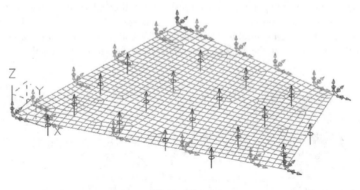

Figure (c)

The solver was invoked. In Figure (e) we show a contour plot of the von Mises stresses on a deformed plate.

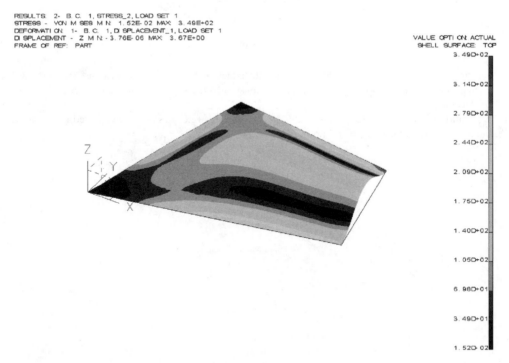

Figure (e)

In Figure (f) we show a contour plot of the displacements. The displacements are greatly exaggerated for visual reasons.

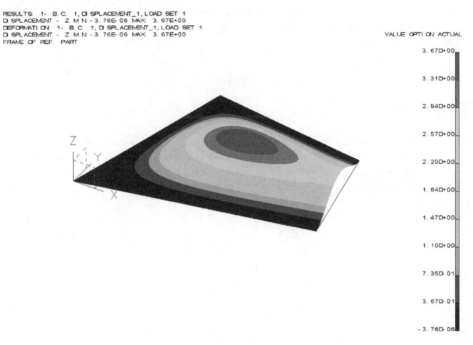

RESULTS 1- B.C. 1, DISPLACEMENT_1, LOAD SET 1
DISPLACEMENT - Z MIN -3.76E-06 MAX 3.67E+00
DEFORMATION: 1- B.C. 1, DISPLACEMENT_1, LOAD SET 1
DISPLACEMENT - Z MIN -3.76E-06 MAX 3.67E+00
FRAME OF REF: PART

VALUE OPTION: ACTUAL

3.67D+00
3.31D+00
2.94D+00
2.57D+00
2.20D+00
1.84D+00
1.47D+00
1.10D+00
7.35D-01
3.67D-01
-3.76D-06

Figure (f)

############

12.6   Shell Structures

The plate bending and plane stress elements are combined into a single element for dealing with both bending and in plane loads, stresses, and displacements. Such is often the case for curved plates and shell structures. These elements are called shell elements and have five components of displacement and force at each node as shown in Figure 12.6.1.

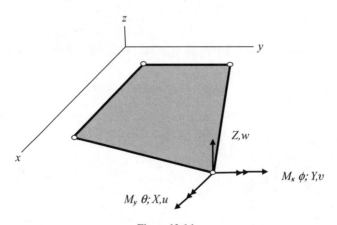

Figure 12.6.1

This element is shown in the *xy* plane. Appropriate coordinate transformations are required to place this in position in 3D space. We shall show in an example how such elements can be used to solve a thin walled beam like structure.

###########

Example 12.6.1

Problem: A cantilever beam with a D shaped cross section such as the one examined in Example 10.7.2 is one meter long and has the cross section dimensions shown in Figure (a). It is loaded by a distributed load of 1 *N/mm* along the vertical web at the free end.
   The wall thickness is 2 *mm*. The material is steel with $E = 206800 \ N/mm^2$.

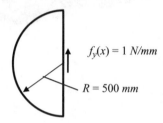

$f_y(x) = 1 \ N/mm$

$R = 500 \ mm$

Figure (a)

Solution: Create a CAD model; add restraints to the edges; add loads to the vertical edge; use the automatic mesh generator to create nodes and elements; invoke the solver; and present contour plots of stresses and displacements.
   A solid model of the beam was created in the I-DEAS software as shown in Figure (b). The shell elements will be created on the surfaces of the model and the thickness entered as a scalar quantity in an element properties table.

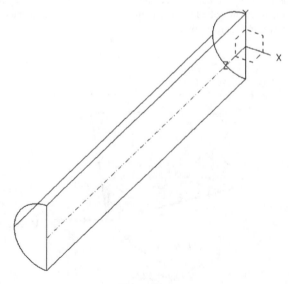

Figure (b)

Restraints and loads are now added to the appropriate edges. These are shown in Figure (c).

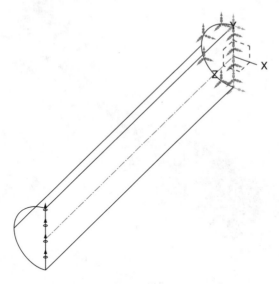

Figure (c)

The nodes and elements are created next. A four noded quadrilateral element is chosen. A nominal size for an element is a 10 *mm* edge length. This is shown in Figure (d).

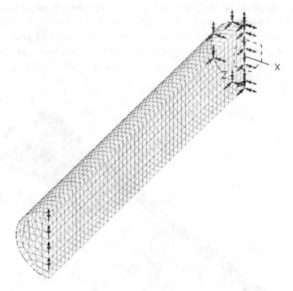

Figure (d)

The solver was invoked next. In Figure (e) we show a contour plot of the von Mises stresses. This is shown in what is called the step shaded format.

Of particular interest is the shear stress. From our study in Chapter 10 we know that the shear stress is tangent to the thin walls. This software only presents stresses in the coordinate directions in contour plots. You can go to the report data and resolve the stress components given in rectangular Cartesian coordinate into components in cylindrical components if necessary. We shall discuss only general conclusions here.

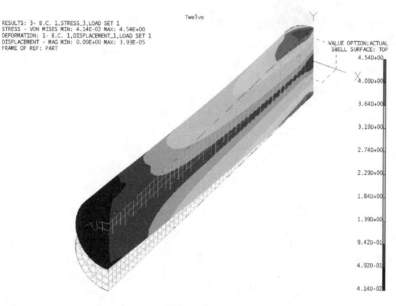

RESULTS: 3- B.C. 1,STRESS_3,LOAD SET 1
STRESS - VON MISES MIN: 4.14E-02 MAX: 4.54E+00
DEFORMATION: 1- B.C. 1,DISPLACEMENT_1,LOAD SET 1
DISPLACEMENT - MAG MIN: 0.00E+00 MAX: 3.93E-05
FRAME OF REF: PART

Twelve

VALUE OPTION:ACTUAL
SHELL SURFACE: TOP
4.54D+00
4.09D+00
3.64D+00
3.19D+00
2.74D+00
2.29D+00
1.84D+00
1.39D+00
9.42D-01
4.92D-01
4.14D-02

Figure (e)

In Figure (f) we show the τ_{yz} stress component. This may be compared with the results obtained in Chapter 10, Example 10.7.2. The gray scale contour plots are difficult to interpret. Looking at the color contour plot shows that the shear stress in the vertical web agrees closely to the value obtain in Chapter 10. A more detailed analysis of the shear stresses in the curved web is needed to confirm that the results agree with those in Chapter 10; however, a look at the τ_{yz} stress in the curved web as shown in Figure (g) confirms that the largest stresses are in the vertical web and it also shows the reversal, that is, negative direction of the shear stresses at the top and bottom of the curved web.

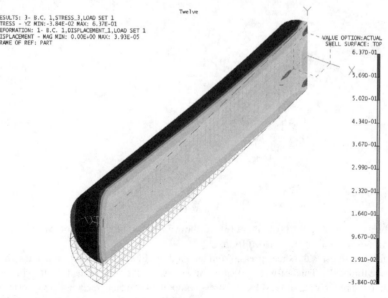

RESULTS: 3- B.C. 1,STRESS_3,LOAD SET 1
STRESS - YZ MIN:-3.84E-02 MAX: 6.37E-01
DEFORMATION: 1- B.C. 1,DISPLACEMENT_1,LOAD SET 1
DISPLACEMENT - MAG MIN: 0.00E+00 MAX: 3.93E-05
FRAME OF REF: PART

Twelve

VALUE OPTION:ACTUAL
SHELL SURFACE: TOP
6.37D-01
5.69D-01
5.02D-01
4.34D-01
3.67D-01
2.99D-01
2.32D-01
1.64D-01
9.67D-02
2.91D-02
-3.84D-02

Figure (f)

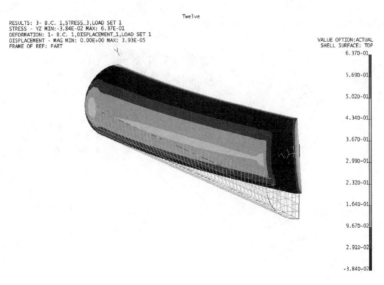

Twelve

VALUE OPTION:ACTUAL
SHELL SURFACE: TOP

6.37D-01
5.69D-01
5.02D-01
4.34D-01
3.67D-01
2.99D-01
2.32D-01
1.64D-01
9.67D-02
2.91D-02
-3.84D-02

Figure (g)

###########

The method is well adapted to multi celled structures as shown in the next example.

###########

Example 12.6.2

Problem: A two celled thin walled beam with a cross section is first analyzed with out stiffeners and then stiffeners are added in the next section. The beam CAD model is shown in Figure (a). It has the same loads and restraints as Example 12.6.1 as shown in Figure (b).

Solution: Create the CAD model, FE model, etc., using exactly the same steps as in Example 12.6.1.

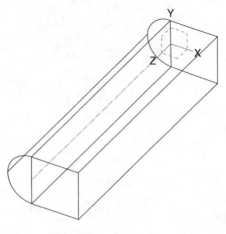

Figure (a)

The loads and restraints are shown in Figure (b).

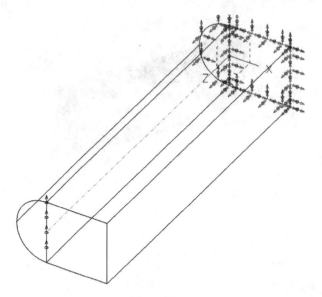

Figure (b)

The mesh is shown in Figure (c).

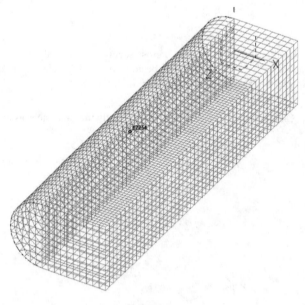

Figure (c)

A contour plot of von Mises stresses is shown in Figure (d).

Twelve

RESULTS: 3- B.C. 1,STRESS_3,LOAD SET 1
STRESS - VON MISES MIN: 7.34E-02 MAX: 1.63E+00
DEFORMATION: 1- B.C. 1,DISPLACEMENT_1,LOAD SET 1
DISPLACEMENT - MAG MIN: 0.00E+00 MAX: 1.32E-05
FRAME OF REF: PART

VALUE OPTION:ACTUAL
SHELL SURFACE: TOP
1.63D+00
1.48D+00
1.32D+00
1.16D+00
1.01D+00
8.52D-01
6.97D-01
5.41D-01
3.85D-01
2.29D-01
7.34D-02

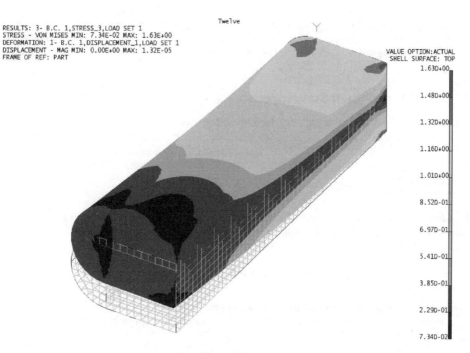

Figure (d)

In Figure (e) we show a contour plot of the τ_{yz} stress component.

Twelve

RESULTS: 3- B.C. 1,STRESS_3,LOAD SET 1
STRESS - YZ MIN:-1.90E-02 MAX: 2.71E-01
DEFORMATION: 1- B.C. 1,DISPLACEMENT_1,LOAD SET 1
DISPLACEMENT - MAG MIN: 0.00E+00 MAX: 1.32E-05
FRAME OF REF: PART

VALUE OPTION:ACTUAL
SHELL SURFACE: TOP
2.71D-01
2.42D-01
2.13D-01
1.84D-01
1.55D-01
1.26D-01
9.71D-02
6.81D-02
3.91D-02
1.00D-02
-1.90D-02

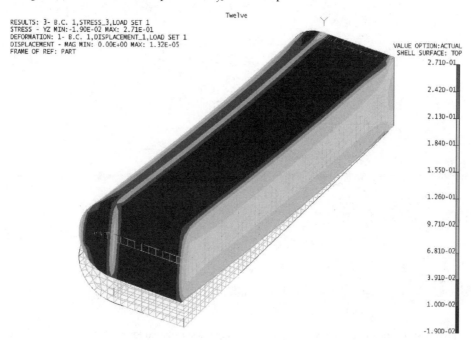

Figure (e)

A comparison of the FEM results with the analytical results from thin walled beam theory establishes good agreement with the exception of the usual local effects at the loadings and the restraints. These local effects are ignored in the thin walled beam theory with the blessing of St. Venant.

12.7   Stiffened Shell Structures

We shall now apply the FEM to stiffened shell structures.

##########

Example 12.7.1

Problem: Solve the cantilever beam with the D shaped cross section and stiffeners examined in Example 10.8.3, that is, repeat Example 12.6.1 with stiffeners added. Let the stiffeners be 10 *mm* by 10 *mm* and the wall thickness is 2 *mm*. Let

$$E = 206800 \, MPa \qquad R = 50 \, mm$$

Solution: Beam elements are added to the model and the problem is resolved.

In Figure (a) we show the FE model with beam elements added. The restraints and loads are the same as in Example 12.6.1

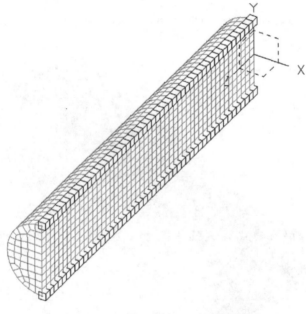

Figure (a)

The von Mises stresses are shown in Figure (b). There is a substantial lowering of the stress by adding the stiffeners as might be expected.

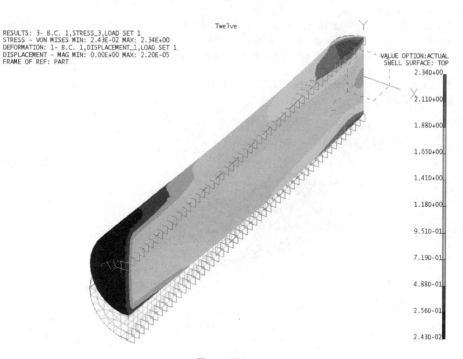

RESULTS: 3- B.C. 1,STRESS_3,LOAD SET 1
STRESS - VON MISES MIN: 2.43E-02 MAX: 2.34E+00
DEFORMATION: 1- B.C. 1,DISPLACEMENT_1,LOAD SET 1
DISPLACEMENT - MAG MIN: 0.00E+00 MAX: 2.20E-05
FRAME OF REF: PART

Twelve

VALUE OPTION:ACTUAL
SHELL SURFACE: TOP

2.34D+00
2.11D+00
1.88D+00
1.65D+00
1.41D+00
1.18D+00
9.51D-01
7.19D-01
4.88D-01
2.56D-01
2.43D-02

Figure (b)

The τ_{yz} shear stresses are shown in Figure (c).

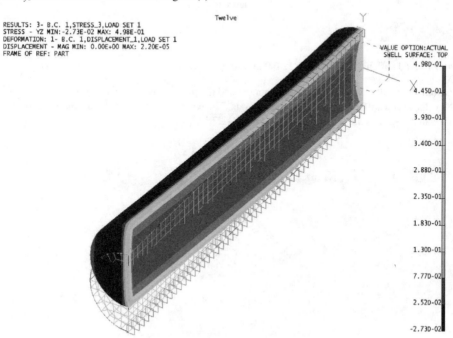

RESULTS: 3- B.C. 1,STRESS_3,LOAD SET 1
STRESS - YZ MIN:-2.73E-02 MAX: 4.98E-01
DEFORMATION: 1- B.C. 1,DISPLACEMENT_1,LOAD SET 1
DISPLACEMENT - MAG MIN: 0.00E+00 MAX: 2.20E-05
FRAME OF REF: PART

Twelve

VALUE OPTION:ACTUAL
SHELL SURFACE: TOP

4.98D-01
4.45D-01
3.93D-01
3.40D-01
2.88D-01
2.35D-01
1.83D-01
1.30D-01
7.77D-02
2.52D-02
-2.73D-02

Figure (c)

The results compare with those obtained in Chapter 10 for the vertical web. A comparison with Example 12.6.1 shows a more nearly uniform value of the shear stress in the vertical web. This supports the assumption of a constant shear stress in the web as explained in Chapter 10 to obtain a first approximation.

A look at the curved web in Figure (d) shows the much lower shear stresses and the stress reversal obtained in Chapter 10.

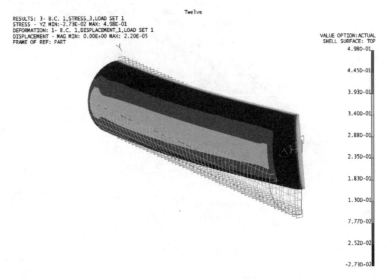

Figure (d)

Example 12.7.2

Problem: Add stiffeners to Example 12.6.2 and solve.

Solution: Beam elements are added to the model and the problem is resolved.

Now let us add stiffeners and repeat the analysis. The beam with stiffeners is shown in Figure (f). The loading and restraints are the same as in Example 12.7.2.

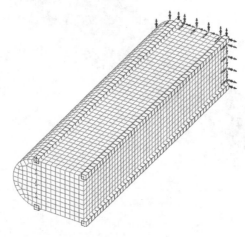

Figure (f)

A contour plot of the von Mises stresses is shown in Figure (g)

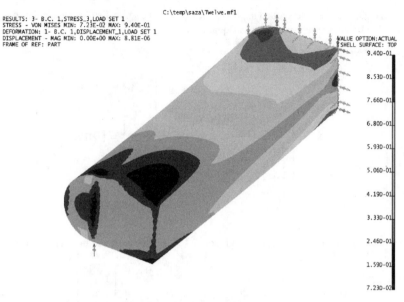

Figure (g)

A contour plot of the τ_{yz} stress component is shown in Figure (h)

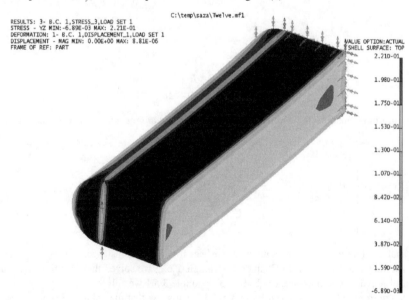

Figure (h)

Again the von Mises stresses are much lower. The shear stress in the vertical web is more closely uniform and we have the usual local effects where the loading and restraints are applied. Comparison with a thin walled beam theory solution would show good agreement.

##########

12.8 Three Dimensional Structures—Classical and FEM Analysis

The classical equations were developed in previous chapters. The strain displacement equations are given in Equation 3.2.10 and are repeated here as Equation 12.8.1.

$$
\{\varepsilon\} =
\begin{bmatrix}
\varepsilon_x \\
\varepsilon_y \\
\varepsilon_z \\
\gamma_{xy} \\
\gamma_{yz} \\
\gamma_{zx}
\end{bmatrix}
=
\begin{bmatrix}
\dfrac{\partial}{\partial x} & 0 & 0 \\
0 & \dfrac{\partial}{\partial y} & 0 \\
0 & 0 & \dfrac{\partial}{\partial z} \\
\dfrac{\partial}{\partial y} & \dfrac{\partial}{\partial x} & 0 \\
0 & \dfrac{\partial}{\partial z} & \dfrac{\partial}{\partial y} \\
\dfrac{\partial}{\partial z} & 0 & \dfrac{\partial}{\partial x}
\end{bmatrix}
\begin{bmatrix}
u \\
v \\
w
\end{bmatrix}
= [D]\{u\}
\tag{12.8.1}
$$

The stress strain equations are given in Equation 3.4.30 and are repeated here as Equation 12.8.2.

$$
\{\sigma\} =
\begin{bmatrix}
\sigma_x \\
\sigma_y \\
\sigma_z \\
\tau_{xy} \\
\tau_{yz} \\
\tau_{zx}
\end{bmatrix}
=
\begin{bmatrix}
\lambda + 2G & \lambda & \lambda & 0 & 0 & 0 \\
\lambda & \lambda + 2G & \lambda & 0 & 0 & 0 \\
\lambda & \lambda & \lambda + 2G & 0 & 0 & 0 \\
0 & 0 & 0 & G & 0 & 0 \\
0 & 0 & 0 & 0 & G & 0 \\
0 & 0 & 0 & 0 & 0 & G
\end{bmatrix}
\begin{bmatrix}
\varepsilon_x \\
\varepsilon_y \\
\varepsilon_z \\
\gamma_{xy} \\
\gamma_{yz} \\
\gamma_{zx}
\end{bmatrix}
= [G]\{\varepsilon\}
\tag{12.8.2}
$$

where

$$
\lambda = \frac{\nu E}{(1+\nu)(1-2\nu)}
$$

The equilibrium equations are given in Equation 2.6.13 and are repeated here as Equation 12.8.3.

$$
[E]\{\sigma\} =
\begin{bmatrix}
\dfrac{\partial}{\partial x} & 0 & 0 & \dfrac{\partial}{\partial y} & 0 & \dfrac{\partial}{\partial z} \\
0 & \dfrac{\partial}{\partial y} & 0 & \dfrac{\partial}{\partial x} & \dfrac{\partial}{\partial z} & 0 \\
0 & 0 & \dfrac{\partial}{\partial z} & 0 & \dfrac{\partial}{\partial y} & \dfrac{\partial}{\partial x}
\end{bmatrix}
\begin{bmatrix}
\sigma_x \\
\sigma_y \\
\sigma_z \\
\tau_{xy} \\
\tau_{yz} \\
\tau_{zx}
\end{bmatrix}
= -
\begin{bmatrix}
f_{bx} \\
f_{by} \\
f_{bz}
\end{bmatrix}
= -\{f_b\}
\tag{12.8.3}
$$

Exact analytical solutions to these equations have been found only in instances of very simple geometry, loading, and material properties. In Chapter 2, Example 2.6.2, we solved them for the stress in a cube under uniform hydrostatic pressure. In Chapter 3, Example 3.5.1 we solved them for a thin rectangular plate with a uniform edge load, but in Example 3.5.2, we show that the solution becomes much more difficult when the loading is not uniform.

Continuing in Chapter 4, Section 4.2, we show that the equations are solved exactly for a slender bar with a rectangular cross section as long as the axial force in uniformly distributed over the cross section. In Chapter 6, Section 6.4, we show that they are solved for a slender shaft with a circular cross section if the loading is a radially linear force per unit area applied at the ends. And in Chapter 7, Section 7.7, we show that they are solved for a beam with a rectangular cross section and a linearly varying normal force per unit area and/or a parabolic shear force per unit area on the ends. In all cases the material is homogenous and isotropic. And in all cases the message is that any change from the simple geometry assumed or any variation from the constant or linearly varying loading makes an exact analytical solution at best difficult and most likely impossible.

There are so few analytical solutions to these equations we turn directly to virtual work as expressed in Equations 12.3.1 and derive the FEM equations.

The shape functions for four elements, the four and ten noded tetrahedrons and the eight and twenty noded hexahedrons, are the ones primarily in use. The four noded tetrahedron is shown in Figure 12.8.1.

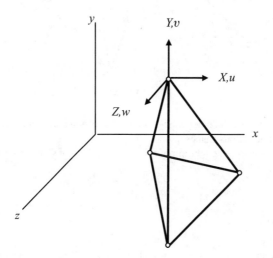

Figure 12.8.1

At node n

$$\{r_m\}_n = \begin{bmatrix} u_n \\ v_n \\ w_n \end{bmatrix} \tag{12.8.4}$$

There are four nodes for element m with three nodal displacements at each node. The shape function matrix is of the form

$$\begin{bmatrix} u(x, y, z) \\ v(x, y, z) \\ w(x, y, z) \end{bmatrix} = [3x12][12x1] = [n]\{r_m\} \tag{12.8.5}$$

The element stiffness matrix is formed from the same virtual work equations as Equations 11.4.20. We enter the order of the matrices rather than type them in full.

$$[p_m] = [12x1] = [12x12][12x1] = [k_m]\{r_m\} \tag{12.8.6}$$

The assembly process to produce the global matrices proceeds are described before.

The ten noded tetrahedral element is shown in Figure 12.8.2. This currently is the most used element because of its utility in automatic mesh generation.

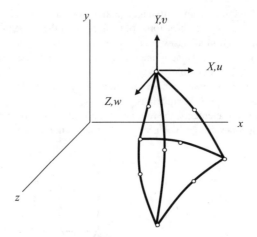

Figure 12.8.2

The hexahedral elements are six sided blocks. Each face is a quadrilateral. Its irregular shape allows it to match solid bodies of various shapes.

Solid elements must be used when the simplifying assumptions of slender bar and shell theory do not apply. Here is an example.

###########

Example 12.8.1

Problem: The bracket shown in Figure (a) is loaded with a vertical force on a pin placed in the hole. The head of the pin transmits the load to the bracket. The bracket is attached to what is assumed to be a rigid wall by an adhesive. Find the stresses and displacements and modify the design to reduce the highest stresses. The material is steel with $E = 206899\ MPa$. The load is $1000\ N$.

Solution: Use a commercial FEM code to find the stresses and displacements.

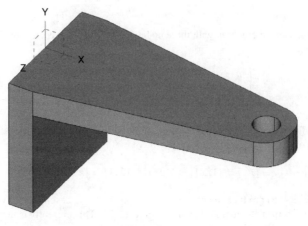

Figure (a)

First we create a surface on which to apply the load as shown in Figure (b).

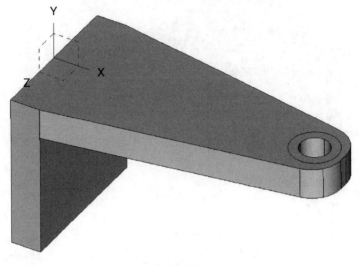

Figure (b)

Now apply the boundary condition restraints and loads as shown in Figure (c).

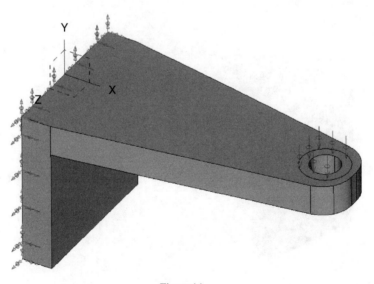

Figure (c)

We now invoke the automatic mesh generator and use ten noded tetrahedral elements as shown in Figure (d).

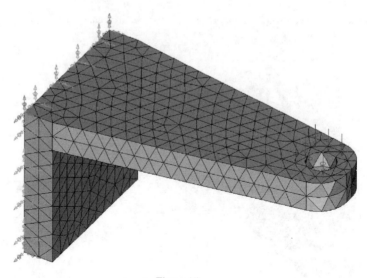

Figure (d)

Next the solver. As might be expected the highest stresses are at the sharp corner as shown in the contour plot of the von Mises stresses in Figure (e).

solid

```
RESULTS: 3- B.C. 1,STRESS_3,LOAD SET 1
STRESS - VON MISES MIN: 3.11E-02 MAX: 2.25E+02
DEFORMATION: 1- B.C. 1,DISPLACEMENT_1,LOAD SET 1
DISPLACEMENT - MAG MIN: 0.00E+00 MAX: 6.45E-01
FRAME OF REF: PART
```

VALUE OPTION:ACTUAL

2.25D+02

2.02D+02

1.80D+02

1.57D+02

1.35D+02

1.12D+02

8.99D+01

6.74D+01

4.50D+01

2.25D+01

3.11D-02

Figure (e)

The maximum displacement is 0.645 *mm*. If the maximum von Mises stress of 225 *MPa* is higher than desired it can be reduced by design changes in the geometry. A common change is to put a fillet in the corner. This has been done as is shown in Figure (f).

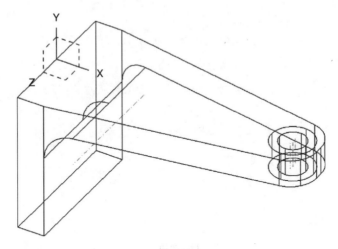

Figure (f)

The boundary conditions are added; the new model is meshed; the solver is run again and the results are displayed in Figure (g).

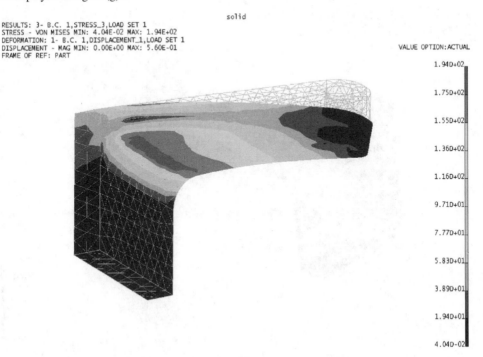

```
                                                           solid
RESULTS: 3- B.C. 1,STRESS_3,LOAD SET 1
STRESS - VON MISES MIN: 4.04E-02 MAX: 1.94E+02
DEFORMATION: 1- B.C. 1,DISPLACEMENT_1,LOAD SET 1
DISPLACEMENT - MAG MIN: 0.00E+00 MAX: 5.60E-01                            VALUE OPTION:ACTUAL
FRAME OF REF: PART

                                                                         1.94D+02
                                                                         1.75D+02
                                                                         1.55D+02
                                                                         1.36D+02
                                                                         1.16D+02
                                                                         9.71D+01
                                                                         7.77D+01
                                                                         5.83D+01
                                                                         3.89D+01
                                                                         1.94D+01
                                                                         4.04D-02
```

Figure (g)

The maximum stress is reduced but perhaps not enough so let us consider another design change. A stiffener is added as shown in Figure (h).

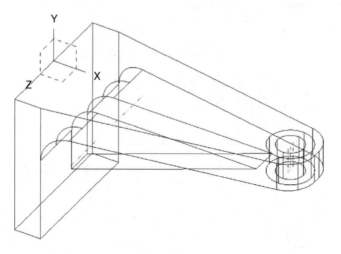

Figure (h)

The results of this analysis are shown in Figure (i).

```
                                                    solid
RESULTS: 3- B.C. 1,STRESS_3,LOAD SET 1
STRESS - VON MISES MIN: 5.57E-02 MAX: 9.09E+01
DEFORMATION: 1- B.C. 1,DISPLACEMENT_1,LOAD SET 1
DISPLACEMENT - MAG MIN: 0.00E+00 MAX: 1.23E-01                    VALUE OPTION:ACTUAL
FRAME OF REF: PART

                                                                   9.09D+01

                                                                   8.18D+01

                                                                   7.27D+01

                                                                   6.36D+01

                                                                   5.45D+01

                                                                   4.55D+01

                                                                   3.64D+01

                                                                   2.73D+01

                                                                   1.82D+01

                                                                   9.14D+00

                                                                   5.57D-02
```

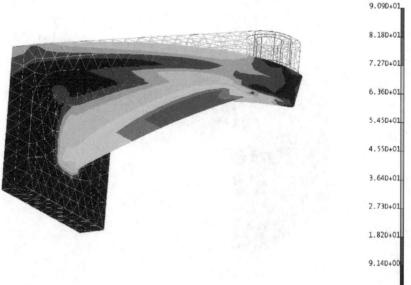

Figure (i)

There is a considerable reduction in both the maximum stress and the maximum displacement.

The purpose of this exercise is not to create an optimum design but rather to illustrate that changes in geometry no longer restrict the ability to get accurate solutions. All three of these brackets would be very difficult to analyze, even with approximate analytic methods, before the arrival of FEM.

12.9   Summary and Conclusions

Exact analytical solutions of the classical differential equations of two and three dimension structures are limited to only a few very simple cases of geometry, loading, and restraint. Several books on the theory of elasticity grapple with this problem and for the most part are forced to go to approximate methods. The finite element method is by far the most productive. Therefore we emphasize this approach and largely skip all others.

13

Analysis of Thin Laminated Composite Material Structures

13.1 Introduction to Classical Lamination Theory

Fiber-reinforced composite laminates are increasingly being used in the aerospace industry and are now being used in a wide variety of aerospace structural applications. The Boeing 787 commercial airplane represents a revolutionary step in the application of thin-walled laminated composites in primary (load bearing) aircraft structural components such as the fuselage, tail and wing. Ongoing developments and advancements in the fiber and polymer manufacturing industries will undoubtedly enhance the performance of future laminated structural components with stronger fibers and tougher resin systems.

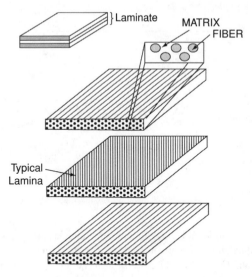

Figure 13.1.1 Schematic of a continuous fiber laminated composite

A laminated structure contains multiple layers that are laminated together with each layer oriented in a specified direction selected by the design engineer, as schematically shown in Figure 13.1.1. Since the

Analysis of Structures: An Introduction Including Numerical Methods, First Edition. Joe G. Eisley and Anthony M. Waas.
© 2011 John Wiley & Sons, Ltd. Published 2011 by John Wiley & Sons, Ltd.

fiber orientation and the manner by which the layers are stacked (usually called the stacking sequence) can be changed as desired, the overall structural properties of a fiber reinforced laminate can be tailored as needed. This type of flexibility is one of the main advantages of using a laminate in a structural application as will be evident in this chapter. Because one can change structural properties, a design engineer can "optimize" the deformation response and dynamic behavior of structural components as desired.

Although there are a wide variety of fiber reinforced laminates in the market today, in this chapter we will focus on analyzing the deformation of continuous, straight fiber reinforced laminates, that are made by stacking laminae, the latter being the basic building block of laminates.

When a laminate is fabricated, they act as one single layer material. The bond between any two laminae in a laminate is assumed to be perfect, that is, infinitesimally thin and not shear deformable. Thus, laminae cannot slip over each other, and the displacements therefore must remain continuous across the bond line (or the interface).

13.2   Strain Displacement Equations for Laminates

Let us consider the deformation of a section of a laminate in the xz plane when we place the coordinate axes such that the xy plane coincides with the geometrical midplane of the laminate, as shown in Figure 13.2.1

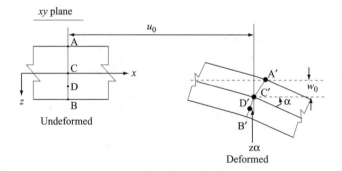

Figure 13.2.1

1. Assume that a line ACDB, originally straight and perpendicular to the x axis, also remains straight and perpendicular to the deformed geometric midplane of the laminate.
2. Further assume that point C undergoes displacements $u_0(x, y)$, $v_0(x, y)$ and $w_0(x, y)$ along the x, y and z directions respectively.

Consider now the deformation of point D, which was located at a distance Z below the midplane in the undeformed configuration. Upon deformation to point D' we can write the following displacement field for point D, assuming small (compared to the total laminate thickness) displacements;

$$u(x, y) = u_0(x, y) - z\alpha(x, y) \qquad (13.2.1)$$

where $\alpha(x, y)$ is the slope of the laminate's midplane is the x-direction. Notice that we are assuming $\alpha(x, y)$ to be small in the sense that $\tan \alpha \approx \alpha$ and $\cos \alpha \approx 1$, that is, small slopes. Then

$$\alpha = \frac{\partial w_0}{\partial x} \quad \rightarrow \quad u = u_0 - z \frac{\partial w_0}{\partial x} \tag{13.2.2}$$

By a similar consideration

$$v = v_0 - z \frac{\partial w_0}{\partial y} \tag{13.2.3}$$

The displacement $w(x, y, z)$ of any point on ACDB is the displacement $w_0(x, y)$ of the midplane ($z = 0$), plus any stretching of ACDB. It is assumed that the stretch of ACDB is insignificant compared to w_0 itself and thus the "w" displacement of any point on ACDB is w_0 of the midplane (point C) itself.

Let us write the infinitesimal strains ε_x, ε_y and ε_z for the displacement field, given by Equation 13.2.4 as shown in Equation 13.2.5

$$u(x, y, z) = u_0(x, y) - z \frac{\partial w_0(x, y)}{\partial x}$$

$$v(x, y, z) = v_0(x, y) - z \frac{\partial w_0(x, y)}{\partial y} \tag{13.2.4}$$

$$w(x, y, z) = w_0(x, y)$$

The strains are therefore,

$$\varepsilon_x = \frac{\partial u}{\partial x} = \frac{\partial u_0(x, y)}{\partial x} - z \frac{\partial^2 w_0(x, y)}{\partial x^2}$$

$$\varepsilon_y = \frac{\partial v}{\partial y} = \frac{\partial v_0(x, y)}{\partial y} - z \frac{\partial^2 w_0(x, y)}{\partial y^2} \tag{13.2.5}$$

$$2\varepsilon_{xy} = \frac{\partial u}{\partial y} + \frac{\partial v}{\partial z} = \frac{\partial u_0(x, y)}{\partial y} + \frac{\partial v_0(x, y)}{\partial x} - 2z \frac{\partial^2 w_0(x, y)}{\partial x \partial y}$$

Notice that γ_{yz} and γ_{zx} are zero and that ε_z is zero on account of the assumption that $w(x, y, z) = w_0(x, y)$.

We can write Equation 13.2.5 as

$$\begin{bmatrix} \varepsilon_x \\ \varepsilon_y \\ \gamma_{xy} \end{bmatrix} = \begin{bmatrix} \varepsilon_x^0 \\ \varepsilon_y^0 \\ \gamma_{xy}^0 \end{bmatrix} + z \begin{bmatrix} \chi_x \\ \chi_y \\ \chi_{xy} \end{bmatrix} \tag{13.2.6}$$

where

$$\begin{bmatrix} \varepsilon_x^0 \\ \varepsilon_y^0 \\ \gamma_{xy}^0 \end{bmatrix} = \begin{bmatrix} \dfrac{\partial u_0}{\partial x} \\ \dfrac{\partial v_0}{\partial y} \\ \dfrac{\partial u_0}{\partial y} + \dfrac{\partial v_0}{\partial x} \end{bmatrix} \qquad \begin{bmatrix} \chi_x \\ \chi_y \\ \chi_{xy} \end{bmatrix} = \begin{bmatrix} -\dfrac{\partial^2 w_0(x, y)}{\partial x^2} \\ -\dfrac{\partial^2 w_0(x, y)}{\partial y^2} \\ -2\dfrac{\partial^2 w_0(x, y)}{\partial x \partial y} \end{bmatrix} \tag{13.2.7}$$

To obtain the stress distribution across the laminate we need to use the strain distribution, Equation 13.2.6, and the lamina constitutive description. This will require using the stress-strain relations for laminae, as presented next.

13.3　Stress-Strain Relations for a Single Lamina

For a transversely isotropic fiber reinforced lamina as shown in Figure 13.3.1, we have,

$$
\begin{bmatrix} \varepsilon_1 \\ \varepsilon_2 \\ \varepsilon_3 \\ \gamma_{23} \\ \gamma_{31} \\ \gamma_{12} \end{bmatrix} =
\begin{bmatrix}
\dfrac{1}{E_1} & -\dfrac{v_{21}}{E_2} & -\dfrac{v_{21}}{E_2} & 0 & 0 & 0 \\[2mm]
-\dfrac{v_{12}}{E_1} & \dfrac{1}{E_2} & -\dfrac{v_{32}(=v_{23})}{E_2} & 0 & 0 & 0 \\[2mm]
-\dfrac{v_{12}}{E_1} & -\dfrac{v_{23}}{E_2} & \dfrac{1}{E_2} & 0 & 0 & 0 \\[2mm]
0 & 0 & 0 & \dfrac{2(1+v_{23})}{E_2} & 0 & 0 \\[2mm]
0 & 0 & 0 & 0 & \dfrac{1}{G_{12}} & 0 \\[2mm]
0 & 0 & 0 & 0 & 0 & \dfrac{1}{G_{12}}
\end{bmatrix}
\begin{bmatrix} \sigma_1 \\ \sigma_2 \\ \sigma_3 \\ \tau_{23} \\ \tau_{31} \\ \tau_{12} \end{bmatrix}
\qquad (13.3.1)
$$

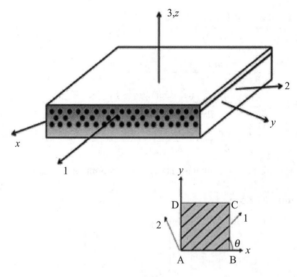

Figure 13.3.1　A single lamina of a fiber reinforced laminate.

For a condition of plane stress (in the 1-2 plane), we can set the out-of-plane stress components to be negligible (owing to the relatively small cross-sectional area of planes with normals in the "1" and "2" directions, because of the small lamina thickness) compared to the in-plane stress components; thus, set

$$
\sigma_3 = \tau_{23} = \tau_{31} = 0 \qquad (13.3.2)
$$

so that

$$
\begin{bmatrix} \varepsilon_1 \\ \varepsilon_2 \\ \gamma_{12} \end{bmatrix} =
\begin{bmatrix} S_{11} & S_{12} & 0 \\ S_{21} & S_{22} & 0 \\ 0 & 0 & S_{66} \end{bmatrix}
\begin{bmatrix} \sigma_1 \\ \sigma_2 \\ \tau_{12} \end{bmatrix} \qquad (13.3.3)
$$

where

$$S_{11} = \frac{1}{E_1} \qquad S_{22} = \frac{1}{E_2} \qquad S_{12} = S_{21} = -\frac{\nu_{21}}{E_2} = -\frac{\nu_{12}}{E_1} \qquad S_{66} = \frac{1}{G_{12}} \qquad (13.3.4)$$

Therfore, stress in terms of strain is

$$\begin{bmatrix} \sigma_1 \\ \sigma_2 \\ \tau_{12} \end{bmatrix} = \begin{bmatrix} Q_{11} & Q_{12} & 0 \\ Q_{21} & Q_{22} & 0 \\ 0 & 0 & 2Q_{66} \end{bmatrix} \begin{bmatrix} \varepsilon_1 \\ \varepsilon_2 \\ \dfrac{\gamma_{12}}{2} \end{bmatrix} = [Q] \begin{bmatrix} \varepsilon_1 \\ \varepsilon_2 \\ \dfrac{\gamma_{12}}{2} \end{bmatrix} \qquad (13.3.5)$$

where

$$Q_{11} = \frac{S_{22}}{S_{11}S_{22} - S_{12}^2} = \frac{E_1}{1 - \nu_{21}\nu_{12}}$$

$$Q_{12} = \frac{S_{12}}{S_{11}S_{22} - S_{12}^2} = \frac{\nu_{12}E_2}{1 - \nu_{21}\nu_{12}} = Q_{21}$$

$$Q_{22} = \frac{S_{11}}{S_{11}S_{22} - S_{12}^2} = \frac{E_2}{1 - \nu_{21}\nu_{12}} \qquad (13.3.6)$$

$$Q_{66} = \frac{1}{S_{66}} = G_{12}$$

Consider a lamina, whose fiber direction (the "1" direction) is at some angle, θ, to the "x" coordinate direction. Using the rules for stress tranformations, the components of stress in the x-y plane can be found from,

$$\begin{bmatrix} \sigma_x \\ \sigma_y \\ \tau_{xy} \end{bmatrix} = \begin{bmatrix} c^2 & s^2 & -2cs \\ s^2 & c^2 & 2cs \\ cs & -cs & c^2 - s^2 \end{bmatrix} \begin{bmatrix} \sigma_1 \\ \sigma_2 \\ \tau_{12} \end{bmatrix} = [T]^{-1} \begin{bmatrix} \sigma_1 \\ \sigma_2 \\ \tau_{12} \end{bmatrix} \qquad (13.3.7)$$

where

$$c = \cos\theta \qquad s = \sin\theta \qquad (13.3.8)$$

It follows that

$$\begin{bmatrix} \sigma_1 \\ \sigma_2 \\ \tau_{12} \end{bmatrix} = [T] \begin{bmatrix} \sigma_x \\ \sigma_y \\ \tau_{xy} \end{bmatrix} \qquad (13.3.9)$$

where

$$[T]^{-1} = \begin{bmatrix} c^2 & s^2 & -2cs \\ s^2 & c^2 & 2cs \\ cs & -cs & c^2 - s^2 \end{bmatrix} \qquad (13.3.10)$$

The tensorial strain transformation is similar.

$$\begin{bmatrix} \varepsilon_1 \\ \varepsilon_2 \\ \dfrac{\gamma_{12}}{2} \end{bmatrix} = [T] \begin{bmatrix} \varepsilon_x \\ \varepsilon_y \\ \dfrac{\gamma_{xy}}{2} \end{bmatrix} \qquad (13.3.11)$$

Note that we have used half of the engineering shear strain, γ_{12}, in the above tensor transformation relations. Therefore

$$\begin{bmatrix} \sigma_x \\ \sigma_y \\ \tau_{xy} \end{bmatrix} = [T]^{-1}[Q][T] \begin{bmatrix} \varepsilon_x \\ \varepsilon_y \\ \dfrac{\gamma_{xy}}{2} \end{bmatrix} \tag{13.3.12}$$

which ultimately provides (now, using engineering shear strain in the final result)

$$\begin{bmatrix} \sigma_x \\ \sigma_y \\ \tau_{xy} \end{bmatrix} = \begin{bmatrix} \overline{Q}_{11} & \overline{Q}_{12} & \overline{Q}_{16} \\ \overline{Q}_{12} & \overline{Q}_{22} & \overline{Q}_{26} \\ \overline{Q}_{16} & \overline{Q}_{26} & \overline{Q}_{66} \end{bmatrix} \begin{bmatrix} \varepsilon_x \\ \varepsilon_y \\ \gamma_{xy} \end{bmatrix} \tag{13.3.13}$$

where

$$\overline{Q}_{11} = Q_{11}c^4 + Q_{22}s^4 + 2(Q_{12} + 2Q_{66})s^2c^2$$
$$\overline{Q}_{12} = (Q_{11} + Q_{22} - 4Q_{66})c^2s^2 + Q_{12}(s^4 + c^4)$$
$$\overline{Q}_{22} = Q_{11}s^4 + Q_{22}c^4 + 2(Q_{12} + 2Q_{66})s^2c^2$$
$$\overline{Q}_{16} = (Q_{11} - Q_{12} - 2Q_{66})c^3s - (Q_{22} - Q_{12} - 2Q_{66})cs^3 \tag{13.3.14}$$
$$\overline{Q}_{26} = (Q_{11} - Q_{12} - 2Q_{66})cs^3 - (Q_{22} - Q_{12} - 2Q_{66})c^3s$$
$$\overline{Q}_{66} = (Q_{11} + Q_{22} - 2Q_{12} - 2Q_{66})c^2s^2 + Q_{66}(s^4 + c^4)$$

There are only four independent constants in the set of equations given above in 13.3.13. Inverting Equation 13.3.12, we obtain

$$\begin{bmatrix} \varepsilon_x \\ \varepsilon_y \\ \gamma_{xy} \end{bmatrix} = \begin{bmatrix} \overline{S}_{11} & \overline{S}_{12} & \overline{S}_{16} \\ \overline{S}_{12} & \overline{S}_{22} & \overline{S}_{26} \\ \overline{S}_{16} & \overline{S}_{26} & \overline{S}_{66} \end{bmatrix} \begin{bmatrix} \sigma_x \\ \sigma_y \\ \tau_{xy} \end{bmatrix} \tag{13.3.15}$$

where

$$\overline{S}_{11} = S_{11}c^4 + (2S_{12} + S_{66})s^2c^2 + S_{22}s^4$$
$$\overline{S}_{12} = S_{12}(s^4 + c^4) + (S_{11} + S_{22} - S_{66})s^2c^2$$
$$\overline{S}_{22} = S_{11}s^4 + (2S_{12} + S_{66})s^2c^2 + S_{22}c^4$$
$$\overline{S}_{16} = (2S_{11} - 2S_{12} - S_{66})sc^3 - (2S_{22} - 2S_{12} - S_{66})s^3c \tag{13.3.16}$$
$$\overline{S}_{26} = (2S_{11} - 2S_{12} - S_{66})s^3c - (2S_{22} - 2S_{12} - S_{66})sc^3$$
$$\overline{S}_{66} = 2(2S_{11} + 2S_{22} - 4S_{12} - S_{66})s^2c^2 + S_{66}(c^4 + s^4)$$

For example, suppose we want the modulus of elasticity associated with uni-axial loading along the x-direction. Then, in Equation 13.3.14, we set,

$$\sigma_x = \sigma \qquad \sigma_y = \tau_{xy} = 0, \tag{13.3.17}$$

so that,

$$E_x = \frac{\sigma_x}{\varepsilon_x} = \frac{\sigma}{\overline{S}_{11}\sigma} = \frac{1}{\overline{S}_{11}} = \frac{1}{S_{11}c^4 + (2S_{12} + S_{66})s^2c^2 + S_{22}s^4}$$

$$= \frac{1}{\dfrac{1}{E_1}c^4 + \left(-2\dfrac{\nu_{12}}{E_1} + \dfrac{1}{G_{12}}\right)s^2c^2 + \dfrac{1}{E_2}s^4}. \tag{13.3.18}$$

Furthermore, the major Poisson's ratio follows as,

$$
v_{xy} = -\frac{\varepsilon_y}{\varepsilon_x} = -\frac{\overline{S}_{12}}{\overline{S}_{11}} = -\frac{S_{12}\left(s^4 + c^4\right) + \left(S_{11} + S_{22} - S_{66}\right)s^2 c^2}{S_{11}c^4 + \left(2S_{12} + S_{66}\right)s^2 c^2 + S_{22}s^4}
$$

$$
= -\frac{\dfrac{v_{12}}{E_1}\left(s^4 + c^4\right) + \left(\dfrac{1}{E_1} + \dfrac{1}{E_2} - \dfrac{1}{G_{12}}\right)s^2 c^2}{\dfrac{1}{E_1}c^4 + \left(-2\dfrac{v_{12}}{E_1} + \dfrac{1}{G_{12}}\right)s^2 c^2 + \dfrac{1}{E_2}s^4} \tag{13.3.19}
$$

We can also define shear coupling coefficients (similar to Poisson's ratios) now. For example,

$$
\eta_{x,xy} = \frac{\gamma_{xy}}{\varepsilon_x} = \frac{\overline{S}_{16}}{\overline{S}_{11}} \tag{13.3.20}
$$

which is a measure of the amount of shear strain generated in the xy plane per unit normal strain along the direction of applied normal stress, σ_x. Clearly $\eta_{x,xy} = 0$ for $\theta = 0°$ and $\theta = 90°$. Another shear coupling constant exists for different types of load states. For example, when $\tau_{xy} \neq 0$ and $\sigma_x = \sigma_y = 0$, then

$$
\eta_{xy,y} = \frac{\varepsilon_y}{\gamma_{xy}} = \frac{\overline{S}_{26}}{\overline{S}_{66}} \tag{13.3.21}
$$

This characterizes the normal strain response along the y-direction due to a shear stress in the xy plane. We can also write $\overline{Q}_{ij}$ in the following convenient form;

$$
\overline{Q}_{11} = u_1 + u_2 \cos 2\theta + u_3 \cos 4\theta
$$

$$
\overline{Q}_{12} = u_4 - u_3 \cos 4\theta
$$

$$
\overline{Q}_{22} = u_1 - u_2 \cos 2\theta + u_3 \cos 4\theta
$$

$$
\overline{Q}_{16} = -\frac{u_2}{2} \sin 2\theta - u_3 \sin 4\theta \tag{13.3.22}
$$

$$
\overline{Q}_{26} = -\frac{u_2}{2} \sin 2\theta + u_3 \sin 4\theta
$$

$$
\overline{Q}_{66} = \frac{1}{2}\left(u_1 - u_4\right) - u_3 \cos 4\theta = u_5 - u_3 \cos 4\theta
$$

where

$$
u_1 = \frac{1}{8}\left(3Q_{11} + 3Q_{22} + 2Q_{12} + 4Q_{66}\right)
$$

$$
u_2 = \frac{1}{2}\left(Q_{11} - Q_{22}\right)
$$

$$
u_3 = \frac{1}{8}\left(Q_{11} + Q_{22} - 2Q_{12} - 4Q_{66}\right) \tag{13.3.23}
$$

$$
u_4 = \frac{1}{8}\left(Q_{11} + Q_{22} + 6Q_{12} - 4Q_{66}\right)
$$

$$
u_5 = \frac{1}{2}\left(u_1 - u_4\right)
$$

Now, u_1, u_2, u_3, u_4, u_5 are invariants and therefore do not change with respect to rotation in the 1-2 plane. All the rotational effects are captured in the trigonometric functions appearing in Equation 13.3.21. Thus these equations can be useful for optimization problems where optimum lay-ups (or stacking sequences of the laminae) are sought for a given application.

Having obtained the lamina stress-strain relations in the x-y coordinate frame, we can now obtain the stresses in the kth lamina. Let any point on the kth lamina be situated between $z = z_k$ and $z = z_{k-1}$. Then, from the constitutive equations, Equations 13.3.12, and Equations 13.2.6

$$
\begin{bmatrix} \sigma_x \\ \sigma_y \\ \tau_{xy} \end{bmatrix}_{kth} = \begin{bmatrix} \overline{Q}_{11} & \overline{Q}_{12} & \overline{Q}_{16} \\ \overline{Q}_{12} & \overline{Q}_{22} & \overline{Q}_{26} \\ \overline{Q}_{16} & \overline{Q}_{26} & \overline{Q}_{66} \end{bmatrix}_{kth} \begin{bmatrix} \varepsilon_x \\ \varepsilon_y \\ \gamma_{xy} \end{bmatrix}
$$

$$
= \begin{bmatrix} \overline{Q}_{11} & \overline{Q}_{12} & \overline{Q}_{16} \\ \overline{Q}_{12} & \overline{Q}_{22} & \overline{Q}_{26} \\ \overline{Q}_{16} & \overline{Q}_{26} & \overline{Q}_{66} \end{bmatrix}_{kth} \begin{bmatrix} \varepsilon_x^0 \\ \varepsilon_y^0 \\ \gamma_{xy}^0 \end{bmatrix} + z \begin{bmatrix} \overline{Q}_{11} & \overline{Q}_{12} & \overline{Q}_{16} \\ \overline{Q}_{12} & \overline{Q}_{22} & \overline{Q}_{26} \\ \overline{Q}_{16} & \overline{Q}_{26} & \overline{Q}_{66} \end{bmatrix}_{kth} \begin{bmatrix} \chi_x \\ \chi_y \\ \chi_{xy} \end{bmatrix}
$$

$$(13.3.24)$$

where

$$ z_{k-1} \leq z \leq z_k \qquad (13.3.25) $$

13.4   Stress Resultants for Laminates

The stresses in a laminate vary from layer to layer, as sketched in Figure 13.4.1.

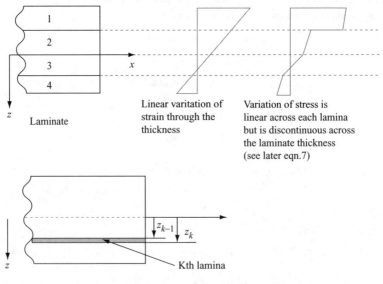

Figure 13.4.1

Thus, it is convenient to introduce the notion of stress resultants and moment resultants, which are thickness integrated quantities, as defined in Equations 13.4.1 and 13.4.2, and acting as shown in Figure 13.4.2

A resultant is defined as one that is obtained by integrating a particular component over the entire thickness of the laminate. Thus they are also referred to as 'thickness average' resultants. Introduce, the in-plane force resultants, N_x, N_y and N_{xy}, as,

$$ N_x = \int_{-\frac{h}{2}}^{\frac{h}{2}} \sigma_x dz \qquad N_y = \int_{-\frac{h}{2}}^{\frac{h}{2}} \sigma_y dz \qquad N_{xy} = \int_{-\frac{h}{2}}^{\frac{h}{2}} \tau_{xy} dz \qquad (13.4.1) $$

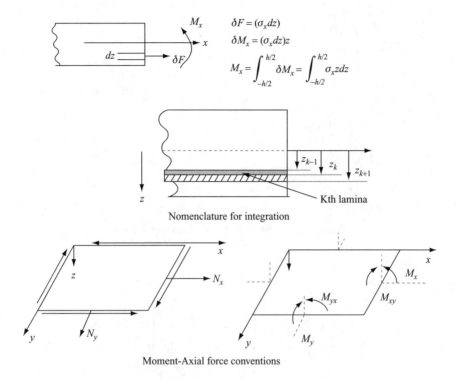

Figure 13.4.2 Stress and Moment Resultants

Similarly the resultant moments can be defined as follows;

$$M_x = \int_{-\frac{h}{2}}^{\frac{h}{2}} \sigma_x z\, dz \qquad M_y = \int_{-\frac{h}{2}}^{\frac{h}{2}} \sigma_y z\, dz \qquad M_{xy} = \int_{-\frac{h}{2}}^{\frac{h}{2}} \tau_{xy} z\, dz \qquad (13.4.2)$$

Positive forces and moments are as indicated in Figure 13.4.2. Since the stresses in each layer are different, the integrals defined in Equations 13.4.1 and 13.4.2 have to be converted to sums over each layer. For example

$$N_x = \int_{-\frac{h}{2}}^{\frac{h}{2}} \sigma_x\, dz = \int_{-h_0}^{-h_1} \sigma_x\, dz + \int_{-h_1}^{-h_2} \sigma_x\, dz \ldots\ldots\ldots + \int_{h_{n-1}}^{h_n} \sigma_x\, dz \qquad (13.4.3)$$

The thickness coordinates used in the integration are as shown in Figure 13.4.3. The force resultants can be written as,

$$\begin{bmatrix} N_x \\ N_y \\ N_{xy} \end{bmatrix} = \int_{-\frac{h}{2}}^{\frac{h}{2}} \begin{bmatrix} \sigma_x \\ \sigma_y \\ \tau_{xy} \end{bmatrix} dz = \sum_{k=1}^{n} \int_{h_{k-1}}^{h_k} \begin{bmatrix} \sigma_x \\ \sigma_y \\ \tau_{xy} \end{bmatrix} dz, \qquad (13.4.4)$$

and the moment resultants, as,

$$\begin{bmatrix} M_x \\ M_y \\ M_{xy} \end{bmatrix} = \int_{-\frac{h}{2}}^{\frac{h}{2}} \begin{bmatrix} \sigma_x \\ \sigma_y \\ \tau_{xy} \end{bmatrix} z\, dz = \sum_{k=1}^{n} \int_{h_{k-1}}^{h_k} \begin{bmatrix} \sigma_x \\ \sigma_y \\ \tau_{xy} \end{bmatrix} z\, dz \qquad (13.4.5)$$

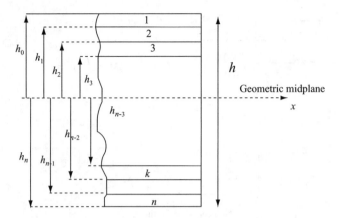

Figure 13.4.3 Nomenclature for thickness integration

By substituting Equation 13.3.23 into Equations 13.4.4 and 13.4.5 we get

$$
\begin{bmatrix} N_x \\ N_y \\ N_{xy} \end{bmatrix} = \sum_{k=1}^{n} \int_{h_{k-1}}^{h_k} \begin{bmatrix} \overline{Q}_{11} & \overline{Q}_{12} & \overline{Q}_{16} \\ \overline{Q}_{12} & \overline{Q}_{22} & \overline{Q}_{26} \\ \overline{Q}_{16} & \overline{Q}_{26} & \overline{Q}_{66} \end{bmatrix}_{k^{th}} \begin{bmatrix} \varepsilon_x^0 \\ \varepsilon_y^0 \\ \gamma_{xy}^0 \end{bmatrix} dz
$$

$$
+ \sum_{k=1}^{n} \int_{h_{k-1}}^{h_k} \begin{bmatrix} \overline{Q}_{11} & \overline{Q}_{12} & \overline{Q}_{16} \\ \overline{Q}_{12} & \overline{Q}_{22} & \overline{Q}_{26} \\ \overline{Q}_{16} & \overline{Q}_{26} & \overline{Q}_{66} \end{bmatrix}_{k^{th}} \begin{bmatrix} \chi_x \\ \chi_y \\ \chi_{xy} \end{bmatrix} z \, dz
$$

(13.4.6)

and

$$
\begin{bmatrix} M_x \\ M_y \\ M_{xy} \end{bmatrix} = \sum_{k=1}^{n} \int_{h_{k-1}}^{h_k} \begin{bmatrix} \overline{Q}_{11} & \overline{Q}_{12} & \overline{Q}_{16} \\ \overline{Q}_{12} & \overline{Q}_{22} & \overline{Q}_{26} \\ \overline{Q}_{16} & \overline{Q}_{26} & \overline{Q}_{66} \end{bmatrix}_{k^{th}} \begin{bmatrix} \varepsilon_x^0 \\ \varepsilon_y^0 \\ \gamma_{xy}^0 \end{bmatrix} z \, dz
$$

$$
+ \sum_{k=1}^{n} \int_{h_{k-1}}^{h_k} \begin{bmatrix} \overline{Q}_{11} & \overline{Q}_{12} & \overline{Q}_{16} \\ \overline{Q}_{12} & \overline{Q}_{22} & \overline{Q}_{26} \\ \overline{Q}_{16} & \overline{Q}_{26} & \overline{Q}_{66} \end{bmatrix}_{k^{th}} \begin{bmatrix} \chi_x \\ \chi_y \\ \chi_{xy} \end{bmatrix} z^2 \, dz
$$

(13.4.7)

Noting that the midplane strains and plate curvatures remain constant not only within a lamina but for all laminae in the laminate, we obtain from Equations 13.4.6 and 13.4.7

$$
\begin{bmatrix} N_x \\ N_y \\ N_{xy} \end{bmatrix} = \begin{bmatrix} A_{11} & A_{12} & A_{16} \\ A_{12} & A_{22} & A_{26} \\ A_{16} & A_{26} & A_{66} \end{bmatrix} \begin{bmatrix} \varepsilon_x^0 \\ \varepsilon_y^0 \\ \gamma_{xy}^0 \end{bmatrix} + \begin{bmatrix} B_{11} & B_{12} & B_{16} \\ B_{12} & B_{22} & B_{26} \\ B_{16} & B_{26} & B_{66} \end{bmatrix} \begin{bmatrix} \chi_x \\ \chi_y \\ \chi_{xy} \end{bmatrix}
$$

(13.4.8)

and

$$
\begin{bmatrix} M_x \\ M_y \\ M_{xy} \end{bmatrix} = \begin{bmatrix} B_{11} & B_{12} & B_{16} \\ B_{12} & B_{22} & B_{26} \\ B_{16} & B_{26} & B_{66} \end{bmatrix} \begin{bmatrix} \varepsilon_x^0 \\ \varepsilon_y^0 \\ \gamma_{xy}^0 \end{bmatrix} + \begin{bmatrix} D_{11} & D_{12} & D_{16} \\ D_{12} & D_{22} & D_{26} \\ D_{16} & D_{26} & D_{66} \end{bmatrix} \begin{bmatrix} \chi_x \\ \chi_y \\ \chi_{xy} \end{bmatrix}
$$

(13.4.9)

where

$$A_{ij} = \sum_{k=1}^{n} \left[\overline{Q}_{ij}\right]_k (h_k - h_{k-1})$$

$$B_{ij} = \frac{1}{2} \sum_{k=1}^{n} \left[\overline{Q}_{ij}\right]_k \left(h_k^2 - h_{k-1}^2\right) \qquad (13.4.10)$$

$$D_{ij} = \frac{1}{3} \sum_{k=1}^{n} \left[\overline{Q}_{ij}\right]_k \left(h_k^3 - h_{k-1}^3\right)$$

13.5 CLT Constitutive Description

Equations 13.4.8 and 13.4.9 can be represented by

$$\begin{bmatrix} \{N\} \\ \{M\} \end{bmatrix} = \begin{bmatrix} [A] & [B] \\ [B] & [D] \end{bmatrix} \begin{bmatrix} \{\varepsilon^0\} \\ \{\chi\} \end{bmatrix}, \qquad (13.5.1)$$

where each of the symbols A, B and D represent a 3×3 matrix. $[A]$ is an extensional stiffness matrix, $[B]$ is a coupling (extensional-bending) stiffness matrix, and $[D]$ is a bending stiffness matrix.

Note that in-plane loads are determined by mid-plane strains and plate curvatures (stretch-bending coupling), that is, both stretching (and compressing) and bending induce mid-plane stress resultants. The bending effects in this case are related to the B matrix and it is apparent that there would be no in-plane/bending coupling if $B = 0$.

The B_{ij} are even functions of the h_k and thus will be zero for laminates which are symmetric about the midplane, that is, for each layer above the midplane there is an identical layer (in properties, thickness, and orientation) located at the same distance below the midplane.

Examination of the A_{ij} shows another form of coupling: in-plane shear and normal forces as represented in Equation 13.5.2.

$$\begin{bmatrix} N_x \\ N_y \\ N_{xy} \end{bmatrix} = \begin{bmatrix} A_{11} & A_{12} & A_{16} \\ A_{12} & A_{22} & A_{26} \\ A_{16} & A_{26} & A_{66} \end{bmatrix} \begin{bmatrix} \varepsilon_x^0 \\ \varepsilon_y^0 \\ \gamma_{xy}^0 \end{bmatrix} \qquad (13.5.2)$$

This particular coupling can be removed if $A_{16} = A_{26} = 0$. From the definition of A_{ij} we see that A_{ij} can be zero only if all the Q_{ij} are zero or if some Q_{ij} are positive and some negative. The Q_{11}, Q_{12}, Q_{22} and Q_{66} are always positive; and so are the corresponding A_{ij}. The Q_{16} and Q_{26} however are zero for 0^0 and 90^0 orientation and can be either positive or negative, since odd parameters c and s appear in the defining expression. So, if for every lamina with positive $+\theta$ orientation, there is another lamina with the identical properties and thickness at negative θ, then $A_{16} = A_{26} = 0$. The z position of the lamina is not important, that is, it need not be midplane symmetric. Of course, if it were, then additionally $B_{ij} = 0$.

From Equation 13.5.1 we have

$$\{M\} = [B]\{\varepsilon^0\} + [D]\{\chi\} \qquad (13.5.3)$$

We see that $[B]$ also relates in-plane and bending effects. Furthermore, note that the moments M_x and M_y are due, in part, to midplane shearing and plate twisting while M_{xy} is partly due to midplane normal strain and normal plate curvature.

The coupling between plate twisting and normal bending is eliminated if $D_{16} = D_{26} = 0$. This will occur if $Q_{16} = Q_{26} = 0$, that is, for 0^0 or 90^0 orientation or if for every layer at $+\theta$ at a given distance above the midplane there is an identical layer, that is, in properties and thickness, at the same distance below the midplane but at negative θ. The $\left(h_k^3 - h_{k-1}^3\right)$ terms are the same for both the layers but the

Q terms are of opposite sign. Unfortunately, such a laminate would not possess midplane symmetry; however, the D_{16} and $D_{26} \to 0$ for thick laminates since the $\pm\theta$ layers tend to cancel each other.

Knowledge of the $[A]$, $[B]$, and $[D]$ matrices allows the stresses within the laminate to be found. The stress resultants will be known from an analysis of the structure, hence the midplane strains and plate curvatures may be found by inverting the constitutive equation. Then, using the transformation matrix $[T]$, for each layer, the lamina stresses in the principal material directions, 1-2, are obtained. Usually a limiting strain would be of interest. For example, it is of interest to know the fiber direction tensile strain, ε_1, to examine if it exceeds the fiber direction maximum allowable tensile strain. This can be obtained from coupon level material tensile tests or from a handbook of materials strength data.

In practice, the situation is slightly more complex than indicated since a thermal strain must be included in the formulation of the constitutive equation to account for stresses that arise during the curing of the laminate. These stresses may be high enough to crack the laminate.

The above equations are commonly known as 'Classical Lamination Theory' and are adequate for predicting laminate stiffness behavior and laminae stresses away from an edge. In the region of an edge a three dimensional analysis is needed to calculate the through the thickness stresses.

############

Example 13.5.1

Problem: A two layer laminate consists of layer (sometimes, the word "ply" is also used to describe a layer) orientation of 0^0 and 45^0 with respect to the x-y axes as shown in Figure (a).

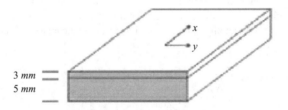

3 mm
5 mm

Figure (a)   Two layer laminate

The 0^0 lamina is 5 mm thick and the 45^0 lamina is 3 mm thick. The $[Q]$matrix for the ply material is given in Equation 13.5.1.

$$[Q] = \begin{bmatrix} 20 & 0.7 & 0 \\ 0.7 & 20 & 0 \\ 0 & 0 & 0.7 \end{bmatrix} \times 10^9 \ N/mm^2 \tag{a}$$

The thickness coordinates are (refer to Figure 13.4.3 for the nomenclature)

$$h_0 = -4 \ mm \qquad h_1 = -1 \ mm \qquad h_2 = 4 \ mm \tag{b}$$

Obtain the laminate constitutive description, that is, the A, B and D matrices.

Solution: The $[Q]$ matrix is defined in Equations 13.3.5 and 13.3.6 and its values are given in the statement of the problem. The $[\bar{Q}]$ matrix is defined in Equation 13.3.12 and 13.3.13. First find the $[\bar{Q}]$ matrices for 0^0 and 45^0 and then use Equations 13.4.10 to find $[A]$, $[B]$, and $[D]$.

Step 1: Find $[Q]_{0°}$ and $[Q]_{45°}$.

The $[Q]_{0°}$ matrix is the same as $[Q]$ since $\cos\theta = 1$ and $\sin\theta = 0$, that is,

$$[Q]_{0°} = \begin{bmatrix} 20 & 0.7 & 0 \\ 0.7 & 20 & 0 \\ 0 & 0 & 0.7 \end{bmatrix} GPa \tag{c}$$

For the $[Q]_{45°}$ matrix we have

$$\overline{Q}_{11} = 20(\cos 45°)^4 + 2(\sin 45°)^4 + 2(0.7 + 2 \times 0.7)(\sin 45°)^2(\cos 45°)^2 = 6.55\ GPa \tag{d}$$

Similarly,

$$\overline{Q}_{22} = 6.55\ GPa \qquad \overline{Q}_{12} = 5.15\ GPa \qquad \overline{Q}_{66} = 5.15\ GPa \qquad \overline{Q}_{16} = 4.50\ GPa \tag{e}$$

and so

$$[Q]_{45°} = \begin{bmatrix} \overline{Q}_{11} & \overline{Q}_{12} & \overline{Q}_{16} \\ \overline{Q}_{12} & \overline{Q}_{22} & \overline{Q}_{26} \\ \overline{Q}_{16} & \overline{Q}_{26} & \overline{Q}_{66} \end{bmatrix} = \begin{bmatrix} 6.55 & 5.15 & 4.50 \\ 5.15 & 6.55 & 4.50 \\ 4.50 & 4.50 & 5.15 \end{bmatrix} GPa \tag{f}$$

Step 2: Find $[A]$, $[B]$, and $[D]$.

Note that the interface does not coincide with the midplane.

$$A_{ij} = \sum_{k=1}^{n} [\overline{Q}_{ij}]_k (h_k - h_{k-1}) = [\overline{Q}_{ij}]_{45°}((-1) - (-4)) + [\overline{Q}_{ij}]_{0°}((4) - (-1))$$
$$= 3[\overline{Q}_{ij}]_{45°} + 5[\overline{Q}_{ij}]_{0°}\ GPa\ mm \tag{g}$$

$$[A] = \begin{bmatrix} 119.65 & 18.95 & 13.50 \\ 18.95 & 29.65 & 13.50 \\ 13.50 & 13.50 & 18.95 \end{bmatrix} GPa\ mm \tag{h}$$

$$B_{ij} = \frac{1}{2}\sum_{k=1}^{n} [\overline{Q}_{ij}]_k (h_k^2 - h_{k-1}^2) = \frac{1}{2}[\overline{Q}_{ij}]_{45°}((-1)^2 - (-4)^2) + \frac{1}{2}[\overline{Q}_{ij}]_{0°}((4)^2 - (-1)^2)$$
$$= -7.5[\overline{Q}_{ij}]_{45°} + 7.5[\overline{Q}_{ij}]_{0°}\ GPa\ mm^2 \tag{i}$$

$$[B] = \begin{bmatrix} 100.9 & -33.4 & -33.75 \\ -33.4 & -34.1 & -33.75 \\ -33.75 & -33.75 & -33.4 \end{bmatrix} GPa\ mm^2 \tag{j}$$

$$D_{ij} = \frac{1}{3}\sum_{k=1}^{n} [\overline{Q}_{ij}]_k (h_k^3 - h_{k-1}^3) = \frac{1}{3}[\overline{Q}_{ij}]_{45°}((-1)^3 - (-4)^3) + \frac{1}{3}[\overline{Q}_{ij}]_{0°}((4)^3 - (-1)^3)$$
$$= 21[\overline{Q}_{ij}]_{45°} + 21.67[\overline{Q}_{ij}]_{0°}\ GPa\ mm^3 \tag{k}$$

$$[D] = \begin{bmatrix} 571 & 123 & 94.5 \\ 123 & 181 & 94.5 \\ 94.5 & 94.5 & 123 \end{bmatrix} GPa\ mm^3 \tag{l}$$

##########

13.6   Determining Laminae Stress/Strains

We know from Equation 13.5.1, repeated here as Equation 13.6.1, that

$$\begin{bmatrix} \{N\} \\ \{M\} \end{bmatrix} = \begin{bmatrix} [A] & [B] \\ [B] & [D] \end{bmatrix} \begin{bmatrix} \{\varepsilon^0\} \\ \{\chi\} \end{bmatrix} \tag{13.6.1}$$

Expanding we have

$$\begin{aligned} \{N\} &= [A]\{\varepsilon^0\} + [B]\{\chi\} \\ \{M\} &= [B]\{\varepsilon^0\} + [D]\{\chi\} \end{aligned} \tag{13.6.2}$$

Solving for $\{\varepsilon^0\}$ from the first equation of 13.6.2, we have,

$$\{\varepsilon^0\} = [A]^{-1}(\{N\} - [B]\{\chi\}) \tag{13.6.3}$$

From the second of Equations 13.6.2, we get

$$\{M\} = [B][A]^{-1}(\{N\} - [B]\{\chi\}) + [D]\{\chi\} \tag{13.6.4}$$

Thus we get a partially inverted system of equations given by

$$\begin{bmatrix} \{\varepsilon^0\} \\ \{M\} \end{bmatrix} = \begin{bmatrix} [A^*] & [B^*] \\ [C^*] & [D^*] \end{bmatrix} \begin{bmatrix} \{N\} \\ \{\chi\} \end{bmatrix} \tag{13.6.5}$$

where

$$\begin{aligned} [A^*] &= [A]^{-1} \\ [B^*] &= -[A]^{-1}[B] \\ [C^*] &= [B][A]^{-1} = -[B^*]^T \\ [D^*] &= [D] - [B][A]^{-1}[B] \end{aligned} \tag{13.6.6}$$

From the second of Equations 13.6.5 we get

$$\{\chi\} = [D^*]^{-1}\{M\} - [D^*]^{-1}[C^*]\{N\} \tag{13.6.7}$$

Using this in the first of Equations 13.6.5 we get

$$\{\varepsilon^0\} = \left([A^*] - [B^*][D^*]^{-1}[C^*]\right)\{N\} + [B^*][D^*]^{-1}\{M\} \tag{13.6.8}$$

From Equations 13.6.7 and 13.6.8 we get the fully inverted form

$$\begin{bmatrix} \{\varepsilon^0\} \\ \{\chi\} \end{bmatrix} = \begin{bmatrix} [A'] & [B'] \\ [C'] & [D'] \end{bmatrix} \begin{bmatrix} \{N\} \\ \{M\} \end{bmatrix} \tag{13.6.9}$$

where

$$\begin{aligned} [A'] &= [A^*] - [B^*][D^*]^{-1}[C^*] \\ [B'] &= [B^*][D^*]^{-1} \\ [C'] &= -[D^*]^{-1}[C^*] = [B']^T \\ [D'] &= [D^*]^{-1} \end{aligned} \tag{13.6.10}$$

This inverted form, Equation 13.6.9, is also very useful for solving certain types of boundary value problems.

13.7 Laminated Plates Subject to Transverse Loads

The theory that was developed (CLT) assumes that each lamina and the laminate as a whole is in a state of plane stress and further it is assumed that plane sections remain-plane and normal to the deformed centerline, as in classical plate theory. Furthermore, with the restriction to linear elastic material response, it is assumed that strains are small.

In Figure 13.7.1 the resultant forces acting on an element of a plate with transverse loading is shown. We can now derive the equilibrium equations for laminated plates, using a direct approach, in which we use summation of forces and summation of moments. Then, with respect to Figure 13.7.1, summation of forces in the x-direction provides

$$N_x dy + \frac{\partial N_x}{\partial x} dy dx + N_{xy} dx + \frac{\partial N_{xy}}{\partial y} dy dx - N_x dy - N_{xy} dx = 0$$

which gives us

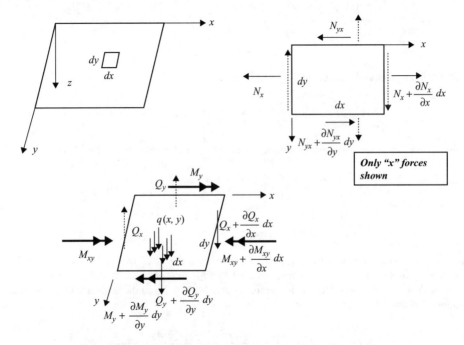

Figure 13.7.1   Force and moment resultants and transverse pressure loads

Force and moment resultants and transverse pressure loads
$q(x, y)$, on an infinitesimal "chunk" of a laminate in the x-y plane

$$\frac{\partial N_x}{\partial x} + \frac{\partial N_{xy}}{\partial y} = 0 \qquad (13.7.1)$$

Similarly in the y-direction

$$N_y dx + \frac{\partial N_y}{\partial y} dy dx + N_{xy} dy + \frac{\partial N_{xy}}{\partial x} dy dx - N_y dx - N_{xy} dy = 0$$

$$\frac{\partial N_y}{\partial y} + \frac{\partial N_{xy}}{\partial x} = 0 \qquad (13.7.2)$$

For force equilibrium in the z- direction (out of plane), the edge shears equilibrate the transverse pressure loading $q(x, y)$.

$$Q_x dy + \frac{\partial Q_x}{\partial x} dydx + Q_y dx + \frac{\partial Q_y}{\partial y} dydx - Q_x dy - Q_y dx + q(x, y) dxdy = 0$$

$$\frac{\partial Q_x}{\partial x} + \frac{\partial Q_y}{\partial y} + q(x, y) = 0$$

(13.7.3)

Moment equilibrium about the x axis, gives,

$$-M_y dx - \frac{\partial M_y}{\partial y} dydx - M_{xy} dy - \frac{\partial M_{xy}}{\partial x} dydx + Q_y dydx + \frac{\partial Q_y}{\partial y} dydxdy$$

$$+ qdxdy\frac{dy}{2} + Q_x dy\frac{dy}{2} + \frac{\partial Q_x}{\partial x} dydx\frac{dy}{2} + M_y dx + M_{xy} dy - Q_x dy\frac{dy}{2} = 0$$

(13.7.4)

$$\frac{\partial M_y}{\partial y} + \frac{\partial M_{xy}}{\partial x} - Q_y = 0$$

Similarly, moment equilibrium about the y axis, provides,

$$\frac{\partial M_x}{\partial x} + \frac{\partial M_{xy}}{\partial y} - Q_x = 0$$

(13.7.5)

Using the last two equations in the third force equation (z-direction) of equilibrium, we obtain,

$$\frac{\partial}{\partial x}\left(\frac{\partial M_x}{\partial x} + \frac{\partial M_{xy}}{\partial y}\right) + \frac{\partial}{\partial y}\left(\frac{\partial M_y}{\partial y} + \frac{\partial M_{xy}}{\partial x}\right) + q = 0$$

$$\frac{\partial^2 M_x}{\partial x^2} + \frac{\partial^2 M_{xy}}{\partial x \partial y} + \frac{\partial^2 M_y}{\partial y^2} + q = 0$$

(13.7.6)

This last equation and the first two equations are in a form that is convenient for solving laminated plate bending problems. These three equations can be cast in terms of centerline displacements, u_0, v_0, and w, by using the strain-displacement equations, Equation 13.2.7, and the laminated plate constitutive equations, Equation 13.5.1, to arrive at,

$$A_{11}\frac{\partial^2 u_0}{\partial x^2} + A_{12}\frac{\partial^2 v_0}{\partial x \partial y} + 2A_{16}\frac{\partial^2 u_0}{\partial x \partial y} + A_{66}\frac{\partial^2 u_0}{\partial y^2} + A_{16}\frac{\partial^2 v_0}{\partial x^2}$$

$$+ A_{66}\frac{\partial^2 v_0}{\partial x \partial y} + A_{26}\frac{\partial^2 v_0}{\partial y^2} - B_{11}\frac{\partial^3 w}{\partial x^3} - 3B_{16}\frac{\partial^3 w}{\partial x^2 \partial y}$$

(13.7.7)

$$- (B_{12} + 2B_{66})\frac{\partial^3 w}{\partial x \partial y^2} - B_{26}\frac{\partial^3 w}{\partial y^3} = 0$$

$$A_{16}\frac{\partial^2 u_0}{\partial x^2} + (A_{12} + A_{66})\frac{\partial^2 u_0}{\partial x \partial y} + A_{26}\frac{\partial^2 u_0}{\partial y^2} + A_{66}\frac{\partial^2 v_0}{\partial x^2} + 2A_{26}\frac{\partial^2 v_0}{\partial x \partial y}$$

$$+ A_{22}\frac{\partial^2 v_0}{\partial y^2} + B_{16}\frac{\partial^3 w}{\partial x^3} - (B_{12} + 2B_{66})\frac{\partial^3 w}{\partial x^2 \partial y} - 3B_{26}\frac{\partial^3 w}{\partial x \partial y^2} - B_{22}\frac{\partial^3 w}{\partial y^3} = 0$$

(13.7.8)

$$D_{11}\frac{\partial^4 w}{\partial x^4} + 4D_{16}\frac{\partial^4 w}{\partial x^3 \partial y} + 2(D_{12} + 2D_{66})\frac{\partial^4 w}{\partial x^2 \partial y^2} + 4D_{26}\frac{\partial^4 w}{\partial x \partial y^3} + D_{22}\frac{\partial^4 w}{\partial y^4}$$

$$- B_{11}\frac{\partial^3 u_0}{\partial x^3} - 3B_{16}\frac{\partial^3 u_0}{\partial x^2 \partial y} - (B_{12} + 2B_{66})\frac{\partial^3 u_0}{\partial x \partial y^2} - B_{26}\frac{\partial^3 u_0}{\partial y^3} - B_{16}\frac{\partial^3 v_0}{\partial x^3} \qquad (13.7.9)$$

$$- (B_{12} + 2B_{66})\frac{\partial^3 v_0}{\partial x^2 \partial y} - 3B_{26}\frac{\partial^3 v_0}{\partial x \partial y^2} - B_{22}\frac{\partial^3 v_0}{\partial y^3} = q(x, y)$$

We will now demonstrate how these equations can be used to solve laminates subjected to transverse loads.

###########

Example 13.7.1

Consider a rectangular, symmetric, cross-ply laminate that is free from any in-plane forces and moments on its edges and, in fact, is free to move in the x and y directions on all edges, but is prevented from out of plane motions at the edges. Suppose such a laminate is subjected to uniform transverse pressure loads. Assume the origin of the xyz coordinate system is at the geometric center of the plate.

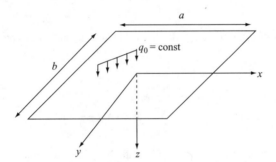

Figure (a)   Cross-ply laminate under transverse pressure loads

Since the plate is symmetric and of cross-ply construction

$$\text{Cross} - \text{ply} \rightarrow \quad A_{16} = A_{26} = D_{16} = D_{26} = 0$$
$$\text{Symmetric} \rightarrow \quad B_{ij} = 0 \quad \text{for all } i \text{ and } j$$

As a result, the governing equations simplify to,

$$A_{11}\frac{\partial^2 u_0}{\partial x^2} + A_{66}\frac{\partial^2 u_0}{\partial y^2} + (A_{11} + A_{66})\frac{\partial^2 v_0}{\partial x \partial y} = 0$$

$$(A_{12} + A_{66})\frac{\partial^2 u_0}{\partial x \partial y} + A_{66}\frac{\partial^2 v_0}{\partial x^2} + A_{22}\frac{\partial^2 v_0}{\partial y^2} = 0 \qquad (a)$$

$$D_{11}\frac{\partial^4 w}{\partial x^4} + 2(D_{12} + 2D_{66})\frac{\partial^4 w}{\partial x^2 \partial y^2} + D_{22}\frac{\partial^4 w}{\partial y^4} = q_0$$

with q_0 being the magnitude of the uniform load. The boundary conditions are as follows:

$$\text{At } x = \pm \frac{a}{2} \rightarrow$$

$$N_x = A_{11} \frac{\partial u_0}{\partial x} + A_{12} \frac{\partial v_0}{\partial y} = 0$$

$$N_{xy} = A_{66} \left(\frac{\partial u_0}{\partial y} + \frac{\partial v_0}{\partial x} \right) = 0 \tag{b}$$

$$M_x = -D_{11} \frac{\partial^2 w}{\partial x^2} - D_{12} \frac{\partial^2 w}{\partial y^2} = 0$$

$$w = 0$$

$$\text{At } y = \pm \frac{b}{2} \rightarrow$$

$$N_y = A_{12} \frac{\partial u_0}{\partial x} + A_{22} \frac{\partial v_0}{\partial y} = 0$$

$$N_{xy} = A_{66} \left(\frac{\partial u_0}{\partial y} + \frac{\partial v_0}{\partial x} \right) = 0 \tag{c}$$

$$M_y = -D_{12} \frac{\partial^2 w}{\partial x^2} - D_{22} \frac{\partial^2 w}{\partial y^2} = 0$$

$$w = 0$$

Sometimes the form of the solution can be determined by inspection. For this problem a solution of the form

$$u_0(x, y) = 0$$
$$v_0(x, y) = 0$$
$$w(x, y) = \sum_{m=1,3,5,\ldots}^{\infty} \sum_{n=1,3,5,\ldots}^{\infty} W_{mn} \cos \left(\frac{m\pi x}{a} \right) \cos \left(\frac{n\pi y}{b} \right) \tag{d}$$

satisfies governing equations, Equation (a), and all the boundary conditions at each edge, Equations (b) and (c). The W_{mn} can be determined by substituting the form for $w(x, y)$ into the third differential equation, expanding the uniform load in the same double cosine series and equating coefficients. For any cross-ply laminate with these same boundary conditions, the following procedure is valid if the load can be expanded in terms of a double cosine Fourier series. The double cosine Fourier series representation of the uniform load is

$$q_0(x, y) = \sum_{m=1,3,5,\ldots}^{\infty} \sum_{n=1,3,5,\ldots}^{\infty} Q_{mn} \cos \left(\frac{m\pi x}{a} \right) \cos \left(\frac{n\pi y}{b} \right) \tag{e}$$

with

$$Q_{mn} = \frac{16 q_0}{mn\pi^2} (-1)^{\frac{m+n}{2} - 1} \tag{f}$$

Substituting the series expressions for $w(x, y)$ and $q_0(x, y)$ into the third differential equation, and equating the like terms in m and n leads to

$$\left(D_{11} \left(\frac{m\pi}{a} \right)^4 + 2(D_{12} + 2D_{66}) \left(\frac{m\pi}{a} \right)^2 \left(\frac{n\pi}{b} \right)^2 + D_{22} \left(\frac{n\pi}{b} \right)^4 \right) W_{mn} = Q_{mn} \tag{g}$$

This results in

$$W_{mn} = \frac{Q_{mn}}{\left(D_{11}\left(\frac{m\pi}{a}\right)^4 + 2\left(D_{12} + 2D_{66}\right)\left(\frac{m\pi}{a}\right)^2\left(\frac{n\pi}{b}\right)^2 + D_{22}\left(\frac{n\pi}{b}\right)^4\right)} \tag{h}$$

or

$$W_{mn} = \frac{\frac{16q_0}{mn\pi^2}(-1)^{\frac{m+n}{2}-1}}{\left(D_{11}\left(\frac{m\pi}{a}\right)^4 + 2\left(D_{12} + 2D_{66}\right)\left(\frac{m\pi}{a}\right)^2\left(\frac{n\pi}{b}\right)^2 + D_{22}\left(\frac{n\pi}{b}\right)^4\right)} \tag{i}$$

With W_{mn} known, the curvatures are

$$\chi_x = -\frac{\partial^2 w}{\partial x^2} = \sum_{m=1,3,5,\ldots}^{\infty}\sum_{n=1,3,5,\ldots}^{\infty} W_{mn}\left(\frac{m\pi}{a}\right)^2 \cos\left(\frac{m\pi x}{a}\right)\cos\left(\frac{n\pi y}{b}\right)$$

$$\chi_y = -\frac{\partial^2 w}{\partial y^2} = \sum_{m=1,3,5,\ldots}^{\infty}\sum_{n=1,3,5,\ldots}^{\infty} W_{mn}\left(\frac{n\pi}{b}\right)^2 \cos\left(\frac{m\pi x}{a}\right)\cos\left(\frac{n\pi y}{b}\right) \tag{j}$$

$$\chi_{xy} = -2\frac{\partial^2 w}{\partial x\partial y} = -2\sum_{m=1,3,5,\ldots}^{\infty}\sum_{n=1,3,5,\ldots}^{\infty} W_{mn}\left(\frac{m\pi}{a}\right)\left(\frac{n\pi}{b}\right)\sin\left(\frac{m\pi x}{a}\right)\sin\left(\frac{n\pi y}{b}\right)$$

The moment resultants are

$$M_x = D_{11}\chi_x + D_{12}\chi_y$$
$$= \sum_{m=1,3,5,\ldots}^{\infty}\sum_{n=1,3,5,\ldots}^{\infty}\left(D_{11}\left(\frac{m\pi}{a}\right)^2 + D_{12}\left(\frac{n\pi}{b}\right)^2\right)W_{mn}\cos\left(\frac{m\pi x}{a}\right)\cos\left(\frac{n\pi y}{b}\right)$$

$$M_y = D_{12}\chi_x + D_{22}\chi_y$$
$$= \sum_{m=1,3,5,\ldots}^{\infty}\sum_{n=1,3,5,\ldots}^{\infty}\left(D_{12}\left(\frac{m\pi}{a}\right)^2 + D_{22}\left(\frac{n\pi}{b}\right)^2\right)W_{mn}\cos\left(\frac{m\pi x}{a}\right)\cos\left(\frac{n\pi y}{b}\right) \tag{k}$$

$$M_{xy} = D_{66}\chi_{xy}$$
$$= -2D_{66}\sum_{m=1,3,5,\ldots}^{\infty}\sum_{n=1,3,5,\ldots}^{\infty}\left(\frac{m\pi}{a}\right)\left(\frac{n\pi}{b}\right)W_{mn}\cos\left(\frac{m\pi x}{a}\right)\cos\left(\frac{n\pi y}{b}\right)$$

The transverse shear force resultants are

$$Q_x = \frac{\partial M_x}{\partial x} + \frac{\partial M_{xy}}{\partial y}$$
$$= -\sum_{m=1,3,5,\ldots}^{\infty}\sum_{n=1,3,5,\ldots}^{\infty}\left(D_{11}\left(\frac{m\pi}{a}\right)^3 + D_{12}\left(\frac{n\pi}{b}\right)^2\left(\frac{m\pi}{a}\right) + 2D_{66}\left(\frac{n\pi}{b}\right)^2\left(\frac{m\pi}{a}\right)\right)$$
$$\times W_{mn}\sin\left(\frac{m\pi x}{a}\right)\cos\left(\frac{n\pi y}{b}\right)$$

$$Q_y = \frac{\partial M_y}{\partial y} + \frac{\partial M_{xy}}{\partial x} \tag{l}$$
$$= -\sum_{m=1,3,5,\ldots}^{\infty}\sum_{n=1,3,5,\ldots}^{\infty}\left(D_{22}\left(\frac{n\pi}{b}\right)^3 + D_{12}\left(\frac{n\pi}{b}\right)\left(\frac{m\pi}{a}\right)^2 + 2D_{66}\left(\frac{n\pi}{b}\right)\left(\frac{m\pi}{a}\right)^2\right)$$
$$\times W_{mn}\cos\left(\frac{m\pi x}{a}\right)\sin\left(\frac{n\pi y}{b}\right)$$

However, these are not the effective transverse shear force resultants discussed earlier, in Chapter 12 (note that V_x and V_y are the effective or Kirchhoff shear resultants). The effective shear force resultants consist of a combination of the transverse shear resultants, Q_x and Q_y, and the gradients along each plate edge, of the twisting moments, M_{xy} and M_{yx}, respectively. That is, with respect to Figure 13.7.1, along the edge where Q_x, acts, the effective transverse Kirchhoff shear force resultant, V_x, is $V_x = Q_x + \frac{\partial M_{xy}}{\partial y}$, and along the edge where Q_y, acts, the effective transverse Kirchhoff shear force resultant, V_y, is $V_y = Q_y + \frac{\partial M_{yx}}{\partial x}$. More details of the derivation of V_x and V_y are given by Timoshenko and Woinowsky-Krieger (1987).

The strains at any point x, y, z within the laminate are

$$\varepsilon_x = z\chi_x = z \sum_{m=1,3,5,\ldots}^{\infty} \sum_{n=1,3,5,\ldots}^{\infty} W_{mn} \left(\frac{m\pi}{a}\right)^2 \cos\left(\frac{m\pi x}{a}\right) \cos\left(\frac{n\pi y}{b}\right)$$

$$\varepsilon_y = z\chi_y = z \sum_{m=1,3,5,\ldots}^{\infty} \sum_{n=1,3,5,\ldots}^{\infty} W_{mn} \left(\frac{n\pi}{b}\right)^2 \cos\left(\frac{m\pi x}{a}\right) \cos\left(\frac{n\pi y}{b}\right) \tag{m}$$

$$\gamma_{xy} = z\chi_{xy} = -2z \sum_{m=1,3,5,\ldots}^{\infty} \sum_{n=1,3,5,\ldots}^{\infty} W_{mn} \left(\frac{m\pi}{a}\right)\left(\frac{n\pi}{b}\right) \sin\left(\frac{m\pi x}{a}\right) \sin\left(\frac{n\pi y}{b}\right)$$

Since the plate deflection, the internal stress and moment resultants and the strains within the laminate are related to the external load, one can find the value of the external load that will cause any of these quantities to be excessive. More examples of composite laminates under different loading conditions are provided in the textbook by Kollar and Springer (2003). The textbooks by Hyer (1996) and Herakovich (1998) are recommended for further reading.

13.8   Summary and Conclusion

Our purpose in this chapter is to provide an introduction to analyze thin fiber reinforced laminates that are increasingly used in the Aerospace industry. The simplest theory to analyze panels, classical lamination thin plate theory (CLT), has been introduced and some example problems have been done to illustrate the solution process of the displacement equations of equilibrium.

14

Buckling

14.1 Introduction

We have assumed up to now that the differential equations for the bending displacement of slender bars subjected to lateral loads are uncoupled from the effects of axial loads. We have made a similar assumption for the effects of lateral and in plane loads in thin plates. This is not always the case. When slender bar and thin plate structures are subjected to low levels of axial and in plane stresses they behave as described by the various equations considered in the previous chapters. When the axial or in plane compressive loads and therefore the compressive stresses become large enough the bending displacements are substantially greater for any given set of lateral loads. What this means is that in an infinitesimal element of the structure, considered in the deformed configuration, the in-plane loads can produce out-of-plane load resultants that are significant and of a magnitude that is non-negligible. This leads to another phenomenon referred to as *geometric instability* or *buckling*. This can cause a catastrophic failure at levels of stress well below the yield stress of the material. We shall study buckling of bars and thin walled structures in this chapter.

14.2 The Equations for a Beam with Combined Lateral and Axial Loading

Consider a beam with both lateral and axial loads as shown in Figure 14.2.1. The loading may include concentrated forces and concentrated moments as well. You will note that in the previous chapters the distributed forces appear in the governing equations while any concentrated forces are included via boundary conditions. In the following development the effect of axial concentrated forces as well as axial distributed forces will appear in the differential equation for the lateral displacement in bending. For this example the loads act only in the xy plane. The cross sectional area is oriented so that $I_{xy} = 0$.

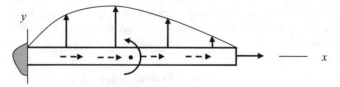

Figure 14.2.1

Analysis of Structures: An Introduction Including Numerical Methods, First Edition. Joe G. Eisley and Anthony M. Waas.
© 2011 John Wiley & Sons, Ltd. Published 2011 by John Wiley & Sons, Ltd.

Referring back to the derivation of the displacement equations for the axially loaded bar in Chapter 4 and the bending of beams in Chapter 7 we can combine the two effects into one set of equations for bending. Let us label the axial displacement of the beam centroidal axis as $u_A(x)$ to distinguish it from the total displacement in the x direction which includes the bending component. Keeping $v(x)$ as the label for the lateral displacement in bending under combined loading we can say that the total displacement in the x direction is

$$u(x, y) = u_A - y\frac{dv}{dx} \tag{14.2.1}$$

where it is understood that $u_A = u_A(x)$ and $v = v(x)$. From the strain displacement equations we get for moderately large rotations and small strains,

$$\varepsilon_x(x, y) = \frac{du}{dx} = \frac{du_A}{dx} - y\frac{d^2v}{dx^2} + \frac{1}{2}\left(\frac{dv}{dx}\right)^2 = \frac{du_A}{dx} + \frac{1}{2}\left(\frac{dv}{dx}\right)^2 - y\frac{d^2v}{dx^2} \tag{14.2.2}$$

Notice that we have now included extra terms (quadratic terms) in Equation 14.2.2 that were neglected before. This is because when the bending deflections are significant, the magnitude of the quadratic term, $\frac{1}{2}\left(\frac{dv}{dx}\right)^2$, is comparable to the other terms in the expression for $\varepsilon_x(x, y)$ in Equation 14.2.2. From Hooke's Law we have

$$\sigma_x(x, y) = E\varepsilon_x = E\left(\frac{du_A}{dx} + \frac{1}{2}\left(\frac{dv}{dx}\right)^2\right) - Ey\frac{d^2v}{dx^2} \tag{14.2.3}$$

Now we must establish equilibrium. Here we show the circumstances under which the axial forces produce significant out-of-plane force resultants. The bar will displace in both the x and and y directions as shown in Figure 14.2.2.

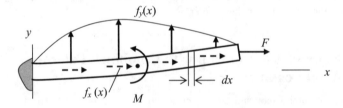

Figure 14.2.2

If we look at the equilibrium of a thin slice of this beam of width dx, that was originally in the undeformed state, then, due to the deformation, the slice gets rotated in the deflected position, as shown in Figure 14.2.3.

We have defined the internal axial force by the symbol N. N receives contributions from both the distributed and concentrated axial applied loads. In previous chapters we used the symbol P but now we want to reserve the symbol P for a compressive axial force to be consistent with the most common notation used in other texts.

From equilibrium in the x direction we get,

$$\sum F_x = 0 \quad \rightarrow \quad N(x + dx)\cos(\theta + d\theta) - N(x)\cos(\theta) + V(x)\sin(\theta)$$
$$- V(x + dx)\sin(\theta + d\theta) + f_x(x)dx = 0 \tag{14.2.4}$$

Where, we have used the Taylor expansion to express $\theta(x + dx) \approx \theta(x) + \theta'(x)dx = \theta + d\theta$. Let θ be small, so that $\cos\theta \approx 1$, and $\sin\theta \approx \theta$. These approximations specify the smallness of θ. Note that

$$\theta(x) = \frac{dv(x)}{dx} \tag{14.2.5}$$

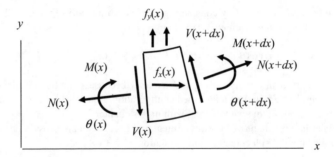

Figure 14.2.3

With these approximations, Equation 14.2.4 reduces to

$$N(x+dx) - N(x) + V(x)\theta - V(x+dx)(\theta + d\theta) = f(x) = 0 \qquad (14.2.6)$$

This further simplifies to

$$\frac{dN}{dx} + f_x(x) - \frac{d(V\theta)}{dx} = 0 \qquad (14.2.7)$$

Now, compared to $N(x)$, $V(x)\theta$ is a second order quantity. Therefore, (14.2.7) finally simplifies to,

$$\frac{dN}{dx} = -f_x(x) \qquad (14.2.8)$$

This is the same as Equations 4.3.6 in Chapter 4 where only axial loads in the x direction are applied. From equilibrium in the y direction we get

$$\sum F_y = 0 \quad \rightarrow \quad V(x+dx)\cos(\theta + d\theta) - V(x)\cos(\theta) - N(x)\sin(\theta)$$
$$+ N(x+dx)\sin(\theta + d\theta) + f_y(x)\,dx = 0$$

Using the same simplifications as in the axial case we get

$$\frac{dV}{dx} + \frac{d(N\theta)}{dx} + f_y(x) = 0 \qquad (14.2.9)$$

This is similar to the first of Equations 7.3.7 in Chapter 7, where only lateral loads in the y direction are applied, but now notice that because we satisfy equilibrium in the deformed configuration, the axial load produces a vertical resultant (the second term in Equation 14.2.9), which would not be there if we considered equilibrium in the undeformed configuration.

Finally, from summation of moments about the z axis and at the left hand face we get

$$\sum M_z = 0 \quad \rightarrow \quad [V(x+dx)\cos(\theta + d\theta)]\,dx + [N(x+dx)\sin(\theta + d\theta)]\,dx$$
$$+ [V(x+dx)\sin(\theta + d\theta) - N(x+dx)\cos(\theta + d\theta)]\frac{dv}{dx}\,dx \qquad (14.2.10)$$
$$+ M(x+dx) - M(x) + [f_y dx]\frac{dx}{2} = 0$$

When simplified this becomes

$$\frac{dM}{dx} = -V(x) \qquad (14.2.11)$$

Unlike in the beam theory presented in Chapter 7 we will now need a new expression for the vertical force resultant on a cross section because the internal resultant forces N and V are now rotated. Such an expression can be easily obtained by considering the forces marked in Figure 14.2.3. If we call the vertical (in the y direction) force resultant $\hat{V}(x)$, then

$$\hat{V} = V(x)\cos\theta + N(x)\sin\theta = V(x) + N(x)\theta = V + N\frac{dv}{dx} \qquad (14.2.12)$$

This term did not appear in the derivations in Chapter 7 because $\hat{V}$ and V are identical in the case where we consider equilibrium in the undeformed configuration. Equations 14.2.8-9 and 14.2.11 are the three equations of equilibrium that we will deal with in this chapter.

Since the moment is a resultant of the stress, integrated over the beam cross-section area, it can be given in terms of the lateral displacement. Using Equation 14.2.3 we get

$$M(x) = -\int_A \sigma_x y\, dA = -\int_A \left(E\left(\frac{du_A}{dx} + \frac{1}{2}\left(\frac{dv}{dx}\right)^2\right) - Ey\frac{d^2v}{dx^2}\right) y\, dA$$
$$= E\frac{d^2v}{dx^2}\int_A y^2\, dA = EI_{zz}\frac{d^2v}{dx^2} \qquad (14.2.13)$$

and the axial force resultant in terms of displacement is

$$N(x) = \int_A \sigma_x\, dA = \int_A \left(E\left(\frac{du_A}{dx} + \frac{1}{2}\left(\frac{dv}{dx}\right)^2\right) - Ey\frac{d^2v}{dx^2}\right) dA$$
$$= EA\left(\frac{du_A}{dx} + \frac{1}{2}\left(\frac{dv}{dx}\right)^2\right) \qquad (14.2.14)$$

The uniform axial stress across the cross section does not contribute to the moment, and the linear stress through the thickness does not contribute to the axial force resultant.

$$\int_A E\left(\frac{du_A}{dx} + \frac{1}{2}\left(\frac{dv}{dx}\right)^2\right) y\, dA = E\left(\frac{du_A}{dx} + \frac{1}{2}\left(\frac{dv}{dx}\right)^2\right)\int_A y\, dA = 0 \quad \rightarrow \quad \int_A y\, dA = 0 \quad (14.2.15)$$

Combining Equations 14.2.11-13 we get

$$V(x) = -\frac{dM}{dx} = -\frac{d}{dx}EI_{zz}\frac{d^2v}{dx^2} \quad \rightarrow \quad \hat{V}(x) = -\frac{d}{dx}EI_{zz}\frac{d^2v}{dx^2} + N\frac{dv}{dx} \qquad (14.2.16)$$

By combining Equation 14.2.9 with Equation 14.2.11 we have

$$-\frac{d^2M}{dx^2} + \frac{d}{dx}\left(N\frac{dv}{dx}\right) + f_y(x) = 0$$
$$\rightarrow \quad \frac{d^2}{dx^2}\left(EI_{zz}\frac{d^2v}{dx^2}\right) - \frac{d}{dx}\left(N\frac{dv}{dx}\right) = f_y(x) \qquad (14.2.17)$$

This shows that the internal axial force does affect the lateral displacement. In the statically determinate case N is found directly by applying equilibrium. For statically indeterminate cases the second order equation for axial displacement must be solved simultaneously with Equation 14.2.17 since the equations for $u_A(x)$ and $v(x)$ are coupled. That is, solve

$$\frac{d}{dx}EA\left(\frac{du_A}{dx} + \frac{1}{2}\left(\frac{dv}{dx}\right)^2\right) = -f_x(x) \quad \rightarrow \quad N(x) = EA\left(\frac{du_A}{dx} + \frac{1}{2}\left(\frac{dv}{dx}\right)^2\right) \qquad (14.2.18)$$

When EI_{zz} is constant Equation 14.2.17 may be written as,

$$EI_{zz}\frac{d^4v}{dx^4} - \frac{d}{dx}\left(N\frac{dv}{dx}\right) = f_y(x) \quad \rightarrow \quad \frac{d^4v}{dx^4} - \frac{1}{EI_{zz}}\frac{d}{dx}\left(N\frac{dv}{dx}\right) = \frac{f_y(x)}{EI_{zz}} \qquad (14.2.19)$$

Further, when the axial distributed load, $f_x(x) = 0$, then, from the axial equilibrium equation, it follows that $N(x) = \text{constant} = N$, say, which further reduces (14.2.19) to

$$EI_{zz}\frac{d^4v}{dx^4} - N\frac{d^2v}{dx^2} = f_y(x) \quad \rightarrow \quad \frac{d^4v}{dx^4} - \frac{N}{EI_{zz}}\frac{d^2v}{dx^2} = \frac{f_y(x)}{EI_{zz}} \tag{14.2.20}$$

In summary

$$M_c, F_c \;\rightarrow\; f_y(x) \quad N(x) \quad V(x) \quad M(x) \quad \sigma_x(x,y) \quad \varepsilon_x(x,y) \quad u(x,y) \quad v(x) \leftarrow \rho$$

$$\frac{dv}{dx} + \frac{d}{dx}\left(N\frac{dv}{dx}\right) + f_y(x) = 0 \qquad \frac{dM}{dx} = -V(x) \qquad\qquad u(x,y) = u_A - y\frac{dv}{dx}$$

$$\frac{dN}{dx} = -f_x(x) \qquad\qquad\qquad\qquad \varepsilon_x(x,y) = \frac{du_A}{dx} + \frac{1}{2}\left(\frac{dv}{dx}\right)^2 - y\frac{d^2v}{dx^2}$$

$$\sigma_x(x,y) = E\varepsilon_x = E\left(\frac{du_A}{dx} + \frac{1}{2}\left(\frac{dv}{dx}\right)^2\right) - Ey\frac{d^2v}{dx^2}$$

$$N(x) = \int_A \sigma_x dA = EA\left(\frac{du_A}{dx} + \frac{1}{2}\left(\frac{dv}{dx}\right)^2\right) \tag{14.2.21}$$

$$M(x) = -\int_A \sigma_x y\,dA = E\frac{d^2v}{dx^2}\int_A y^2\,dA = EI_{zz}\frac{d^2v}{dx^2} \quad\rightarrow\quad \sigma_x = \frac{N(x)}{A} - \frac{M(x)y}{I_{zz}}$$

$$\frac{dM}{dx} = \frac{d}{dx}EI_{zz}\frac{d^2v}{dx^2}; \quad \hat{V}(x) = -\frac{d}{dx}EI_{zz}\frac{d^2v}{dx^2} + N\frac{dv}{dx} \quad\rightarrow\quad \tau_{xy}(x,y) = \frac{\hat{V}}{I_{zz}h}\int_{A_y} y\,dA$$

$$\frac{d^2}{dx^2}EI_{zz}\frac{d^2v}{dx^2} - \frac{d}{dx}\left(N\frac{dv}{dx}\right) = f_y(x)$$

The summary in Equation 14.2.21 looks much the same as Equation 7.3.17 in Chapter 7 but with the changes in force equilibrium. This results in an additional term in the definition of shear force (vertical force resultant), $\hat{V}(x)$, Equation 14.2.16, and in the final equation for displacements, Equation 14.2.17. Note that the distributed applied lateral load, $f_y(x)$, is explicit in the displacement equation but the distributed applied axial loads, $f_x(x)$ are implicit in the value of N. Furthermore, concentrated applied lateral loads appear in the boundary conditions while concentrated applied axial loads are also implicit in the value of N.

For the next few examples we will limit discussion to applied axial forces that are not functions of x, that is, $f_x(x) = 0$; however, concentrated axial forces do exist. Distributed applied axial forces would result in distributed internal axial forces that are function of x, that is, $N(x)$. This would create differential equations with non constant coefficients and this presents exceptional difficulty in attempting solutions. For the same reason we shall consider only cases where EI_{zz} is constant. We deal with situations in which $N(x)$ and/or $EI_{zz}(x)$ by other means later in this chapter. In those cases where distributed axial forces are absent (but concentrated forces can still be present) and when EI_{zz} is constant, we find that the equations of equilibrium can be uncoupled and the solution becomes much easier, because the internal axial force, N, is constant. This is easily seen if we put $f_x(x) = 0$, in Equation 14.2.8, then the axial equilibrium equation admits the solution, $N = \text{constant}$, and using this fact in Equation 14.2.19, with EI_{zz} constant, we have an ordinary differential equation with constant coefficients.

$$\frac{d^4v}{dx^4} - \frac{N}{EI_{zz}}\frac{d^2v}{dx^2} = \frac{f_y(x)}{EI_{zz}} \tag{14.2.22}$$

This has a well known solution. Of special interest is the case where the lateral load, $f_y(x)$, is zero and the axial load is compressive. Such a structure is commonly called a *column*. We consider this in the next section.

14.3   Buckling of a Column

Consider the case of a uniform beam with no lateral load but a constant axial compressive load that produces a constant internal force $N = -P$. Equation 14.2.22 becomes

$$\frac{d^4v}{dx^4} + \frac{P}{EI_{zz}} \frac{d^2v}{dx^2} = 0 \tag{14.3.1}$$

For convenience let

$$\lambda^2 = \frac{P}{EI_{zz}} \tag{14.3.2}$$

then

$$\frac{d^4v}{dx^4} + \lambda^2 \frac{d^2v}{dx^2} = 0 \tag{14.3.3}$$

This equation has a very interesting solution. Either $v(x) = 0$ or λ takes on a series of discrete non zero values for which a solution other than $v(x) = 0$ is possible. In the study of differential equations this is called an *eigenvalue* equation. The discrete values of λ imply discrete values of P. The lowest non zero value of P is called the *buckling load*. Furthermore for each discrete value of λ there is a non zero solution for $v(x)$ which is called the *eigenvector* or *mode shape*. We shall now develop this solution.

Ordinary differential equations with constant coefficients have solutions of the form

$$v = ce^{\alpha x} \tag{14.3.4}$$

If we substitute this into Equation 14.3.3 we get

$$(\alpha^4 + \lambda^2\alpha^2) ce^{\alpha x} = 0 \quad \rightarrow \quad \alpha^4 + \lambda^2\alpha^2 = 0 \tag{14.3.5}$$

This equation has four roots as follows

$$\alpha_1 = i\lambda \qquad \alpha_2 = -i\lambda \qquad \alpha_3 = 0 \qquad \alpha_4 = 0 \tag{14.3.6}$$

where $i = \sqrt{-1}$.

The solution to Equation 14.3.3 is

$$v(x) = ae^{i\lambda x} + be^{-i\lambda x} + cx + d \tag{14.3.7}$$

or since

$$e^{i\lambda x} = \cos\lambda x + i\sin\lambda x \qquad e^{-i\lambda x} = \cos\lambda x - i\sin\lambda x$$

this may be rewritten as

$$\begin{aligned}
v(x) &= ae^{i\lambda x} + be^{-i\lambda x} + cx + d \\
&= a(\cos\lambda x + i\sin\lambda x) + b(\cos\lambda x - i\sin\lambda x) + cx + d \\
&= (a+b)\cos\lambda x + i(a-b)\sin\lambda x + cx + d \\
&= A\sin\lambda x + B\cos\lambda x + Cx + D
\end{aligned} \tag{14.3.8}$$

The values of these constants of integration are obtained from the boundary conditions.

For a typical column with two boundary conditions at each end the result is a set of four homogeneous algebraic equations containing the four constants of integration. These can be written in matrix form where the coefficients of the constants of integrations are just shown as a 4×4 matrix.

$$[4x4] \begin{bmatrix} A \\ B \\ C \\ D \end{bmatrix} = \begin{bmatrix} 0 \\ 0 \\ 0 \\ 0 \end{bmatrix} \tag{14.3.9}$$

One possible solution is that

$$\begin{bmatrix} A \\ B \\ C \\ D \end{bmatrix} = \begin{bmatrix} 0 \\ 0 \\ 0 \\ 0 \end{bmatrix} \tag{14.3.10}$$

but this means there is no lateral displacement. It so happens that there is another possibility under certain circumstances. If the determinant of the coefficient matrix is zero then it is possible that the constants of integration are not all zero and a lateral displacement occurs, that is, if

$$|[4x4]| = 0 \tag{14.3.11}$$

If we set the determinant to zero we can expand it to obtain an equation in terms of the unknown λ. The roots of this equation have interesting meaning, namely, the values of the applied axial compressive load for which a solution other than no displacement is possible. As noted above problems of this type are called *eigenvalue problems*. The roots are called *eigenvalues* and the corresponding loads are called *critical buckling loads*.

We shall find that three of the four constants of integration can be found but a fourth one is always indeterminate. Thus we can define the shape of the displacement but not its amplitude.

This is best explored by examples.

###########

Example 14.3.1

Problem: A uniform column has both ends pinned, or simply supported, and a constant axial compressive load as shown in Figure (a). Find the buckling load and mode.

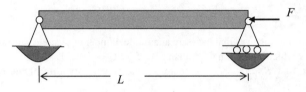

Figure (a)

Solution: Insert Equation 14.3.8 into the four boundary conditions to obtain the four homogenius equations. Solve for the lowest buckling load and its mode shape.

The four required boundary conditions are

$$v(0) = 0$$

$$M(0) = EI_{zz}\frac{d^2v(0)}{dx^2} = 0$$

$$v(L) = 0 \tag{a}$$

$$M(L) = EI_{zz}\frac{d^2v(L)}{dx^2} = 0$$

Substitute the solution, Equation 14.3.8, into each of the boundary conditions and you get

$$v(0) = A\sin\lambda 0 + B\cos\lambda 0 + C\cdot 0 + D = B + D = 0 \tag{b}$$

$$\frac{d^2v(0)}{dx^2} = -A\lambda^2\sin\lambda 0 - B\lambda^2\cos\lambda 0 = -B\lambda^2 = 0 \tag{c}$$

$$v(L) = A\sin\lambda L + B\cos\lambda L + CL + D = 0 \tag{d}$$

$$\frac{d^2v(L)}{dx^2} = -A\lambda^2\sin\lambda L - B\lambda^2\cos\lambda L = 0 \tag{e}$$

In matrix form

$$\begin{bmatrix} 0 & 1 & 0 & 1 \\ 0 & -\lambda^2 & 0 & 0 \\ \sin\lambda L & \cos\lambda L & L & 1 \\ -\lambda^2\sin\lambda L & -\lambda^2\cos\lambda L & 0 & 0 \end{bmatrix}\begin{bmatrix} A \\ B \\ C \\ D \end{bmatrix} = \begin{bmatrix} 0 \\ 0 \\ 0 \\ 0 \end{bmatrix} \tag{f}$$

so we examine under what conditions a non zero displacement is possible by setting the determinant of the coefficient matrix equal to zero.

$$\begin{vmatrix} 0 & 1 & 0 & 1 \\ 0 & -\lambda^2 & 0 & 0 \\ \sin\lambda L & \cos\lambda L & L & 1 \\ -\lambda^2\sin\lambda L & -\lambda^2\cos\lambda L & 0 & 0 \end{vmatrix} = 0 \tag{g}$$

We can expand the determinant; however, in this case you can simplify it by noting that from Equation (c) you get $B = 0$; then from the Equation (b) you get $D = 0$. The Equations (d) and (e) reduce to

$$A\sin\lambda L + CL = 0 \qquad A\sin\lambda L = 0 \tag{h}$$

If $A = 0$ then $C = 0$ and there would be no displacement; but there is another possibility, namely, that $\sin\lambda L = 0$. If this is true then $C = 0$; however, A is indeterminate. Under what circumstances is $\sin\lambda L = 0$? When

$$\lambda L = n\pi \quad \rightarrow \quad P = \left(\frac{n\pi}{L}\right)^2 EI \tag{i}$$

where n is an integer. This suggests that there are an infinite number of discrete values of P, one for each integer value of n, for which a solution other than $A = B = C = D = 0$ exists.

We did determine that $B = C = D = 0$ therefore the general solution of the equation for these boundary conditions is

$$v(x) = A\sin\frac{n\pi}{L}x \tag{j}$$

We know the shape of the deflection but the amplitude A is indeterminate. This shape is called the *mode shape*. We simply do not have enough information to determine the amplitude of the displacement.

In practice the lowest positive non zero value, that is, $n = 1$, is of most interest. As a load is slowly applied once it reaches the lowest value of the buckling load the damage has been done. So we usually just say that the *critical buckling load* and the corresponding *critical buckling stress* are

$$P_{cr} = \frac{\pi^2}{L^2} EI \qquad \sigma_{cr} = -\frac{P}{A} = -\frac{\pi^2}{AL^2} EI \tag{k}$$

Experiments bear this out. Real columns will deflect catastrophically when this load is reached. The mode shape is a simple half sine wave.

$$v(x) = A \sin \frac{\pi}{L} x \tag{l}$$

and is plotted in Figure (b). We arbitrarily assign the value $A = 1$ for plotting purposes.

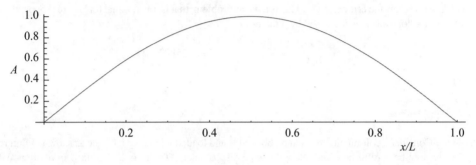

Figure (b)

Example 14.3.2

Problem: Pin jointed trusses are made up of an assembly of slender bars simply supported at the ends. Those members in compression may be candidates for buckling. Consider the truss in Example 4.10.1. What value of F would cause one of the members of the truss to buckle? The truss is shown here in Figure (a). The cross section is a 20 *mm* × 20 *mm* square.

$$A_1 = A_2 = 400 \ mm^2 \qquad F = 10000 \ N \qquad E = 206840 \ MPa. \tag{a}$$

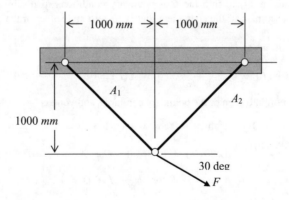

Figure (a)

Solution: Increase the load until a member buckles.

We know from Example 4.10.1 that member 2 is in compression. The critical buckling load in member 2 would be

$$P_{cr} = \frac{\pi^2}{L^2}EI = \left(\frac{\pi}{1414}\right)^2 \cdot 206800 \cdot \frac{20^4}{12} = 13611N \qquad (b)$$

The corresponding compressive stress, called the *critical stress*, σ_{cr}, is

$$\sigma_{cr} = -\frac{P_{cr}}{A} = -34.03 \, MPa \qquad (c)$$

The actual stress found in Example 4.10.1 is

$$\sigma_2 = -6.471 \, MPa \qquad (d)$$

This is well below the critical stress. This is a linear problem, that is, the internal forces are proportional to the external loads. If F is increased by a factor of

$$\frac{34.03}{6.471} = 5.259 \quad \text{or} \quad F = 52588N \qquad (e)$$

the member will buckle.

Example 14.3.3

Problem: Consider a column that is fixed on both ends and loaded with an axial force as shown in Figure (a). We have to be careful here and allow the axial displacement. The right end is fixed in rotation and lateral deflection but axial displacement is permitted. Find the buckling load and mode.

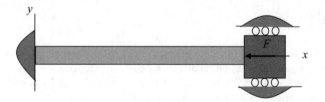

Figure (a)

Solution: Insert Equation 14.3.8 into the four boundary conditions to obtain the four homogenius equations. Set the determinant of the coefficient matrix to zero and solve for the critical buckling load and mode.

The boundary conditions are

$$v(0) = 0 \qquad \frac{dv(0)}{dx} = 0 \qquad v(L) = 0 \qquad \frac{dv(L)}{dx} = 0 \qquad (a)$$

Substitute the solution into each of the boundary condtions and you get

$$v(0) = A \sin \lambda 0 + B \cos \lambda 0 + C \cdot 0 + D = B + D = 0 \qquad (b)$$

$$\frac{dv(0)}{dx} = A\lambda \cos \lambda 0 - B\lambda \sin \lambda 0 + C = A\lambda + C = 0 \qquad (c)$$

$$v(L) = A \sin \lambda L + B \cos \lambda L + CL + D = 0 \qquad (d)$$

$$\frac{dv(0)}{dx} = A\lambda \cos \lambda L - B\lambda \sin \lambda L + C = 0 \qquad (e)$$

This can be presented as a set of four equations in four unknowns in matrix equation form.

$$
\begin{bmatrix}
0 & 1 & 0 & 1 \\
\lambda & 0 & 1 & 0 \\
\sin \lambda L & \cos \lambda L & L & 1 \\
\lambda \cos \lambda L & -\lambda \sin \lambda L & 1 & 0
\end{bmatrix}
\begin{bmatrix}
A \\ B \\ C \\ D
\end{bmatrix}
=
\begin{bmatrix}
0 \\ 0 \\ 0 \\ 0
\end{bmatrix}
\tag{f}
$$

Again an obvious solution is

$$
\begin{bmatrix}
A \\ B \\ C \\ D
\end{bmatrix}
=
\begin{bmatrix}
0 \\ 0 \\ 0 \\ 0
\end{bmatrix}
\tag{g}
$$

Another possible solution is that

$$
\begin{vmatrix}
0 & 1 & 0 & 1 \\
\lambda & 0 & 1 & 0 \\
\sin \lambda L & \cos \lambda L & L & 1 \\
\lambda \cos \lambda L & -\lambda \sin \lambda L & 1 & 0
\end{vmatrix}
= 0
\tag{h}
$$

We can simplify the equations and reduce this to a 2×2 determinant by noting the $D = -B$ and $C = -A\lambda$ from Equations (b) and (c). Equations (d) and (e) can then be written

$$
A (\sin \lambda L - \lambda L) + B (\cos \lambda L - 1) = 0 \tag{i}
$$
$$
A\lambda (\cos \lambda L - 1) - B\lambda \sin \lambda L = 0 \tag{j}
$$

and the determinant becomes

$$
\begin{vmatrix}
\sin \lambda L - \lambda L & \cos \lambda L - 1 \\
\cos \lambda L - 1 & \sin \lambda L
\end{vmatrix}
= 0
\tag{k}
$$

If we expand the determinant we get

$$
\lambda L \sin \lambda L + 2(\cos \lambda L - 1) = 0 \tag{l}
$$

By inspection the lowest positive non zero root is found to be

$$
\lambda L = 2\pi \quad \rightarrow \quad P_{cr} = \left(\frac{2\pi}{L} \right)^2 EI \tag{m}
$$

Substituting $\lambda L = 2\pi$ back into the equations we get

$$
\begin{aligned}
A (\sin 2\pi - 2\pi) + B (\cos 2\pi - 1) = 0 &\quad \rightarrow \quad -2\pi A + B \cdot 0 = 0 \quad \rightarrow \quad A = 0 \\
A\lambda (\cos 2\pi - 1) - B\lambda \sin 2\pi = 0 &\quad \rightarrow \quad 0 = 0
\end{aligned}
\tag{n}
$$

If $A = 0$ then from Equation (c) $C = 0$ and since $D = -B$ from Equation (b) the mode shape is found to be

$$
v(x) = B \left(\cos \frac{2\pi x}{L} - 1 \right) \tag{o}
$$

The mode shape is shown in Figure (b) with a value $B = -0.5$ assigned for plotting purposes. This plots the maximum amplitude as 1.0. Remember that the actual amplitude is indeterminate.

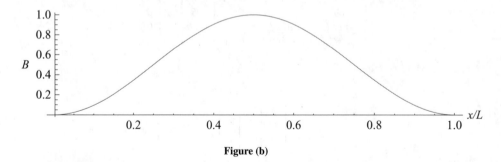

Figure (b)

Example 14.3.4

Problem: Now consider a column fixed on one end and free on the other as shown in Figure (a). Find the buckling load and mode.

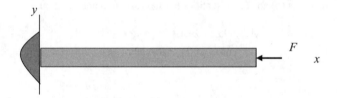

Figure (a)

Solution: Insert Equation 14.3.8 into the four boundary conditions to obtain the four homogenius equations. Expand the determinant of the coefficients and solve for the lowest buckling load and mode.

With one exception the boundary conditions are the same as discussed in Chapter 7 for a cantilever beam. We specify the displacement and slope on the left end and the moment and shear force on the right end. The exception is that the shear force applied at the right end is now given by

$$EI_{zz}\frac{d^3 v}{dx^3} + P\frac{dv}{dx} = -V \tag{a}$$

The boundary conditions are

$$v(0) = 0$$

$$\frac{dv(0)}{dx} = 0$$

$$M(L) = EI_{zz}\frac{d^2 v(L)}{dx^2} = 0 \tag{b}$$

$$-V(L) = EI_{zz}\frac{d^3 v(L)}{dx^3} + P\frac{dv(L)}{dx} = 0$$

Substituting we get

$$v(0) = A \sin \lambda 0 + B \cos \lambda 0 + C \cdot 0 + D = B + D = 0 \tag{c}$$

$$\frac{dv(0)}{dx} = A\lambda \cos \lambda 0 - B\lambda \sin \lambda 0 + C = A\lambda + C = 0 \tag{d}$$

$$\frac{d^2 v(L)}{dx^2} = -A\lambda^2 \sin \lambda L - B\lambda^2 \cos \lambda L = 0 \tag{e}$$

$$\frac{d^3 v(L)}{dx^3} + \lambda^2 \frac{dv(L)}{dx}$$
$$= -A\lambda^3 \cos \lambda L + B\lambda^3 \sin \lambda L + \lambda^2 (A\lambda \cos \lambda L - B\lambda \sin \lambda L + C) \tag{f}$$
$$= C\lambda^2 = 0$$

We can set up the determinant of the coefficients or in this case note that $C = 0$ from Equation (f) and therefore $A = 0$ from Equation (d). We are left with

$$B\lambda^2 \cos \lambda L = 0 \qquad B + D = 0 \tag{g}$$

from Equations (c) and (e). Either $B = 0$ and thus $D = 0$ or $\cos \lambda L = 0$. The condition for $\cos \lambda L = 0$ to be true is for

$$\lambda L = \frac{(2n + 1)\pi}{2} \qquad n = 0, 1, 2, \ldots \tag{h}$$

In this case the lowest critical buckling load is for $n = 0$, or

$$P_{cr} = \left(\frac{\pi}{2L}\right)^2 EI_{zz} \tag{i}$$

The mode shape is

$$v(x) = B \cos \frac{\pi x}{2L} + D = B \left(\cos \frac{\pi x}{2L} - 1\right) \tag{j}$$

The mode shape is shown in Figure (b) with a value $B = -1$ assigned for plotting purposes.

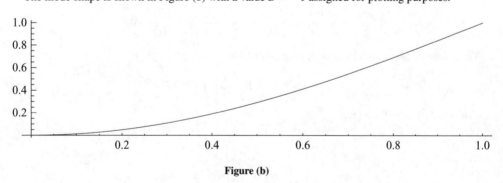

Figure (b)

###########

We can go on giving examples, including multiple compressive loads at different points along the beam and variable cross sections. As the loading gets more complicated there is a better way. This will be shown in Section 14.5.

14.4   The Beam Column

When both axial and lateral loads are present and the compressive axial load is well below the buckling load the axial and bending effects are nearly uncoupled. As the axial load approaches the buckling load we can expect the behavior to be quite different. This structure is sometimes called a *beam column*. The differential equation for the case with both a compressive axial load and a lateral load is

$$\frac{d^4v}{dx^4} + \lambda^2 \frac{d^2v}{dx^2} = \frac{f_y(x)}{EI_{zz}}$$ (14.4.1)

Let us examine what happens with a simple example.

###########

Example 14.4.1

Problem: Find the displacement and stresses in the beam in Figure (a). Let the axial load vary from a small to a large value approaching the buckling load.

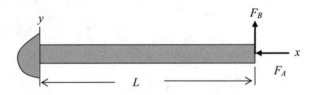

Figure (a)

Solution: Solve Equation 14.2.22 for a range of values of the axial load. Adapt Equations 14.2.20 to this problem.

$$\frac{d^4v}{dx^4} - \frac{N}{EI}\frac{d^2v}{dx^2} = \frac{f_y(x)}{EI} \quad \rightarrow \quad \frac{d^4v}{dx^4} + \lambda^2 \frac{d^2v}{dx^2} = 0$$ (a)

The solution is given in Section 14.3, Equation 14.3.8.

$$v(x) = A \sin \lambda x + B \cos \lambda x + Cx + D$$ (b)

The boundary conditions are

$$v(0) = 0$$

$$\frac{dv(0)}{dx} = 0$$

$$M(L) = EI\frac{d^2v(L)}{dx^2} = 0$$ (c)

$$-V(L) = \frac{d^3v(L)}{dx^3} + \lambda^2\frac{dv(L)}{dx} = -\frac{F_B}{EI}$$

Substituting we get

$$v(0) = A \sin \lambda 0 + B \cos \lambda 0 + C \cdot 0 + D = B + D = 0 \tag{d}$$

$$\frac{dv(0)}{dx} = A\lambda \cos \lambda 0 - B\lambda \sin \lambda 0 + C = A\lambda + C = 0 \tag{e}$$

$$\frac{d^2 v(L)}{dx^2} = -A\lambda^2 \sin \lambda L - B\lambda^2 \cos \lambda L = 0 \tag{f}$$

$$\frac{d^3 v(L)}{dx^3} + \lambda^2 \frac{dv(L)}{dx} = -\frac{V(L)}{EI}$$

$$= -A\lambda^3 \cos \lambda L + B\lambda^3 \sin \lambda L + \lambda^2 (A\lambda \cos \lambda L - B\lambda \sin \lambda L + C)$$

$$= C\lambda^2 = -\frac{F_B}{EI} \tag{g}$$

From Equation (g) we have

$$C = -\frac{F_B}{\lambda^2 EI} = -\frac{F_B EI}{PEI} = -\frac{F_B}{P} \tag{h}$$

From the Equation (e) we have

$$A\lambda + C = A\lambda - \frac{F_B}{P} = 0 \quad \rightarrow \quad A = \frac{F_B}{\lambda P} \tag{i}$$

From the Equation (f) we have

$$A \sin \lambda L + B \cos \lambda L = B \cos \lambda L + \frac{F_B}{\lambda P} \sin \lambda L = 0 \quad \rightarrow \quad B = -\frac{F_B}{\lambda P} \tan \lambda L \tag{j}$$

Finally from Equation (d) we have

$$B + D = 0 \quad \rightarrow \quad D = -B = \frac{F_B}{\lambda P} \tan \lambda L \tag{k}$$

And so the solution is

$$v(x) = A \sin \lambda x + B \cos \lambda x + Cx + D$$

$$= \frac{F_B}{\lambda P} \sin \lambda x - \frac{F_B}{\lambda P} \tan \lambda L \cos \lambda x - \frac{F_B}{P} x + \frac{F_B}{\lambda P} \tan \lambda L \tag{l}$$

$$= \frac{F_B}{\lambda P} \sin \lambda x + \frac{F_B}{\lambda P} \tan \lambda L \, (1 - \cos \lambda x) - \frac{F_B}{P} x$$

The lateral tip deflection is

$$v(L) = \frac{F_B}{\lambda P} \sin \lambda L + \frac{F_B}{\lambda P} \tan \lambda L \, (1 - \cos \lambda L) - \frac{F_B}{P} L$$

$$= \frac{F_B}{\lambda P} \sin \lambda L + \frac{F_B}{\lambda P} \tan \lambda L - \frac{F_B}{\lambda P} \sin \lambda L - \frac{F_B}{P} L \tag{m}$$

$$= \frac{F_B}{P} \left(\frac{1}{\lambda} \tan \lambda L - L \right)$$

The value of λ and the buckling load when F_B is zero are (see Example 14.3.4)

$$\lambda = \frac{\pi}{2L} \quad \rightarrow \quad P_{cr} = \left(\frac{\pi}{2L}\right)^2 EI_{zz} \tag{n}$$

At $P = 0$, that is, when there is no axial load, the value of the tip deflection as presented in Equation (m) is indeterminate. Through a limiting process you can find the value, but, of course, from Example 7.3.2, Equation (d) we know the deflection is

$$v(x) = \frac{F_B}{EI}\left(\frac{Lx^2}{2} - \frac{x^3}{6}\right) \quad \rightarrow \quad v(L) = \frac{F_B L^3}{3EI} \tag{o}$$

So now let us plot the value of the tip deflection versus the value of λL from zero to the buckling load. For plotting purposes we format the tip deflection as follows.

$$v(L) = \frac{F_B}{P}\left(\frac{1}{\lambda}\tan\lambda L - L\right) = \frac{F_B}{\lambda P}(\tan\lambda L - \lambda L) = \frac{F_B L^3}{EI}\frac{(\tan\lambda L - \lambda L)}{(\lambda L)^3} \tag{p}$$

Plot $v(L)$ versus λL in Figure (b).

It is quite clear that as the axial load approaches the buckling load the lateral displacement grows rapidly.

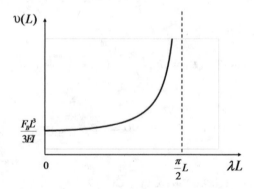

Figure (b)

To complete the plot we ask what happens when the axial force is tension. Then the equation is

$$\frac{d^4v}{dx^4} - \lambda^2\frac{d^2v}{dx^2} = 0 \tag{q}$$

The solution is straight forward following the same steps that we just presented. We shall not go through them here. Suffice it to say that the effect is a small reduction in the amplitude of the lateral displacement caused by the tensile axial force. The moment generated by the axial force times the lateral displacement is very small compared to the moment generated by the lateral force and its moment arm. The plot in Figure (b) has been extended downward to show this.

##########

We can continue to consider other loading and boundary conditions as we have done for beams in Chapter 7. We can also extend the analysis to beams with different internal forces in different regions by writing equations for each region and matching boundary conditions where regions meet. This becomes quite time consuming and tedious so we will defer these problems to the next section.

We can continue examples of beam column analysis to include more complicated cases including frames; however, the solutions become quite laborious. There is a better way which we consider in the next section.

14.5   The Finite Element Method for Bending and Buckling

We shall extend the FEM for cases that include the effect of internal axial forces on the lateral bending of beams. First we show the beam column form of these equations and then we simplify for buckling analysis.

When both axial and lateral loads are present the virtual work consists of the external work done by the loads minus the strain energy of bending. Again we look at the bar in the deflected position. Then

$$\delta W = \delta W_a^e + \delta W_l^e + \delta W^i = \delta W^e - \delta U \tag{14.5.1}$$

where δW_a^e is the external virtual work of the axial loads and δW_l^e is the external virtual work of the lateral loads.

Consider an element m with an axial compressive load and a set of lateral loads. The lateral bending displacement is represented by nodal displacements and a shape function.

$$v(s_m) = [n]\{r_m\} \tag{14.5.2}$$

When these loads are applied to an element m the strain energy of bending is precisely what we found before in Chapter 11, Equation 11.3.33.

$$U_m = \frac{1}{2}\int_V (\varepsilon_{xm}\sigma_{xm})dV = \frac{EI_m}{2}\int_0^{l_m}\left(\frac{d^2v}{ds^2}\right)^2 ds$$

$$= \frac{EI_m}{2}\int_0^{l_m}\left(\{r_m\}^T \frac{d^2}{ds^2}[n]^T\right)\left(\frac{d^2}{ds^2}[n]\{r_m\}\right)ds \tag{14.5.3}$$

and the virtual strain energy is given in Equation 11.3.34.

$$\delta U_m = EI_m\int_0^{l_m}\frac{d^2\delta v}{ds^2}\frac{d^2v}{ds^2}ds = EI_m\int_0^{l_m}\left(\{\delta r_m\}^T \frac{d^2}{ds^2}[n]^T\right)\left(\frac{d^2}{ds^2}[n]\{r_m\}\right)ds \tag{14.5.4}$$

Note that we do not include the strain energy of the axial stress. It is negligible when compared to that of bending and so has little effect on the solution.

Now consider a beam in the deflected position with an axial load as shown in Figure 14.5.1.

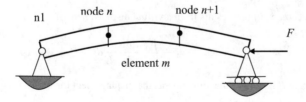

Figure 14.5.1

The applied axial force F provides an internal axial force P on element m. To find the external virtual work consider an axial compressive load on element m. The work done by the axial force P is equal to the force times the distance it moves as a result of the lateral displacement. For example in Figure 14.5.2 we can identify the shortening of the distance between the two ends of an element due to the lateral displacement. The undeflected element is shown with a dashed outline.

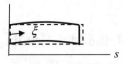

Figure 14.5.2

We isolate the element m and assign a coordinate ξ which follows the centerline of the curved beam and we retain coordinate s of the undeflected element.

We express the shortening of the distance on a small slice of the element as

$$du = ds - d\xi \tag{14.5.5}$$

where

$$d\xi = \sqrt{ds^2 + dv^2} = ds\sqrt{1 + \left(\frac{dv}{ds}\right)^2} \tag{14.5.6}$$

For small values of the slope this can be approximated by expanding in a series and neglecting higher order terms.

$$d\xi = ds\left[1 + \frac{1}{2}\left(\frac{dv}{ds}\right)^2\right] + \text{higher order terms} \tag{14.5.7}$$

Thus

$$du = ds - d\xi = ds - ds\left[1 + \frac{1}{2}\left(\frac{dv}{ds}\right)^2\right] = -\frac{1}{2}\left(\frac{dv}{ds}\right)^2 ds \tag{14.5.8}$$

The total axial displacement due to the lateral displacement for the element m is

$$u(l_m) = -\frac{1}{2}\int_0^{l_m}\left(\frac{dv}{ds}\right)^2 ds \tag{14.5.9}$$

The virtual displacement δu that results from a virtual displacement δv is then

$$\delta u(l_m) = -\int_0^{l_m}\frac{d\delta v}{ds}\frac{dv}{ds}ds \tag{14.5.10}$$

Since the axial force P is in the same direction as the displacement the external virtual work done by the axial force is

$$\delta W_{am}^e = P\delta u = P\int_0^{l_m}\frac{d\delta v}{ds}\frac{dv}{ds}ds = \{\delta r_m\}^T P\int_0^{l_m}\frac{d[n]^T}{ds}\frac{d[n]}{ds}ds\{r_m\} \tag{14.5.11}$$

The external virtual work done by the lateral forces is

$$\delta W_{lm}^e = \{\delta r_m\}^T \{p_m\} \tag{14.5.12}$$

The total virtual work is

$$\delta W = \{\delta r_m\}^T \{p_m\} + \{\delta r_m\}^T P \int_0^{lm} \frac{d\,[n]^T}{ds} \frac{d\,[n]}{ds} ds \{r_m\}$$

$$- \{\delta r_m\}^T EI_m \int_0^{lm} \frac{d^2\,[n]}{ds^2}^T \frac{d^2\,[n]}{ds^2} ds \{r_m\} = 0 \tag{14.5.13}$$

or

$$\{p_m\} = -P \int_0^{lm} \frac{d\,[n]^T}{ds} \frac{d\,[n]}{ds} ds \{r_m\}$$

$$+ EI_m \int_0^{lm} \frac{d^2\,[n]}{ds^2}^T \frac{d^2\,[n]}{ds^2} ds \{r_m\} = 0 \tag{14.5.14}$$

Note again that we can neglect the strain energy contributed by the axial stress and strain because its small compared to the bending strain energy, and we can neglect the work done by compressive axial displacement, but we include the external work done by the axial force due to lateral displacement.

This leaves us with

$$\left([k_m] - P\left[k_m^d\right]\right) \{r_m\} = \{p_m\} \tag{14.5.15}$$

where

$$[k_m] = EI_m \int_0^{lm} \frac{d^2\,[n]}{ds^2}^T \frac{d^2\,[n]}{ds^2} \{r_m\}\, ds \qquad \left[k_m^d\right] = \int_0^{lm} \frac{d\,[n]^T}{ds} \frac{d\,[n]}{ds}\, ds \tag{14.5.16}$$

For a finite element solution we could consider shape functions based on the general homogenous solution of the fourth order differential equation. It has been found that a simpler approach, namely, using the same shape functions that were used for the beam with lateral loading only, is more convenient. It is easier to obtain a desired level of accuracy by using more elements than it is to fuss over the accuracy of each element.

The shape functions for bending developed in Chapter 8 and summarized in Equation 10.2.30, repeated here, are

$$v(s_m) = [n]\{r_m\}$$

$$= \frac{1}{l_m^3}\left[l_m^3 - 3l_m s_m^2 + 2s_m^3 \quad l_m^3 s_m - 2l_m^2 s_m^2 + l_m s_m^3 \quad 3l_m s_m^2 - 2s_m^3 \quad -l_m^2 s_m^2 + l_m s_m^3\right] \begin{bmatrix} w_n \\ \theta_n \\ w_{n+1} \\ \theta_{n+1} \end{bmatrix} \tag{14.5.17}$$

These are not the solution to the differential equation for buckling but they meet the criteria for kinematically admissible shape functions.

As demonstrated in Chapter 11, Equations 11.3.38-41 we get for $[k_m]$ for element m

$$[k_m] = EI \int_0^{l_m} \frac{d^2[n]}{ds^2}^T \frac{d^2[n]}{ds^2} ds = \frac{E_m I_m}{l_m^3} \begin{bmatrix} 12 & 6l_m & -12 & 6l_m \\ 6l_m & 4l_m^2 & -6l_m & 2l_m^2 \\ -12 & -6l_m & 12 & -6l_m \\ 6l_m & 2l_m^2 & -6l_m & 4l_m^2 \end{bmatrix} \quad (14.5.18)$$

The matrix $\left[k_m^d\right]$ has various names. It is called the *differential stiffness matrix*, or the *geometric stiffness matrix*, or the *initial stress matrix,* among others.

Using the same shape functions we get

$$[k_m^d] = \int_0^{l_m} \frac{d}{ds}[n]^T \frac{d}{ds}[n] ds = \frac{1}{30 l_m} \begin{bmatrix} 36 & -3l_m & -36 & -3l_m \\ -3l_m & 4l_m^2 & 3l_m & -l_m^2 \\ -36 & 3l_m & 36 & 3l_m \\ -3l_m & -l_m^2 & 3l_m & 4l_m^2 \end{bmatrix} \quad (14.5.19)$$

For the element m we get

$$([k_m] - P[k_m^d])\{r_m\} = \left(\frac{E_m I_m}{l_m^3} \begin{bmatrix} 12 & 6l_m & -12 & 6l_m \\ 6l_m & 4l_m^2 & -6l_m & 2l_m^2 \\ -12 & -6l_m & 12 & -6l_m \\ 6l_m & 2l_m^2 & -6l_m & 4l_m^2 \end{bmatrix} \right.$$

$$\left. - \frac{P}{30 L_m} \begin{bmatrix} 36 & -3l_m & -36 & -3l_m \\ -3l_m & 4l_m^2 & 3l_m & -l_m^2 \\ -36 & 3l_m & 36 & 3l_m \\ -3l_m & -l_m^2 & 3l_m & 4l_m^2 \end{bmatrix} \right) \begin{bmatrix} v_n \\ \theta_n \\ v_{n+1} \\ \theta_{n+1} \end{bmatrix} \quad (14.5.20)$$

Next we must assemble the matrices.

$$[K] = \sum [k_m] \qquad [K^d] = \sum [k_m^d] \quad (14.5.21)$$

The global equations are

$$([K] - P[K^d])\{r_m\} = \{F\} \quad (14.5.22)$$

So using the same shape functions from bending theory and assuming a constant EI we can summarize the process of finding the global matrix equations. The summary provided here in Equation 14.5.22 is very similar to that in Chapter 8, Equation 8.4.5 with the addition of the $\left[k_m^d\right]$ element matrix.

In summary

$$\{F\} \rightarrow f_{y_m}(s_m) \quad P_m \quad V_m(s_m) \quad M_{z_m}(s_m) \quad \sigma_{s_m}(s_m, y) \quad \varepsilon_{s_m}(s_m, y) \quad u(s_m, y) \quad v(s_m) \quad \{r_m\} \leftarrow \rho$$

$$F_{em} = \int_0^{l_m} f_y(s_m)\, ds \quad \rightarrow \quad \sum F_{em} \qquad\qquad v(s_m) = [n(s_m)]\{r_m\}$$

$$F = F_c + F_e \qquad\qquad\qquad\qquad u_m(s_m) = -y\frac{d}{ds}[n]\{r_m\}$$

$$\varepsilon_{s_m}(s_m) = -y\frac{d^2}{ds^2}[n]\{r_m\}$$

$$\sigma_{s_m}(s_m) = -yE\frac{d^2}{ds^2}[n]\{r_m\} \qquad\qquad (14.5.23)$$

$$\delta W = \{\delta r_m\}^T P \int_0^{l_m} \frac{d[n]^T}{ds}\frac{d[n]}{ds}\,ds\,\{r_m\} - \{\delta r_m\}^T EI_m \int_0^{l_m} \frac{d^2[n]^T}{ds^2}\frac{d^2[n]}{ds^2}\,ds\,\{r_m\} = 0$$

$$[k_m] = EI_m \int_0^{l_m} \frac{d^2[n]^T}{ds^2}\frac{d^2[n]}{ds^2}\,ds \qquad\qquad \left[k_m^d\right] = P \int_0^{l_m} \frac{d[n]^T}{ds}\frac{d[n]}{ds}\,ds\,\{r_m\}$$

$$\frac{1}{l_m^3}\left[l_m^3 - 3l_m s_m^2 + 2s_m^3 \quad l_m^3 s_m - 2l_m^2 s_m^2 + l_m s_m^3 \quad 3l_m s_m^2 - 2s_m^3 \quad -l_m^2 s_m^2 + l_m s_m^3 \right]$$

$$\left([k_m] - P\left[k_m^d\right]\right)\{r_m\} = \left(\frac{E_m I_m}{l_m^3} \begin{bmatrix} 12 & 6l_m & -12 & 6l_m \\ 6l_m & 4l_m^2 & -6l_m & 2l_m^2 \\ -12 & -6l_m & 12 & -6l_m \\ 6l_m & 2l_m^2 & -6l_m & 4l_m^2 \end{bmatrix} \right.$$

$$\left. -\frac{P}{30L_m} \begin{bmatrix} 36 & -3l_m & -36 & -3l_m \\ -3l_m & 4l_m^2 & 3l_m & -l_m^2 \\ -36 & 3l_m & 36 & 3l_m \\ -3l_m & -l_m^2 & 3l_m & 4l_m^2 \end{bmatrix} \right) \begin{bmatrix} v_n \\ \theta_n \\ v_{n+1} \\ \theta_{n+1} \end{bmatrix}$$

$$[K] = \sum [k_m] \qquad \left[K^d\right] = \sum \left[k_m^d\right]$$

$$\left([K] - P\left[K^d\right]\right)\{r\} = \{F\}$$

We shall show how the element stiffness matrices are assembled to solve a buckling problem in an example. In such cases the applied lateral loads are zero.

##########

Example 14.5.1

Problem: Reconsider Example 14.3.3. This time do it by FEA.

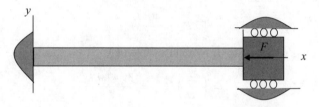

Figure (a)

Solution: Divide the column into elements and label the nodes and elements. Then assemble the elements stiffness and differential stiffness matrices, add boundary conditions, partition, and solve.

Let us divide the column into two elements and let $L_1 = L_2 = \frac{L}{2} = l$ as shown in Figure (b).

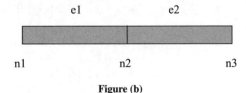

Figure (b)

We assemble the element equations to obtain the global equations, Equation 14.5.22, add the boundary restraints, and partition.

$$
\begin{bmatrix} R_1 \\ Q_1 \\ 0 \\ 0 \\ R_3 \\ Q_3 \end{bmatrix} = \left(\frac{EI_{zz}}{l^3} \begin{bmatrix} 12 & -6l & -12 & -6l & 0 & 0 \\ -6l & 4l^2 & 6l & 2l^2 & 0 & 0 \\ -12 & 6l & 24 & 0 & -12 & -6l \\ -6l & 2l^2 & 0 & 8l^2 & 6l & 2l^2 \\ 0 & 0 & -12 & 6l & 12 & 6l \\ 0 & 0 & -6l & 2l^2 & 6l & 4l^2 \end{bmatrix} - \frac{P}{30l} \begin{bmatrix} 36 & -3l & -36 & -3l & 0 & 0 \\ -3l & 4l^2 & 3l & -l^2 & 0 & 0 \\ -36 & 3l & 72 & 0 & -36 & -3l \\ -3l & -l^2 & 0 & 8l^2 & 3l & -l^2 \\ 0 & 0 & -36 & 3l & 36 & 3l \\ 0 & 0 & -3l & -l^2 & 3l & 4l^2 \end{bmatrix} \right) \begin{bmatrix} 0 \\ 0 \\ v_2 \\ \theta_2 \\ 0 \\ 0 \end{bmatrix}
$$

(a)

This quickly reduces the problem to just two equations in two unknowns.

$$
([K] - P[K^d]) \{r\} = \left(\frac{EI_{zz}}{l^3} \begin{bmatrix} 24 & 0 \\ 0 & 8l^2 \end{bmatrix} - \frac{P}{30l} \begin{bmatrix} 72 & 0 \\ 0 & 8l^2 \end{bmatrix} \right) \begin{bmatrix} v_2 \\ \theta_2 \end{bmatrix} = \begin{bmatrix} 0 \\ 0 \end{bmatrix}
$$

(b)

Either $v_2 = \theta_2 = 0$ or the determinant of the coefficients is zero. To simplify the notation let

$$
\beta = \frac{Pl^2}{30EI_{zz}}
$$

(c)

Then the determinant of the coefficients reduces to

$$
\begin{vmatrix} 24 - 72\beta & 0 \\ 0 & 8l^2 (1 - \beta) \end{vmatrix} = 0
$$

(d)

There are two possibilities:

$$
24 - 72\beta = 0 \quad \rightarrow \quad \beta_1 = \frac{1}{3} = \frac{Pl^2}{30EI_{zz}} \quad \rightarrow \quad P_1 = \frac{10EI_{zz}}{l^2} = \frac{40EI_{zz}}{L^2}
$$

$$
1 - \beta = 0 \quad \rightarrow \quad \beta_2 = 1 = \frac{Pl^2}{30EI_{zz}} \quad \rightarrow \quad P_2 = \frac{30EI_{zz}}{l^2} = \frac{120EI_{zz}}{L^2}
$$

(e)

The lowest value is the critical value.

$$
P_{cr} = \frac{40EI_{zz}}{L^2}
$$

(f)

We can compare this with the exact answer in Example 14.3.3.

$$
P_{cr} = \left(\frac{2\pi}{L} \right)^2 EI_{zz} = \frac{39.478EI_{zz}}{L^2}
$$

(g)

The error is 1.32%. This is quite a good approximation for such a small number of elements.

The mode shape is obtained by substituting the critical value back into the equations.

$$(24 - 72\beta_1)\, v_2 = \left(24 - 72\frac{1}{3}\right) v_2 = 0 \cdot v_2 = 0 \qquad \left(1 - \frac{1}{3}\right)\theta_2 = \frac{2}{3}\theta_2 = 0 \tag{h}$$

It follows that $\theta_2 = 0$ and v_2 is indeterminate. The mode shape is

$$v(x) = [n_1]\,\{r_1\} = \frac{1}{l^3}\left[3ls^2 - 2s^3\right]v_2 = \frac{8}{L^3}\left[\frac{3L}{2}x^2 - 2x^3\right]v_2 \qquad 0 \le x \le \frac{L}{2}$$

$$v(x) = [n_2]\,\{r_2\}$$

$$= \frac{1}{l^3}\left[l^3 - 3ls^2 + 2s^3\right]v_2 = \frac{8}{L^3}\left[\frac{L^3}{8} - \frac{3L}{2}\left(x - \frac{L}{2}\right)^2 + 2\left(x - \frac{L}{2}\right)^3\right]v_2 \qquad \frac{L}{2} \le x \le L$$

$$\tag{i}$$

The mode shape from the analytical solution in Example 14.3.3 is

$$v(x) = B\left(\cos\frac{2\pi x}{L} - 1\right) \tag{j}$$

What we have done is approximate a mode shape that is exactly a cosine function with two piecewise cubic polynomials.

Let us compare the exact and approximate shapes. Remember that mode shapes do not have a defined amplitude. The coefficients v_2 in Equation (i) and B in Equation (j) are not determined. In order to compare we assign a value to each coefficient so that the maximum displacements, in this case at $L/2$, are the same amplitude. This is called *normalizing*. In this case we assign

$$B = -\frac{1}{2} \qquad v_2 = 1 \tag{k}$$

We shall evaluate the mode shapes at intervals of $L/8$ and present the results in Table (a)

Table (a)

x	Equation (j)	Equation (i)
0	0.000	0.000
$L/8$	0.146	0.156
$L/4$	0.500	0.500
$3L/8$	0.854	0.844
$L/2$	1.000	1.000
$5L/8$	0.854	0.844
$3L/4$	0.500	0.500
$7L/8$	0.146	0.156
L	0.000	0.000

This shows that even a crude approximation of just two elements provides excellent correlation. Maybe there is something to this finite element stuff after all.

Both the exact analytical and the FEM approximate modes shapes are shown in Figure (c). They are so close a single line is shown.

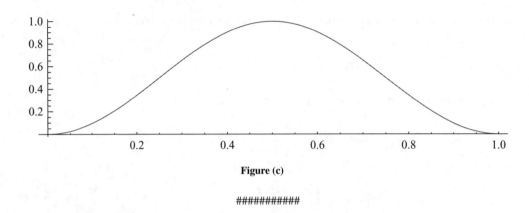

Figure (c)

###########

In more complicated problems involving three or more equations it is helpful to use one of the numerical equation solvers. Some of the solvers can deal with the format we have here, that is

$$\left([K] - P\left[K^d\right]\right)\{r\} = 0 \quad \text{or} \quad [K]\{r\} = P\left[K^d\right]\{r\} \tag{14.5.24}$$

Other solvers often require that the set of equations be put in standard form for solving for the eigenvalues and eigenfunctions. That form requires the inversion of $\left[K^d\right]$, that is, $\left[K^d\right]^{-1}$. Then

$$\left[K^d\right]^{-1}\left([K] - P\left[K^d\right]\right)\{r\} = 0 \quad \rightarrow \quad \left(\left[K^d\right]^{-1}[K] - P[1]\right)\{r\} = 0 \tag{14.5.25}$$

and the eigenvalue determinant becomes

$$\left|\left[K^d\right]^{-1}[K] - P[1]\right| = 0 \tag{14.5.26}$$

where $[K]^{-1}[K] = [1]$, the unit matrix that has ones on the diagonal and zeros elsewhere.

Since for large determinants the solver may use an iterative procedure that finds the highest eigenvalue first it is sometimes desirable to invert the matrix $[K]$ instead and format the problem as follows.

$$[K]^{-1}\left([K] - P\left[K^d\right]\right)\{r\} = 0 \quad \rightarrow \quad \left([K]^{-1}\left[K^d\right] - \frac{1}{P}[1]\right)\{r\} = 0 \tag{14.5.27}$$

This way we get the lowest critical load first and can stop the iteration.

###########

Example 14.5.2

Problem: Here is a problem that lends itself to solution by one of the equation solvers. Consider the column with multiple loads. Let $L_1 = L_2 = L_3 = \frac{L}{3} = l$ as shown in Figure (a). Set up the global equations for buckling analysis.

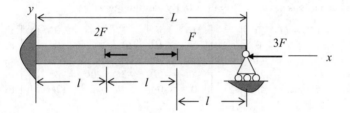

Figure (a)

Solution: Use three elements. Please note how tedious this would be to solve by direct analytical means. We number the nodes and elements in Figure (b).

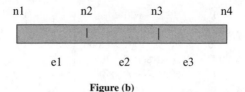

Figure (b)

The assembled stiffness matrix is

$$[K] = \frac{EI_{zz}}{l^3}\begin{bmatrix}
12 & -6l & -12 & -6l & 0 & 0 & 0 & 0 \\
-6l & 4l^2 & 6l & 2l^2 & 0 & 0 & 0 & 0 \\
-12 & 6l & 24 & 0 & -12 & -6l & 0 & 0 \\
-6l & 2l^2 & 0 & 8l^2 & 6l & 2l^2 & 0 & 0 \\
0 & 0 & -12 & 6l & 24 & 0 & -12 & -6l \\
0 & 0 & -6l & 2l^2 & 0 & 8l^2 & 6l & 2l^2 \\
0 & 0 & 0 & 0 & -12 & 6l & 12 & 6l \\
0 & 0 & 0 & 0 & -6l & 2l^2 & 6l & 4l^2
\end{bmatrix} \qquad (a)$$

Next we must find the axial load in each element.

$$P_1 = 4F \qquad P_2 = 2F \qquad P_3 = 3F \qquad (b)$$

The assembled differential stiffness or initial stress matrix is

$$[K^d] = \frac{F}{30l}\begin{bmatrix}
144 & -12l & -144 & -12l & 0 & 0 & 0 & 0 \\
-12l & 16l^2 & 12l & -4l^2 & 0 & 0 & 0 & 0 \\
-144 & 12l & 144+72 & 12l-6l & -72 & -6l & 0 & 0 \\
-12l & -4l^2 & 12l-6l & 16l^2+8l^2 & 6l & -2l^2 & 0 & 0 \\
0 & 0 & -72 & 6l & 72+108 & 6l-9l & -108 & -9l \\
0 & 0 & -6l & -2l^2 & 6l-9l & 8l^2+12l^2 & 9l & -3l^2 \\
0 & 0 & 0 & 0 & -108 & 9l & 108 & 9l \\
0 & 0 & 0 & 0 & -9l & -3l^2 & 9l & 12l^2
\end{bmatrix}$$

$$(c)$$

The boundary conditions are $v_1 = \theta_1 = v_4 = 0$. By partitioning we get

$$
\left(\begin{bmatrix} 24 & 0 & -12 & -6l & 0 \\ 0 & 8l^2 & 6l & 2l^2 & 0 \\ -12 & 6l & 24 & 0 & -6l \\ -6l & 2l^2 & 0 & 8l^2 & 2l^2 \\ 0 & 0 & -6l & 2l^2 & 4l^2 \end{bmatrix} - \beta \begin{bmatrix} 216 & 6l & -72 & -6l & 0 \\ 6l & 24l^2 & 6l & -2l^2 & 0 \\ -72 & 6l & 180 & -3l & -9l \\ -6l & -2l^2 & -3l & 20l^2 & -3l^2 \\ 0 & 0 & -9l & -3l^2 & 12l^2 \end{bmatrix} \right) \begin{bmatrix} v_2 \\ \theta_2 \\ v_3 \\ \theta_3 \\ \theta_4 \end{bmatrix} = \begin{bmatrix} 0 \\ 0 \\ 0 \\ 0 \\ 0 \end{bmatrix}
$$

(d)

where

$$
\beta = \frac{Fl^2}{30EI_{zz}}
$$

(e)

To proceed numerically we need only define the length of the elements. Let it be 100 *mm*; we shall add numbers for the other quantities later. Equation (d) becomes

$$
\left(\begin{bmatrix} 24 & 0 & -12 & -600 & 0 \\ 0 & 80000 & 600 & 20000 & 0 \\ -12 & 600 & 24 & 0 & -600 \\ -600 & 20000 & 0 & 80000 & 20000 \\ 0 & 0 & -600 & 20000 & 40000 \end{bmatrix} \right.
$$

$$
\left. -\beta \begin{bmatrix} 216 & 600 & -72 & -600 & 0 \\ 600 & 240000 & 600 & -20000 & 0 \\ -72 & 600 & 180 & -300 & -900 \\ -600 & -20000 & -300 & 200000 & -30000 \\ 0 & 0 & -900 & -30000 & 120000 \end{bmatrix} \right) \begin{bmatrix} v_2 \\ \theta_2 \\ v_3 \\ \theta_3 \\ \theta_4 \end{bmatrix} = \begin{bmatrix} 0 \\ 0 \\ 0 \\ 0 \\ 0 \end{bmatrix}
$$

(f)

This can be inserted into an equation solver for eigenvalue extraction.

By this example we have established the validity of the FEM for buckling. As the problems become more complicated the commercial codes are used.

###########

The FEM is especially good when the column is tapered and/or the axial force is not a constant. Then the differential equation has non constant coefficients and is very difficult to solve. The FEM equations can be divided into constant cross section elements and we can use concentrated equivalent nodal loads.

14.6 Buckling of Frames

To extend the FEM analysis to frames requires applying the transformation matrix presented in Section 10.4, Equations 10.4.3-9, to the $\left[k_m^d\right]$ matrix as well as to the $[k_m]$ matrix. As you can imagine when there are several elements requiring the transformation this can become time consuming and tedious. It is best to go directly to commercial FEM codes that have this process nicely programmed. We shall not pursue this further here.

14.7 Buckling of Thin Plates and Other Structures

Thin flat plates are normally the next component to consider for being susceptible to buckling. By an argument similar to that used for columns the plate equations can be extended to include the effect of both lateral and in plane loads on the deflected plate.

Consider a flat plate with in plane loads on its edges. Prior to buckling this is a plane stress problem and the equations of Chapter 12, Section 12.2, apply. Assume we have solved the problem and the internal stresses have been found. If we multiply these stresses by the plate thickness h we have the internal forces per unit length.

$$N_x(x, y) = \sigma_x(x, y)h \qquad N_y(x, y) = \sigma_y(x, y)h \qquad N_{xy}(x, y) = \tau_{xy}(x, y)h \qquad (14.7.1)$$

When we apply equilibrium in the deflected position the plate equations with both lateral and in plane loads become

$$D_h\left(\frac{\partial^4 w}{\partial x^4} + 2\frac{\partial^4 w}{\partial x^2 \partial y^2} + \frac{\partial^4 w}{\partial y^4}\right) = p(x, y) + N_x\frac{\partial^2 w}{\partial x^2} + 2N_{xy}\frac{\partial^2 w}{\partial x \partial y} + N_y\frac{\partial^2 w}{\partial y^2} \qquad (14.7.2)$$

where

$$D_h = \frac{Eh^3}{12(1 - v^2)} \qquad (14.7.3)$$

For those who wish to examine the derivation of this equation we refer you to books on the subject. We will restrict our discussion to a brief look at the subject.

When $p(x,y) = 0$ and there are compressive loads present the plate is a candidate for buckling.

Exact analytical solutions for buckling of plates have been found for a few rare cases of geometry, restraint, and loading. One such case is a rectangular plate with uniform edge loads and simply supported on all four edges. In practice plates are not often rectangular, edge loads are seldom uniform, and truly simply supported edges do not exist. Even in tightly controlled laboratory experiments it is virtually impossible to build a rectangular plate with truly simply supported edges. Nevertheless the solution of this problem provides great insight into the nature of plate buckling and exists as a measure for accuracy of approximate solutions.

In Section 12.2 we found the internal plane stress for a plate with uniform edge loads to be uniform internal stress and hence uniform internal forces per unit length. Let us use the solution from Example 12.2.1 to examine the buckling of a rectangular plate.

###########

Example 14.7.1

Problem: A rectangular plate with uniform compressive edge loads is shown in Figure (a). It is simply supported on all four edges. The edge loads have the units of force per unit area. There are no lateral loads. Find the buckling load.

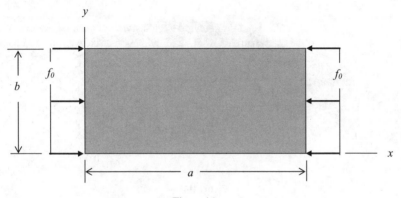

Figure (a)

Solution: An exact analytical solution has been found that we shall adopt.

The plane stress solution from Example 12.2.1 is

$$N_x(x, y) = \sigma_x(x, y)h = f_0 h = N_0 \qquad N_y(x, y) = 0 \qquad N_{xy}(x, y) = 0 \tag{a}$$

Equation 14.7.2 becomes

$$\frac{\partial^4 w}{\partial x^4} + 2\frac{\partial^4 w}{\partial x^2 \partial y^2} + \frac{\partial^4 w}{\partial y^4} = -\frac{N_0}{D_H}\frac{\partial^2 w}{\partial x^2} \tag{b}$$

It may be noted that the simply supported boundary conditions require that

$$w(0, y) = w(a, y) = \frac{\partial^2 w}{\partial^2 x}(0, y) = \frac{\partial^2 w}{\partial^2 x}(a, y) = 0$$

$$w(x, 0) = w(x, b) = \frac{\partial^2 w}{\partial^2 y}(x, 0) = \frac{\partial^2 w}{\partial^2 y}(x, b) = 0 \tag{c}$$

These are the conditions for no lateral displacement and no moments on all four edges. There is no restraint of in plane displacement in order for the in plane load to be applied.

Try a solution of the form

$$w(x, y) = C_{mn}\sin\frac{m\pi x}{a}\sin\frac{n\pi y}{b} \tag{d}$$

This satisfies the boundary conditions, so let us see if it satisfies the governing equation. Upon substitution we obtain the value of N_0 for the critical buckling load.

$$\left[\frac{m^2}{a^2} + \frac{n^2}{b^2}\right]^2 = \frac{m^2}{a^2}\frac{N_0}{\pi^2 D_H} \quad \rightarrow \quad N_0 = \left(m\frac{b}{a} + \frac{n^2}{m}\frac{a}{b}\right)^2\frac{\pi^2 D_H}{b^2} \tag{e}$$

Take, for example, a square plate for which $a = b$. Then the lowest buckling load is for $m = n = 1$ or

$$N_0 = 4\frac{\pi^2 D_h}{b^2} \tag{f}$$

and the mode shape is

$$w(x, y) = C\sin\frac{\pi x}{b}\sin\frac{\pi y}{b} \tag{g}$$

This is plotted in Figure (b).

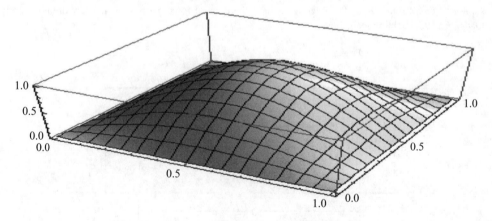

Figure (b)

Other combinations of m and n can be found for other ratios of a/b. For example, if $a/b = 2$ the lowest buckling load is for $m = 2$ and $n = 1$, or

$$N_0 = 4\frac{\pi^2 D_h}{b^2} \tag{h}$$

and the mode is

$$w(x, y) = C \sin\frac{2\pi x}{a} \sin\frac{\pi y}{b} \tag{i}$$

This is plotted in Figure (c).

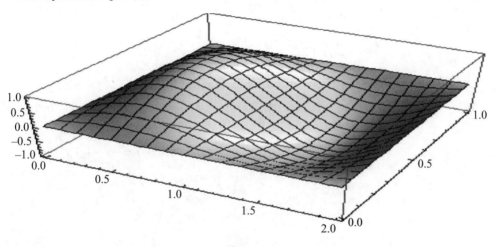

Figure (c)

Unfortunately, the slightest changes in plate shape, edge loading, or edge support make it impossible to find an exact analytical solution and quite difficult to find an approximate analytical solution. But fortunately, there is help on hand—FEM to the rescue.

Please accept that appropriate shape function has been found and element stiffness and differential stiffness matrices have been derived. It is beyond our efforts in this introductory text to carry this further.

###########

14.8   Summary and Conclusions

Compressive loads can cause catastrophic failure at loads below those which achieve yield stress when the bars are slender or plates are thin. For slender bars in summary we have the following equations for classical buckling analysis.

$$M_c, F_c \;\to\; f_y(x) \quad N(x) \quad V(x) \quad M(x) \quad \sigma_x(x,y) \quad \varepsilon_x(x,y) \quad u(x,y) \quad v(x) \leftarrow \rho$$

$$\frac{dv}{dx} + \frac{d}{dx}\left(N\frac{dv}{dx}\right) + f_y(x) = 0 \qquad \frac{dM}{dx} = -V(x) \qquad\qquad u(x,y) = u_A - y\frac{dv}{dx}$$

$$\frac{dN}{dx} = -f_x(x) \qquad\qquad \varepsilon_x(x,y) = \frac{du_A}{dx} + \frac{1}{2}\left(\frac{dv}{dx}\right)^2 - y\frac{d^2v}{dx^2}$$

$$\sigma_x(x,y) = E\varepsilon_x = E\left(\frac{du_A}{dx} + \frac{1}{2}\left(\frac{dv}{dx}\right)^2\right) - Ey\frac{d^2v}{dx^2} \qquad\qquad (14.8.1)$$

$$N(x) = \int_A \sigma_x dA = EA\left(\frac{du_A}{dx} + \frac{1}{2}\left(\frac{dv}{dx}\right)^2\right)$$

$$M(x) = -\int_A \sigma_x y\,dA = E\frac{d^2v}{dx^2}\int_A y^2 dA = EI_{zz}\frac{d^2v}{dx^2} \;\;\to\;\; \sigma_x = \frac{N(x)}{A} - \frac{M(x)y}{I_{zz}}$$

$$\frac{dM}{dx} = \frac{d}{dx}EI_{zz}\frac{d^2v}{dx^2}; \quad \hat{V}(x) = -\frac{d}{dx}EI_{zz}\frac{d^2v}{dx^2} + N\frac{dv}{dx} \;\;\to\;\; \tau_{xy}(x,y) = \frac{\hat{V}}{I_{zz}h}\int_{A_y} y\,dA$$

$$\frac{d^2}{dx^2}EI_{zz}\frac{d^2v}{dx^2} - \frac{d}{dx}\left(N\frac{dv}{dx}\right) = f_y(x)$$

The equivalent finite element equations are summarized as follows.

$$\{F\} \;\to\; f_{ym}(s) \quad P_m \quad V_m(s) \quad M_{zm}(s) \quad \sigma_{xm}(s,y) \quad \varepsilon_{xm}(s,y) \quad u_m(s,y) \quad v_m(s) \quad \{r_m\} \leftarrow \rho$$

$$v_m(s) = [n(s)]\{r_m\}$$

$$u_m(s) = -y\frac{d}{ds}[n]\{r_m\}$$

$$\varepsilon_{xm}(s) = -y\frac{d^2}{ds^2}[n]\{r_m\}$$

$$\sigma_{sm}(s) = -yE\frac{d^2}{ds^2}[n]\{r_m\}$$

$$\delta W = \{\delta r_m\}^T P \int_0^{l_m} \frac{d[n]^T}{ds}\frac{d[n]}{ds}ds\{r_m\} - \{\delta r_m\}^T EI_m \int_0^{l_m} \frac{d^2[n]^T}{ds^2}\frac{d^2[n]}{ds^2}ds\{r_m\} = 0$$

$$[k_m] = EI_m \int_0^{l_m} \frac{d^2[n]^T}{ds^2}\frac{d^2[n]}{ds^2}ds \quad [k_m^d] P \int_0^{l_m} \frac{d[n]^T}{ds}\frac{d[n]}{ds}ds\{r_m\} \qquad (14.8.2)$$

$$[n] = \frac{1}{l_m^3}\left[\,l_m^3 - 3l_m s^2 + 2s^3 \quad l_m^3 s - 2l_m^2 s^2 + l_m s^3 \quad 3l_m s^2 - 2s^3 \quad -l_m^2 s^2 + l_m s^3\,\right]$$

$$\left([k_m] - P[k_m^d]\right)\{r_m\} = \left(\frac{E_m I_m}{l_m^3}\begin{bmatrix} 12 & 6l_m & -12 & 6l_m \\ 6l_m & 4l_m^2 & -6l_m & 2l_m^2 \\ -12 & -6l_m & 12 & -6l_m \\ 6l_m & 2l_m^2 & -6l_m & 4l_m^2 \end{bmatrix}\right.$$

$$\left.-\frac{P}{30L_m}\begin{bmatrix} 36 & -3l_m & -36 & -3l_m \\ -3l_m & 4l_m^2 & 3l_m & -l_m^2 \\ -36 & 3l_m & 36 & 3l_m \\ -3l_m & -l_m^2 & 3l_m & 4l_m^2 \end{bmatrix}\right)\begin{bmatrix} v_n \\ \theta_n \\ v_{n+1} \\ \theta_{n+1} \end{bmatrix}$$

$$[K] = \sum[k_m] \qquad\qquad [K^d] = \sum[k_m^d]$$

Similar equations can be derived for plates and other structures.

15

Structural Dynamics

15.1 Introduction

In the analysis of structures up to this point we have assumed that loads are applied slowly, so slowly, in fact, that all the loads and the resulting stresses and displacements are independent of time. We are all aware, however, that loads are often applied rapidly and time dependent motion of the structure occurs. We are also aware that time dependent motion can be induced by imposing a displacement with a force or a restraint and then releasing the force or restraint. We speak here of time dependent displacement with no net rigid body motion. The study of such motion is called *structural dynamics* or *vibration of structures*.

When structures are subjected to dynamic loads inertia terms must be added to the equations. To set the stage for the study of structural dynamics we introduce the inertia term in a simple case of mass/spring systems in the next section. In this way the nature of vibration, the form of the governing equations, and the methods of analysis can be introduced in their simplest forms.

We then examine the axial motion of a slender bar. This is the simplest of the structural forms. We follow this with the torsional motion of shafts, the dynamic bending of beams, plates and shells, and dynamic motion of general three dimensions structures. In each case we shall consider first the exact analytical solution based on the differential equations of motion and then we shall solve the same problems by the finite element method. It just so happens that the equations for the dynamics of continuous structures such as slender bars, plates and shells, and general solid bodies are reduced by the finite element method to equations of the same form as mass/spring systems.

For each structural form we shall first consider *free motion*, that is, motion induced by an initial displacement, or an initial velocity, or both. The next major concern is a phenomenon called *resonance*, that is, a displacement driven by an oscillatory load. The final concern is the *dynamic response* due to a time dependent but non oscillating load.

15.2 Dynamics of Mass/Spring Systems

15.2.1 Free Motion

Insight into the nature of structural dynamics or vibration is quickly gained by looking at a one degree of freedom mass/spring system as shown in Figure 15.2.1 where k is the spring stiffness with units of Newtons per millimeter (N/mm), M is the value of the mass in kilograms (kg), and $F(t)$ the applied load in Newtons (N). The displacement is $v(t)$ in millimeters (mm). We allow the mass to stretch the

Analysis of Structures: An Introduction Including Numerical Methods, First Edition. Joe G. Eisley and Anthony M. Waas.
© 2011 John Wiley & Sons, Ltd. Published 2011 by John Wiley & Sons, Ltd.

spring under the action of gravity until it is in static equilibrium and measure the displacement from that position. The spring is assumed to be massless. This is a reasonable assumption when the mass of the spring is small compared to the attached mass.

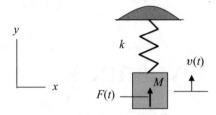

Figure 15.2.1

From D'Alembert's principle the magnitude of the inertia force, F_I, is equal to the mass times the acceleration, and is directed opposite to the direction of motion (y-direction). We can indicate this force in the positive y-direction as

$$F_I = -M\frac{d^2v}{dt^2} = -M\ddot{v} \tag{15.2.1}$$

This is often called the *reversed effective force*. We have introduced the common practice of using dots over the symbol to indicate differentiation with respect to time.

A free body diagram of the mass is shown in Figure 15.2.2.

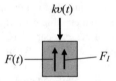

Figure 15.2.2

The equation of motion is then the sum of the applied load, the inertia force, and the spring force set equal to zero.

$$F(t) + F_I - kv = 0 \quad \rightarrow \quad F(t) - M\ddot{v} - kv = 0 \quad \rightarrow \quad \ddot{v} + \omega^2 v = \frac{F(t)}{M} \tag{15.2.2}$$

where $\omega^2 = \frac{k}{M}$ for convenience of notation. When there is no applied load, that is, $F(t) = 0$, we have what is called *free motion* or *free vibration*. Motion is induced by an initial displacement or velocity of both. The equation for free vibration is

$$\ddot{v} + \omega^2 v = 0 \tag{15.2.3}$$

From the methods for the solution of ordinary differential equations with constant coefficients the homogeneous equation has a solution of the form

$$v(t) \approx Ce^{\alpha t} \tag{15.2.4}$$

If we substitute this into Equation 14.2.3 we get

$$C\alpha^2 e^{\alpha t} + C\omega^2 e^{\alpha t} = (\alpha^2 + \omega^2)Ce^{\alpha t} = 0 \tag{15.2.5}$$

If $C = 0$ there is no motion. Since $e^{\alpha t}$ is not zero it is possible that

$$\alpha^2 + \omega^2 = 0 \quad \rightarrow \quad \alpha^2 = -\omega^2 \quad \rightarrow \quad \alpha = \pm i\omega \tag{15.2.6}$$

where $i = \sqrt{-1}$. There are two possible solutions to Equation 14.2.3

$$v(t) = C_1 e^{i\omega t} \quad \text{and} \quad v(t) = C_2 e^{-i\omega t} \tag{15.2.7}$$

For a linear equation the principle of superposition holds and so the sum of these two solutions is also a solution. It is usually more convenient to convert the solution from its exponential form to its trigonometric form.

$$\begin{aligned} v(t) &= C_1 e^{i\omega t} + C_2 e^{-i\omega t} \\ &= C_1(\cos \omega x + i \sin \omega x) + C_2(\cos \omega x - i \sin \omega x) \\ &= A \sin \omega t + B \cos \omega t \end{aligned} \tag{15.2.8}$$

To find the motion in a specific case we must assign values to the constants A and B. This is done by providing *initial conditions*. The most general statement of initial conditions is to assign values to an initial displacement and an initial velocity.

$$v(0) = v_0 \qquad \dot{v}(0) = \dot{v}_0 \tag{15.2.9}$$

When the initial conditions are applied to find the constants of integration we get

$$\begin{aligned} v(t) &= A \sin \omega t + B \cos \omega t \quad \rightarrow \quad v(0) = B = v_0 \\ \dot{v}(t) &= A\omega \cos \omega t - B\omega \sin \omega t \quad \rightarrow \quad \dot{v}(0) = A\omega = \dot{v}_0 \quad \rightarrow \quad A = \frac{\dot{v}_0}{\omega} \end{aligned} \tag{15.2.10}$$

or

$$v(t) = \frac{\dot{v}_0}{\omega} \sin \omega t + v_0 \cos \omega t \tag{15.2.11}$$

where

$$\omega = \sqrt{\frac{k}{M}} \; rad/s \tag{15.2.12}$$

is known as the *natural frequency* of the system. The frequency has units of radians per second, rad/s. The standard unit of frequency is cycles per second or Hertz, Hz. We shall use the symbol f for frequency measured in Hertz, thus,

$$f = \frac{\omega}{2\pi} \; Hz \tag{15.2.13}$$

The natural period, T, that is, the length in time of one cycle of the motion is

$$T = \frac{2\pi}{\omega} \tag{15.2.14}$$

The motion can be excited by any initial displacement, initial velocity, or combination of initial displacement and velocity. In any case the frequency of oscillation will always be $\omega = \sqrt{k/M}$ and the motion will always be sinusoidal.

##########

| Example 15.2.1 |

Problem: A single degree of freedom mass/spring system has the following properties.

$$M = 2 \ kg \qquad k = 32 \ N/mm \qquad \text{(a)}$$

From its static equilibrium position it is given an initial displacement of 4 *mm*, no initial velocity, and released. Find the natural frequency and resulting motion.

Solution: Use Equations 15.2.12 and 15.2.13r.
The natural frequency is

$$\omega = \sqrt{\frac{k}{M}} = \sqrt{\frac{32}{2}} = 4 \ rad/s \quad \rightarrow \quad f = \frac{\omega}{2\pi} = \frac{4}{2\pi} = 0.6366 \ Hz \qquad \text{(b)}$$

The resulting motion is

$$v(t) = \frac{\dot{v}_0}{\omega} \sin \omega t + v_0 \cos \omega t = 0 \cdot \sin 4t + 4 \cos 4t = 4 \cos 4t \qquad \text{(c)}$$

A plot of the first few cycles of the motion is given in Figure (a).

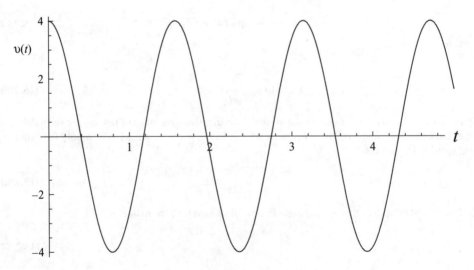

Figure (a)

##########

Now consider a system with two masses and two springs as shown in Figure 15.2.3.

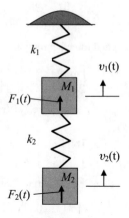

Figure 15.2.3

To derive the equations of motion we set up a free body diagram for each mass as shown in Figure 15.2.4.

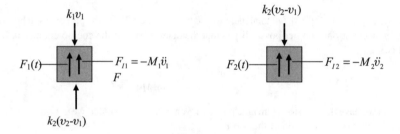

Figure 15.2.4

From summation of forces

$$F_1(t) - k_1 v_1 + k_2(v_2 - v_1) - M_1 \ddot{v}_1 = 0 \quad \rightarrow \quad M_1 \ddot{v}_1 + k_1 v_1 - k_2(v_2 - v_1) = F_1(t)$$
$$F_2(t) - k_2(v_2 - v_1) - M_2 \ddot{v}_2 = 0 \quad \rightarrow \quad M_2 \ddot{v}_2 + k_2(v_2 - v_1) = F_2(t)$$

(15.2.15)

Collecting terms we get

$$M_1 \ddot{v}_1 + (k_1 + k_2)v_1 - k_2 v_2 = F_1(t)$$
$$M_2 \ddot{v}_2 - k_2 v_1 + k_2 v_2 = F_2(t)$$

(15.2.16)

These may be put in the matrix form

$$\begin{bmatrix} M_1 & 0 \\ 0 & M_2 \end{bmatrix} \begin{bmatrix} \ddot{v}_1 \\ \ddot{v}_2 \end{bmatrix} + \begin{bmatrix} k_1 + k_2 & -k_2 \\ -k_2 & k_2 \end{bmatrix} \begin{bmatrix} v_1 \\ v_2 \end{bmatrix} = \begin{bmatrix} F_1(t) \\ F_2(t) \end{bmatrix}$$

(15.2.17)

This may be written as

$$[M]\{\ddot{v}\} + [K]\{v\} = \{F(t)\}$$

(15.2.18)

where

$$\{v\} = \begin{bmatrix} v_1 \\ v_2 \end{bmatrix} \quad [M] = \begin{bmatrix} M_1 & 0 \\ 0 & M_2 \end{bmatrix} \quad [K] = \begin{bmatrix} k_1 + k_2 & -k_2 \\ -k_2 & k_2 \end{bmatrix} \quad \{F(t)\} = \begin{bmatrix} F_1(t) \\ F_2(t) \end{bmatrix} \quad (15.2.19)$$

For free motion, that is, there is no applied force but motion is excited by initial conditions, we have

$$\begin{bmatrix} M_1 & 0 \\ 0 & M_2 \end{bmatrix} \begin{bmatrix} \ddot{v}_1 \\ \ddot{v}_2 \end{bmatrix} + \begin{bmatrix} k_1 + k_2 & -k_2 \\ -k_2 & k_2 \end{bmatrix} \begin{bmatrix} v_1 \\ v_2 \end{bmatrix} = \begin{bmatrix} 0 \\ 0 \end{bmatrix} \quad (15.2.20)$$

From our knowledge of the solution for the one degree of freedom system let us try a solution of the form

$$\begin{bmatrix} v_1 \\ v_2 \end{bmatrix} \approx \begin{bmatrix} C_1 \\ C_2 \end{bmatrix} e^{i\omega t} \quad (15.2.21)$$

By substituting Equation 15.2.21 into Equation 15.2.20 we obtain

$$\left(\begin{bmatrix} k_1 + k_2 & -k_2 \\ -k_2 & k_2 \end{bmatrix} - \omega^2 \begin{bmatrix} M_1 & 0 \\ 0 & M_2 \end{bmatrix} \right) \begin{bmatrix} C_1 \\ C_2 \end{bmatrix} e^{i\omega t} = \begin{bmatrix} 0 \\ 0 \end{bmatrix} \quad (15.2.22)$$

Since $e^{i\omega t} \neq 0$

$$\left(\begin{bmatrix} k_1 + k_2 & -k_2 \\ -k_2 & k_2 \end{bmatrix} - \omega^2 \begin{bmatrix} M_1 & 0 \\ 0 & M_2 \end{bmatrix} \right) \begin{bmatrix} C_1 \\ C_2 \end{bmatrix} = \begin{bmatrix} 0 \\ 0 \end{bmatrix} \quad (15.2.23)$$

One possibility is that $\{C\} = 0$, but then there would be no motion. From the properties of homogeneous linear algebraic equations another possibility is that the determinant of the coefficient matrix is equal to zero, that is,

$$\begin{vmatrix} (k_1 + k_2) - \omega^2 M_1 & -k_2 \\ -k_2 & k_2 - \omega^2 M_2 \end{vmatrix} = 0 \quad (15.2.24)$$

Once again we have the classic eigenvalue problem. When this is expanded we get a quadratic equation in terms of ω^2, that is, an equation of the form

$$\alpha_1(\omega^2)^2 + \alpha_2\omega^2 + \alpha_3 = 0 \quad (15.2.25)$$

This equation has two roots ω_1^2 and ω_2^2. These are squares of the two natural frequencies of the system. To find the values for $\{C\}$ we substitute each natural frequency in turn back into Equation 15.2.23. For each natural frequency there will be non zero values for $\{C\}$.

$$\left(\begin{bmatrix} k_1 + k_2 & -k_2 \\ -k_2 & k_2 \end{bmatrix} - \omega_1^2 \begin{bmatrix} M_1 & 0 \\ 0 & M_2 \end{bmatrix} \right) \begin{bmatrix} C_{11} \\ C_{12} \end{bmatrix} = \begin{bmatrix} 0 \\ 0 \end{bmatrix}$$

$$\left(\begin{bmatrix} k_1 + k_2 & -k_2 \\ -k_2 & k_2 \end{bmatrix} - \omega_2^2 \begin{bmatrix} M_1 & 0 \\ 0 & M_2 \end{bmatrix} \right) \begin{bmatrix} C_{21} \\ C_{22} \end{bmatrix} = \begin{bmatrix} 0 \\ 0 \end{bmatrix} \quad (15.2.26)$$

We learned in Chapter 14 that for homogenous equations we cannot solve for all the constants of integration, but we can solve for all but one of the constants in terms of that one. We know the shape of the displacement but not its magnitude. These solutions are of the form

$$\begin{bmatrix} C_{11} \\ C_{12} \end{bmatrix} = C_{11} \begin{bmatrix} 1 \\ \dfrac{C_{12}}{C_{11}} \end{bmatrix} = C_{11}\{\varphi\}_1 = C_{11} \begin{bmatrix} 1 \\ \varphi_1 \end{bmatrix} \qquad \begin{bmatrix} C_{21} \\ C_{22} \end{bmatrix} = C_{21} \begin{bmatrix} 1 \\ \dfrac{C_{22}}{C_{21}} \end{bmatrix} = C_{21}\{\varphi\}_2 = C_{21} \begin{bmatrix} 1 \\ \varphi_2 \end{bmatrix}$$

$$(15.2.27)$$

The resulting values are called the *modes* or *mode shapes*. The quantities $\{\varphi\}_1$ and $\{\varphi\}_2$ are called the *normal modes*. Note that we could have chosen to factor C_{12} or C_{22} and normalize that way. The actual amplitude of any normal mode is arbitrary.

Equations 15.2.26 also can be written in terms of the normal modes since the factored constant can be divided out of each equation.

$$\begin{aligned}
([K] - \omega_1^2[M]) C_{11}\{\varphi\}_1 = 0 \quad &\rightarrow \quad ([K] - \omega_1^2[M])\{\varphi\}_1 = 0 \\
([K] - \omega_2^2[M]) C_{21}\{\varphi\}_2 = 0 \quad &\rightarrow \quad ([K] - \omega_2^2[M])\{\varphi\}_2 = 0
\end{aligned} \tag{15.2.28}$$

Since each mode is a solution to its respective equation in Equation 15.2.28, these modes can be used to find the complete solution to the original equations, Equation 15.2.20. We take our cue from the one degree of system solution and note that there is a solution to the equations based on the first frequency and mode

$$\{v(t)\}_1 = \begin{bmatrix} v_1(t) \\ v_2(t) \end{bmatrix}_1 = (a_1 \sin \omega_1 t + b_1 \cos \omega_1 t) C_{11} \begin{bmatrix} 1 \\ \varphi_1 \end{bmatrix} = (A_1 \sin \omega_1 t + B_1 \cos \omega_1 t) \begin{bmatrix} 1 \\ \varphi_1 \end{bmatrix} \tag{15.2.29}$$

A second solution is based on the second frequency and mode.

$$\{v(t)\}_2 = \begin{bmatrix} v_1(t) \\ v_2(t) \end{bmatrix}_2 = (a_2 \sin \omega_2 t + b_2 \cos \omega_2 t) C_{21} \begin{bmatrix} 1 \\ \varphi_2 \end{bmatrix} = (A_2 \sin \omega_2 t + B_2 \cos \omega_2 t) \begin{bmatrix} 1 \\ \varphi_2 \end{bmatrix} \tag{15.2.30}$$

If each is a solution, then from the principle of superposition for linear systems, the sum of the two is a solution

$$\begin{aligned}
\{v(t)\} = \{v(t)\}_1 + \{v(t)\}_2 &= \begin{bmatrix} v_1(t) \\ v_2(t) \end{bmatrix}_1 + \begin{bmatrix} v_1(t) \\ v_2(t) \end{bmatrix}_2 \\
&= \begin{bmatrix} v_1(t) \\ v_2(t) \end{bmatrix} = (A_1 \sin \omega_1 t + B_1 \cos \omega_1 t) \begin{bmatrix} 1 \\ \varphi_1 \end{bmatrix} + (A_2 \sin \omega_2 t + B_2 \cos \omega_2 t) \begin{bmatrix} 1 \\ \varphi_2 \end{bmatrix}
\end{aligned} \tag{15.2.31}$$

Initial conditions are used to find the values of the constants A_1, B_1, A_2, B_2. Just how this is done will be demonstrated with specific examples. In Example 15.2.2 we shall find the frequencies and modes. In Example 15.2.3 we shall find the complete solution for the motion for specific initial conditions.

###########

Example 15.2.2

Problem: The mass spring system shown in Figure (a) has the following values assigned to its properties.

$$M_1 = 2M \qquad M_2 = M \qquad k_1 = 2k \qquad k_2 = k \tag{a}$$

Find the natural frequencies and modes.

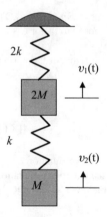

Figure (a)

Solution: Use Equations 15.2.17-28.
   The coefficient matrices are

$$[M] = M\begin{bmatrix} 2 & 0 \\ 0 & 1 \end{bmatrix} \qquad [K] = \begin{bmatrix} k_1 + k_2 & -k_2 \\ -k_2 & k_2 \end{bmatrix} = k\begin{bmatrix} 3 & -1 \\ -1 & 1 \end{bmatrix} \qquad (b)$$

The equations of motion are

$$M\begin{bmatrix} 2 & 0 \\ 0 & 1 \end{bmatrix}\begin{bmatrix} \ddot{v}_1(t) \\ \ddot{v}_2(t) \end{bmatrix} + k\begin{bmatrix} 3 & -1 \\ -1 & 1 \end{bmatrix}\begin{bmatrix} v_1 \\ v_2 \end{bmatrix} = \begin{bmatrix} 0 \\ 0 \end{bmatrix} \qquad (c)$$

The frequencies are obtained from

$$\left|[K] - \omega^2[M]\right| = \left|k\begin{bmatrix} 3 & -1 \\ -1 & 1 \end{bmatrix} - \omega^2 M\begin{bmatrix} 2 & 0 \\ 0 & 1 \end{bmatrix}\right| = \begin{vmatrix} 3k - 2M\omega^2 & -k \\ -k & k - M\omega^2 \end{vmatrix} = 0 \qquad (d)$$

Expanding the determinant we get

$$(\omega^2 M)^2 - \frac{5}{2}k(\omega^2 M) + k^2 = 0 \quad \rightarrow \quad (\omega^2)^2 - \frac{5}{2}\frac{k}{M}(\omega^2) + \left(\frac{k}{M}\right)^2 = 0 \qquad (e)$$

This is a quadratic equation in terms of ω^2 whose two roots are

$$\omega_1^2 = \frac{k}{2M} \qquad \omega_2^2 = \frac{2k}{M} \qquad (f)$$

or

$$\omega_1 = \sqrt{\frac{k}{2M}} \qquad \omega_2 = \sqrt{\frac{2k}{M}} \qquad (g)$$

The equations for the first normal mode are then

$$([K] - \omega_1^2[M])\{\varphi\}_1 = \left(k\begin{bmatrix} 3 & -1 \\ -1 & 1 \end{bmatrix} - \frac{k}{2}\begin{bmatrix} 2 & 0 \\ 0 & 1 \end{bmatrix}\right)\begin{bmatrix} 1 \\ \varphi_1 \end{bmatrix} = \begin{bmatrix} 0 \\ 0 \end{bmatrix}$$
$$\rightarrow \begin{bmatrix} 2 & -1 \\ -1 & 0.5 \end{bmatrix}\begin{bmatrix} 1 \\ \varphi_1 \end{bmatrix} = \begin{bmatrix} 0 \\ 0 \end{bmatrix} \qquad (h)$$

We can solve for φ_1. Use either one of the two equations; they give the same result.

$$2 - \varphi_1 = 0 \qquad -1 + 0.5\varphi_1 = 0 \quad \rightarrow \quad \varphi_1 = 2 \qquad (i)$$

The first normal mode can be written

$$\{\varphi\}_1 = \begin{bmatrix} 1 \\ 2 \end{bmatrix}$$ (j)

Follow the same procedure for the second mode.

$$([K] - \omega_2^2[M])\{\varphi\}_2 = \left(k\begin{bmatrix} 3 & -1 \\ -1 & 1 \end{bmatrix} - 2k\begin{bmatrix} 2 & 0 \\ 0 & 1 \end{bmatrix} \right)\begin{bmatrix} 1 \\ \varphi_2 \end{bmatrix} = \begin{bmatrix} 0 \\ 0 \end{bmatrix}$$

$$\rightarrow \begin{bmatrix} -1 & -1 \\ -1 & -1 \end{bmatrix}\begin{bmatrix} 1 \\ \varphi_2 \end{bmatrix} = \begin{bmatrix} 0 \\ 0 \end{bmatrix}$$ (k)

We can solve for φ_2.

$$-1 - \varphi_2 = 0 \qquad -1 - \varphi_2 = 0 \qquad \rightarrow \qquad \varphi_2 = -1$$ (l)

The second normal mode can be written

$$\{\varphi\}_2 = \begin{bmatrix} 1 \\ -1 \end{bmatrix}$$ (m)

We pause here to note an interesting property of normal modes that we will use to advantage later on. Note that

$$\{\varphi\}_1^T[M]\{\varphi\}_2 = \begin{bmatrix} 1 & 2 \end{bmatrix} M\begin{bmatrix} 2 & 0 \\ 0 & 1 \end{bmatrix}\begin{bmatrix} 1 \\ -1 \end{bmatrix} = \begin{bmatrix} 1 & 2 \end{bmatrix} M\begin{bmatrix} 2 \\ -1 \end{bmatrix} = 0$$

$$\{\varphi\}_1^T[k]\{\varphi\}_2 = \begin{bmatrix} 1 & 2 \end{bmatrix} k\begin{bmatrix} 3 & -1 \\ -1 & 1 \end{bmatrix}\begin{bmatrix} 1 \\ -1 \end{bmatrix} = \begin{bmatrix} 1 & 2 \end{bmatrix} k\begin{bmatrix} 4 \\ -2 \end{bmatrix} = 0$$ (n)

In fact

$$\{\varphi\}_2^T[M]\{\varphi\}_1 = \begin{bmatrix} 1 & -1 \end{bmatrix} M\begin{bmatrix} 2 & 0 \\ 0 & 1 \end{bmatrix}\begin{bmatrix} 1 \\ 2 \end{bmatrix} = \begin{bmatrix} 1 & -1 \end{bmatrix} M\begin{bmatrix} 2 \\ 2 \end{bmatrix} = 0$$

$$\{\varphi\}_2^T[k]\{\varphi\}_1 = \begin{bmatrix} 1 & -1 \end{bmatrix} k\begin{bmatrix} 3 & -1 \\ -1 & 1 \end{bmatrix}\begin{bmatrix} 1 \\ 2 \end{bmatrix} = \begin{bmatrix} 1 & -1 \end{bmatrix} k\begin{bmatrix} 1 \\ 1 \end{bmatrix} = 0$$ (o)

We also note that

$$\{\varphi\}_1^T[M]\{\varphi\}_1 = \begin{bmatrix} 1 & 2 \end{bmatrix} M\begin{bmatrix} 2 & 0 \\ 0 & 1 \end{bmatrix}\begin{bmatrix} 1 \\ 2 \end{bmatrix} = \begin{bmatrix} 1 & 2 \end{bmatrix} M\begin{bmatrix} 4 \\ 2 \end{bmatrix} = 8M \neq 0$$

$$\{\varphi\}_2^T[M]\{\varphi\}_2 = \begin{bmatrix} 1 & -1 \end{bmatrix} M\begin{bmatrix} 2 & 0 \\ 0 & 1 \end{bmatrix}\begin{bmatrix} 1 \\ -1 \end{bmatrix} = \begin{bmatrix} 1 & -1 \end{bmatrix} M\begin{bmatrix} 2 \\ -1 \end{bmatrix} = 3M \neq 0$$ (p)

$$\{\varphi\}_1^T[k]\{\varphi\}_1 = \begin{bmatrix} 1 & 2 \end{bmatrix} k\begin{bmatrix} 3 & -1 \\ -1 & 1 \end{bmatrix}\begin{bmatrix} 1 \\ 2 \end{bmatrix} = \begin{bmatrix} 1 & 2 \end{bmatrix} k\begin{bmatrix} 1 \\ 1 \end{bmatrix} = 3k \neq 0$$

$$\{\varphi\}_2^T[k]\{\varphi\}_2 = \begin{bmatrix} 1 & -1 \end{bmatrix} k\begin{bmatrix} 3 & -1 \\ -1 & 1 \end{bmatrix}\begin{bmatrix} 1 \\ -1 \end{bmatrix} = \begin{bmatrix} 1 & -1 \end{bmatrix} k\begin{bmatrix} 4 \\ -2 \end{bmatrix} = 6k \neq 0$$ (q)

This property is known as *orthogonality*. It can be shown that in general

$$\{\varphi\}_i^T[M]\{\varphi\}_j = 0 \quad i \neq j \qquad \{\varphi\}_i^T[k]\{\varphi\}_j = 0 \quad i \neq j$$
$$\qquad\qquad\qquad \neq 0 \quad i = j \qquad\qquad\qquad\qquad \neq 0 \quad i = j$$ (r)

We shall make good use of orthogonality when we study forced motion.

##########

We have found the natural frequencies and modes but now we must find the actual motion. Let us first find the motion for a set of initial conditions. For two degrees of freedom there are four initial conditions.

$$v_1(0) = v_{10} \qquad \dot{v}_1(0) = \dot{v}_{10} \qquad v_2(0) = v_{20} \qquad \dot{v}_2(0) = \dot{v}_{20} \qquad (15.2.32)$$

The solution given in Equation 15.2.31, repeated here as Equation 15.2.33, has four constants to be determined.

$$v(t) = \begin{bmatrix} v_1(t) \\ v_2(t) \end{bmatrix} = (A_1 \sin \omega_1 t + B_1 \cos \omega_1 t) \begin{bmatrix} 1 \\ \varphi_1 \end{bmatrix} + (A_2 \sin \omega_2 t + B_2 \cos \omega_2 t) \begin{bmatrix} 1 \\ \varphi_2 \end{bmatrix} \qquad (15.2.33)$$

The initial conditions provide us with the necessary information to find the four constants of integration.

##########

Example 15.2.3

Problem: The system in Example 15.2.2 has the following initial conditions.

$$v_1(0) = v_{10} \qquad \dot{v}_1(0) = 0 \qquad v_2(0) = 0 \qquad \dot{v}_2(0) = 0 \qquad (a)$$

Find the resulting motion.

Solution: Find a solution based on normal modes.
Equation 15.2.33 for this case is

$$v(t) = \begin{bmatrix} v_1(t) \\ v_2(t) \end{bmatrix} = (A_1 \sin \omega_1 t + B_1 \cos \omega_1 t) \begin{bmatrix} 1 \\ 2 \end{bmatrix} + (A_2 \sin \omega_2 t + B_2 \cos \omega_2 t) \begin{bmatrix} 1 \\ -1 \end{bmatrix} \qquad (b)$$

Substituting Equation (b) into the four initial conditions we get

$$v_1(0) = (A_1 \sin \omega_1 0 + B_1 \cos \omega_1 0) 1$$
$$\qquad + (A_2 \sin \omega_2 0 + B_2 \cos \omega_2 0) 1 = B_1 + B_2 = v_{10}$$
$$\dot{v}_1(0) = (A_1 \omega_1 \cos \omega_1 0 - B_1 \omega_1 \sin \omega_1 0) 1$$
$$\qquad + (A_2 \omega_2 \cos \omega_2 0 - B_2 \omega_2 \sin \omega_2 0) 1 = A_1 \omega_1 + A_2 \omega_2 = 0$$
$$v_2(0) = (A_1 \sin \omega_1 0 + B_1 \cos \omega_1 0) 2 \qquad \qquad (c)$$
$$\qquad - (A_2 \sin \omega_2 0 + B_2 \cos \omega_2 0) 1 = 2B_1 - B_2 = 0$$
$$\dot{v}_2(0) = (A_1 \omega_1 \cos \omega_1 0 - B_1 \omega_1 \sin \omega_1 0) 2$$
$$\qquad - (A_2 \omega_2 \cos \omega_2 0 - B_2 \omega_2 \sin \omega_2 0) 1 = 2A_1 \omega_1 - A_2 \omega_2 = 0$$

From the second and fourth equations you find that

$$A_1 = 0 \qquad A_2 = 0 \qquad (d)$$

From the first and third equations we have

$$B_1 = \frac{v_{10}}{3} \qquad B_2 = \frac{2v_{10}}{3} \qquad (e)$$

The resulting motion is

$$\{v(t)\} = \begin{bmatrix} v_1(t) \\ v_2(t) \end{bmatrix} = \left(\frac{v_{10}}{3} \cos \omega_1 t \right) \begin{bmatrix} 1 \\ 2 \end{bmatrix} + \left(\frac{2v_{10}}{3} \cos \omega_2 t \right) \begin{bmatrix} 1 \\ -1 \end{bmatrix} \qquad (f)$$

Let us add some numerical values and plot the displacements. Let

$$M = 2 \, kg \qquad k = 64 \, N/mm \qquad v_{10} = 3 \, mm \tag{g}$$

Then

$$\omega_1 = \sqrt{\frac{k}{2M}} = \sqrt{\frac{64}{2 \cdot 2}} = 4 \, rad/s \qquad \omega_2 = \sqrt{\frac{2k}{M}} = \sqrt{\frac{2 \cdot 64}{2}} = 8 \, rad/s \tag{h}$$

The displacements of the two masses are

$$v_1(t) = \cos 4t + 2 \cos 8t \qquad v_2(t) = 2 \cos 4t - 2 \cos 8t \tag{i}$$

These are plotted in Figure (a).

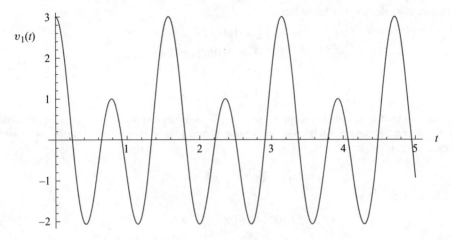

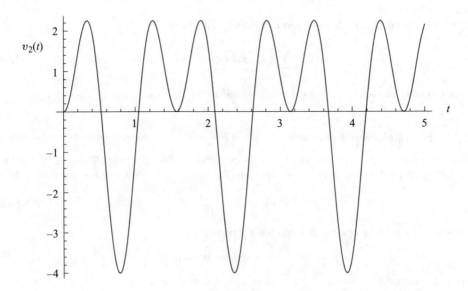

Figure (a)

##########

When there are more masses and springs and hence more degrees of freedom the equations for the free motion are of the form

$$[M]\{\ddot{v}\} + [K]\{v\} = \{0\} \tag{15.2.34}$$

where the number of equations is equal to the number of degrees of freedom. The solution is found by first finding the natural frequencies. These are determined by assuming a solution of the form

$$\{v(t)\} \approx \{C\}e^{i\omega t} \tag{15.2.35}$$

which results in

$$\left([K] - \omega_i^2[M]\right)\{C\}e^{i\omega t} = 0 \quad \rightarrow \quad \left|[K] - \omega^2[M]\right| = 0 \tag{15.2.36}$$

The natural modes are then found from

$$\begin{aligned}
([K] - \omega_1^2[M])\{C\}_1 &= 0 \\
([K] - \omega_2^2[M])\{C\}_2 &= 0 \\
&\vdots \\
([K] - \omega_n^2[M])\{C\}_n &= 0
\end{aligned} \tag{15.2.37}$$

The normal modes are found in shape but their amplitude is not determined. As noted it is customary to factor out one of the constants so the remaining matrix $\{\varphi\}_i$ is known, for example, as shown in Equation 15.2.27.

$$\{C\}_i = C_{i1}\{\varphi\}_i \tag{15.2.38}$$

It follows that

$$\left([K] - \omega_i^2[M]\right)\{\varphi\}_i = 0 \tag{15.2.39}$$

The constant C_{i1} is often assigned an arbitrary value, say, $C_{i1} = 1$. The resulting matrix $\{\varphi\}_i$ is then called a *normal mode*.

To satisfy the initial conditions the normal modes $\{\varphi\}_i$ are summed.

$$\{v(t)\} = \sum_{1}^{n} (A_i \sin \omega_i t + B_i \cos \omega_i t)\{\varphi\}_i \tag{15.2.40}$$

The constants are evaluated using the initial conditions.

15.2.2   Forced Motion—Resonance

When an applied force is present this is known as *forced motion*. For a single degree of freedom mass/spring system the equation of motion is given by Equation 15.2.2 repeated here as Equation 15.2.41.

$$\ddot{v} + \omega^2 v = \frac{F(t)}{M} \tag{15.2.41}$$

Its solution is made up of a homogenous and a particular part

$$v(t) = A \sin \omega t + B \cos \omega t + v_{particular} \tag{15.2.42}$$

Of major interest is the solution for a periodic applied force, for example,

$$F(t) = F_0 \sin \Omega t \tag{15.2.43}$$

We try a particular solution of the form

$$v_p = P \sin \Omega t \qquad (15.2.44)$$

Substitute this into Equation 15.2.41.

$$\ddot{v} + \omega^2 v = \frac{F_0}{M} \sin \Omega t$$

$$\rightarrow \frac{d^2}{dt^2}(P \sin \Omega t) + \omega^2 P \sin \Omega t = \frac{F_0}{M} \sin \Omega t$$

$$\rightarrow (-\Omega^2 + \omega^2) P \sin \Omega t = \frac{F_0}{M} \sin \Omega t \qquad (15.2.45)$$

$$\rightarrow P = \frac{F_0}{(\omega^2 - \Omega^2)M}$$

The complete solution is then

$$v(t) = A \sin \omega t + B \cos \omega t + \frac{F_0}{(\omega^2 - \Omega^2)M} \sin \Omega t \qquad (15.2.46)$$

Initial conditions are used to find the values of the constants A and B. For initial conditions

$$v(0) = v_0 \qquad \dot{v}(0) = \dot{v}_0 \qquad (15.2.47)$$

we get

$$v(0) = A \sin \omega 0 + B \cos \omega 0 + \frac{F_0}{(\omega^2 - \Omega^2)M} \sin \Omega 0 = v_0 \quad \rightarrow \quad B = v_0$$

$$\dot{v}(0) = A\omega \cos \omega 0 - B\omega \sin \omega 0 + \frac{F_0}{(\omega^2 - \Omega^2)M} \Omega \cos \Omega 0 \qquad (15.2.48)$$

$$= A\omega + \frac{F_0}{(\omega^2 - \Omega^2)M} \Omega = \dot{v}_0 \quad \rightarrow \quad A = \frac{\dot{v}_0}{\omega} - \frac{F_0}{(\omega^2 - \Omega^2)M} \frac{\Omega}{\omega}$$

When these values of A and B are inserted into Equation 15.2.46 the complete solution becomes

$$v(t) = A \sin \omega t + B \cos \omega t + \frac{F_0}{(\omega^2 - \Omega^2)M} \sin \Omega t$$

$$= \frac{\dot{v}_0}{\omega} \sin \omega t - \frac{F_0}{(\omega^2 - \Omega^2)M} \frac{\Omega}{\omega} \sin \omega t + v_0 \cos \omega t + \frac{F_0}{(\omega^2 - \Omega^2)M} \sin \Omega t \qquad (15.2.49)$$

$$= v_0 \cos \omega t + \frac{\dot{v}_0}{\omega} \sin \omega t + \frac{F_0}{(\omega^2 - \Omega^2)M} \left(\sin \Omega t - \frac{\Omega}{\omega} \sin \omega t \right)$$

We note that when the applied frequency approaches the natural frequency the particular solution grows rapidly and when they are equal the displacement is infinite for any and all initial conditions.

Note that the amplitude of the motion is determined by the coefficient of the particular solution which can be reformulated.

$$P = \frac{F_0}{(\omega^2 - \Omega^2)M} = \frac{F_0}{\omega^2 M} \frac{1}{\left(1 - \dfrac{\Omega^2}{\omega^2}\right)} = \frac{F}{\left(1 - \dfrac{\Omega^2}{\omega^2}\right)} \quad \rightarrow \quad \frac{P}{F} = \frac{1}{\left(1 - \dfrac{\Omega^2}{\omega^2}\right)} \qquad (15.2.50)$$

where

$$F = \frac{F_0}{\omega^2 M} \qquad (15.2.51)$$

The plot of the absolute value of $\left|\dfrac{P}{F}\right|$ versus $\dfrac{\Omega}{\omega}$ in Figure 15.2.5 shows the dependence of amplitude on the relation between the applied frequency and the natural frequency. When the applied frequency equals the natural frequency, that is, when $\dfrac{\Omega}{\omega} = 1$, the amplitude is infinite.

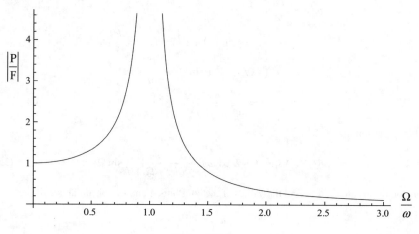

Figure 15.2.5

########

Example 15.2.4

Problem: Find the forced motion for a single degree of freedom mass/spring system when

$$F(t) = F_0 \sin \Omega t \tag{a}$$

and the initial conditions are

$$v(0) = v_0 = 0 \qquad \dot{v}(0) = \dot{v}_0 = 0 \tag{b}$$

Solution: Use initial conditions to find the constants in Equation 15.2.49.
   Applying the initial conditions we get

$$\begin{aligned} v(t) &= v_0 \cos \omega t + \frac{\dot{v}_0}{\omega} \sin \omega t + \frac{F_0}{(\omega^2 - \Omega^2)M}\left(\sin \Omega t - \frac{\Omega}{\omega}\sin \omega t\right) \\ &= \frac{F_0}{(\omega^2 - \Omega^2)M}\left(\sin \Omega t - \frac{\Omega}{\omega}\sin \omega t\right) \end{aligned} \tag{c}$$

If we insert numbers from Example 15.2.1

$$\omega = 4 \; rad/s \qquad M = 2 \; kg \tag{d}$$

and assign numerical values to the others

$$F_0 = 10 \; N \qquad \Omega = 2 \; rad/s \tag{e}$$

the motion is

$$v(t) = \frac{F_0}{(\omega^2 - \Omega^2)M} \left(\sin \Omega t - \frac{\Omega}{\omega} \sin \omega t \right) = \frac{10}{(16-4)\,2} \left(\sin 2t - \frac{2}{4} \sin 4t \right)$$

$$= \frac{5}{12} \sin 2t - \frac{5}{24} \sin 4t = 0.4167 \sin 2t - 0.2083 \sin 4t$$

(f)

If we change the applied frequency to twice the value of the natural frequency, that is, to $\Omega = 8\ rad/s$, we see a decrease in the amplitude of the motion.

$$v(t) = \frac{F_0}{(\omega^2 - \Omega^2)M} \left(\sin \Omega t - \frac{\Omega}{\omega} \sin \omega t \right) = \frac{10}{(16-64)\,2} \left(\sin 8t - \frac{8}{4} \sin 4t \right)$$

$$= -\frac{5}{48} \sin 8t + \frac{5}{24} \sin 4t = -0.1042 \sin 8t + 0.2083 \sin 4t$$

(g)

If we change the applied frequency to a value much closer to the natural frequency, say, $\Omega = 3.6\ rad/s$, we see a large increase in the amplitude of the motion.

$$v(t) = \frac{F_0}{(\omega^2 - \Omega^2)M} \left(\sin \Omega t - \frac{\Omega}{\omega} \sin \omega t \right) = \frac{10}{(16-12.96)\,2} \left(\sin 3.6t - \frac{3.6}{4} \sin 4t \right)$$

$$= 1.645 \sin 3.6t - 1.480 \sin 4t$$

(h)

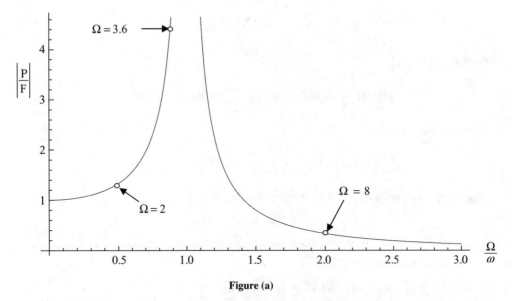

Figure (a)

For this mass/spring system with a natural frequency of 4 rad/s the amplitudes on Figure 15.2.5 would be as shown in Figure (a) for each of the applied frequencies given in Equations (f), (g), and (h).

########

We see that when the applied frequency Ω approaches the natural frequency ω the motion grows quite large. This is a phenomenon called *resonance* that can be quite dangerous and, at times, destructive. Thus an important design consideration is to tune the values of M and k to avoid any natural frequencies

ω that might be near values of applied frequencies Ω. Alternatively, we can place operating restrictions so that there is no applied frequency at or near the natural frequency.

Let us extend this to two or more degrees of freedom. The equations of motion are of the form

$$[M]\{\ddot{v}\} + [K]\{v\} = \{F(t)\} \tag{15.2.52}$$

The homogeneous equations can be solved for the frequencies and modes as just described. Once we have found the frequencies and modes we can note that

$$\left([K] - \omega_i^2[M]\right)\{\varphi\}_i = 0 \quad \rightarrow \quad [K]\{\varphi\}_i = \omega_i^2[M]\{\varphi\}_i \tag{15.2.53}$$

We noted in a two degree of freedom system the complete solution is the sum of the solution based on the two normal modes as expressed in Equation 15.2.33 and this can be extended to as many degrees of freedom as there are shown in Equation 15.2.40. This can be extended to forced motion and an advantage taken by the orthogonality property of normal modes mentioned briefly for the special case of two degrees of freedom in Example 15.2.2. This property allows us to simplify the solution for a large number of degrees of freedom.

For many degrees of freedom a solution is found by assuming a series of the normal modes.

$$\{v\} = \sum_{i=1}^{n} \xi_i(t)\{\varphi\}_i \tag{15.2.54}$$

where $\xi_i(t)$ is to be determined. Insert this into Equation 15.2.51

$$[M] \sum_{i=1}^{n} \ddot{\xi}_i(t)\{\varphi\}_i + [K] \sum_{i=1}^{n} \xi_i(t)\{\varphi\}_i = \{F(t)\} \tag{15.2.55}$$

and premultiply by $\{\varphi\}_j^T$

$$\{\varphi\}_j^T[M] \sum_{i=1}^{n} \ddot{\xi}_i(t)\{\varphi\}_i + \{\varphi\}_j^T[K] \sum_{i=1}^{n} \xi_i(t)\{\varphi\}_i = \{\varphi\}_j^T\{F(t)\} \tag{15.2.56}$$

Now rearrange the terms

$$\sum_{i=1}^{n} \ddot{\xi}_i(t)\{\varphi\}_j^T[M]\{\varphi\}_i + \sum_{i=1}^{n} \xi_i(t)\{\varphi\}_j^T[K]\{\varphi\}_i = \{\varphi\}_j^T\{F(t)\} \tag{15.2.57}$$

The orthogonality relation applies to all normal modes.

$$\begin{aligned} \{\varphi\}_j^T[M]\{\varphi\}_i &= 0 \quad i \neq j \\ &= M_j \quad i = j \end{aligned} \tag{15.2.58}$$

From Equations 15.2.53 and 15.2.58 we can say

$$\begin{aligned} \{\varphi\}_j^T[K]\{\varphi\}_i = \{\varphi\}_j^T \omega_i^2[M]\{\varphi\}_i &= 0 \quad i \neq j \\ &= \omega_j^2 M_j \quad i = j \end{aligned} \tag{15.2.59}$$

We can rewrite Equation 15.2.57 as follows

$$M_j\ddot{\xi}_j + \omega_j^2 M_j \xi_j = f_j(t) \quad \rightarrow \quad \ddot{\xi}_j + \omega_j^2 \xi_j = \frac{f_j(t)}{M_j} \tag{15.2.60}$$

where

$$f_j(t) = \{\varphi_j\}^T\{F(t)\} \tag{15.2.61}$$

Each equation contains only one dependent variable. This is the same equation as for the one degree of freedom system and the solution is the same.

By expanding the displacement in a series of normal modes and taking advantage of the orthogonality of modes we have transformed the set of coupled equations, Equations 15.2.52, into a set of uncoupled equations, Equations 15.2.60.

It is time for an example.

###########

Example 15.2.5

Problem: The mass/spring system in Example 15.2.2 is initially at rest and is then excited by a sinusoidal force applied to the first mass equal to

$$F_1(t) = F_{10} \sin \Omega t \tag{a}$$

The initial conditions are

$$v(0) = v_0 = 0 \qquad \dot{v}(0) = \dot{v}_0 = 0 \tag{b}$$

Find the resulting motion.

Solution: Use orthogonality of modes to find the uncoupled equations. Solve the uncoupled equations and sum according to Equations 15.2.54.

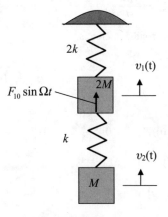

Figure (a)

The equations of motion are

$$M \begin{bmatrix} 2 & 0 \\ 0 & 1 \end{bmatrix} \begin{bmatrix} \ddot{v}_1(t) \\ \ddot{v}_2(t) \end{bmatrix} + k \begin{bmatrix} 3 & -1 \\ -1 & 1 \end{bmatrix} \begin{bmatrix} v_1 \\ v_2 \end{bmatrix} = \begin{bmatrix} F_0 \sin \Omega t \\ 0 \end{bmatrix} \tag{c}$$

We use the natural frequencies and normal modes found in Example 15.2.2 to convert the equations to the uncoupled form.

To convert to the form of Equations 15.2.60 we find

$$M_1 = \{\varphi\}_1^T [M]\{\varphi\}_1 = [1 \quad 2] M \begin{bmatrix} 2 & 0 \\ 0 & 1 \end{bmatrix} \begin{bmatrix} 1 \\ 2 \end{bmatrix} = 6M$$

$$M_2 = \{\varphi\}_2^T [M]\{\varphi\}_2 = [1 \quad -1] M \begin{bmatrix} 2 & 0 \\ 0 & 1 \end{bmatrix} \begin{bmatrix} 1 \\ -1 \end{bmatrix} = 3M$$

$$f_1(t) = \{\varphi\}_1^T \{F(t)\} = [1 \quad 2] \begin{bmatrix} F_{10} \sin \Omega t \\ 0 \end{bmatrix} = F_{10} \sin \Omega t$$

$$f_2(t) = \{\varphi\}_2^T \{F(t)\} = [1 \quad -1] \begin{bmatrix} F_{10} \sin \Omega t \\ 0 \end{bmatrix} = F_{10} \sin \Omega t$$

(d)

The first uncoupled equation is

$$\ddot{\xi}_1 + \omega_1^2 \xi_1 = \frac{f_1(t)}{M_1} \quad \rightarrow \quad \ddot{\xi}_1 + \omega_1^2 \xi_1 = \frac{F_{10}}{6M} \sin \Omega t$$

(e)

This equation if the same for as the single mass/spring system and so the particular solution is

$$\xi_{1\,particular} = \frac{F_0}{(\omega_1^2 - \Omega^2) 6M} \sin \Omega t$$

(f)

The complete solution for the first uncoupled equation is

$$\xi_1(t) = a_1 \sin \omega_1 t + b_1 \cos \omega_1 t + \frac{F_0}{(\omega_1^2 - \Omega^2) 6M} \sin \Omega t$$

(g)

By the same procedure the second uncoupled equation is

$$\ddot{\xi}_2 + \omega_2^2 \xi_2 = \frac{f_2(t)}{M_2} \quad \rightarrow \quad \ddot{\xi}_2 + \omega_2^2 \xi_2 = \frac{F_{10}}{3M} \sin \Omega t$$

(h)

and the complete solution for the second equation is

$$\xi_2(t) = a_2 \sin \omega_2 t + b_2 \cos \omega_2 t + \frac{F_0}{(\omega_2^2 - \Omega^2) 3M} \sin \Omega t$$

(i)

The initial conditions are

$$\{v(0)\} = \sum_{i=1}^{n} [\xi_i(0)] \{\varphi\}_i = 0 \quad \rightarrow \quad \xi_1(0) \begin{bmatrix} 1 \\ 2 \end{bmatrix} + \xi_2(0) \begin{bmatrix} 1 \\ -1 \end{bmatrix} = 0$$

$$\rightarrow \xi_1(0) = 0 \quad \xi_2(0) = 0$$

$$\{\dot{v}(0)\} = \sum_{i=1}^{n} [\dot{\xi}_i(0)] \{\varphi\}_i = 0 \quad \rightarrow \quad \dot{\xi}_1(0) \begin{bmatrix} 1 \\ 2 \end{bmatrix} + \dot{\xi}_2(0) \begin{bmatrix} 1 \\ -1 \end{bmatrix} = 0$$

$$\rightarrow \dot{\xi}_1(0) = 0 \quad \dot{\xi}_2(0) = 0$$

(j)

From the solution for the single degree of freedom system given in Equation 15.2.49 for initial conditions at rest the solutions for the uncoupled equations are

$$\xi_1(t) = \frac{F_{10}}{(\omega_1^2 - \Omega^2) 6M} \left(\sin \Omega t - \frac{\Omega}{\omega_1} \sin \omega_1 t \right)$$

$$\xi_2(t) = \frac{F_{10}}{(\omega_2^2 - \Omega^2) 3M} \left(\sin \Omega t - \frac{\Omega}{\omega_2} \sin \omega_2 t \right)$$

(k)

It follows that

$$\{v\} = \begin{bmatrix} v_1(t) \\ v_2(t) \end{bmatrix} = \sum_{i=1}^{n} \xi_i(t)\{\varphi\}_i = \left(\frac{F_{10}}{(\omega_1^2 - \Omega^2) 6M} \left(\sin \Omega t - \frac{\Omega}{\omega_1} \sin \omega_1 t \right) \right) \begin{bmatrix} 1 \\ 2 \end{bmatrix}$$

$$+ \left(\frac{F_{10}}{(\omega_2^2 - \Omega^2) 3M} \left(\sin \Omega t - \frac{\Omega}{\omega_2} \sin \omega_2 t \right) \right) \begin{bmatrix} 1 \\ -1 \end{bmatrix} \tag{1}$$

Note that each time the applied frequency approaches a natural frequency the motion grows quite large in amplitude.

############

For multi degree of freedom systems with the same applied load on the first mass and the same initial conditions the complete solution is

$$v(t) = \sum_{i=1}^{n} \frac{F_{10}}{(\omega_i^2 - \Omega^2) M_i} \left(\sin \Omega t - \frac{\Omega}{\omega_i} \sin \omega_i t \right) \{\varphi\}_i \tag{15.2.62}$$

15.2.3 Forced Motion—Response

For multi degree of freedom systems when $f_j(t)$ is not periodic we get

$$\xi_j = a_j \sin \omega_j t + b_j \cos \omega_j t + \xi_{jparticular} \tag{15.2.63}$$

The particular solutions are not periodic and so resonance does not occur. In this case finding the maximum amplitude is often a determining design factor.

Again let us see an example.

############

Example 15.2.6

Problem: The one degree of freedom mass/spring system is at rest at $t = 0$, that is,

$$v(0) = \dot{v}(0) = 0 \tag{a}$$

and then a constant force is applied to the mass.

$$\begin{aligned} F(t) &= 0 & t < 0 \\ F(t) &= F_0 & t \geq 0 \end{aligned} \tag{b}$$

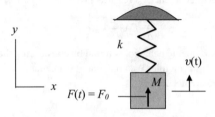

Figure (a)

Find the resulting motion.

Solution: Find the particular solution for this case.
   The governing equation is

$$\ddot{v} + \omega^2 v = \frac{F_0}{M} \tag{c}$$

Try a particular solution of the form

$$v_p = P \tag{d}$$

Then

$$\omega^2 P = \frac{F_0}{M} \tag{e}$$

The complete solution is

$$v(t) = a \sin \omega t + b \cos \omega t + \frac{F_0}{\omega^2 M} \tag{f}$$

Apply initial conditions.

$$v(0) = a \sin \omega 0 + b \cos \omega 0 + \frac{F_0}{\omega^2 M} = 0 \quad \rightarrow \quad b = -\frac{F_0}{\omega^2 M}$$
$$\dot{v}(0) = a\omega \cos \omega 0 - b\omega \sin \omega 0 = 0 \quad \rightarrow \quad a = 0 \tag{g}$$

The complete solution is

$$v(t) = \frac{F_0}{\omega^2 M} (1 - \cos \omega t) \tag{h}$$

This is a steady state oscillation of amplitude $\dfrac{F_0}{\omega^2 M}$.

###########

   The extension to multiple degree of freedom systems is a matter of finding the appropriate particular solutions. We shall explore this further later on in this chapter.
   Next let us consider the vibration of various structural forms starting with the axial motion of a slender bar

15.3   Axial Vibration of a Slender Bar

15.3.1   Solutions Based on the Differential Equation

We shall see that the axial bar can have a very similar behavior to that of a mass/spring system. The equation for the static axial displacement of a bar was given in Chapter 4, Section 4.3, Equation 4.3.7.

$$\frac{d}{dx} EA \frac{du}{dx} = -f_x(x) \tag{15.3.1}$$

   The governing equation for dynamic behavior is found by considering the inertia force as an additional applied load using D'Alembert's principle. We note that $u(x, t)$, that is, u is a function of both space and time and so partial derivatives are needed.
   The governing equation is

$$\frac{\partial}{\partial x} EA \frac{\partial u}{\partial x} = -f_x(x, t) - F_I = -f(x, t) + m\ddot{u} \tag{15.3.2}$$

where m is the mass per unit length of the bar.

For a uniform bar this becomes

$$EA\frac{\partial^2 u}{\partial x^2} - m\ddot{u} = -f_x(x, t) \tag{15.3.3}$$

First we consider what is called the free vibration, that is, vibration induced by initial conditions but no applied load.

$$EA\frac{\partial^2 u}{\partial x^2} - m\ddot{u} = 0 \tag{15.3.4}$$

To find a solution we use a method called *separation of variables*, that is, we assume

$$u(x, t) = U(x)T(t) \tag{15.3.5}$$

and substitute it into Equation 15.3.4. The result can be separated into two distinct equations.

$$EA\frac{\partial^2 U}{\partial x^2}T - mU\ddot{T} = 0 \quad \rightarrow \quad \frac{EA\dfrac{d^2 U}{dx^2}}{mU} = \frac{1}{T}\frac{d^2 T}{dt^2} = \text{constant} = -\omega^2 \tag{15.3.6}$$

On the left side of the equation all the quantities are functions of x only, while those to the right are functions of t only. This can be true only if both are equal to the same constant. The choice of the constant as $-\omega^2$ is done by hindsight. It turns out that ω is the *natural frequency* of the system.

Thus we obtain two separate equations.

$$\ddot{T} + \omega^2 T = 0 \qquad \frac{d^2 U}{dx^2} + \frac{m\omega^2}{EA}U = \frac{d^2 U}{dx^2} + \alpha^2 U = 0 \tag{15.3.7}$$

where

$$\alpha^2 = \frac{m\omega^2}{EA} \tag{15.3.8}$$

The solutions are

$$T(t) = a\sin\omega t + b\cos\omega t \qquad U(x) = c\sin\alpha x + d\cos\alpha x \tag{15.3.9}$$

The natural frequencies and modes are found by applying boundary conditions to the second of equations 15.3.9. We shall demonstrate this with a specific example.

##########

Example 15.3.1

Problem: Consider the bar in Figure (a) with a cross sectional area A (mm^2), Young's modulus E (N/mm^2), length L (mm), and mass per unit length m (kg/mm). It is restrained on the left end as shown in Figure (a). Find the natural frequencies and modes.

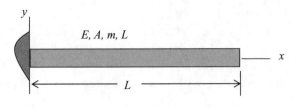

Figure (a)

Solution: Separate variables and apply boundary conditions to the second of Equations 15.3.9. Given

$$U(x) = c \sin \alpha x + d \cos \alpha x \tag{a}$$

we apply boundary conditions of no displacement at $x = 0$ and no load at $x = L$.

$$u(0, t) = T(t)U(0) = 0 \quad \rightarrow \quad U(0) = c \sin \alpha \cdot 0 + d \cos \alpha \cdot 0 = 0 \quad \rightarrow \quad d = 0$$

$$EA\frac{du(L, t)}{dx} = EA\frac{dU(L)}{dx}T(t) = 0 \quad \rightarrow \quad \frac{dU(L)}{dx} = c\alpha \cos \alpha L = 0 \tag{b}$$

It must be true that either $c = 0$, or $\alpha = 0$, or $\cos \alpha L = 0$. If either c or α is zero then nothing happens, that is, there is no displacement; however, if

$$\cos \alpha L = 0 \quad \rightarrow \quad \alpha_i L = \frac{(2i - 1)}{2}\pi = L\sqrt{\frac{m\omega^2}{EA}} \quad \rightarrow \quad \omega_i = \frac{(2i - 1)\pi}{2}\frac{1}{L}\sqrt{\frac{EA}{m}} \tag{c}$$

where i is any positive integer, we can have displacement. The first four frequencies, for example, are

$$\omega_1 = \frac{1}{2}\frac{\pi}{L}\sqrt{\frac{EA}{m}} \qquad \omega_2 = \frac{3}{2}\frac{\pi}{L}\sqrt{\frac{EA}{m}} \qquad \omega_3 = \frac{5}{2}\frac{\pi}{L}\sqrt{\frac{EA}{m}} \qquad \omega_4 = \frac{7}{2}\frac{\pi}{L}\sqrt{\frac{EA}{m}} \tag{d}$$

From the solution to the time dependent part of Equation 15.3.9

$$T(t) = a \sin \omega t + b \cos \omega t \tag{e}$$

we can recognize that the ω's are the natural frequencies and that there are an infinite number of them. As noted if $c \neq 0$ then

$$U_i = c_i \varphi_i(x) = c_i \sin \alpha_i x \tag{f}$$

Since the c_i are indeterminate we assign them the value of 1 and the normal modes are

$$\varphi_i(x) = \sin \alpha_i x = \sin \frac{(2i - 1)}{2L}\pi x \tag{g}$$

There are an infinite number of them. The first four normal modes are

$$\varphi_1(x) = \sin \alpha_1 x = \sin \frac{\pi}{2}\frac{x}{L}$$

$$\varphi_2(x) = \sin \alpha_2 x = \sin \frac{3\pi}{2}\frac{x}{L}$$

$$\varphi_3(x) = \sin \alpha_3 x = \sin \frac{5\pi}{2}\frac{x}{L} \tag{h}$$

$$\varphi_4(x) = \sin \alpha_4 x = \sin \frac{7\pi}{2}\frac{x}{L}$$

These modes are plotted in Figure (b).

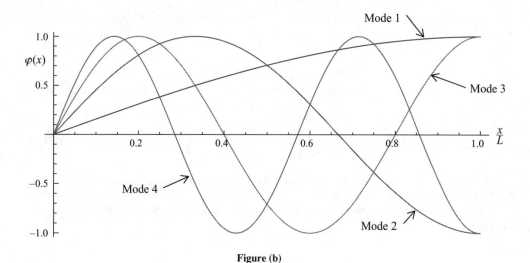

Figure (b)

###########

Taking our cue from the mass/spring systems it is suggested that the motion can be described by a sum of the normal modes as follows

$$u(x, t) = \sum (A_i \sin \omega_i t + B_i \cos \omega_i t) \, \varphi_i(x) \tag{15.3.10}$$

The initial conditions would then be satisfied by

$$u(x, 0) = \sum (A_i \sin \omega_i 0 + B_i \cos \omega_i 0) \, \varphi_i(x) = \sum B_i \varphi_i(x)$$
$$\dot{u}(x, 0) = \sum (A_i \omega_i \cos \omega_i 0 - B_i \omega_i \sin \omega_i 0) \, \varphi_i(x) = \sum A_i \omega_i \varphi_i(x) \tag{15.3.11}$$

The actual initial conditions may not be the same functional shape as the normal modes but an interesting property of the normal modes saves the day for us. Similar to what happens in the mass/spring systems it just so happens that

$$\int_0^L \varphi_i(x) m \varphi_j(x) dx(x) = 0, \qquad j \neq i$$
$$= M_i \qquad j = i \tag{15.3.12}$$

This is the property called *orthogonality* as it applies in this case. In this case it is weighted by the distributed mass.

If we multiply both sides of Equations 15.3.7 by $m\varphi_j(x)$ and integrate over the length of the bar as follows:

$$\int_0^L u(x, 0)\varphi_j(x) m dx = \int_0^L \sum B_i \varphi_i(x)\varphi_j(x) m dx = \sum B_i \int_0^L \varphi_i(x)\varphi_j(x) m dx$$
$$\int_0^L \dot{u}(x, 0)\varphi_j(x) m dx = \int_0^L \sum A_i \omega_i \varphi_i(x)\varphi_j(x) m dx = \sum A_i \omega_i \int_0^L \varphi_i(x)\varphi_j(x) m dx \tag{15.3.13}$$

We note that all the terms in the summation on the right side are zero except when $i = j$. We then can solve for A_i and B_i.

$$B_i = \frac{1}{M_i} \int_0^L u(x, 0)\varphi_i(x)m\,dx$$

$$A_i = \frac{1}{M_i\omega_i} \int_0^L \dot{u}(x, 0)\varphi_i(x)m\,dx \qquad (15.3.14)$$

$$M_i = \int_0^L (\varphi_i(x))^2 \, m\,dx \qquad (15.3.15)$$

Now there are an infinite number of frequencies and modes and therefore and infinite number of terms in the series solution given in Equation 15.3.10. Fortunately this is a highly convergent series and only a few terms need be evaluated to achieve an accurate representation of the ensuing motion in most cases. Let us look at an example.

############

Example 15.3.2

Problem: The bar in Example 15.3.1, shown here in Figure (a), is given an initial displacement and no initial velocity.

$$u(x, 0) = u_0 \frac{x}{L} \qquad \dot{u}(x, 0) = 0 \qquad (a)$$

From Chapter 4 we know that

$$u(L) = \frac{FL}{EA} = u_0 \quad \rightarrow \quad F = \frac{u_0 EA}{L} \qquad (b)$$

The force is applied to provide an initial displacement and then at $t = 0$ is suddenly released. Find the resulting motion.

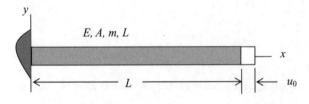

Figure (a)

Solution: Find a series solution using Equation 15.3.10.
The dynamic displacement based on the normal modes is

$$u(x, t) = \sum (A_i \sin \omega_i t + B_i \cos \omega_i t) \sin \alpha_i x \qquad (c)$$

Use the initial conditions to find the values of the constants.

$$B_i = \frac{1}{M_i} \int_0^L u(x,0)\varphi_i(x)\,mdx = \frac{1}{M_i} \int_0^L u_0\frac{x}{L}(\sin\alpha_i x)\,mdx$$

$$= \frac{u_0 m}{M_i L} \int_0^L x(\sin\alpha_i x)\,dx = \frac{u_0 m}{\alpha_i^2 M_i L}(\sin\alpha_i x - \alpha_i x\cos\alpha_i x)\Big|_0^L$$

$$= \frac{u_0 m}{\alpha_i^2 M_i L}(\sin\alpha_i L - \sin\alpha_i 0 - \alpha_i L\cos\alpha_i L + \alpha_i 0\cos\alpha_i 0)$$

$$= \frac{u_0 m}{\alpha_i^2 M_i L}(\sin\alpha_i L)$$

$$A_i = \frac{1}{M_i\omega_i} \int_0^L \dot{u}(x,0)\varphi_i(x)\,mdx = \frac{1}{M_i\omega_i} \int_0^L 0(\sin\alpha_i x)\,mdx = 0 \qquad (d)$$

where

$$M_i = \int_0^L (\sin\alpha_i x)^2 mdx = \frac{m}{2\alpha_i}(\alpha_i x + \sin\alpha_i x\cos\alpha_i x)\Big|_0^L$$

$$= \frac{m}{2\alpha_i}(\alpha_i L - \alpha_i\cdot 0 + \sin\alpha_i L\cos\alpha_i L - \sin\alpha_i\cdot 0\cos\alpha_i\cdot 0) = \frac{mL}{2} \qquad (e)$$

The displacement is

$$u(x,t) = \sum B_i \sin\alpha_i x\cos\omega_i t = \sum \frac{u_0 m}{\alpha_i^2 M_i L}(\sin\alpha_i L)\sin\alpha_i x\cos\omega_i t \qquad (f)$$

Remember from Example 15.3.1 we found that

$$\cos\alpha_i L = 0 \quad\rightarrow\quad \alpha_i L = \frac{(2i-1)}{2}\pi \qquad (g)$$

and therefore

$$\sin\alpha_i L = 1 \qquad i = odd$$
$$= -1 \qquad i = even \qquad (h)$$

The first few values of B_i are

$$B_i = \frac{u_0 m}{\alpha_i^2 M_i L}(\sin\alpha_i L) = \frac{2u_0 m}{\alpha_i^2 mL^2}(\sin\alpha_i L) = \frac{2u_0}{(\alpha_i L)^2}(\sin\alpha_i L)$$

$$\rightarrow B_1 = \frac{8u_0}{\pi^2}, \quad B_2 = -\frac{8u_0}{9\pi^2}, \quad B_3 = \frac{8u_0}{25\pi^2}, \quad B_4 = -\frac{8u_0}{49\pi^2} \qquad (i)$$

The displacement can now be written as

$$u(x,t) = \sum B_i \sin\alpha_i x\cos\omega_i t$$

$$= \frac{8u_0}{\pi^2}\sin\frac{\pi}{2L}x\cos\omega_1 t$$

$$- \frac{8u_0}{9\pi^2}\sin\frac{3\pi}{2L}x\cos\omega_2 t$$

$$+ \frac{8u_0}{25\pi^2}\sin\frac{5\pi}{2L}x\cos\omega_3 t$$

$$- \frac{8u_0}{49\pi^2}\sin\frac{7\pi}{2L}x\cos\omega_4 t$$

$$+ \cdots \qquad (j)$$

##########

Let us demonstrate the rate of convergence of the series solution given by Equation 15.3.10. A numerical example will help.

############

Example 15.3.3

Problem: The bar in Example 15.3.2 is given the following properties.

$$L = 300 \ mm \qquad A = 100 \ mm^2 \qquad E = 68950 \ N/mm^2 \qquad m = 0.271 \ kg/mm$$

Show how many terms are needed to demonstrate convergence of the series solution for the motion given by Equation (j) in Example 15.3.2.

Solution: Evaluate Equation (j) for the first four modes in the series solution and compare one term, two term, three term, and four term solutions.

A solution based on the first four terms, that is, summing the first four terms of the infinite series would be

$$
\begin{aligned}
u(x,t) = \sum B_i \sin \alpha_i x \cos \omega_i t \\
= \frac{8u_0}{\pi^2} \sin \frac{\pi}{2L} x \cos \omega_1 t \\
- \frac{8u_0}{9\pi^2} \sin \frac{3\pi}{2L} x \cos \omega_2 t \\
+ \frac{8u_0}{25\pi^2} \sin \frac{5\pi}{2L} x \cos \omega_3 t \\
- \frac{8u_0}{49\pi^2} \sin \frac{7\pi}{2L} x \cos \omega_4 t
\end{aligned}
\tag{a}
$$

The natural frequencies found in Example 15.3.1 are

$$
\begin{aligned}
\omega_1 = \frac{1}{2} \frac{\pi}{L} \sqrt{\frac{EA}{m}} = \frac{1}{2} \frac{\pi}{300} \sqrt{\frac{68950 \cdot 100}{0.271}} = 26.411 \ rad/sec \\
\omega_2 = \frac{3}{2} \frac{\pi}{L} \sqrt{\frac{EA}{m}} = 3\omega_1 = 79.232 \ rad/sec \\
\omega_3 = \frac{5}{2} \frac{\pi}{L} \sqrt{\frac{EA}{m}} = 5\omega_1 = 132.054 \ rad/sec \\
\omega_4 = \frac{7}{2} \frac{\pi}{L} \sqrt{\frac{EA}{m}} = 7\omega_1 = 184.875 \ rad/sec
\end{aligned}
\tag{b}
$$

For plotting purposes let $u_0 = 1$ and let us evaluate the motion at $x = L$.

$$
\begin{aligned}
u(L,t) = \frac{8}{\pi^2} \cos 26.411t \\
+ \frac{8}{9\pi^2} \cos 79.232t \\
+ \frac{8}{25\pi^2} \cos 132.054t \\
+ \frac{8}{49\pi^2} \cos 184.875t
\end{aligned}
\tag{c}
$$

Note that the amplitude of the second term is 1/9 of the first term and the frequency is three times as great. The amplitudes get smaller and the frequencies higher for each successive term.

One term, two term, three term, and four term solutions are plotted in order in Figures (a), (b), (c), and (d).

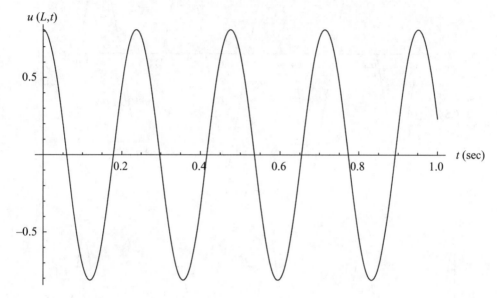

Figure (a)

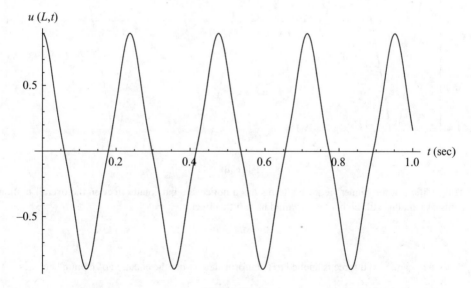

Figure (b)

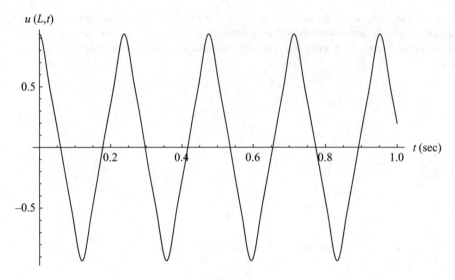

Figure (c)

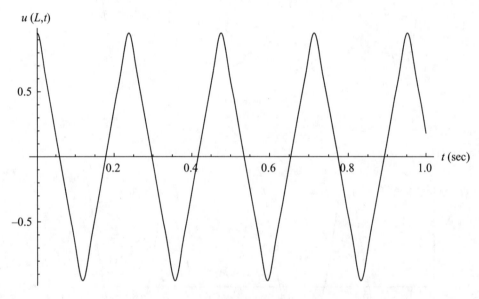

Figure (d)

The addition of the higher frequency terms has a noticeable but small effect on the overall motion. For this case adding still higher terms would have little effect.

###########

When a dynamic axial force is applied to a uniform slender bar the equation of motion becomes

$$EA\frac{\partial^2 u}{\partial x^2} - m\ddot{u} = -f(x, t) \tag{15.3.16}$$

In pursuit of a solution we first find the natural frequencies and modes as just presented. Having found the frequencies and modes we assume a solution of the form

$$u(x, t) = \sum_{i=1}^{\infty} \xi_i(t)\varphi_i(x) \tag{15.3.17}$$

We insert this form into Equation 14.3.12

$$EA\frac{\partial^2}{\partial x^2} \sum_{i=1}^{\infty} \xi_i(t)\varphi_i(x) - m \sum_{i=1}^{\infty} \ddot{\xi}_i(t)\varphi_i(x) = -f(x, t) \tag{15.3.18}$$

We can rearrange the terms.

$$EA\sum_{i=1}^{\infty} \xi_i \frac{\partial^2 \varphi_i}{\partial x^2} - m \sum_{i=1}^{\infty} \ddot{\xi}\varphi_i = -f(x, t) \tag{15.3.19}$$

Now multiply all terms by $\varphi_j(x)$ and integrate all terms over the length of the bar.

$$\int_0^L \varphi_j \left[EA\sum_{i=1}^{\infty} \xi_i \frac{\partial^2 \varphi_i}{\partial x^2} - m \sum_{i=1}^{\infty} \ddot{\xi}\varphi_i = -f(x, t) \right] dx$$

$$= \sum_{i=1}^{\infty} \xi_i \int_0^L EA\varphi_j \frac{\partial^2 \varphi_i}{\partial x^2} dx - \sum_{i=1}^{\infty} \ddot{\xi}_i \int_0^L \varphi_j m\varphi_i dx = - \int_0^L \varphi_j f(x, t) dx \tag{15.3.20}$$

The normal modes are solutions to the separated equation

$$EA\frac{d^2 U}{dx^2} + m\omega^2 U = 0 \tag{15.3.21}$$

It follows that

$$EA\frac{d^2 \varphi_i}{dx^2} + m\omega_i^2 \varphi_i = 0 \tag{15.3.22}$$

Thus we can set

$$EA\frac{d^2 \varphi_i}{dx^2} = -m\omega_i^2 \varphi_i \tag{15.3.23}$$

By replacing $EA\frac{d^2 \varphi_i}{dx^2}$ with $-m\omega_i^2 \varphi_i$ Equation 15.3.18 becomes

$$-\sum_{i=1}^{\infty} \xi_i \omega_i^2 \int_0^L \varphi_j m\varphi_i dx - \sum_{i=1}^{\infty} \ddot{\xi}_i \int_0^L \varphi_j m\varphi_i dx = - \int_0^L \varphi_j f(x, t) dx \tag{15.3.24}$$

We designate

$$f_j(t) = - \int_0^L \varphi_j f(x, t) dx \tag{15.3.25}$$

We note that from orthogonality

$$\int_0^L \varphi_j m\varphi_i dx = 0 \qquad i \neq j$$
$$= M_j \qquad i = j \tag{15.3.26}$$

Finally the equation for forced motion may be written

$$\ddot{\xi}_j + \omega_j^2 \xi = \frac{f_j(t)}{M_j} \tag{15.3.27}$$

We now solve for $\xi_j(t)$ and insert it in the series of Equation 15.3.17.
Once again an example will help.

##########

Example 15.3.4

Problem: A periodic force is applied to the right end of the bar as shown in Figure (a).

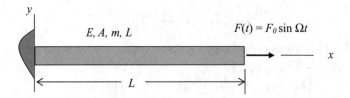

Figure (a)

Solution: Solve Equations 15.3.27, repeated here as Equation (a).

$$\ddot{\xi}_j + \omega_j^2 \xi = \frac{f_j(t)}{M_j} \tag{a}$$

Then sum the results as given by Equation 15.3.17, repeated here as Equation (b).

$$u(x, t) = \sum_{i=1}^{\infty} \xi_i(t)\varphi_i(x) \tag{b}$$

From Example 15.3.1, Equation (c), we have the natural frequencies and the normal modes

$$\cos \alpha L = 0 \quad \rightarrow \quad \alpha_i L = \frac{(2i - 1)}{2}\pi = L\sqrt{\frac{m\omega^2}{EA}} \quad \rightarrow \quad \omega_i = \frac{(2i - 1)\,\pi}{2}\frac{1}{L}\sqrt{\frac{EA}{m}} \tag{c}$$

$$\varphi_i(x) = \sin \alpha_i x$$

First evaluate

$$f_j(t) = -\int_0^L \varphi_j f(x, t)dx = -\int_0^L F_0 \sin \Omega t \sin \alpha_j x dx$$

$$= -F_0 \sin \Omega t \int_0^L \sin \alpha_j x dx = -\frac{F_0}{\alpha_i} \sin \Omega t \left(-\cos \alpha_i x|_0^L\right)$$

$$= \frac{F_0}{\alpha_i} \sin \Omega t(\cos \alpha_i L - \cos \alpha_i 0) = \frac{F_0}{\alpha_i} \sin \Omega t(0 - 1) = -\frac{F_0}{\alpha_i} \sin \Omega t \tag{d}$$

$$M_j = \int_0^L \varphi_j m \varphi_j dx = \int_0^L m(\sin \alpha_j x)^2 dx$$

$$= \frac{m}{2\alpha_j}(\alpha_j x - \sin \alpha_j x \cos \alpha_j x)|_0^L$$

$$= \frac{m}{2\alpha_j}(\alpha_j L - \sin \alpha_j L \cos \alpha_j L - \alpha_j 0 - \sin \alpha_j 0 \cos \alpha_j 0)$$

$$= \frac{m}{2\alpha_j}\alpha_j L = \frac{mL}{2}$$

Insert these values in Equation (a).

$$\ddot{\xi}_j + \omega_j^2 \xi_j = \frac{f_j(t)}{M_j} = -\frac{2F_0}{\alpha_j mL} \sin \Omega t \tag{e}$$

When you compare this equation with the single degree of freedom mass/spring equation in Equation 15.2.41 and its solution in Equation 15.2.46 you see immediately that

$$\xi_j(t) = A_j \sin \omega_j t + B_j \cos \omega_j t - \frac{2F_0}{\alpha_j \left(\omega_j^2 - \Omega^2\right) mL} \sin \Omega t \tag{f}$$

Apply initial conditions to solve for A_j and B_j. The complete solution is given by Equation (b). Resonance will occur when an applied frequency approaches any one of the infinite number of natural frequencies.

##########

The response to a non periodic applied force is reduced to finding the particular solutions to Equations 15.3.27.

##########

Example 15.3.5

Problem: Instead of a periodic force a step function force is applied at the right end of the bar in Example 15.3.4.

$$\begin{aligned} F(t) &= 0 & t &< 0 \\ &= F_0 & t &\geq 0 \end{aligned} \tag{a}$$

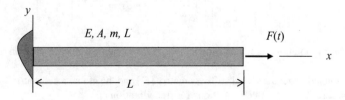

Figure (a)

Solution: Find $f_j(t)$ and then the particular solutions.
The applied force becomes

$$f_j(t) = -\int_0^L \varphi_j f(x,t)dx = -\int_0^L F_0 \sin \alpha_i x \, dx = -\frac{F_0}{\alpha_i}(\cos \alpha_i L - 1) = -\frac{F_0}{\alpha_i} \tag{b}$$

The uncoupled equations of motion are then

$$\ddot{\xi}_j + \omega_j^2 \xi_j = -\frac{F_0}{M_j \alpha_j} \tag{c}$$

This has a solution of the same form as that obtained in Example 15.2.6.

$$\xi_j(t) = A_j \sin \omega_j t + B_j \cos \omega_j t - \frac{F_0}{\omega_j^2 M_j \alpha_j} \tag{d}$$

The same steps as in Example 15.2.6 are followed and the final motion is given by Equation 15.3.17.

$$u(x, t) = \sum_{i=1}^{\infty} \xi_i(t)\varphi_i(x) \qquad\qquad (e)$$

###########

15.3.2   Solutions Based on FEM

From Chapter 5, Section 5.3, we note that the stiffness matrix of an axial member for element m is

$$\{p_m\} = [k_m]\{r_m\} = \frac{E_m A_m}{l_m} \begin{bmatrix} 1 & -1 \\ -1 & 1 \end{bmatrix} \begin{bmatrix} r_{m,n} \\ r_{m,n+1} \end{bmatrix} \qquad (15.3.28)$$

When inertia forces are assigned the equivalent nodal loads resulting from the inertia forces for element m are

$$\{F_I\}_{enl} = \int_0^l [n]^T(-m\ddot{u})\,dA = -\int_0^l [n]^T[n]m\,ds \begin{bmatrix} \ddot{r}_n \\ \ddot{r}_{n+1} \end{bmatrix} = -[m]\{\ddot{r}_m\} \qquad (15.3.29)$$

We have just defined a mass matrix

$$[m_m] = \int_0^l [n]^T[n]m\,ds \qquad\qquad (15.3.30)$$

If m is a constant value the element mass matrix is

$$[m_m] = \int_0^l \begin{bmatrix} 1-\dfrac{s}{l} \\[2mm] \dfrac{s}{l} \end{bmatrix} \begin{bmatrix} 1-\dfrac{s}{l} & \dfrac{s}{l} \end{bmatrix} m\,ds = \frac{m_m l_m}{6} \begin{bmatrix} 2 & 1 \\ 1 & 2 \end{bmatrix} \qquad (15.3.31)$$

There is confusing notation here. The subscript m refers to element m. The other m is the mass per unit length.

These are assembled to get the global stiffness matrix $[K]$ and the global mass matrix $[M]$ in the same way as the static case. The assembled equations for free vibration are

$$[M]\{\ddot{r}\} + [K]\{r\} = \{0\} \qquad\qquad (15.3.32)$$

This is best seen by example.

###########

Example 15.3.6

Problem: Solve for the free vibration of the same problem posed in Example 15.3.2 by the FEM. The initial conditions are

$$u(x, 0) = u_0\frac{x}{L} \qquad \dot{u}(x, 0) = 0 \qquad\qquad (a)$$

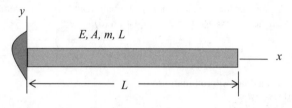

Figure (a)

Solution: Assemble the FE equations and solve using the same methods as those used for the mass/spring system.

Let us use just one element. The node and element numbering is shown in Figure (b).

e1

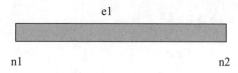

n1   n2

Figure (b)

Then the equation of motion is

$$[M]\{\ddot{r}\} + [K]\{r\} = \frac{mL}{6}\begin{bmatrix} 2 & 1 \\ 1 & 2 \end{bmatrix}\begin{bmatrix} 0 \\ \ddot{u}_2 \end{bmatrix} + \frac{EA}{L}\begin{bmatrix} 1 & -1 \\ -1 & 1 \end{bmatrix}\begin{bmatrix} 0 \\ u_2 \end{bmatrix} = \begin{bmatrix} R_1 \\ 0 \end{bmatrix} \qquad (b)$$

After partitioning the remaining equation for the displacement is

$$\frac{mL}{3}\ddot{u}_2 + \frac{EA}{L}u_2 = 0 \quad \rightarrow \quad \ddot{u}_2 + \frac{3EA}{mL^2}u_2 = \ddot{u}_2 + \omega^2 u_2 = 0 \qquad (c)$$

This is exactly the same form as Equation 15.2.3 for the mass/spring system and has a solution in exactly the same form. The natural frequency is

$$\omega^2 = \frac{3EA}{mL^2} \quad \rightarrow \quad \omega = \sqrt{\frac{3EA}{mL^2}} \qquad (d)$$

Assigning initial conditions

$$u(x, 0) = u_0\frac{x}{L} \quad \rightarrow \quad u(L, 0) = u_2(0) = u_{20} \qquad (e)$$

$$\dot{u}(x, 0) = 0 \quad \rightarrow \quad \dot{u}(L, 0) = \dot{u}_2(0) = 0$$

we obtain

$$u(t) = a\sin\omega t + b\cos\omega t \quad \rightarrow \quad u(0) = b = u_{20} \qquad \dot{u}(0) = a\omega = 0 \qquad (f)$$

The final solution for the motion of node 2 is

$$u_2(t) = u_{20}\cos\omega t \qquad (g)$$

It follows that the complete motion for the bar is approximated by

$$u(s, t) = \begin{bmatrix} 1 - \dfrac{s}{l} & \dfrac{s}{l} \end{bmatrix}\begin{bmatrix} u_1(t) \\ u_2(t) \end{bmatrix}$$

$$= u(x, t) = \begin{bmatrix} 1 - \dfrac{x}{L} & \dfrac{x}{L} \end{bmatrix}\begin{bmatrix} 0 \\ u_2(t) \end{bmatrix} = \frac{x}{L}(u_{20}\cos\omega t) \qquad (h)$$

Only one frequency results for the one element approximation and it may be compared with the exact value for the lowest frequency in Example 15.3.1, Equation (d).

$$\text{FE} \rightarrow \omega_1 = \frac{\sqrt{3}}{L}\sqrt{\frac{EA}{m}} = \frac{1.732}{L}\sqrt{\frac{EA}{m}} \qquad \text{Exact} \rightarrow \omega_1 = \frac{\pi}{2L}\sqrt{\frac{EA}{m}} = \frac{1.571}{L}\sqrt{\frac{EA}{m}} \qquad \text{(i)}$$

This is not bad for such a crude approximation.
The normal mode shape is

$$\varphi_1 = \frac{x}{L} \qquad \text{(j)}$$

where we have arbitrarily assigned $u_{20} = 1$. We can compare this mode with the exact value. The two are plotted in Figure (c). Note that this very crude approximation for the mode shape, namely, a linear function, compared to the exact shape, a sine function, still gives a ball park value for the frequency.

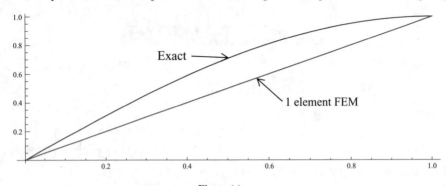

Exact

1 element FEM

Figure (c)

We shall improve on the FEM solution by using more elements. Let us do free vibration with two elements, each of length $L/2$. The nodal and element numbers are shown in Figure (d)

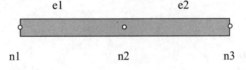

e1   e2

n1   n2   n3

Figure (d)

The assemble global FE equations for free vibration with restraints inserted are

$$\frac{mL}{12}\begin{bmatrix} 2 & 1 & 0 \\ 1 & 4 & 1 \\ 0 & 1 & 2 \end{bmatrix}\begin{bmatrix} 0 \\ \ddot{u}_2 \\ \ddot{u}_3 \end{bmatrix} + \frac{2EA}{L}\begin{bmatrix} 1 & -1 & 0 \\ -1 & 2 & -1 \\ 0 & -1 & 1 \end{bmatrix}\begin{bmatrix} 0 \\ u_2 \\ u_3 \end{bmatrix} = \begin{bmatrix} R_1 \\ 0 \\ 0 \end{bmatrix} \qquad \text{(k)}$$

The effect of the partitioned equations is to obtain

$$\left(\frac{2EA}{L}\begin{bmatrix} 2 & -1 \\ -1 & 1 \end{bmatrix} - \omega^2\frac{mL}{12}\begin{bmatrix} 4 & 1 \\ 1 & 2 \end{bmatrix} \right)\begin{bmatrix} u_2 \\ u_3 \end{bmatrix}e^{i\omega t} = \begin{bmatrix} 0 \\ 0 \end{bmatrix} \qquad \text{(l)}$$

The resulting equations have a solution other than

$$\begin{bmatrix} u_2 \\ u_3 \end{bmatrix} = \begin{bmatrix} 0 \\ 0 \end{bmatrix} \qquad \text{(m)}$$

only if the determinant of the coefficients is equal to zero, that is,

$$\left| \frac{2EA}{L} \begin{bmatrix} 2 & -1 \\ -1 & 1 \end{bmatrix} - \omega^2 \frac{mL}{12} \begin{bmatrix} 4 & 1 \\ 1 & 2 \end{bmatrix} \right| = 0 \tag{n}$$

Expanding the determinant we get a polynomial in ω^2. To simplify things a bit let

$$\lambda = \frac{\omega^2 mL^2}{24EA} \tag{o}$$

Then the determinant becomes

$$\begin{vmatrix} 2 - 4\lambda & -1 - \lambda \\ -1 - \lambda & 1 - 2\lambda \end{vmatrix} = 0 \tag{p}$$

Expanded we get

$$7\lambda^2 - 10\lambda + 1 = 0 \tag{q}$$

The two roots of this polynomial and the natural frequencies are

$$\lambda_1 = 0.1082 \qquad\qquad \lambda_2 = 1.320$$

$$\omega_1^2 = \frac{\lambda_1 24EA}{mL^2} = 2.5968 \frac{EA}{mL^2} \qquad \omega_2^2 = \frac{\lambda_2 24EA}{mL^2} = 31.68 \frac{EA}{mL^2} \tag{r}$$

The frequencies compared to exact values are

$$\begin{array}{llll}
\text{FE} & \rightarrow & \omega_1 = \dfrac{1.611}{L}\sqrt{\dfrac{EA}{m}} & \text{Exact} \rightarrow \omega_1 = \dfrac{\pi}{2L}\sqrt{\dfrac{EA}{m}} = \dfrac{1.571}{L}\sqrt{\dfrac{EA}{m}} \\[4mm]
\text{FE} & \rightarrow & \omega_2 = \dfrac{5.629}{L}\sqrt{\dfrac{EA}{m}} & \text{Exact} \rightarrow \omega_2 = \dfrac{3\pi}{2L}\sqrt{\dfrac{EA}{m}} = \dfrac{4.712}{L}\sqrt{\dfrac{EA}{m}}
\end{array} \tag{s}$$

As we increase the number of elements we get more frequencies and improve the accuracy of the lower frequencies.

Note that this time we shall normalize by factoring c_{13} from the equation. This will provide a unit value for the normal mode displacement at the free end and will simplifying comparing the FEM modes with the exact analytical modes which all have unit values at the free end. Remember that the amplitude of the normal modes can be arbitrarily assigned.

The first mode is found from these equations.

$$\left(\begin{bmatrix} 2 & -1 \\ -1 & 1 \end{bmatrix} - \lambda_1 \begin{bmatrix} 4 & 1 \\ 1 & 2 \end{bmatrix} \right) \begin{bmatrix} c_{12} \\ c_{13} \end{bmatrix} = \begin{bmatrix} 0 \\ 0 \end{bmatrix}$$

$$\rightarrow \left(\begin{bmatrix} 2 & -1 \\ -1 & 1 \end{bmatrix} - 0.1082 \begin{bmatrix} 4 & 1 \\ 1 & 2 \end{bmatrix} \right) c_{13} \begin{bmatrix} c_{12}/c_{13} \\ 1 \end{bmatrix} \tag{t}$$

$$\rightarrow \begin{bmatrix} 1.5672 & -1.1082 \\ -1.1082 & 0.7836 \end{bmatrix} \begin{bmatrix} \varphi_{12} \\ 1 \end{bmatrix} = \begin{bmatrix} 0 \\ 0 \end{bmatrix}$$

From the first equation in Equation (s) we obtain

$$1.5672\varphi_{12} - 1.1082 = 0 \quad \rightarrow \quad \varphi_{12} = 0.707 \tag{u}$$

The first normal mode is then

$$\{\varphi_1\} = \begin{bmatrix} 0.707 \\ 1 \end{bmatrix} \tag{v}$$

The second normal mode is found from these equations.

$$
\left(\begin{bmatrix} 2 & -1 \\ -1 & 1 \end{bmatrix} - \lambda_2 \begin{bmatrix} 4 & 1 \\ 1 & 2 \end{bmatrix} \right) \begin{bmatrix} c_{22} \\ c_{23} \end{bmatrix} = \begin{bmatrix} 0 \\ 0 \end{bmatrix}
$$

$$
\rightarrow \left(\begin{bmatrix} 2 & -1 \\ -1 & 1 \end{bmatrix} - 1.320 \begin{bmatrix} 4 & 1 \\ 1 & 2 \end{bmatrix} \right) c_{23} \begin{bmatrix} c_{22}/c_{23} \\ 1 \end{bmatrix} = \begin{bmatrix} 0 \\ 0 \end{bmatrix} \tag{w}
$$

$$
\rightarrow \begin{bmatrix} -3.28 & -2.32 \\ -2.32 & -1.64 \end{bmatrix} \begin{bmatrix} \varphi_{22} \\ 1 \end{bmatrix} = \begin{bmatrix} 0 \\ 0 \end{bmatrix}
$$

From the first equation in Equation (v) we obtain

$$
- 3.28\varphi_{22} - 2.32 = 0 \quad \rightarrow \quad \varphi_{22} = -0.707 \tag{x}
$$

The second normal mode is

$$
\{\varphi_2\} = \begin{bmatrix} -0.707 \\ 1 \end{bmatrix} \tag{y}
$$

We compare the first two exact analytical normal modes with those obtained by the two element FEM solution. In Figure (e) we compare the first modes.

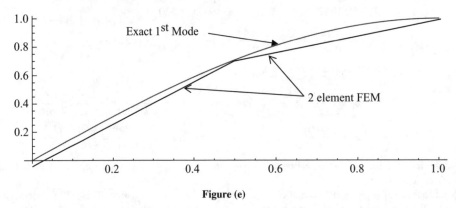

Figure (e)

In Figure (f) we compare the second modes.

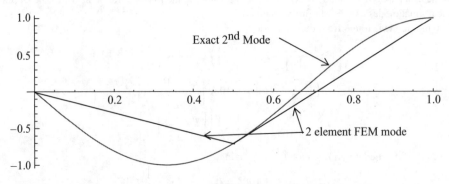

Figure (f)

The complete solution is found by normal modes exactly as in the analytical solution. We just have approximate values for the modes and frequencies and a finite number of modes according to the number of nodes and elements.

The solution in terms of the two normal modes is

$$\{u(t)\} = \begin{bmatrix} u_2(t) \\ u_3(t) \end{bmatrix} = (A_1 \sin \omega_1 t + B_1 \cos \omega_1 t) \begin{bmatrix} 1 \\ 1.414 \end{bmatrix} + (A_2 \sin \omega_2 t + B_2 \cos \omega_2 t) \begin{bmatrix} 1 \\ -1.414 \end{bmatrix} \quad (z)$$

The initial conditions are

$$u_2(0) = \frac{u_0}{2} \qquad \dot{u}_2(0) = 0 \qquad u_3(0) = u_0 \qquad \dot{u}_3(0) = 0 \qquad (aa)$$

Lo and behold, we now have four constants to satisfy the four initial conditions.

$$u_2(0) = (A_1 \sin \omega_1 0 + B_1 \cos \omega_1 0)1$$
$$+ (A_2 \sin \omega_2 0 + B_2 \cos \omega_2 0)1 = B_1 + B_2 = \frac{u_0}{2}$$
$$\dot{u}_2(0) = (A_1 \omega_1 \cos \omega_1 0 - B_1 \omega_1 \sin \omega_1 0)1$$
$$+ (A_2 \omega_2 \cos \omega_2 0 - B_2 \omega_2 \sin \omega_2 0)1 = A_1 \omega_1 + A_2 \omega_2 = 0$$
$$u_3(0) = (A_1 \sin \omega_1 0 + B_1 \cos \omega_1 0)1.414 \qquad\qquad (bb)$$
$$- (A_2 \sin \omega_2 0 + B_2 \cos \omega_2 0)1.414 = 1.414B_1 - 1.414B_2 = u_0$$
$$\dot{u}_3(0) = (A_1 \omega_1 \cos \omega_1 0 - B_1 \omega_1 \sin \omega_1 0)1.414$$
$$- (A_2 \omega_2 \cos \omega_2 0 - B_2 \omega_2 \sin \omega_2 0)1.414 = 1.414A_1 \omega_1 - 1.414A_2 \omega_2 = 0$$

Solving we get

$$A_1 = 0 \qquad A_2 = 0 \qquad B_1 = \frac{1 + \sqrt{2}}{4} u_0 \qquad B_2 = \frac{1 - \sqrt{2}}{4} u_0 \qquad (cc)$$

The resulting motion at the nodes is

$$\{u(t)\} = \begin{bmatrix} u_2(t) \\ u_3(t) \end{bmatrix} = \frac{1 + \sqrt{2}}{4} u_0 \cos \omega_1 t \begin{bmatrix} 1 \\ \sqrt{2} \end{bmatrix} + \frac{1 - \sqrt{2}}{4} u_0 \cos \omega_2 t \begin{bmatrix} 1 \\ -\sqrt{2} \end{bmatrix} \qquad (dd)$$

The distributed displacement is found by using the shape functions for each element.

Increasing the number of elements increases the accuracy of both frequencies and modes. For free motion only a few modes are needed.

###########

The finite element equations for free motion, Equation 15.3.32, are modified to add the applied load.

$$[M]\{\ddot{r}\} + [K]\{r\} = \{F(t)\} \qquad (15.3.33)$$

The coefficient matrices are assembled from the element matrices. For element m the stiffness matrix is

$$[k_m] = \frac{EA}{L} \begin{bmatrix} 1 & -1 \\ -1 & 1 \end{bmatrix} \qquad (15.3.34)$$

and the mass matrix is

$$\{m_m\} = \frac{mL}{6} \begin{bmatrix} 2 & 1 \\ 1 & 2 \end{bmatrix} \qquad (15.3.35)$$

Concentrated loads at nodes can be entered directly. Distributed applied loads must be converted to equivalent nodal loads. From Section 4.4

$$\{F_{enl}\} = \int [n]^T F(t) ds = \int \begin{bmatrix} 1 - \dfrac{s}{l} \\[2mm] \dfrac{s}{l} \end{bmatrix} F(t) ds \tag{15.3.36}$$

############

Example 15.3.7

Problem: Find the response of the bar in Example 15.3.5 to a sinusoidal load in time.

Solution: Use just one element.

For just one element our FE equations become

$$[M]\{\ddot{r}\} + [K]\{r\} = \frac{mL}{6}\begin{bmatrix} 2 & 1 \\ 1 & 2 \end{bmatrix}\begin{bmatrix} 0 \\ \ddot{u}_2 \end{bmatrix} + \frac{EA}{L}\begin{bmatrix} 1 & -1 \\ -1 & 1 \end{bmatrix}\begin{bmatrix} 0 \\ u_2 \end{bmatrix} = \begin{Bmatrix} R_1 \\ F_0 \sin \Omega t \end{Bmatrix} \tag{a}$$

and so the partitioned equation for the motion is

$$\frac{mL}{3}\ddot{u}_2 + \frac{EA}{L}u_2 = F_0 \sin \Omega t \quad \rightarrow \quad \ddot{u}_2 + \frac{3EA}{mL^2}u_2 = \frac{3F_0}{mL}\sin \Omega t \quad \rightarrow \quad \ddot{u}_2 + \omega^2 u_2 = \frac{3F_0}{mL}\sin \Omega t \tag{b}$$

Compare Equation (b) with Equation 15.2.41 with the periodic force applied as shown here

$$\frac{d^2 w}{dt^2} + \omega^2 w = \frac{F_0}{m}\sin \Omega t \tag{c}$$

With the same applied load and the same initial conditions it has the same solution as given in Equation 15.2.49, or

$$u_2(t) = \frac{3F_0}{(\omega^2 - \Omega^2)mL}\left(\sin \Omega t - \frac{\Omega}{\omega}\sin \omega t \right) \tag{d}$$

############

Clearly the same phenomenon of resonance is observed. Anytime an applied frequency is at or near any one of the natural frequencies of a structure there is danger of catastrophic resonance failure. This can be alleviated by adding *damping* to the system, but for now that is another subject.

In our analytic solution we noted an infinite number of frequencies. In the FE solution the number of frequencies you obtain is equal to the number of unrestrained nodal displacements. The lowest frequency is found and the others are in ascending order.

Those needing evidence of the validity of this approach can turn to the mathematical and engineering literature on the classical *eigenvalue* problem.

A majority of problems in vibration are solved by finding the natural frequencies and then adjusting the mass and stiffness of the structure to avoid likely applied frequencies. Alternatively, operating limits can be enforced to ensure that no applied frequency is allowed that is at or near any natural frequency.

Of course, non periodic applied forces also are important, for example, when an airplane enters a gust, performs a maneuver or, say, on landing. Then the mass and stiffness must be adjusted to provide acceptable displacements and stresses or operating constraints must be imposed to provide the acceptable responses.

15.4 Torsional Vibration

15.4.1 Torsional Mass/Spring Systems

Torsional motion can also be described in terms of the rotation of masses and the resistance of springs. Consider the shaft with masses as shown in Figure 15.4.1. The rotation is denoted by $\beta(\tau)$ with units of radians or degrees, the shaft stiffness is denoted by a torsion spring constant k_β with units of moment/radian or degrees (*Nmm/rad Nmm/deg,*), and the polar mass moment of inertia I_M of the disk with units of kilograms-millimeters squared (*kg-mm²*).

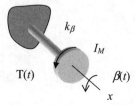

Figure 15.4.1

From summation of moments about the x axis the equation of motion is

$$I_M\ddot{\beta} + k_\beta\beta = M(t) \quad \rightarrow \quad \ddot{\beta} + \omega^2\beta = \frac{T(t)}{I_M} \tag{15.4.1}$$

where

$$\omega^2 = \frac{k_\beta}{I_M} \tag{15.4.2}$$

This equation is exactly the same form as Equation 15.2.3 for the linear mass/spring system. The solutions for free and forced motion are easily obtained by analogy.

For free motion, that is, when $T(t) = 0$, the solution is

$$\beta(t) = A\sin\omega t + B\cos\omega t \tag{15.4.3}$$

and initial conditions are applied in the same way. Multiple degrees of freedom torsional systems are also represented by the matrix equations of the form

$$[I_M]\{\ddot{\beta}\} + [K_\beta]\{\beta\} = \{0\} \tag{15.4.4}$$

The natural frequencies are found from

$$|[K_\beta] - \omega^2[I_M]| = 0 \tag{15.4.5}$$

and the modes from

$$([K_\beta] - \omega_i^2[I_M])\{C\}_i = 0 \tag{15.4.6}$$

The complete solution is

$$\beta(t) = \sum_{i=1}^{n}(A_i\sin\omega_i t + B_i\cos\omega_i t)\varphi_i \tag{15.4.7}$$

where φ_i are the normal modes and the constants A_i and B_i are evaluated using the initial conditions.

Forced motion, that is, when $\{M(t)\} \neq \{0\}$, is defined by the equations

$$[I_M]\{\ddot{\beta}\} + [K_\beta]\{\beta\} = \{\text{T}(t)\} \tag{15.4.8}$$

To find the solution the frequencies and normal modes are found as just discussed for the axial case and a solution is assumed of the form

$$\{\beta\} = \sum_{i=1}^{n} \xi_i(t)\{\varphi_i\} \tag{15.4.9}$$

As in the case of axial vibration

$$\left([K_\beta] - \omega_i^2[I_M]\right)\{\varphi_i\} = 0 \quad \rightarrow \quad [K_\beta]\{\varphi_i\} = \omega_i^2[I_M]\{\varphi_i\} \tag{15.4.10}$$

and the modes are orthogonal

$$\{\varphi_i\}^T[I_M]\{\varphi_j\} = 0 \qquad i \neq j$$
$$= I_{M_j} \qquad i = j \tag{15.4.11}$$

If we substitute Equation 15.4.9 into Equation 15.4.8 we get

$$[I_M]\sum_{i=1}^{n}\ddot{\xi}_i(t)\{\varphi_i\} + [K_\beta]\sum_{i=1}^{n}\xi_i(t)\{\varphi_i\} = \{\text{T}(t)\} \tag{15.4.12}$$

Now use Equation 15.4.10 and pre multiply by φ_j

$$\{\varphi_j\}^T \left([I_M]\sum_{i=1}^{n}\ddot{\xi}_i(t)\{\varphi_i\} + \omega_i^2[I_M]\sum_{i=1}^{n}\xi_i(t)\{\varphi_i\} = \{\text{T}(t)\}\right) \tag{15.4.13}$$

We then rearrange terms and take advantage of orthogonality to obtain

$$I_{M_j}\ddot{\xi}_j + \omega_j^2 I_{M_j}\xi_j = T_j(t) \quad \rightarrow \quad \ddot{\xi}_j + \omega_j^2\xi_j = \frac{T_j(t)}{I_{M_j}} \tag{15.4.14}$$

where

$$\text{T}_j(t) = \{\varphi_j\}^T\{\text{T}(t)\} \tag{15.4.15}$$

The solutions to these equations are obtained in the same way as for the linear mass spring systems.

15.4.2 Distributed Torsional Systems

The differential equation for the torsional vibration of a uniform bar with a circular cross section is obtained from Equation 6.3.10 by including the rotational inertia as part of the applied load.

$$GJ\frac{\partial^2\beta}{\partial x^2} = -t(x) - I_m\frac{\partial^2\beta}{\partial t^2} \quad \rightarrow \quad GJ\frac{\partial^2\beta}{\partial x^2} + I_m\frac{\partial^2\beta}{\partial t^2} = -t(x) \tag{15.4.16}$$

where I_m is the mass polar moment of inertia per unit length about the x axis.

This is the same differential equation form as the axial case and all we learned there carries over with just a change in coefficients. As in the axial case the first step is to find the natural frequencies and this requires separation of variables.

$$\beta(x, t) = \text{B}(x)T(t) \tag{15.4.17}$$

resulting in

$$\frac{d^2B}{dx^2} + \frac{I_m\omega^2}{GJ}B = 0 \qquad \ddot{T} + \omega^2 T = 0 \tag{15.4.18}$$

The solution proceeds exactly as presented in Section 15.3 for the axial case.

In a finite element solution the shape functions and element stiffness matrix are the same as used in Chapter 6, Section 6.6.

$$[n] = \begin{bmatrix} 1 - \dfrac{s}{l_m} & \dfrac{s}{l_m} \end{bmatrix} \qquad [k_{\beta m}] = \frac{GJ}{l_m}\begin{bmatrix} 1 & -1 \\ -1 & 1 \end{bmatrix} \tag{15.4.19}$$

To this we add the mass matrix which by analogy to the axial case is

$$[I_m] = \frac{I_m l_m}{6}\begin{bmatrix} 2 & 1 \\ 1 & 2 \end{bmatrix} \tag{15.4.20}$$

The assembled matrix equations are of the form

$$[I]\{\ddot{r}\} + [K]\{r\} = \{F\} \tag{15.4.21}$$

The element matrices are assembled exactly as presented in Section 15.3. Again it should be noted that the shape functions are not exact in this case since the inertia forces are a distributed, that is, not concentrated, load.

15.5 Vibration of Beams in Bending

15.5.1 Solutions of the Differential Equation

For the free vibration of uniform beams we have

$$EI\frac{\partial^4 v}{\partial x^4} = f_y(x, t) = -m\ddot{v} \quad \rightarrow \quad EI\frac{\partial^4 v}{\partial x^4} + m\ddot{v} = 0 \tag{15.5.1}$$

Separate variables

$$v(x, t) = V(x)T(t) \tag{15.5.2}$$

to obtain

$$EI\frac{d^4 V}{dx^4} - m\omega^2 V = 0 \qquad \ddot{T} + \omega^2 T = 0 \tag{15.5.3}$$

The first of these equations may be rewritten

$$\frac{d^4 V}{dx^4} - \frac{m\omega^2}{EI}V = 0 \quad \rightarrow \quad \frac{d^4 V}{dx^4} - \alpha^4 V = 0 \tag{15.5.4}$$

The general solution for $V(x)$ is found by assuming an exponential form.

$$V \approx Ce^{ax} \tag{15.5.5}$$

Substituting this into Equation 15.5.4 gives us

$$a^4 Ce^{ax} - \alpha^4 Ce^{ax} = 0 \quad \rightarrow \quad a^4 = \alpha^4 \tag{15.5.6}$$

There are two pairs of roots. One pair $\pm\alpha$ is real and the other $\pm i\alpha$ is imaginary. Thus

$$V(x) = C_1 e^{\alpha x} + C_2 e^{-\alpha x} + C_3 e^{i\alpha x} C_4 e^{-l\alpha x} \tag{15.5.7}$$

which can be put in the more convenient form of

$$V(x) = c \sinh \alpha x + d \cosh \alpha x + e \sin \alpha x + f \cos \alpha x \qquad (15.5.8)$$

To find the natural frequencies substitute Equation 15.5.8 into the boundary conditions. An example will help.

##########

| Example 15.5.1 |

Problem: Find the natural frequencies and modes of the uniform simply supported beam shown here in Figure (a).

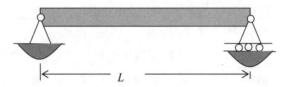

Figure (a)

Solution: Insert Equation 15.5.8 into the boundary conditions to obtain the eigenvalue matrix. Find the frequencies and mode shapes.

The boundary conditions are that the displacements and moments are zero at both ends.

$$v(0, t) = V(0)T(t) = 0 \quad \rightarrow \quad V(0) = 0$$

$$EI \frac{\partial^2 v(0, t)}{\partial x^2} = EI \frac{d^2 V(0)}{dx^2} T(t) = 0 \quad \rightarrow \quad \frac{d^2 V(0)}{dx^2} = 0$$

$$v(L, t) = V(L)T(t) = 0 \quad \rightarrow \quad V(L) = 0 \qquad (a)$$

$$EI \frac{\partial^2 v(L, t)}{\partial x^2} = EI \frac{d^2 V(L)}{dx^2} T(t) = 0 \quad \rightarrow \quad \frac{d^2 V(L)}{dx^2} = 0$$

Upon substitution

$$V(0) = c \sinh \alpha 0 + d \cosh \alpha 0 + e \sin \alpha 0 + f \cos \alpha 0 = d + f = 0$$

$$\frac{d^2 V(0)}{dx^2} = c\alpha^2 \sinh \alpha 0 + d\alpha^2 \cosh \alpha 0 - e\alpha^2 \sin \alpha 0 - f\alpha^2 \cos \alpha 0 = \alpha^2(d - f) = 0$$

$$V(L) = c \sinh \alpha L + d \cosh \alpha L + e \sin \alpha L + f \cos \alpha L = 0 \qquad (b)$$

$$\frac{d^2 V(L)}{dx^2} = c\alpha^2 \sinh \alpha L + d\alpha^2 \cosh \alpha L - e\alpha^2 \sin \alpha L - f\alpha^2 \cos \alpha L = 0$$

When we have a set of homogeneous linear algebraic equations of the form

$$[\Gamma]\{C\} = 0 \qquad (c)$$

either

$$\{C\} = \{0\} \quad \text{or} \quad |\Gamma| = 0 \qquad (d)$$

that is, the unknowns are zero or the determinant of the coefficients is zero.

When we have an eigenvalue problem the trivial solution $\{C\} = \{0\}$ would proclaim that we had no motion. But our experience says this is not so; thus, we look for the conditions where $|\Gamma| = 0$.

We can immediately reduce the size of the set of equations by noting that from the first two equations

$$d = f = 0 \tag{e}$$

The remaining equations are

$$c \sinh \alpha L + e \sin \alpha L = 0$$
$$c \sinh \alpha L - e \sin \alpha L = 0 \tag{f}$$

Adding the two we get

$$c \sinh \alpha L = 0 \quad \rightarrow \quad c = 0 \tag{g}$$

which leaves us with

$$e \sin \alpha L = 0 \tag{h}$$

and either $e = 0$, the trivial solution, or

$$\sin \alpha L = 0 \quad \rightarrow \quad \alpha L = i\pi \quad \rightarrow \quad \alpha = \frac{i\pi}{L} \tag{i}$$

where i is an integer. And there we have our natural frequencies.

$$\alpha^4 = \left(\frac{i\pi}{L}\right)^4 = \frac{m\omega^2}{EI} \quad \rightarrow \quad \omega = \left(\frac{i\pi}{L}\right)^2 \sqrt{\frac{EI}{m}} \tag{j}$$

The mode shapes and the normal modes are

$$V(x) = e \sin \frac{i\pi x}{L} \quad \rightarrow \quad \varphi_i = \sin \frac{i\pi x}{L} \tag{k}$$

We know the shape of the mode but the amplitude is unknown so we arbitrarily assign $e = 1$ for plotting purposes.

############

For forced motion the equation of motion is

$$EI \frac{\partial^4 v}{\partial x^4} + m\ddot{v} = f_y(x, t) \tag{15.5.9}$$

Just as we did in the axial and torsional cases we assume a solution of the form

$$v(x, t) = \sum_{i=1}^{\infty} \xi_i(t)\varphi_i(x) \tag{15.5.10}$$

We insert this form in Equation 15.7.9

$$EI \frac{\partial^4}{\partial x^4} \sum_{i=1}^{\infty} \xi_i(t)\varphi_i(x) - m \sum_{i=1}^{\infty} \ddot{\xi}_i(t)\varphi_i(x) = -f_y(x, t) \tag{15.5.11}$$

We can rearrange the terms.

$$EI \sum_{i=1}^{\infty} \xi_i \frac{\partial^4 \varphi_i}{\partial x^4} - m \sum_{i=1}^{\infty} \ddot{\xi}\varphi_i = -f_y(x, t) \tag{15.5.12}$$

Now multiply all terms by $\varphi_j(x)$ and integrate all terms over the length of the bar.

$$\int_0^L \varphi_j \left[EI \sum_{i=1}^\infty \xi_i \frac{d^4\varphi_i}{dx^4} - m \sum_{i=1}^\infty \ddot{\xi}\varphi_i = -f_y(x,t) \right] dx$$

$$= \sum_{i=1}^\infty \xi_i \int_0^L EI\varphi_j \frac{d^4\varphi_i}{dx^4}dx - \sum_{i=1}^\infty \ddot{\xi}_i \int_0^L \varphi_j m\varphi_i dx = -\int_0^L \varphi_j f_y(x,t)dx \tag{15.5.13}$$

The normal modes are solutions to the separated equation

$$EI\frac{d^4V}{dx^4} + m\omega^2 V = 0 \tag{15.5.14}$$

It follows that

$$EI\frac{d^4\varphi_i}{dx^4} + m\omega_i^2\varphi_i = 0 \quad \rightarrow \quad EI\frac{d^4\varphi_i}{dx^4} = -m\omega_i^2\varphi_i \tag{15.5.15}$$

By replacing $EI\frac{d^4\varphi_i}{dx^4}$ with $-m\omega_i^2\varphi_i$ Equation 15.5.13 becomes

$$-\sum_{i=1}^\infty \xi_i\omega_i^2 \int_0^L \varphi_j m\varphi_i dx - \sum_{i=1}^\infty \ddot{\xi}_i \int_0^L \varphi_j m\varphi_i dx = \int_0^L \varphi_j f_y(x,t)dx \tag{15.5.16}$$

We designate

$$f_{yj}(t) = -\int_0^L \varphi_j f_y(x,t)dx \tag{15.5.17}$$

We note that from orthogonality

$$\int_0^L \varphi_j m\varphi_i dx = 0 \qquad i \neq j$$
$$= M_j \qquad i = j \tag{15.5.18}$$

Finally the equation for forced motion may be written

$$\ddot{\xi}_j + \omega_j^2\xi = \frac{f_{yj}(t)}{M_j} \tag{15.5.19}$$

We now solve for $\xi_j(t)$ and insert it in the series of Equation 15.7.10.
Once again an example will help.

###########

Example 15.5.2

Problem: The simple supported beam in Example 15.5.1 is initially at rest and then is subjected to a dynamic lateral load as shown in Figure (a).

$$f_y(t) = f_0 \sin \Omega t \tag{a}$$

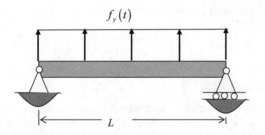

Figure (a)

Solution: Find a solution based on a series of normal modes.

The governing equation is

$$\ddot{\xi}_j + \omega_j^2 \xi = \frac{f_{yj}(t)}{M_j} \tag{b}$$

where

$$f_{yj}(t) = \int_0^L \varphi_j f_y(t)dx = \int_0^L \sin\frac{j\pi x}{L} f_0 \sin\Omega t\, dx = f_0 \frac{L}{j\pi}(1 - \cos j\pi)\sin\Omega t \tag{c}$$

When j is an even integer $f_{yj}(t) = 0$. This is logical since there is symmetry of geometry, restraint, and loading about the mid length of the beam. All modes for j even are anti symmetric about the mid length.

Thus

$$f_{yj}(t) = 2f_0\frac{L}{j\pi}\sin\Omega t \qquad j = \text{odd}$$
$$= 0 \qquad\qquad\qquad j = \text{even} \tag{d}$$

We also note that

$$M_j = \int_0^L \varphi_j^2 m\,dx = \int_0^L m\sin^2\frac{j\pi}{L}x\,dx = \frac{mL}{2} \tag{e}$$

The solution to Equation (b) is

$$\xi_j(t) = A_j\sin\omega_j t + B_j\cos\omega_j t + \xi_{jparticular} \tag{f}$$

The particular solution for an equation of this form was found in Section 15.2, Equation 15.2.46. We just need to use the correct coefficients.

$$\xi_{jparticular} = \frac{f_{yj}(t)}{(\omega_j^2 - \Omega^2)M_j}\sin\Omega t = \frac{4f_0}{j\pi(\omega_j^2 - \Omega^2)m}\sin\Omega t \qquad j = \text{odd} \tag{g}$$

From Equation 15.5.10

$$v(x,t) = \sum_{j=1}^{\infty}\xi_j(t)\varphi_j(x) = \sum_{j=1}^{\infty}\left(A_j\sin\omega_j t + B_j\cos\omega_j t + \frac{4f_0}{j\pi(\omega_j^2 - \Omega^2)m}\sin\Omega t\right)\sin\frac{j\pi}{L}x \tag{h}$$

Apply initial conditions

$$v(0) = 0 \quad \dot{v}(0) = 0 \tag{i}$$

And things proceed just as they have done in other cases.

$$\#\#\#\#\#\#\#\#\#\#$$

15.5.2   Solutions Based on FEM

For the finite element formulation of the beam vibration problem we use the same shape functions developed in Chapter 8, Section 8.2, Equation 8.2.5. The shape function matrix for element m is

$$[n] = \frac{1}{l_m^3}\left[\, l_m^3 - 3l_m s^2 + 2s^3 \quad l_m^3 s - 2l_m^2 s^2 + l_m s^3 \quad 3l_m s^2 - 2s^3 \quad -l_m^2 s^2 + l_m s^3 \,\right] \tag{15.5.20}$$

The stiffness matrix for element m was found to be

$$[k_m] = \frac{E_m I_m}{l_m^3}\begin{bmatrix} 12 & 6l_m & -12 & 6l_m \\ 6l_m & 4l_m^2 & -6L & 2l_m^2 \\ -12 & -6l_m & 12 & -6l_m \\ 6l_m & 2l_m^2 & -6L & 4l_m^2 \end{bmatrix} \tag{15.5.21}$$

If we use D'Alembert's principle the equivalent nodal load for the mass term is

$$\{F\} = \int_0^{l_m} [n]^T \{F_I\}ds = -\int_0^{l_m} [n]^T m\ddot{v}\,ds = -\int [n]^T [n]m ds \begin{bmatrix} \ddot{v}_1 \\ \ddot{\theta}_1 \\ \ddot{v}_2 \\ \ddot{\theta}_2 \end{bmatrix} \tag{15.5.22}$$

We define a mass matrix for element m.

$$[m_m] = \int_0^{l_m} [n]^T [n]m ds \tag{15.5.23}$$

Then the equation of motion for an element becomes

$$[m_m]\{\ddot{r}_m\} + [k_m]\{r_m\} = \{0\} \tag{15.5.24}$$

where

$$\{\ddot{r}_m\} = \begin{bmatrix} \ddot{v}_n \\ \ddot{\theta}_n \\ \ddot{v}_{n+1} \\ \ddot{\theta}_{n+1} \end{bmatrix} \tag{15.5.25}$$

When evaluated the mass matrix becomes

$$[m_m] = \int_0^{l_m} [n]^T [n]m ds = \frac{m l_m}{420}\begin{bmatrix} 156 & 22l_m & 54 & -13l_m \\ 22l_m & 4l_m^2 & 13l_m & -3l_m^2 \\ 54 & 13l_m & 156 & -22l_m \\ -13l_m & -3l_m^2 & -22l_m & 4l_m^2 \end{bmatrix} \tag{15.5.26}$$

These are assembled to get the global stiffness matrix $[K]$ and the global mass matrix $[M]$ in the same way as the static case to form the global equations.

$$[M]\{\ddot{r}\} + [K]\{r\} = \{0\} \tag{15.5.27}$$

###########

Example 15.5.3

Problem: Let us find the first two natural frequencies of a uniform beam with fixed supports on both ends as shown in Figure (a) suing the FEM.

Figure (a)

Solution: Use two elements.

We divide the beam into two elements with three nodes and assemble the global mass and stiffness matrices. The mass matrix is

$$[M] = \frac{mL}{840} \begin{bmatrix} 156 & 22\left(\frac{L}{2}\right) & 54 & -13\left(\frac{L}{2}\right) & 0 & 0 \\ 22\left(\frac{L}{2}\right) & 4\left(\frac{L}{2}\right)^2 & 13\left(\frac{L}{2}\right) & -3\left(\frac{L}{2}\right)^2 & 0 & 0 \\ 54 & 13\left(\frac{L}{2}\right) & 312 & 0 & 54 & -13\left(\frac{L}{2}\right) \\ -13\left(\frac{L}{2}\right) & -3\left(\frac{L}{2}\right)^2 & 0 & 8\left(\frac{L}{2}\right)^2 & 13\left(\frac{L}{2}\right) & -3\left(\frac{L}{2}\right)^2 \\ 0 & 0 & 54 & 13\left(\frac{L}{2}\right) & 156 & -22\left(\frac{L}{2}\right) \\ 0 & 0 & -13\left(\frac{L}{2}\right) & -3\left(\frac{L}{2}\right)^2 & -22\left(\frac{L}{2}\right) & 4\left(\frac{L}{2}\right)^2 \end{bmatrix} \tag{a}$$

and the stiffness matrix is

$$[K] = \frac{8EI}{L^3} \begin{bmatrix} 12 & 6\left(\frac{L}{2}\right) & -12 & 6\left(\frac{L}{2}\right) & 0 & 0 \\ 6\left(\frac{L}{2}\right) & 4\left(\frac{L}{2}\right)^2 & -6\left(\frac{L}{2}\right) & 2\left(\frac{L}{2}\right)^2 & 0 & 0 \\ -12 & -6\left(\frac{L}{2}\right) & 24 & 0 & -12 & 6\left(\frac{L}{2}\right) \\ 6\left(\frac{L}{2}\right) & 2\left(\frac{L}{2}\right)^2 & 0 & 8\left(\frac{L}{2}\right)^2 & -6\left(\frac{L}{2}\right) & 2\left(\frac{L}{2}\right)^2 \\ 0 & 0 & -12 & -6\left(\frac{L}{2}\right) & 12 & -6\left(\frac{L}{2}\right) \\ 0 & 0 & 6\left(\frac{L}{2}\right) & 2\left(\frac{L}{2}\right)^2 & -6\left(\frac{L}{2}\right) & 4\left(\frac{L}{2}\right)^2 \end{bmatrix} \tag{b}$$

The boundary conditions are

$$v_1 = \theta_1 = v_3 = \theta_3 = 0 \tag{c}$$

The partitioned equations reduce to just two.

$$[M]\{\ddot{r}\} + [K]\{r\} = \frac{mL}{840}\begin{bmatrix} 312 & 0 \\ 0 & 8\left(\frac{L}{2}\right)^2 \end{bmatrix}\begin{bmatrix} \ddot{v}_2 \\ \ddot{\theta}_2 \end{bmatrix} + \frac{8EI}{L^3}\begin{bmatrix} 24 & 0 \\ 0 & 8\left(\frac{L}{2}\right)^2 \end{bmatrix}\begin{bmatrix} v_2 \\ \theta_2 \end{bmatrix} = \begin{bmatrix} 0 \\ 0 \end{bmatrix} \tag{d}$$

Assuming

$$\{r\} = \{r_c\}e^{i\omega t} \tag{e}$$

and substituting

$$\left(\begin{bmatrix} 24 & 0 \\ 0 & 8\left(\frac{L}{2}\right)^2 \end{bmatrix} - \frac{\omega^2 mL^4}{840 \cdot 8 \cdot EI}\begin{bmatrix} 312 & 0 \\ 0 & 8\left(\frac{L}{2}\right)^2 \end{bmatrix}\right)\begin{bmatrix} v_{c2} \\ \theta_{c2} \end{bmatrix} = \begin{bmatrix} 0 \\ 0 \end{bmatrix} \tag{f}$$

Let

$$\gamma = \frac{\omega^2 mL^4}{840 \cdot 8 \cdot EI} \tag{g}$$

The determinant of the coefficients is

$$\left| \begin{bmatrix} 24 & 0 \\ 0 & 8\left(\frac{L}{2}\right)^2 \end{bmatrix} - \gamma\begin{bmatrix} 312 & 0 \\ 0 & 8\left(\frac{L}{2}\right)^2 \end{bmatrix} \right| = 0 \tag{h}$$

This expands to

$$(24 - 312\gamma)(1 - \gamma)8\left(\frac{L}{2}\right)^2 = 0 \tag{i}$$

For the lowest frequency we get

$$(24 - 312\gamma) = 0 \quad \rightarrow \quad \gamma = \frac{1}{13} = 0.0769 \quad \rightarrow \quad \omega_1^2 = 516.92\frac{EI}{mL^4} \tag{j}$$

The second frequency is

$$(1 - \gamma) = 0 \quad \rightarrow \quad \gamma = 1 \quad \rightarrow \quad \omega_2^2 = 6720\frac{EI}{mL^4} \tag{k}$$

Thus

$$\omega_1 = 22.74\sqrt{\frac{EI}{mL^4}} \qquad \omega_2 = 81.98\sqrt{\frac{EI}{mL^4}} \tag{l}$$

The exact values from a differential equation solution

$$\omega_1 = 22.37\sqrt{\frac{EI}{mL^4}} \qquad \omega_2 = 61.67\sqrt{\frac{EI}{mL^4}} \tag{m}$$

Not bad for such a crude approximation.

##########

15.6   The Finite Element Method for all Elastic Structures

The element FE equations for the vibration of all elastic structures are of the form

$$[m_m]\{\ddot{r}_m\} + [k_m]\{r_m\} = \{F_m\} \tag{15.6.1}$$

These are assembled into global equations of the form

$$[M]\{\ddot{r}\} + [K]\{r\} = \{F\} \tag{15.6.2}$$

This is exactly the same form as the equations for multi degree of freedom mass/spring systems described in detail in Section 15.2.

The natural frequencies ω_i and the corresponding modes $\{\varphi\}_i$ are found from the homogeneous equations.

$$[M]\{\ddot{r}\} + [K]\{r\} = \{0\} \tag{15.6.3}$$

We assume harmonic motion (a fast way to separate variables) of the form

$$\{r(t)\} = \{r_c\}e^{i\omega t} \tag{15.6.4}$$

and obtain

$$([K] - \omega^2[M])\{r_c\}e^{i\omega t} = \{0\} \tag{15.6.5}$$

The frequencies are found by setting the determinant of the coefficients to zero.

$$|[K] - \omega^2[M]| = 0 \tag{15.6.6}$$

By expanding this determinant we get a polynomial in ω^2 and the roots of this polynomial are the natural frequencies. The natural frequencies are reinserted into the homogeneous equations and the normal modes are found. All that is different for various structures, such as, beams, shafts, plates, shells, and 3D solid elastic bodies are the specific forms of the mass and stiffness matrices.

Solutions for specific initial conditions and applied forces are found by assuming a series solution based on the normal modes (see Equation 15.2.54)

$$\{r\} = \sum_{i=1}^{n} \xi_i(t)\{\varphi\}_i \tag{15.6.7}$$

and proceeding as presented in Equations 15.2.55-61.

In all cases resonance is encountered whenever an applied frequency is at or near a natural frequency.

15.7   Addition of Damping

The motion of structures as described in the previous sections of this chapter continue indefinitely once started by imposing initial conditions and/or applied loads. In practice this motion often is modified or even dies out as a result of forces that resist the motion. These are called the *damping forces*. The actual physical phenomena that cause damping are complex and generally hard to model mathematically. The actual damping mechanisms are beyond the space available and the scope of this text; however, one idealization of the true damping forces that has a certain amount of validity and mathematical simplicity is called *viscous damping*. This type generates a force proportional to the velocity and in the opposite direction of the velocity.

An additional device called a *viscous damper* is often added to mass spring systems. It is shown schematically in Figure 15.7.1.

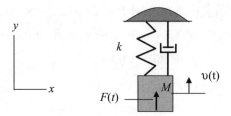

Figure 15.7.1

This device exerts a force opposing the motion proportional to the velocity, that is,

$$F_D = -\mu \dot{v} \tag{15.7.1}$$

where μ is the *damping coefficient* and has the units of kilograms per second (kg/s) or Newton-seconds per millimeter (Ns/mm). Add this to the free body diagram as shown in Figure 15.7.2.

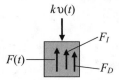

Figure 15.7.2

The equation of motion becomes

$$F(t) + F_I + F_D - kv = 0$$
$$\rightarrow \; F(t) - M\ddot{v} - \mu\dot{v} - kv = 0$$
$$\rightarrow \; M\ddot{v} + \mu\dot{v} + kv = F(t) \tag{15.7.2}$$
$$\rightarrow \; \ddot{v} + \beta\dot{v} + \omega^2 v = \frac{F(t)}{M}$$

where

$$\beta = \frac{\mu}{M} \qquad \omega^2 = \frac{k}{M} \tag{15.7.3}$$

To solve this equation we try

$$v \approx C e^{\lambda t} \tag{15.7.4}$$

The result is

$$(\lambda^2 + \beta\lambda + \omega^2)C e^{\lambda t} = 0 \tag{15.7.5}$$

Since $C e^{\lambda t} \neq 0$ it follows that

$$\lambda^2 + \beta\lambda + \omega^2 = 0 \tag{15.7.6}$$

Using the quadratic formula we obtain

$$\lambda_1 = -\frac{\beta}{2} + \sqrt{\frac{\beta^2}{4} - \omega^2} \qquad \lambda_2 = -\frac{\beta}{2} - \sqrt{\frac{\beta^2}{4} - \omega^2} \tag{15.7.7}$$

When

$$\beta^2 \geq 4\omega^2 \tag{15.7.8}$$

the roots are real and negative. The solution to the homogenous equation is then

$$v = Ae^{\lambda_1 t} + Be^{\lambda_2 t}$$

Both the terms represent an exponential decay with time. When

$$\beta^2 \leq 4\omega^2 \tag{15.7.9}$$

both roots are complex

$$\lambda_1 = -\frac{\beta}{2} + i\Phi \qquad \lambda_2 = -\frac{\beta}{2} + i\Phi \qquad \Phi = \sqrt{\omega^2 - \frac{\beta^2}{4}} \tag{15.7.10}$$

and the homogenous solution for free motion may be written as

$$v(t) = e^{-\frac{\beta}{2}t}(ae^{i\Phi t} + be^{-i\Phi t}) = e^{-\frac{\beta}{2}t}(A \sin \Phi t + B \cos \Phi t) \tag{15.7.11}$$

This is an oscillatory motion of frequency Φ multiplied by an exponential decay.

##########

Example 15.7.1

Problem: A single degree of freedom mass/spring system with viscous damping has the following properties.

$$M = 2 \, kg \qquad k = 32 \, \frac{N}{mm} \qquad \mu = 0.8 \, \frac{kg\text{-}sec}{mm} \tag{a}$$

From its static equilibrium position it is given an initial displacement of 4 mm, no initial velocity, and released. Find the natural frequency and resulting motion.

Solution: Use Equations 15.7.11 to find the motion.
The natural frequency for the undamped system, as found in Example 15.2.1, is

$$\omega = \sqrt{\frac{k}{M}} = \sqrt{\frac{32}{2}} = 4 \, rad/s \quad \rightarrow \quad f = \frac{\omega}{2\pi} = \frac{4}{2\pi} = 0.6366 \, Hz \tag{b}$$

With damping it is

$$\Phi = \sqrt{\omega^2 - \frac{\beta^2}{4}} = \sqrt{16 - 0.04} = \sqrt{15.96} = 3.995 \, rad/s \tag{c}$$

Apply initial conditions to Equation 15.7.11 we get

$$v(t) = e^{-\frac{\beta}{2}t}(A \sin \Phi t + B \cos \Phi t)$$

$$\rightarrow \quad v(0) = e^0 (A \sin 0 + B \cos 0) = 4 \quad \rightarrow \quad B = 4$$

$$\dot{v}(t) = e^{-\frac{\beta}{2}t}(A\Phi \cos \Phi t - B\Phi \sin \Phi t) - (A \sin \Phi t + B\Phi \cos \Phi t)\frac{\beta}{2}e^{-\frac{\beta}{2}t} = 0 \tag{d}$$

$$\rightarrow \quad \dot{v}(0) = e^0(A\Phi \cos 0 - B\Phi \sin 0) - (A\Phi \sin 0 + B\Phi \cos 0)\frac{\beta}{2}e^0 = 0$$

$$\rightarrow \quad A\Phi - \frac{B\Phi\beta}{2} = 0 \quad \rightarrow \quad A = \frac{B}{2}\beta = \frac{4}{2}0.4 = 0.8$$

Thus

$$v(t) = e^{-\frac{\beta}{2}t}(A \sin \Phi t + B \cos \Phi t) = e^{-0.2t}(0.8 \sin 3.995\, t + 4 \cos 3.995\, t) \tag{e}$$

A plot of the first few cycles of the motion is given in Figure (a).

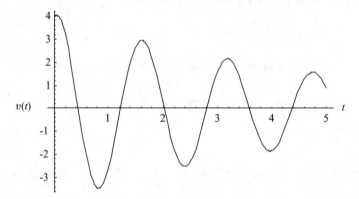

Figure (a)

############

Now let us consider an oscillatory applied force

$$F(t) = F_0 \sin \Omega t \tag{15.7.12}$$

that is,

$$\ddot{v} + \beta\dot{v} + \omega^2 v = \frac{F(t)}{M} = \frac{F_0 \sin \Omega t}{M} \tag{15.7.13}$$

A particular solution has been found by assuming the following form for the trial solution

$$v_p = P(\sin \Omega t - \psi) \tag{15.7.14}$$

When this is substituted into Equation 15.7.13 and the results collected we find that

$$P = \frac{F}{\sqrt{\left(1 - \frac{\Omega^2}{\omega^2}\right)^2 + \frac{\beta^2 \Omega^2}{\omega^2}}} \qquad \tan \psi = \frac{\frac{\beta \Omega}{\omega^2}}{\left(1 - \frac{\Omega^2}{\omega^2}\right)} \qquad F = \frac{F_0}{\omega^2 M} \qquad (15.7.15)$$

We note that the homogenous part of the solution dies out in time for all cases of positive damping leaving only the particular solution, which for this reason is called the *steady-sate solution*. It is instructive to plot the amplitude of the steady-state solution, or more precisely, the absolute value of $\frac{P}{F}$ versus $\frac{\Omega}{\omega}$, for several values of damping β. The effects of damping are very clear. The resonance peaks are now finite.

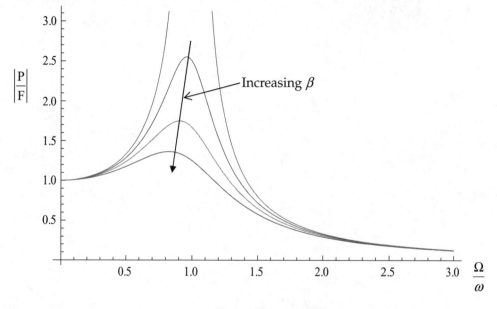

Figure 15.7.3

This can be expanded to multi degree of freedom systems forming the equations

$$[M]\{\ddot{v}\} + [C]\{\dot{v}\} + [K]\{v\} = \{F(t)\} \qquad (15.7.16)$$

Again the natural frequencies and normal modes are found and a solution based on a series of normal modes is sought.

As noted, the dynamics of structures can be formed into equations that mathematically are the same as mass/spring/damper systems via FEM and the same solution methods used. The finite element equations are

$$[M]\{\ddot{r}\} + [C]\{\dot{r}\} + [K]\{r\} = \{F(t)\} \qquad (15.7.17)$$

where $[C]$ is the damping matrix.

15.8   Summary and Conclusions

A brief introduction to the subject of structural dynamics has been completed. It places much attention on discrete mass spring systems because this is the clearest and simplest way to explain and demonstrate the process of solution. The equations of motion of continuous structures are first solved for the natural frequencies and modes. Then a solution is sought by forming a series which is the sum of the natural modes. In both classical and FEM analysis this reduces the problem to solving equations which are the same mathematically as mass/spring and mass/spring/damper systems.

The addition of inertia forces introduces time as a variable and requires a new process called separation of variables. Part of the problem reduces to solving an eigenvalue equation. This solution method is familiar from the study of buckling. Advanced treatment of structural dynamics can be found in many modern textbooks, and the reader is recommended to consult Craig (2006)) for a broader coverage of the subject.

16

Evolution in the (Intelligent) Design and Analysis of Structural Members

16.1   Introduction

In this chapter we shall follow a series of steps taken in designing and analyzing, then repeated re-designing and analyzing, until a final design and analysis of a part is achieved for two specific cases. In each case the goal will be to reduce high stresses and deformations as appropriate, always keeping in mind the reduction of weight or amount of material. A commercial FEM computer program will be used.

The subject of design as distinct from analysis was introduced briefly in Chapter 4, Section 4.9, for an axially loaded bar, and in Section 4.10 for a pin jointed truss. Then in Chapter 7, Section 7.10, the design of beams was considered. As is noted in those sections, in analysis we specify exactly the geometry, materials, loads, and restraints and ask for the displacements, strains, and stresses. We wish to verify that the structure meets minimum requirements on load carrying ability.

In design, we are given a task to perform by a structural member, and we inquire what geometry and materials should be used to achieve a certain acceptable value of stress and displacement. The loads and restraints are determined as suitable to the task. Any extensive discussion of design is well beyond the scope of this text, and the interested student can follow on by taking a modern course in structural design optimization that introduces the student to the ideas of finding solutions to problems with constraints. Since we have developed analysis tools, in particular the finite element method, to include two and three dimensional structures, we take this opportunity to extend our understanding of design methods.

Again as previously noted, design is a much broader subject than that just stated. In addition to the geometric restraints we may have restraints on weight, cost, and ease of manufacture. Safety and margins of safety are ever present. Aesthetics are an additional consideration in many cases, such as the body panels on an automobile and the interior of an airliner. Because of the broad nature of design it is largely a team effort. We cannot begin to study all of these concerns so we shall limit ourselves to geometry, geometric restraints, materials, and weight for a given state of loading. These are the immediate concerns of the structural analyst member of the team.

Analysis of Structures: An Introduction Including Numerical Methods, First Edition. Joe G. Eisley and Anthony M. Waas.
© 2011 John Wiley & Sons, Ltd. Published 2011 by John Wiley & Sons, Ltd.

For an axially loaded slender bar the geometry consists of the length and the cross sectional area. The exact shape of the cross section might be determined by considerations of manufacturing, fasteners and restraints, points or regions of load application, or other considerations. We shall restrict our discussion to the role of modifying the geometry and material to improve the design.

The first example is a truss member. It will be analyzed first by 1D slender bar theory, then by 2D plane stress theory, and finally by 3D solid body theory. There will follow a plane stress example.

These examples are limited to just a few of the cases considered in the previous chapters; however, they illustrate the process of improving design through analysis. The same process can be applied to many other examples.

16.2 Evolution of a Truss Member

Suppose a pin jointed truss has several members. In the first analysis of a truss the axial load in each member is found; however, the geometry of the joints where the members are pinned together is not modeled. At first the truss will be analyzed by the methods used in Chapter 3, Section 3.10 and Chapter 4, Section 4.4. One member of the truss has a known tensile load on it and is isolated for further design changes and analysis. The member we choose is 1000 *mm* long and has an axial load of 100000 *N*. We shall now account for the way the truss member is pinned to the base or to other members and take the truss member through several design changes as shown in Figure 16.2.1.

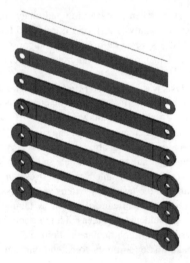

Figure 16.2.1

16.2.1   Step 1. Slender Bar Analysis

At the top in Figure 16.2.1 we represent a slender bar with just a line. This line is the centroidal axis of the axially loaded member. In slender bar analysis the length and the cross sectional area are the only geometric quantities of interest. The actual shape of the cross section and any details on how this member is attached to a base or to other truss members are ignored. For the given load we shall assign a cross section area to achieve a desired stress level and a satisfactory displacement.

We shall give it an initial cross sectional area of 1000 *mm²* without specifying a shape for the cross section. The load is a total force of 100000 *N*. From slender bar theory, as explained in Chapter 4, the stress is,

$$\sigma_x = \frac{100000}{1000} = 100 \, N/mm^2 \text{ or } Mpa \tag{16.2.1}$$

If the bar is made of steel with a Young's modulus of 206800 N/mm^2 the total elongation of the bar is

$$u(L) = \frac{FL}{AE} = \frac{100000 \cdot 1000}{1000 \cdot 206800} = 0.484 \, mm \tag{16.2.2}$$

If we choose a steel alloy with a yield stress of approximately 350 Mpa the factor of safety based on the yield stress is

$$\text{F.S.} = \frac{350}{100} = 3.5 \tag{16.2.3}$$

At this point the large factor of safety is to account in part for the fact that we don't know how the part is attached to a base or to other truss members. We do not need to know the actual shape of the cross section, since the area of the cross section is all we need to know in this analysis.

16.2.2 Step 2. Rectangular Bar—Plane Stress FEM

The second one down in Figure 16.2.1 is a member with a specific cross section. In this case a rectangular section with a height of 100 mm and a depth of 10 mm. It is modeled as a surface; the thickness is entered as a scalar quantity in the plane stress equations. If a uniform stress is applied on the edges at each end, from the plane stress equations, we obtain a constant uniform stress throughout the member, and the same displacement as in Step 1. We also can solve this by FEA using shell elements as shown in Figure 16.2.2, but this only confirms what we already know, that is, it will return a uniform stress of 100 MPa

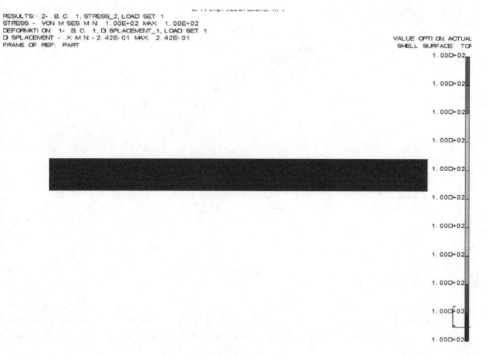

Figure 16.2.2

and a total elongation of 0.484 *mm*. Note that in this analysis the displacement is measured from the centroid of the model on each end face and is 0.242 *mm* in each of the positive and negative directions for a total elongation of 0.484.

16.2.3   Step 3. Rectangular Bar with Pin Holes—Plane Stress Analysis

The third one down in Figure 16.2.1 is more representative of an actual member that is attached to a base and to other members with pins, that is, it is a truly pin-jointed truss member. Material is added at the ends so that the distance between the centers of the holes is 1000 *mm*. This one has no neat analytical solution, so we need to look at FEA. We need to simulate the application of the load which in reality is the pin in contact with the edge of the hole. In this case it is done with a rigid element bearing against the side of the hole. This is an approximation since the pin in reality is a flexible body and will deform to some extent under contact with the edge of the hole thus modifying the distribution of load. The rigid element does give a reasonable first approximation.

We show the results in Figure 16.2.3 with the stress contours shown on the displaced model. The displacements are greatly exaggerated to be made more visible.

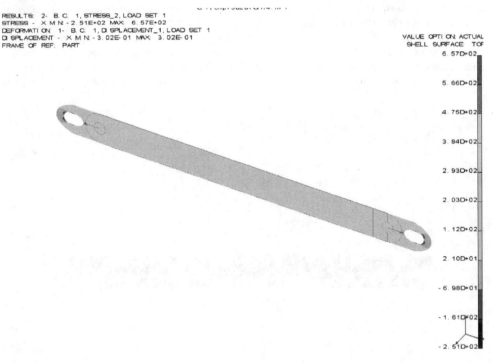

Figure 16.2.3

In Figure 16.2.4 we present a close up of the right end with the stress contours shown on the undeformed model. The result is a large increase in the maximum stress due to the stress concentration effect at the

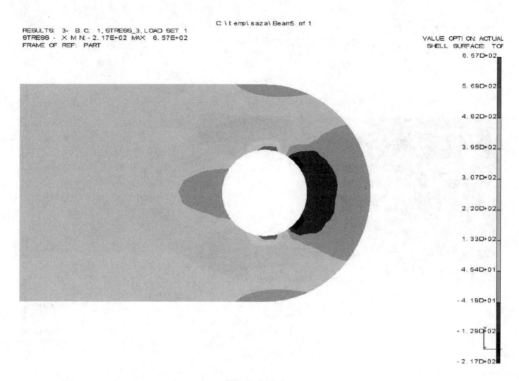

RESULTS: 3- B. C. 1, STRESS_3, LOAD SET 1
STRESS - X M N: - 2. 17E+02 MAX 6. 57E+02
FRAME OF REF: PART

C. \ t emp\ s az a\ Beam5. rrf 1

VALUE OPTI ON: ACTUAL
SHELL SURFACE: TOP
6. 57D+02

5. 69D+02

4. 82D+02

3. 95D+02

3. 07D+02

2. 20D+02

1. 33D+02

4. 54D+01

- 4. 19D+01

- 1. 29D+02

- 2. 17D+02

Figure 16.2.4

hole. The stress of 657 *MPa* exceeds the yield stress by a large factor and is unacceptable. Away from the holes the stress is the same as in our simple models. This is an excellent example of St. Venant's principle that was discussed earlier.

The total displacement of 0.604 *mm* results from the elongation of the region in the vicinity of the holes. The elongation of the mid section is the same as that obtained from the simple analysis.

16.2.4   Step 4. Rectangular Bar with Pin Holes—Solid Body Analysis

Now we shall repeat the solution in Step 3 using a solid body and solid elements.

The model is meshed using a nominal element length of 2.5 *mm* and a solution is found. This is essentially a repeat of the plane stress solution with small differences in the maximum stress due to modeling differences. If a greater number of elements were used in both the plane stress and solid cases the results would converge. The results show that something dramatic is happening in the vicinity of the holes. And it confirms that over most of the length the stress is uniform at 100 *MPa*. This too is an excellent example of St. Venant's principle.

For a better understanding, let us take a closer look at a contour plot of the region around the hole as shown in Figure 16.2.5. In this case we show the stress contours on a deformed model. The deformation is greatly exaggerated of course.

We note that in both the plane stress and solid cases, when loaded by the pins bearing on the hole surfaces, we get much higher stress levels in the vicinity of the hole but essentially the same stress as in

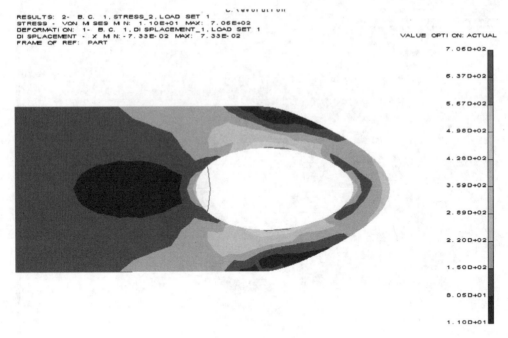

Figure 16.2.5

the first two cases in the middle portion. The values of the several stress components and displacement are

$$
\begin{aligned}
\text{Von Mises} &= 706\,Mpa \\
\text{Maximum Principle} &= 761 \\
\sigma_x &= 753 \\
\tau_{xy} &= 245 \\
\text{Maximum shear} &= 357 \\
\text{Displancement in } x &= 0.1882 \text{ (total elongation)}
\end{aligned}
\tag{16.2.4}
$$

The maximum von Mises stress reported at the top and bottom of the hole is 706 *MPa*. This is much higher than the yield stress of 350 *MPa* and would produce a negative factor of safety. We need to modify the design.

16.2.5   Step 5. Add Material Around the Hole—Solid Element Analysis

We note that the stresses are too high at the hole and we must change the design to reduce the highest stresses to prevent yielding. The first attempt is to place more material in the vicinity of the hole by enlarging the radius surrounding the hole as shown in the fourth figure down and in Figure 16.2.6. The outside diameter is increased to 30 *mm*.

Figure 16.2.6

This is solved with the same density of mesh and the same loading as in the previous case. The results are shown in Figure 16.2.7.

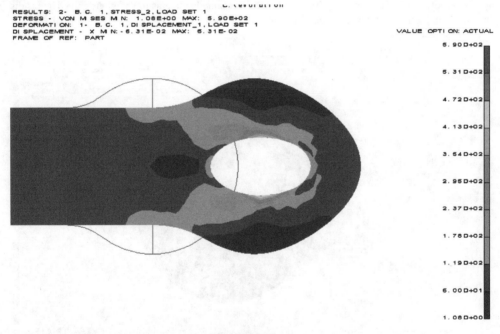

```
RESULTS:  2-  B. C.  1, STRESS_2, LOAD SET 1
STRESS -  VON M SES M N: 1.08E+00  MAX:  5.90E+02
DEFORMATI ON:  1-  B. C.  1, DI SPLACEMENT_1, LOAD SET  1
DI SPLACEMENT  -  X  M N: - 6.31E- 02  MAX:  6.31E- 02
FRAME OF  REF:  PART
```

VALUE OPTI ON: ACTUAL

```
5.90 D+02
5.31 D+02
4.72 D+02
4.13 D+02
3.54 D+02
2.95 D+02
2.37 D+02
1.78 D+02
1.19 D+02
6.00 D+01
1.08 D+00
```

Figure 16.2.7

The values of the several stress components and displacement are

$$
\begin{aligned}
\text{Von Mises} &= 590\,Mpa \\
\text{Maximum Principle} &= 625 \\
\sigma_x &= 619 \\
\tau_{xy} &= 182 \\
\text{Maximum shear} &= 310 \\
\text{Displacement in } x &= 0.1262\,(\text{total elongation})
\end{aligned}
\tag{16.2.5}
$$

The maximum von Mises stress is reduced to 590 MPa. That is still too high. More changes are needed.

16.2.6   Step 6. Bosses Added—Solid Element Analysis

We shall add bosses to the material surrounding the hole as shown in the fifth figure and in Figure 16.2.8. The thickness in this region is now 12 *mm*.

Figure 16.2.8

This is solved with the same density of mesh and the same loading as the previous case. The results are shown in Figure 16.2.9 on an undeformed model.

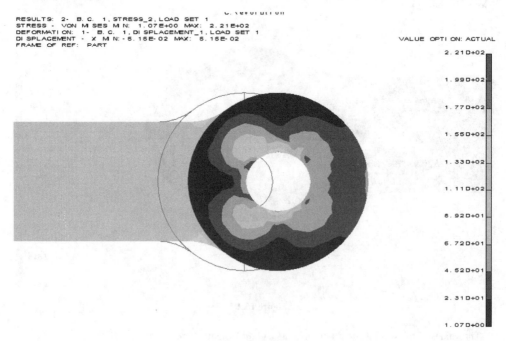

```
RESULTS:  2-  B. C.  1, STRESS_2, LOAD SET 1
STRESS -  VON M SES  M N:  1.07E+00  MAX:  2.21E+02
DEFORMATION:  1-  B. C.  1, DI SPLACEMENT_1, LOAD SET 1
DI SPLACEMENT -  X  M N: -5.15E-02  MAX:  5.15E-02
FRAME OF REF:  PART
```

VALUE OPTI ON: ACTUAL

```
2.21D+02

1.99D+02

1.77D+02

1.55D+02

1.33D+02

1.11D+02

8.92D+01

6.72D+01

4.52D+01

2.31D+01

1.07D+00
```

Figure 16.2.9

The values of the several stress components and displacement are

$$
\begin{aligned}
\text{Von Mises} \quad & = 221\,Mpa \\
\text{Maximum Principle} & = 246 \\
\sigma_x \quad & = 244 \\
\tau_{xy} \quad & = 78 \\
\text{Maximum shear} \quad & = 119 \\
\text{Displacement in } x \quad & = 0.1030\,(\text{total elongation})
\end{aligned}
\tag{16.2.6}
$$

The maximum von Mises stress is now 221 *MPa*. This is well within the range providing a safety factor of

$$F.S. = \frac{350}{221} = 1.58 \tag{16.2.7}$$

But wait! This meets the design requirement on stress and displacement, but is it an efficient design? Much of the material is operating at a safety factor of 3.5. This may be unnecessarily high, producing a heavier member than desired.

16.2.7   Step 7. Reducing the Weight—Solid Element Analysis

Finally, we can reduce the amount of material and reduce the weight by removing material in the regions of low stress as shown in the sixth figure and in Figure 16.2.10. We can remove some of the material in the mid region by narrowing the height of the mid cross section by half.

Figure 16.2.10

This should provide a stress of 200 *Mpa* in the mid region and a safety factor of.

$$F.S. = \frac{350}{200} = 1.75 \tag{16.2.8}$$

With that change a contour plot of the von Mises stresses is given in Figure 16.2.11.

The region of highest stress has now shifted to where the rectangular middle bar joins the curved boss. The stress values are now

$$\begin{aligned}
\text{Von Mises} &= 240 \, Mpa \\
\text{Maximum Principle} &= 257 \\
\sigma_x &= 225 \\
\tau_{xy} &= 88 \\
\text{Maximum shear} &= 122 \\
\text{Displancement in } x &= 0.1882
\end{aligned} \tag{16.2.9}$$

This provides a safety factor based on the von Mises stresses of

$$F.S. = \frac{350}{240} = 1.45 \tag{16.2.10}$$

Further changes, such as a larger fillet between the mid bar and the boss, will refine the design further.

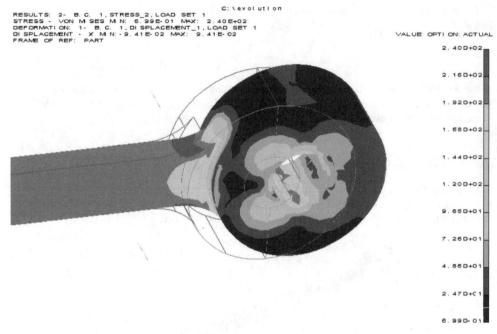

Figure 16.2.11

16.2.8   Step 8. Buckling Analysis

Under certain conditions the loading on the truss member may be reversed subjecting the member to a compressive load of 10000 N. Can this member withstand that load without buckling?

First use slender column theory. In the yx plane it acts as a column simply supported on both ends; in the zx plane it approximates a column with both ends fixed. From Example 8.3.1, for the simply supported case,

$$P_{cr} = \frac{\pi^2}{L^2} E I_{zz} = \frac{\pi^2}{200^2} 206800 \frac{5 \cdot 10^3}{12} = 21261 \, N \tag{16.2.11}$$

From Example 8.3.3, for the fixed end case,

$$P_{cr} = \frac{4\pi^2}{L^2} E I_{yy} = \frac{4\pi^2}{200^2} 206800 \frac{10 \cdot 5^3}{12} = 21261 \, N \tag{16.2.12}$$

Coincidentally, both buckling loads turn out to have the same magnitude, and both are sufficiently high to not warrant a FEA for buckling.

16.3   Evolution of a Plate with a Hole—Plane Stress

Consider the plate in Example 12.5.1. We found the maximum von Mises stress to be 354 N/mm^2 at the upper and lower diameter of the hole. This is the classical example of stress concentration. It is 3.54 times the value of the maximum stress if no hole were present. A typical solution is to reinforce the plate around the hole. In Figure 11.2.1 we take the surface which represents the plate and split into a ring

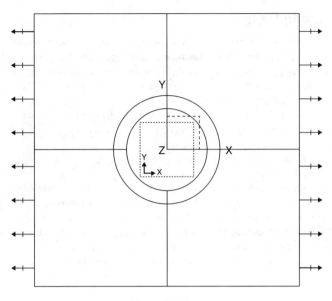

Figure 16.3.1

surface around the hole with a radius of 200 *mm* – the hole has a radius of 150 *mm*. For that surface we specify a thickness of 30 *mm* while the rest of the plate remains at 10 *mm*.

We repeat the analysis with the new geometry. A contour plot of the results is shown in Figure 16.3.2. The maximum stress is now only 116 *N/mm²*.

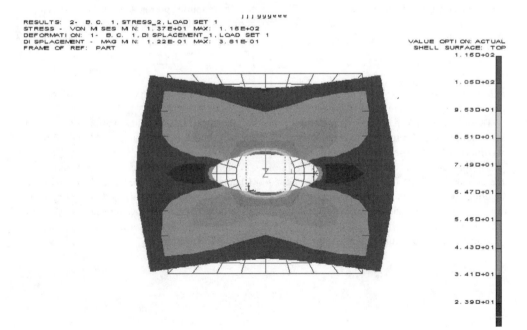

Figure 16.3.2

16.4   Materials in Design

In the last century materials used for structures generally were homogeneous and isotropic. In high stress devices and machines, such as aircraft and automobiles, the materials were dominantly those listed in Chapter 3, Section 3.7, along with other metals. In less high stress areas, isotropic molded plastics were and are used. Orthotropic materials such as plywood, layered metal plates with honeycomb cores, and layered fiber glass were used in selected applications. Then toward the end of the twentieth century new materials became available. Foremost among these for high technical application are fiber reinforced composite laminates. We introduced fiber reinforced laminate material properties in Chapter 3, Section 3.8, and used them in structural applications in Chapter 13.

The significance of such laminates in design is truly immense. In the past, we accepted certain material properties and built the design around certain geometry consistent with the necessary boundary conditions, that is, restraints and applied loads. We can now design a material that best meets the requirements. Thus in the examples given earlier in this chapter we can both modify geometry *and* materials to best meet the design requirements. No longer must the material be homogenous nor isotropic. The challenge to do this with classical analysis tools, that is direct solutions of the differential equations would be an insurmountable task, except for trivial configurations and trivial cases of external loading; however, the finite element method is clearly adaptable to the task and will be the analysis method of choice now and in the future.

16.5   Summary and Conclusions

Our purpose in this chapter is to show how increased analysis capability inherent in the finite element method is used to refine the design of a structural member. We do not pretend that the steps we have taken achieve the most efficient design. Prior to the development of FEA sometimes extreme simplifications of the geometry, restraints, and loadings were necessary to obtain solutions for stress and displacement. Now such simplifications are important in selecting initials values of geometry but are superceded by the more exact methods of analysis provided by FEA and modern computing power. As shown here structural parts often start out with simple geometry, idealized restraints, and simple loading using approximate theories, such as

$$\sigma_x = \frac{F}{A} \tag{16.5.1}$$

was used to start the design of the truss member. Then more detailed geometry, restraints, and loading are added requiring more capable analysis equations, such as FEA.

These same ideas can be used in the evolution of a design to meet buckling and vibration criteria. Start with the simple geometry and material of a structure that meets buckling or vibration criteria and make modification to improve the design.

Often a new product is the derivative of a previous design and the evolutionary process starts with the previous product. The Boeing 787, mentioned in Chapter 13, Section 13.1, has a shape and form determined by aerodynamics that is very much like previous airliners, its engines are moderately improved versions of previous ones; however, it structure is truly a quantum jump in performance. These new materials are the future of design.

Your world of opportunity as a structural analyst is beyond description.

Appendix A

Matrix Definitions and Operations

A.1 Introduction

A brief review of basic matrix definitions and operation is given here. The notation is consistent with that in the main text.

A.2 Matrix Definitions

A *rectangular matrix of order* $m \times n$ is an array of quantities in m rows and n columns as follows:

$$[a] = \begin{bmatrix} a_{11} & a_{12} & \cdots & a_{1n} \\ a_{21} & a_{22} & \cdots & a_{2n} \\ \cdot & \cdot & \cdots & \cdot \\ \cdot & \cdot & \cdots & \cdot \\ \cdot & \cdot & \cdots & \cdot \\ a_{m1} & a_{m2} & \cdots & a_{mn} \end{bmatrix} \tag{A.2.1}$$

An element in the mth row and nth column is represented by the notation a_{mn}. A set of square brackets [ ] will denote a rectangular matrix of any order. For convenience we shall define some special cases and use some special notation as explained in the following paragraphs.

A *column* matrix has m rows and 1 column, or order $m \times 1$. Curved braces { } are used to designate a column matrix. The element in the mth row is designated a_m. For example, if $a_1 = 1, a_2 = 3, a_3 = 2$, the matrix is of order 3×1 and given by

$$\{a\} = \begin{bmatrix} a_1 \\ a_2 \\ a_3 \end{bmatrix} = \begin{bmatrix} 1 \\ 3 \\ 2 \end{bmatrix} \tag{A.2.2}$$

A *row* matrix has 1 row and n columns, for example,

$$[a] = \begin{bmatrix} a_1 & a_2 & a_3 \end{bmatrix} = \begin{bmatrix} 1 & 3 & 2 \end{bmatrix} \tag{A.2.3}$$

and the element of the nth column is designated a_n.

Analysis of Structures: An Introduction Including Numerical Methods, First Edition. Joe G. Eisley and Anthony M. Waas.
© 2011 John Wiley & Sons, Ltd. Published 2011 by John Wiley & Sons, Ltd.

A *square* matrix has the same number of rows and columns, or $m = n$. For example a square matrix of order 3×3 might look like this

$$[a] = \begin{bmatrix} 1 & 3 & 2 \\ 4 & 8 & 1 \\ 7 & 5 & 2 \end{bmatrix} \qquad (A.2.4)$$

A *symmetrical* matrix is a square matrix that has exactly the same value when rows and columns are interchanged, or

$$a_{mn} = a_{nm} \qquad (A.2.5)$$

For example,

$$[a] = \begin{bmatrix} 1 & 3 & 2 \\ 3 & 8 & 1 \\ 2 & 1 & 2 \end{bmatrix} \qquad (A.2.6)$$

A *diagonal* matrix is a square matrix in which all elements are zero except those on the principal diagonal, or

$$a_{mn} = 0, \quad \text{if } m \neq n \qquad (A.2.7)$$

For example,

$$[a] = \begin{bmatrix} 1 & 0 & 0 \\ 0 & 8 & 0 \\ 0 & 0 & 2 \end{bmatrix} \qquad (A.2.8)$$

A diagonal matrix is always symmetrical. The *identity* matrix is denoted by $[I]$ and is a special case of the diagonal matrix for which

$$a_{mn} = 1 \quad \text{when } m = n \qquad (A.2.9)$$

For example

$$[I] = \begin{bmatrix} 1 & 0 & 0 \\ 0 & 1 & 0 \\ 0 & 0 & 1 \end{bmatrix} \qquad (A.2.10)$$

It is also frequently called the *unit* matrix. A *single element* matrix is a *scalar* and may be written with or without brackets.

The *determinant of* $[a]$ is formed from the elements of a square matrix and is designated $|a|$.

The *transpose* of a matrix is formed by interchanging rows and columns and is denoted by the superscript T. The elements of the transpose of a matrix are found by setting

$$(a_{mn})^T = a_{nm} \qquad (A.2.11)$$

For example,

$$[a] = \begin{bmatrix} 1 & 2 \\ 7 & 9 \\ 6 & 3 \end{bmatrix} \qquad [a]^T = \begin{bmatrix} 1 & 7 & 6 \\ 2 & 9 & 3 \end{bmatrix} \qquad (A.2.12)$$

The transpose of a row matrix is a column matrix and vice versa. A symmetrical matrix is identical to its transpose.

A.3   Matrix Algebra

We now define certain rules of matrix mathematics that make it especially useful.

1. *Equality.* Two matrices are equal if they are of the same order and all corresponding elements are equal. That is,

$$[a] = [b] \qquad \text{if} \quad a_{mn} = b_{mn} \tag{A.3.1}$$

2. *Addition and subtraction.* Matrices must be of the same order to add and subtract. Addition is performed by adding corresponding elements and subtraction by subtracting corresponding elements. That is,

$$[a] \pm [b] = [c] \qquad \text{where} \quad c_{mn} = a_{mn} \pm b_{mn} \tag{A.3.2}$$

For example,

$$\begin{bmatrix} 2 & 1 & 7 \\ 3 & 6 & 9 \\ 1 & -4 & 0 \end{bmatrix} + \begin{bmatrix} 1 & 6 & 7 \\ 1 & -2 & 0 \\ 0 & 0 & 1 \end{bmatrix} = \begin{bmatrix} 3 & 7 & 14 \\ 4 & 4 & 9 \\ 1 & -4 & 1 \end{bmatrix} \tag{A.3.3}$$

3. *Multiplication by a scalar.* Any matrix may be multiplied by a scalar by multiplying each element by the scalar. That is,

$$a\,[b] = [c] \qquad \text{where} \quad c_{mn} = a b_{mn} \tag{A.3.4}$$

4. *Multiplication.* We define multiplication of two matrices, say, $[a]$ and $[b]$, provided certain conditions exist. The number of columns in $[a]$ must equal the number of rows in $[b]$. Each element in the product $[c]$ is obtained by multiplying the elements of the corresponding row in $[a]$ by the elements of the corresponding column in $[b]$ and adding the results according to the rule

$$c_{mk} = \sum a_{mn} b_{nk} \tag{A.3.5}$$

For example,

$$\begin{bmatrix} 3 & 2 \\ 1 & 1 \\ 7 & -1 \end{bmatrix} \begin{bmatrix} 1 & 2 & 1 \\ 6 & -3 & 1 \end{bmatrix} = \begin{bmatrix} 3 \cdot 1 + 2 \cdot 6 & 3 \cdot 2 - 2 \cdot 3 & 3 \cdot 1 + 2 \cdot 1 \\ 1 \cdot 1 + 1 \cdot 6 & 1 \cdot 2 - 1 \cdot 3 & 1 \cdot 1 + 1 \cdot 1 \\ 7 \cdot 1 - 1 \cdot 6 & 7 \cdot 2 + 1 \cdot 3 & 7 \cdot 1 - 1 \cdot 1 \end{bmatrix} = \begin{bmatrix} 15 & 0 & 5 \\ 7 & -1 & 2 \\ 1 & 17 & 6 \end{bmatrix} \tag{A.3.6}$$

likewise,

$$\begin{bmatrix} 1 & 2 & 1 \\ 6 & -3 & 1 \end{bmatrix} \begin{bmatrix} 3 & 2 \\ 1 & 1 \\ 7 & -1 \end{bmatrix} = \begin{bmatrix} 1 \cdot 3 + 2 \cdot 1 + 1 \cdot 7 & 1 \cdot 2 + 2 \cdot 1 - 1 \cdot 1 \\ 6 \cdot 3 - 3 \cdot 1 + 1 \cdot 7 & 6 \cdot 2 - 3 \cdot 1 - 1 \cdot 1 \end{bmatrix} = \begin{bmatrix} 12 & 3 \\ 22 & 8 \end{bmatrix} \tag{A.3.7}$$

Note the order of the product in each case.

Matrix multiplication is *associative,* thus,

$$[a]\,([b]\,[c]) = ([a]\,[b])\,[c] \tag{A.3.8}$$

distributive, thus,

$$[a]\,([b] + [c]) = [a]\,[b] + [a]\,[c] \tag{A.3.9}$$

but, in general, <u>not</u> *commutative,* thus,

$$[a][b] \neq [b][a] \tag{A.3.10}$$

5. *Inversion.* Division, as such, is not defined for matrices but is replaced by something called inversion. We denote the inverted matrix with the symbolic form $[a]^{-1}$, and it is defined so that

$$[a]^{-1}[a] = [I] \tag{A.3.11}$$

The elements of an inverse matrix can be obtained algebraically. In practice this is seldom done. Instead, the inverse is found numerically. Numerically, inverting matrices, by developing special algorithms, is a specialized subject which is becoming increasingly important with the speeding up of calculations.

A.4 Partitioned Matrices

A useful operation is the *partitioning* into *submatrices.* These submatrices may be treated as elements of the parent matrix and manipulated by the rules just reviewed. For example,

$$[a] = \begin{bmatrix} 7 & 5 & 9 & 4 & 3 \\ 3 & 9 & 2 & 7 & 8 \\ 1 & 2 & 8 & 6 & 5 \end{bmatrix} = \begin{bmatrix} [A^{11}] & [A^{12}] \\ [A^{21}] & [A^{22}] \end{bmatrix} \tag{A.4.1}$$

where

$$[A^{11}] = \begin{bmatrix} 7 & 5 & 9 \\ 3 & 0 & 2 \end{bmatrix} \qquad [A^{12}] = \begin{bmatrix} 4 & 3 \\ 7 & 8 \end{bmatrix}$$
$$[A^{21}] = \begin{bmatrix} 1 & 2 & 8 \end{bmatrix} \qquad [A^{22}] = \begin{bmatrix} 6 & 5 \end{bmatrix} \tag{A.4.2}$$

A.5 Differentiating and Integrating a Matrix

Differentiate each element in the conventional manner. For example,

$$\text{if} \quad [a] = \begin{bmatrix} x & x^2 & 3x \\ x^2 & x^4 & 2x \\ 3x & 2x & x^3 \end{bmatrix} \quad \text{then} \quad \frac{d}{dx}[a] = \begin{bmatrix} 1 & 2x & 3 \\ 2x & 4x^3 & 2 \\ 3 & 2 & 3x^2 \end{bmatrix} \tag{A.5.1}$$

Integrate each element in the conventional manner. For example,

$$\text{if} \quad [a] = \begin{bmatrix} 1 & 2x & 3 \\ 2x & 4x^3 & 2 \\ 3 & 2 & 3x^2 \end{bmatrix} \quad \text{then} \quad \int [a]dx = \begin{bmatrix} x & x^2 & 3x \\ x^2 & x^4 & 2x \\ 3x & 2x & x^3 \end{bmatrix} + [C] \tag{A.5.2}$$

where $[C]$ are the constants of integration. For definite integrals each term is evaluated for the limits of integration present. Below is a summary of useful relations for following the main text, but the interested reader is alerted to specialized texts on the subject of matrix algebra for wider coverage.

A.6   Summary of Useful Matrix Relations

$$[a][I] = [I][a] = [a]$$

$$[a]([b] + [c]) = [a][b] + [a][c]$$

$$a([b] + [c]) = a[b] + a[c]$$

$$[a]([b][c]) = ([a][b])[c] = [a][b][c]$$

$$[a] + ([b] + [c]) = ([a] + [b]) + [c] = [a] + [b] + [c]$$

$$[a] + [b] = [b] + [a]$$

$$[a][b] \neq [b][a]$$

$$([a][b])^T = [b]^T[a]^T$$

$$([a][b])^{-1} = [b]^{-1}[a]^{-1}$$

$$\left([a]^T\right)^{-1} = \left([a]^{-1}\right)^T$$

Appendix B

Area Properties of Cross Sections

B.1   Introduction

The area, the centroid of area, and the area moments of inertia of the cross sections are needed in slender bar calculations for stress and deflection. To simplify the problem we place the x axis so that it coincides with the loci of centroids of all cross sections of the bar. In our examples the cross sections lie in the yz plane. Furthermore, for beam bending analysis in these chapters we orient the y and z axes so that they are *principal axes of inertia* of the cross section area. This simplifies the equations for stress and displacement. Just what this means is explained in the following sections.

B.2   Centroids of Cross Sections

Consider a cross section with a general shape such as shown in Figure B.2.1 with the x axis normal to the cross section.

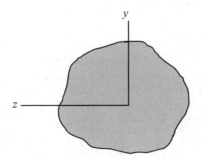

Figure B.2.1

The x axis is a *centroidal* axis if

$$\int_A y\,dA = 0 \qquad \int_A z\,dA = 0 \tag{B.2.1}$$

Analysis of Structures: An Introduction Including Numerical Methods, First Edition. Joe G. Eisley and Anthony M. Waas.
© 2011 John Wiley & Sons, Ltd. Published 2011 by John Wiley & Sons, Ltd.

If the cross sectional is symmetrical the centroid is easily found since it will always lie on the axis of symmetry. For sections with double symmetry, that is, symmetry about both the y and z axes, such as those sections in Figure B.2.2, the location is obvious.

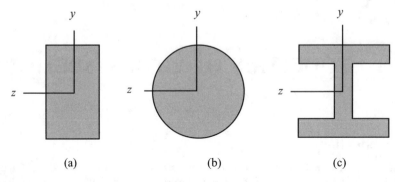

| (a) | (b) | (c) |

Figure B.2.2

For sections with symmetry about just one axis we know the centroid lies on that axis but we must locate just where on that axis. For the sections in Figure B.2.3 we use the formulae

$$\bar{y} = \frac{\int\limits_A ydA}{\int\limits_A dA} \qquad \bar{z} = \frac{\int\limits_A zdA}{\int\limits_A dA} \qquad\qquad (B.2.2)$$

as appropriate.

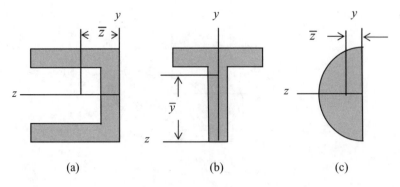

| (a) | (b) | (c) |

Figure B.2.3

When an area can be divided into sub areas with simple geometry so that the centroid of the sub area is easily identified the process of finding the centroid of the original area is simplified to

$$\bar{y} = \frac{\sum y_s A_s}{\sum A_s} \qquad \bar{z} = \frac{\sum z_s A_s}{\sum A_s} \qquad\qquad (B.2.3)$$

where y_s and z_s represent the distances from base axes to the centroids of the sub areas and A_s represents the areas of the sub areas. An example will help.

############

Example B.2.1

Consider the cross section in Figure (a) to which some dimensions have be added.

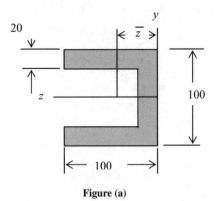

Figure (a)

We place the y axis on the axis of symmetry where we know the centroid lies and the z axis conveniently at the right edge. The cross section is divided into three rectangular areas for which their centroids are known. To find $\bar{z}$

$$\bar{z} = \frac{z_1 A_1 + z_2 A_2 + z_3 A_3}{A_1 + A_2 + A_3} = \frac{50 \cdot 100 \cdot 20 + 10 \cdot 60 \cdot 20 + 50 \cdot 100 \cdot 20}{100 \cdot 20 + 60 \cdot 20 + 100 \cdot 20} = 40.77 \qquad \text{(a)}$$

For slender bar analysis the y axis is moved to the new location. Centroids of some common shapes are given in the last section of this appendix.

############

B.3 Area Moments and Product of Inertia

The *area moments of inertia* are

$$I_{zz} = \int_A y^2 dA \qquad I_{yy} = \int_A z^2 dA \qquad \text{(B.3.1)}$$

and the *area product of inertia* is

$$I_{yz} = \int_A yz\,dA \qquad \text{(B.3.2)}$$

We are interested primarily in values referred to centroidal axes. In many cases $I_{yz} = 0$. This occurs when either the xy or the xz axes plane is a plane of symmetry or the yz axes are oriented so that $I_{yz} = 0$. Then the axes are called *principal axes of inertia*. For sections with double symmetry the integration is often straight forward. An example will help.

###########

Example B.3.1

Consider first the rectangular cross section with double symmetry as shown in Figure (a).

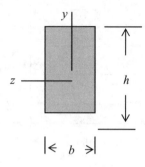

Figure (a)

Given a height h and a width b we have

$$I_{zz} = \int_{-\frac{h}{2}}^{\frac{h}{2}} y^2 b \, dy = \frac{1}{12}bh^3 \qquad I_{yy} = \int_{-\frac{b}{2}}^{\frac{b}{2}} z^2 h \, dz = \frac{1}{12}b^3 h \qquad I_{yz} = \int_{-\frac{b}{2}}^{\frac{b}{2}} \int_{-\frac{h}{2}}^{\frac{h}{2}} yz \, dy \, dz = 0 \quad \text{(a)}$$

Moments of inertia of typical double and single symmetry sections are given in the last section of this appendix.

###########

For sections made up of subsections with known moments of inertia about the centroids of the sub sections there is a transfer process. It is known as the *parallel axis theorem*.

Let the $y_c z_c$ axes be centroid axes for an area whose moments and product of inertia are known. We wish to find the moments of inertia of this area with respect to a yz set of axes. Let $\bar{y}_c$ and $\bar{z}_c$ be the distances from the yz axes to the $y_c z_c$ axes as shown in Figure B.3.1.

$$I_{zz} = \int_A (y_c + \bar{y}_c)^2 \, dA = \int_A y_c^2 \, dA + 2\bar{y}_c \int_A y_c \, dA + y_c^2 \int_A dA \qquad \text{(B.3.3)}$$

Since

$$\int_A y_c \, dA = 0 \qquad \int_A dA = A \qquad \text{(B.3.4)}$$

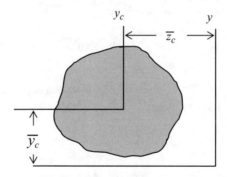

Figure B.3.1

this becomes

$$I_{zz} = I_{z_c z_c} + A\bar{y}_c^2 \tag{B.3.5}$$

Likewise

$$I_{yy} = I_{y_c y_c} + A\bar{z}_c^2 \tag{B.3.6}$$

and

$$I_{yz} = I_{y_c z_c} + A\bar{y}_c\bar{z}_c \tag{B.3.7}$$

###########

Example B.3.2

Find the area moments of inertia with respect to centroidal axes for the cross section in Example B.2.1. Units are millimeters.

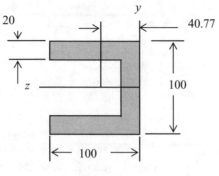

Figure (a)

Transferring the three sections from top to bottom:

$$I_{zz} = \frac{1}{12} \cdot 100 \cdot (20)^3 + 100 \cdot 20 \cdot (40)^2 + \frac{1}{12} \cdot 20 \cdot (60)^3 + \frac{1}{12} \cdot 100 \cdot (20)^3 + 100 \cdot 20 \cdot (40)^2 \tag{a}$$

$$= 6{,}893{,}333 \ mm^4$$

$$I_{yy} = \frac{1}{12} \cdot 20 \cdot (100)^3 + 100 \cdot 20 \cdot (9.23)^2 + \frac{1}{12} \cdot 60 \cdot (20)^3 + 60 \cdot 20 \cdot (30.77)^2$$

$$+ \frac{1}{12} \cdot 20 \cdot (100)^3 + 100 \cdot 20 \cdot (9.23)^2 \tag{b}$$

$$= 4{,}850{,}256 \ mm^4$$

From symmetry the product of intertia $I_{yz} = 0$.

###########

For a section with no symmetry the process requires also finding the product of inertia. We show an example next.

###########

Example B.3.3

Consider the following section. Find the centroid and the area moments and product of inertial with respect to the centroidal axes.

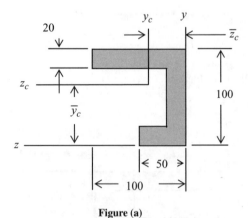

Figure (a)

First find the centroid:

$$\bar{z}_c = \frac{z_1 A_1 + z_2 A_2 + z_3 A_3}{A_1 + A_2 + A_3} = \frac{100 \cdot 20 \cdot 50 + 60 \cdot 20 \cdot 10 + 50 \cdot 20 \cdot 25}{100 \cdot 20 + 60 \cdot 20 + 50 \cdot 20} = 32.62 \ mm$$

$$\bar{y}_c = \frac{y_1 A_1 + y_2 A_2 + y_3 A_3}{A_1 + A_2 + A_3} = \frac{100 \cdot 20 \cdot 90 + 60 \cdot 20 \cdot 50 + 50 \cdot 20 \cdot 10}{100 \cdot 20 + 60 \cdot 20 + 50 \cdot 20} = 59.52 \ mm \tag{a}$$

Now the moments of inertia:

$$I_{zz} = \frac{1}{12} \cdot 100 \cdot (20)^3 + 100 \cdot 20 \cdot (30.48)^2 + \frac{1}{12} \cdot 20 \cdot (60)^3 + 60 \cdot 20 \cdot (9.52)^2$$

$$+ \frac{1}{12} \cdot 50 \cdot (20)^3 + 50 \cdot 20 \cdot (49.52)^2$$

$$= 4{,}879{,}048 \ mm^4$$

$$I_{yy} = \frac{1}{12} \cdot 20 \cdot (100)^3 + 100 \cdot 20 \cdot (17.38)^2 \qquad\qquad\qquad \text{(b)}$$

$$+ \frac{1}{12} \cdot 60 \cdot (20)^3 + 60 \cdot 20 \cdot (22.62)^2 + \frac{1}{12} \cdot 20 \cdot (50)^3 + 50 \cdot 20 \cdot (7.62)^2$$

$$= 3{,}191{,}190 \ mm^4$$

$$I_{yz} = 100 \cdot 20 \cdot 17.38 \cdot 30.48 + 20 \cdot 60 \cdot 9.52 \cdot 22.62 + 50 \cdot 20 \cdot 7.62 \cdot 49.52$$

$$= 1{,}695{,}238 \ mm^4$$

##########

Since in the main text we do all analysis with respect to principle axes, that is, axes for which $I_{yz} = 0$, we must reorient the axes to apply those methods. Consider the rotated $y'z'$ axes in Figure B.3.2.

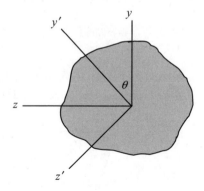

Figure B.3.2

A point with the coordinates y and z with respect to the yz axes have the coordinates y' and z' with respect to the $y'z'$ axes. The transformation equations are

$$y' = y \cos \theta + z \sin \theta$$
$$z' = z \cos \theta - y \sin \theta \qquad\qquad\qquad \text{(B.3.8)}$$

From this we obtain

$$I_{z'z'} = \int_A (y')^2 \, dA = \int_A (y \cos \theta + z \sin \theta)^2 \, dA$$

$$= I_{zz} \cos^2 \theta + I_{yy} \sin^2 \theta + 2I_{yz} \sin \theta \cos \theta \qquad\qquad \text{(B.3.9)}$$

$$= \frac{I_{yy} + I_{zz}}{2} - \frac{I_{yy} - I_{zz}}{2} \cos 2\theta + I_{yz} \sin 2\theta$$

$$I_{y'y'} = \int_A (z')^2 \, dA = \int_A (z \cos\theta - y \sin\theta)^2 \, dA$$
$$= I_{yy} \cos^2\theta + I_{zz} \sin^2\theta - 2I_{yz} \sin\theta \cos\theta \qquad (B.3.10)$$
$$= \frac{I_{yy} + I_{zz}}{2} + \frac{I_{yy} - I_{zz}}{2} \cos 2\theta - I_{yz} \sin 2\theta$$

$$I_{y'z'} = \int_A y'z' \, dA = \int_A (y \cos\theta + z \sin\theta)(z \cos\theta - y \sin\theta) \, dA$$
$$= I_{yy} \sin\theta \cos\theta \, I_{zz} \sin\theta \cos\theta + I_{yz} \sin^2\theta - \cos^2\theta \qquad (B.3.11)$$
$$= \frac{I_{yy} - I_{zz}}{2} \sin 2\theta + I_{yz} \cos 2\theta$$

There is a value of θ for which $I_{xy} = 0$. I may be found by setting

$$\frac{I_{yy} - I_{zz}}{2} \sin 2\theta + I_{yz} \cos 2\theta = 0 \rightarrow \tan 2\theta = \frac{2I_{yz}}{I_{zz} - I_{yy}} \qquad (B.3.12)$$

It may be noted that when $I_{xy} = 0$ then I_{yy} is either a maximum or a minimum and I_{zz} is a corresponding minimum or a maximum. These values are

$$I_{\substack{max \\ min}} = \frac{I_{yy} + I_{zz}}{2} \pm \sqrt{\left(\frac{I_{yy} - I_{zz}}{2}\right)^2 + I_{yz}^2} \qquad (B.3.13)$$

###########

Example B.3.4

Find the rotation angle of the axes to obtain principal axes of inertia and the resulting values for the cross section in Example B.3.2.

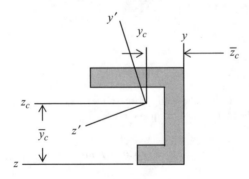

Figure (a)

The angle of rotation is

$$\tan 2\theta = \frac{I_{yz}}{I_{zz} - I_{yy}} = \frac{1,695,238}{4,879,048 - 3,191,190} = 1.00437 \rightarrow \theta = 22.56\,^\circ \qquad (a)$$

The principal moments of inertia are

$$I_{\max} = 5,928,805 \ mm^4 \quad I_{\min} = 2,141,433 \ mm^4 \tag{b}$$

###########

B.4 Properties of Common Cross Sections

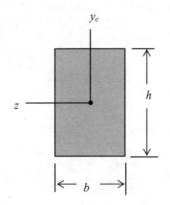

$$A = bh$$

$$I_{yy} = bh^3/12$$

$$I_{zz} = hb^3/12$$

$$I_{yz} = 0$$

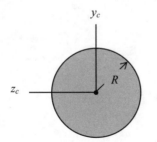

$$A = \pi R^2$$

$$I_{yy} = I_{zz} = \pi R^4/4$$

$$J = \pi R^4/2$$

$$I_{yz} = 0$$

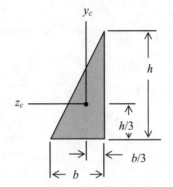

$$A = bh/2$$

$$I_{zz} = bh^3/36$$

$$I_{yy} = hb^3/36$$

$$I_{yz} = b^2h^2/72$$

Appendix C

Solving Sets of Linear Algebraic Equations with Mathematica

C.1 Introduction

You will have the opportunity to solve sets of linear algebraic equations of increasing complexity as the text progresses. It will be advantageous for you to learn the use of one or more software packages as soon as possible. Among the software packages that may be available are Mathematica, Maple, MATLAB®, and Mathcad. There may be others. Any will do. For those who do not already know a package here is a very brief introduction to Mathematica.

C.2 Systems of Linear Algebraic Equations

Simply stated the problem is to solve the equations

$$[A]\{q\} = \{B\} \tag{C.2.1}$$

where $[A]$ and $\{B\}$ are known and $\{q\}$ is to be found. Several software packages can conveniently solve these equations either symbolically or numerically. Here are some simple instructions for using Mathematica. If you are already familiar with Maple or other software that does the job please feel free to use it instead.

Our first interest is in numerical solutions, that is, where both $[A]$ and $\{B\}$ contain only numerical values. There are circumstances, however, when symbolic solutions may be desired so we will cover that as well, but first, numerical solutions.

C.3 Solving Numerical Equations in Mathematica

The following equations are used to illustrate the numerical solution.

$$[A]\{q\} = \begin{bmatrix} 5.25 & -4.1 & 1.0 & 0 \\ -4.1 & 6.05 & -4.1 & 1.0 \\ 1.0 & -4.1 & 6.05 & -4.1 \\ 0 & 1.0 & -4.1 & 5.25 \end{bmatrix} \begin{bmatrix} q_1 \\ q_2 \\ q_3 \\ q_4 \end{bmatrix} = \{B\} = \begin{bmatrix} 0.01 \\ 1.1 \\ 0.0 \\ 0.1 \end{bmatrix} \tag{C.3.1}$$

Analysis of Structures: An Introduction Including Numerical Methods, First Edition. Joe G. Eisley and Anthony M. Waas.
© 2011 John Wiley & Sons, Ltd. Published 2011 by John Wiley & Sons, Ltd.

The brackets and braces are not needed with the symbols A, q, and B. That they represent matrices is declared by the format you use when you enter the values.

Open Mathematica. Enter the data on your Mathematica worksheet. What you enter is given in boldface. That not in boldface is supplied by the program. The symbol <cr> means press the enter key.

A={{5.25,-4.1,1.0,0},{-4.1,6.05,-4.1,1.0},{1.0,-4.1,6.05,-4.1},{0,1.0,-4.1,5.25}}   <cr>
q={{q1},{q2},{q3},{q4}}   <cr>
B={{0.01},{1.1},{0},{0.1}}   <cr>
Solve[A.q==B]   <shift>+<cr>

<shift>+<cr>is the instruction to do the calculation. The software will respond with the following output.

$Out[1] = \{\{5.25,-4.1,1.0,0\},\{-4.1,6.05,-4.1,1.0\},\{1.0,-4.1,6.05,-4.1\},\{0,1.0,-4.1,5.25\}\}$

$Out[2] = \{\{q1\},\{q2\},\{q3\},\{q4\}\}$

$Out[3] = \{\{0.01\},\{1.1\},\{0\},\{0.1\}\}$

$Out[4] = \{\{q1->2.97414,q2->4.97075,q3->4.77586,q4->2.80196\}\}$

There is also a command called LinearSolve. For Solve the format is [A.q = =B]; for LinearSolve the format is [A,B]. Try it.

C.4 Solving Symbolic Equations in Mathematica

Let us try a simple set of equations right out of introductory algebra.

$$1x + 2y + 3z = R$$
$$2x + 4y + 5z = S \quad \text{or} \quad [A\{q\}] = \begin{bmatrix} 1 & 2 & 3 \\ 2 & 4 & 5 \\ 3 & 5 & 6 \end{bmatrix} \begin{bmatrix} x \\ y \\ z \end{bmatrix} = \{B\} = \begin{bmatrix} R \\ S \\ T \end{bmatrix} \quad (C.4.1)$$
$$3x + 5y + 6z = T$$

Enter in Mathematica:

A = {{1,2,3},{2,4,5},{3,5,6}}   <cr>
q = {{p1},{p2},{p3}}   <cr>
B = {{R},{S},{T}}   <cr>
Solve[A.B==q]   <shift>+<cr>

$Out[5]=\{\{1,2,3\},\{2,4,5\},\{3,5,6\}\}$

$Out[6]=\{\{q1\},\{q2\},\{q3\}\}$

$Out[7]=\{\{R\},\{S\},\{T\}\}$

$Out[8]=\{\{q1->R-3S+2T,q2->-3R+3S-T,q3->2R-S\}\}$

C.5   Matrix Multiplication

You will often need to multiply matrices to complete a solution. Take a case which is similar to one that you will encounter.

$$\begin{bmatrix} R_1 \\ R_2 \end{bmatrix} = 12 \begin{bmatrix} 2 & 3 & 4.5 \\ 7 & 1 & -6 \end{bmatrix} \begin{bmatrix} 0.52 \\ 1.20 \\ 0.26 \end{bmatrix} \tag{C.5.1}$$

Enter in Mathematica:

12 {{2,3,4.5},{7,1,-6}} . {0.52,1.20,0.26}   <**shift**>+<**cr**>

Out[1]={69.72, 39.36}

This is the dot or inner product of two matrices.

Appendix D

Orthogonality of Normal Modes

D.1 Introduction

In Chapter 15, Equation 15.2.54, we note the matrix form of the equations for multi degree of freedom mass/spring systems. It is repeated here are as Equation D.1.1.

$$[M]\{\ddot{v}\} + [K]\{v\} = \{F(t)\} \tag{D.1.1}$$

In the preceding pages the homogenous form of these is equations are solved for the natural frequencies, ω_i, and for the normal modes, $\{\varphi\}_i$. In the ensuing paragraphs the equations for finding the forced motion of the system are developed. In the process, the orthogonality of normal modes, Equation 15.2.58 repeated here as Equation D.1.2, is used without proof.

$$\begin{aligned} \{\varphi\}_j^T [M]\{\varphi\}_i &= 0 \qquad i \neq j \\ &= M_j \qquad i = j \end{aligned} \tag{D.1.2}$$

We should note that the equations for FEM analysis are exactly the same form as Equation D.1. In the case of the mass/spring system the mass matrix, $[M]$, contains the concentrated masses while for the FEM case the mass matrix contains equivalent nodal masses depending upon the particular elements used in the derivation. Similarly, the stiffness matrix contains the spring constants for the mass/spring system and the equivalent stiffnesses for the FEM case.

The proof which follows in Section D.2, then, applies to all discrete systems such as mass/spring and FEM formulations.

In Section D.3 we extend the proof to continuous systems.

D.2 Proof of Orthogonality for Discrete Systems

Consider two dissimilar frequencies and modes of Equations D.1

$$\left([K] - \omega_i^2 [M]\right)\{\varphi\}_i = 0 \qquad \left([K] - \omega_j^2 [M]\right)\{\varphi\}_j = 0 \tag{D.2.1}$$

If we premultiply the first by $\{\varphi\}_j^T$ and the second by $\{\varphi\}_i^T$, we get

$$\begin{aligned} \{\varphi\}_j^T \left([K] - \omega_i^2 [M]\right)\{\varphi\}_i &= 0 \\ \{\varphi\}_i^T \left([K] - \omega_j^2 [M]\right)\{\varphi\}_j &= 0 \end{aligned} \tag{D.2.2}$$

Analysis of Structures: An Introduction Including Numerical Methods, First Edition. Joe G. Eisley and Anthony M. Waas.
© 2011 John Wiley & Sons, Ltd. Published 2011 by John Wiley & Sons, Ltd.

Since in matrix multiplications

$$[a]^T [b] [c] = [c]^T [b] [a] \tag{D.2.3}$$

when $[b]$ is symmetric, subtracting the first of Equation D.2.2 from the second, we have

$$\left(\omega_j^2 - \omega_i^2\right)\left(\{\varphi\}_j^T [M] \{\varphi\}_i\right) = \{\varphi\}_j^T [K] \{\varphi\}_i - \{\varphi\}_j^T [K] \{\varphi\}_i = 0 \tag{D.2.4}$$

From this we conclude that

$$\begin{aligned} \{\varphi\}_j^T [M] \{\varphi\}_i &= 0 \qquad \omega_i \neq \omega_j \\ &= M_j \qquad \omega_i = \omega_j \end{aligned} \tag{D.2.5}$$

Thus Equation 15.2.58 is justified.

D.3   Proof of Orthogonality for Continuous Systems

Consider first the differential equation for the forced motion of a uniform axial bar, Equation 15.3.3 repeated here as Equation D.3.1.

$$EA \frac{\partial^2 u}{\partial x^2} - m\ddot{u} = -f_x(x, t) \tag{D.3.1}$$

Once the natural frequencies, ω_i, and normal modes, φ_i, are found we can consider two different normal modes as follows.

$$EA\varphi_i'' - \omega_i^2 m \varphi_i = 0 \qquad EA\varphi_j'' - \omega_j^2 m \varphi_j = 0 \tag{D.3.2}$$

If we premultiply the first by φ_j and the second by φ_i and integrate over the length of the bar, we get

$$\begin{aligned} \int EA\varphi_i'' \varphi_j dx - \omega_i^2 \int m\varphi_i \varphi_j dx = 0 \\ \int EA\varphi_j'' \varphi_i dx - \omega_i^2 \int m\varphi_j \varphi_i dx = 0 \end{aligned} \tag{D.3.3}$$

when $[b]$ is symmetric, subtracting the first of Equations D.2.2 from the second, we have

$$\left(\omega_j^2 - \omega_i^2\right)\left(\int m\varphi_i \varphi_j dx\right) = \int EA\varphi_j'' \varphi_i dx - \int EA\varphi_i'' \varphi_j dx \tag{D.3.4}$$

By integrating the right hand side of Equation D.3.4 by parts, we obtain

From this we conclude that

$$\begin{aligned} \int m\varphi_i \varphi_j dx &= 0 \qquad \omega_i \neq \omega_j \\ &= M_j \qquad \omega_i = \omega_j \end{aligned} \tag{D.3.5}$$

Thus Equation 15.3.8 is justified.

This can be extended to any and all continuous elastic structures.

References

Bathias, C. and A. Pineau (2010) *Fatigue of Materials and Structures*, John Wiley & Sons Ltd.

Craig, R. and A. Kurdila (2006) *Fundamentals of Structural Dynamics*, 2nd edition, John Wiley & Sons Ltd.

Crandall, S.H., N.C. Dahl, and T.S. Lardner (1972) *An Introduction to the Mechanics of Solids*, 2nd edition, McGraw-Hill: New York.

Grandt, A. (2004) *Fundamentals of Structural Integrity*, John Wiley & Sons Ltd.

Herakovich, C.T. (1998) *Mechanics of Fibrous Composites*, John Wiley & Sons Ltd.

Hyer, M.W. (1998) *Stress Analysis of Fiber-Reinforced Composite Materials*, McGraw-Hill: New York.

Kollar, L. and G. Springer (2003) *Mechanics of Composite Structures*, Cambridge University Press.

Matthews, F. (1981) *Course Notes on Aircraft Structures*, AY 1981–1982, Imperial College, London, UK.

Timoshenko, S. and S. Woinowsky-Krieger (1987) *Theory of Plates and Shells*, McGraw-Hill: New York.

Analysis of Structures: An Introduction Including Numerical Methods, First Edition. Joe G. Eisley and Anthony M. Waas.
© 2011 John Wiley & Sons, Ltd. Published 2011 by John Wiley & Sons, Ltd.

Index

Analysis of Structures: An Introduction Including Numerical Methods, First Edition. Joe G. Eisley and Anthony M. Waas.
© 2011 John Wiley & Sons, Ltd. Published 2011 by John Wiley & Sons, Ltd.